COMBINATORIAL MATHEMATICS, OPTIMAL DESIGNS AND THEIR APPLICATIONS

annals of discrete mathematics

NORTH-HOLLAND PUBLISHING COMPANY – AMSTERDAM • NEW YORK • OXFORD

ANNALS OF DISCRETE MATHEMATICS 6

COMBINATORIAL MATHEMATICS, OPTIMAL DESIGNS AND THEIR APPLICATIONS

Edited by
J. SRIVASTAVA
Colorado State University, Fort Collins, USA

1980

NORTH-HOLLAND PUBLISHING COMPANY – AMSTERDAM • NEW YORK • OXFORD

Reprinted from the journal *Annals of Discrete Mathematics*, Volume 6

North-Holland ISBN for this Volume 0 444 86048 7

Published by:

NORTH-HOLLAND PUBLISHING COMPANY
AMSTERDAM • NEW YORK • OXFORD

Sole distributors for the U.S.A. and Canada:

ELSEVIER NORTH-HOLLAND, INC.
52 VANDERBILT AVENUE
NEW YORK, NY 10017

Library of Congress Cataloging in Publication Data

Symposium on Combinatorial Mathematics and Optimal
Design, Colorado State University, 1978.
Combinatorial mathematics, optimal designs, and
their applications.

(Annals of discrete mathematics ; 6)
Bibliography: p.
1. Combinatorial designs and configurations--
Congresses. 2. Combinatorial analysis--Congresses.
I. Srivastava, Jaya. II. Title. III. Series.
QA166.25.S95 1978 511'.6 80-24006
ISBN 0-444-86048-7

PRINTED IN THE NETHERLANDS

PREFACE

A Symposium on Combinatorial Mathematics and Optimal Design was held at Colorado State University (CSU), Fort Collins, Colorado, on June 5–9, 1978. The symposium was international in scope. Both the speakers and the audience ranged from all over the world.

The present volume contains the contributions of the invited speakers. The papers are both of survey and research types.

This symposium was actually a "State of the Art" conference, and was similar to the one held here in September, 1971. The purpose was to help disseminate knowledge and stimulate research by bringing together top ranking workers from diverse areas of the above fields. These include Foundations, Enumerative Techniques, various branches of Graph Theory, Coding Theory, Combinatorial Problems of Designs, Optimal Design Theory, Finite Geometries, Number Theory, Combinatorial games, Computer problems, etc.

The conference was jointly sponsored by the U.S. Air Force Office of Scientific Research, and the Office of Naval Research. Dr. I.N. Shimi, of the Air Force, particularly helped in the same. On behalf of the Organizing Committee, the participants, and the scientific community, I wish to express my deep appreciation and gratitude to them.

The Organizing Committee of the Symposium consisted of Professors R.C. Bose (Colorado State University), Paul Erdös (Hungarian Academy of Sciences), Frank Harary (University of Michigan), G.C. Rota (Massachusetts Institute of Technology), Esther Seiden (Michigan State University), W.T. Tutte (University of Waterloo), and myself. The presence of these people on the organizing committee helped a great deal towards the success of the conference. I am deeply grateful to each and every one of them for being on the committee, and for the tremendous cooperation that I always received from them.

Professors Erdös, Harary, Rota and Tutte were particularly helpful in developing the program of the Conference. This time money was not available for payment for overseas travel. However, these people, particularly Professor Erdös, helped find many outstanding people from abroad who were planning to visit the United States on their own. I am thankful to them for their help in this regard. Several other distinguished foreign scientists were invited by me with the request that they arrange for their overseas travel. I am happy that almost all of them had success. Most sincere thanks go to these foreign governments and organizations for their cooperation.

As is well known now, I have been passing through severe personal problems for the past 15 years, which finally culminated in the events during the last two years. In this connection, a very difficult sequence of situations began just six weeks before the Conference. During these times, the university authorities (including Dr. Williams, Chairman of the Statistics Department, Dean Cook of the College of Natural Sciences, Vice-Presidents Olson and Neidt, and President Chamberlain) extended their understanding and support. Professor Bose informed the other Organizing Committee members about me and they joined in. I am extremely thankful to all of these people; without their encouragement the Conference would not have been held.

Thanks also go to many local people for their help. Among these, particular mention must be made of (i) the secretaries Waydene Casey and Joanne Moynihan, (ii) my then student W. Ariyaratna, and (iii) my esteemed colleagues Professors Manvel and Bose. Finally, to this list, must be added the name of Usha Srivastava, now my sister-in-law. Along with me, Usha also was going through agonies. In spite of this she helped me run things smoothly thus making a great (though indirect) contribution to the success of the Conference.

I am thankful to the authors for the many excellent papers in this volume, and also to the referees for their help.

As in the earlier conferences, the various local arrangements were made by the C.S.U. Department of Conferences and Institutes. This time, unfortunately, some participants suffered inconveniences. I wish to apologize for the same.

I am thankful to North Holland (particularly, the desk editor Aad Thoen) for their promptness in handling the manuscripts, and producing this volume.

Last, but not the least, my thanks go to all the participants in the Symposium for it was their participation which truly made it a success.

Jaya Srivastava
Symposium Director

CONTENTS

PREFACE v

A. BARLOTTI, Results and problems in Galois geometry 1

R.C. BOSE, Combinatorial problems of experimental design II: Factorial designs 7

P.J. CAMERON and D.A. DRAKE, Partial λ-geometries of small nexus 19

C.-S. CHENG and L.J. GRAY, A characterization of group-divisible designs and some related results 31

M. DEZA, On permutation cliques 41

M. DEZA and N.M. SINGHI, Some properties of perfect matroid designs 57

D. DUMONT and G. VIENNOT, A combinatorial interpretation of the Seidel generation of Genocchi numbers 77

P. ERDÖS, A survey of problems in combinatorial number theory 89

P. ERDÖS, F. HARARY and M. KLAWE, Residually-complete graphs 117

D. FOATA, A combinatorial proof of Jacobi's identity 125

A.S. FRAENKEL, From Nim to Go 137

M. HALL, Jr., Configurations in a plane of order ten 157

N. HAMADA and F. TAMARI, Construction of optimal codes and optimal fractional factorial designs using linear programming 175

A. HEDAYAT and S.-Y. R. LI, Combinatorial topology and the trade off method in BIB designs 189

F. HERING, Block designs with cyclic block structure 201

S.A. JONI and G.-C. ROTA, On restricted bases for finite fields 215

A.K. KEVORKIAN, Partitioning of the minimum essential set construction problem 219

J. KIEFER, Optimal design theory in relation to combinatorial design 225

D.J. KLEITMAN and K.J. WINSTON, The asymptotic number of lattices 243

A. LASCOUX and M.P. SCHÜTZENBERGER, A new statistics on words 251

R.C. Mullin, P.J. Schellenberg, D.R. Stinson and S.A. Vanstone, Some results on the existence of squares 257

R. Naik, S.B. Rao, S.S. Shrikhande and N.M. Singhi, Intersection graphs of k-uniform hypergraphs 275

P.N. Rathie, On some enumeration problems in graph theory 281

J. Seberry, Some remarks on amicable orthogonal designs 289

E. Seiden, On the number of nonisomorphic designs in a repeated measurement experiment 293

M. Simonovits and V.T. Sós, Intersection theorems on structures 301

J.N. Srivastava and S. Ghosh, Enumeration and representation of nonisomorphic bipartite graphs 315

R.P. Stanley, Decompositions of rational convex polytopes 333

W.T. Tutte, Rotors in graph theory 343

S.G. Williamson, Embedding graphs in the plane—algorithmic aspects 349

S. Yamamoto and S. Tazawa, Hyperclaw decomposition of complete hypergraphs 385

Annals of Discrete Mathematics 6 (1980) 1–5

RESULTS AND PROBLEMS IN GALOIS GEOMETRY

Adriano BARLOTTI

University of Bologna, Bologna, Italy

1. Introduction

Galois geometry, in its broader sense, is the study of the nonlinear sets of points in finite spaces (including here also finite spaces over nonfield structures). References on the early history of Galois geometry can be found in [29]. In this survey the main research areas in which Galois geometry may be divided were clearly indicated:

(a) To offer pure geometric interpretations of algebraic and number theoretic properties.

(b) To give estimates for the number of points lying on certain algebraic varieties. Here the solution may follow from algebraic results or from purely combinatorial and geometric methods.

(c) To present graphic characterizations of algebraic varieties.

(d) To study $(k;n)$-arcs, $(k;n)$-caps and more generally $(k;n)$-sets. Of particular interest in this study is the "packing problem", i.e. the problem of finding in a given space the maximum number of points which can belong to a $(k;n)$-set, for a given value of n.

In the last 20 years the study of Galois geometry has developed in a quite remarkable way. In 1974 in some lectures we presented a brief survey on this field (see [2, n. 3]). We shall exhibit here some of the progress done since 1974, particularly in the above section (d), and some (old and new) open problems.

2. Basic notions on arcs and caps

PG(r, q) will denote (if $r>2$) a finite r-dimensional projective space of order q. If $r=2$, the symbol PG$(2, q)$ will be used only for a desarguesian plane, whereas $\pi(q)$ will denote any projective plane of order q.

A $(k;n)$-arc of $\pi(q)$ is a set of k points of $\pi(q)$ such that n is the largest number of them which are collinear. The $(k;2)$-arcs are simply called k-arcs. In a given plane a $(k;n)$-arc is "complete" if there does not exist a $(k';n)$-arc which contains it (with $k'>k$).

A $(k;n)$-cap of PG(r, q), where $r\geqslant 3$, is a set of k points of PG(r, q) such that n

is the largest number of them which are collinear. The $(k;2)$-caps are simply called k-caps.

A line g is an s-secant of a $(k;n)$-arc, K, if g contains s points of K; the 1-secants and 0-secants will be called respectively tangent and external lines to K. Let t_s denote the total number of s-secants to K; the numbers t_s are called the characters of K, and K is said to have p characters if exactly p among its characters are different from zero. The arc K is of type $(s_1, s_2, \ldots, s_h)$, where $s_1 < s_2 < \cdots < s_h = n$, if only the characters $t_{s_1}, t_{s_2}, \ldots, t_{s_h}$ are different from zero.

It has been proved that if there exists a k-arc in $\pi(q)$, then $k \leq q+1$ when q is odd and $k \leq q+2$ when q is even (see [9]).

In $\pi(q)$, with q even, every $(q+1)$-arc is incomplete and can be uniquely completed to form a $(q+2)$-arc.

3. A generalization of a theorem by Buekenhout

In 1966 Buekenhout proved the following theorem (see [10]):

Theorem. *If in a projective plane $\pi(q)$ there is a $(q+1)$-arc K such that every hexagon whose vertices are points of K is pascalian, then $\pi(q)$ is pappian and K is a conic.*

This theorem shows that a geometrical property of a single arc can determine the type of the plane in which the arc can be embedded. Alternative proofs for this theorem were given by several authors (see [1, 14, 19, 20, 26]).

In a paper due to appear (see [21]) Korchmaros has obtained a generalization of the above theorem for finite planes by weakening the hypothesis in the following way: the hexagons inscribed in K which are required to be pascalian are only those for which at least one of the lines joining two by two the three diagonal points is a chord or a tangent of the oval.

4. On abstract arcs of type (m, n)

Abstract ovals were defined by Buekenhout [11] and their study led to many interesting problems of classification and existence.

An example of an infinite abstract oval which is not embeddable in a projective plane has been given by Krier [22]. It is still an open question whether there exist or not finite abstract ovals which are not embeddable in a plane.

Barnabei and Zucchini [6] defined abstract arcs of type (m, n) which generalize Buekenhout abstract ovals. Their definition includes as particular cases the $(k;n)$-arcs of types $(0, m, n)$, $(0, n)$ or (m, n) of a finite projective plane.

The notion of abstract arc of type (m, n) is founded on the definition of generalized involution of type (m, n). Let K be a finite set, whose elements we shall call points. A generalized involution of type (m, n) on K (with $1 \leq m < n$) is any partition of the points of K in subsets of cardinality m or n, which are called respectively the blocks of type m or of type n.

An abstract arc of type (m, n) is given by a pair $(K, \mathcal{I})$, where K is a finite set and $\mathcal{I}$ is a set of generalized involutions of type (m, n) on K such that if $\mathcal{V}$ is the set of blocks of all involutions of $\mathcal{I}$, the following axioms hold:

(1) two distinct points of K belong to exactly one block of $\mathcal{V}$;

(2) two blocks of $\mathcal{V}$, with empty intersection, belong to exactly one involution of $\mathcal{I}$;

(3) in $\mathcal{V}$ there is at least one block of type n.

Clearly, (k, n)-arcs of types $(0, m, n)$, $(0, n)$ or (m, n) of any finite projective plane can be seen as abstract arcs of type (m, n) if we consider as elements of I the partitions induced on the points of the arc by the lines of the pencils whose centers do not belong to the arc.

It can be proved that if $|K|:= k>2$ every point of K belongs to the same number, say d, of blocks.

Let $(K, \mathcal{I})$ be any abstract arc of type (m, n). To this we can associate an incidence structure $(\mathcal{P}, \mathcal{L})$ as follows:

(i) $\mathcal{P} = K \cup \mathcal{I}$;

(ii) $\mathcal{L} = \mathcal{V}$;

(iii) incidence is defined in a natural way: a point A of K is incident with $v \in \mathcal{V}$ when $A \in v$ and a point A of $\mathcal{I}$ is incident with $v \in \mathcal{V}$ when $v \in A$.

The problem arises to decide whether or not the above structure can be completed in a suitable way to a projective plane. Examples of sufficient conditions for embeddability are the following:

(A) every involution of $\mathcal{I}$ has a number d of blocks (where d is equal to the number of blocks through any point of K);

(B) $|\mathcal{V}| = d^2 - d + 1$.

5. The packing problem

We denote by $m(r, q)$ the maximum number of points in PG(r, q) that belong to a k-cap if $r \geq 3$ or to a k-arc if $r = 2$. The problem of finding the values of $m(r, q)$, known as the "packing problem", seems to be very difficult and only the following results are known to date:

$m(2, q) = q+1$,	q odd,	Bose [9],
$m(2, q) = q+2$,	q even,	Bose [9],
$m(3, q) = q^2+1$,	q odd,	Bose [9],
$m(3, q) = q^2+1$,	q even, $\neq 2$,	Qvist [25],
$m(r, 2) = 2^r$,	$r \geq 2$,	Bose [9],
$m(4, 3) = 20$,		Pellegrino [24]
$m(5, 3) = 56$,		Hill [16].

The knowledge of $m(r, q)$ is important for applications to statistic and to coding theory, and a considerable amount of work has also been done to provide upper and lower bounds on $m(r, q)$ for $q \geqslant 4$ (see [15, 16, 17, 18, 23]).

In [24] is proved the existence of only two non-isomorphic 20-caps in PG(4, 3).

The packing problem arises (and in general is far from being solved) also for non desarguesian planes, for k-sets of kind s and for (k, n)-arcs and (k, n)-caps.

References

[1] R. Artzy, Pascals theorem on an oval, Am. Math. Monthly 75 (1968) 143–146.

[2] A. Barlotti, Combinatorics of finite planes and other finite geometric structures, in: P. Scherk (ed.), Foundations of Geometry (Univ. of Toronto Press, Toronto, Ont., 1976) 3–15.

[3] M. Barnabei, Alcuni risultati sui $(k; n)$-archi con tre caratteri, Ist. Lombardo Accad. Sci. Lett. Rend. A. 110 (1976) 351–358.

[4] M. Barnabei, On arcs with weighted points, to appear.

[5] M. Barnabei, D. Searby and C. Zucchini, On small $(k; q)$-arcs in planes of order q^2, J. Combin. Theory 24(A) (1978) 241–246.

[6] M. Barnabei and C. Zucchini, Archi astratti di tipo (m, n), Boll. Un. Mat. Ital. 15-A (5) (1978) 585–591

[7] A. Basile and P. Brutti, Proprietà delle ovali di un piano desarguesiano finito, Rend. Circ. Mat. Palermo 24 (2) (1978) 223–243.

[8] A. Basile and P. Brutti, Planes and ovals, to appear.

[9] R.C. Bose, Mathematical theory of the symmetrical factorial design, Sankhya 8 (1947) 107–166.

[10] F. Buekenhout, Plans projectifs à ovoides pascaliens, Arch. Math. 17 (1966) 89–93.

[11] F. Buekenhout, Etude intrinsèque des ovals, Rend. Mat. e Appl. 25 (5) (1966) 333–393.

[12] F. Buekenhout, Existence of unitals in finite translation planes of order q^2 with a kernel of order q, Geometriae Dedicata 5 (1976) 189–194.

[13] F. Buekenhout, Characterizations of semi quadrics. A survey, In: Colloq. Int. Teorie Combinatorie, Roma 1973, Tomo I (1976) 394–417.

[14] G. Conti, Piani proiettivi dotati di un ovale pascaliano, Boll. Un. Mat. Ital. 11 (4) (1975) 330–338.

[15] N. Hamada and F. Tamari, On a geometrical method of construction of maximal t-linearly independent sets, J. Combin. Theory 25 (A) (1978) 14–28.

[16] R. Hill, On the largest size of cap in $S_{5,3}$, Rend. Accad. Naz. Lincei 54 (8) (1973) 378–384.

[17] R. Hill, (1978). Caps and codes, to appear.

[18] R. Hill, (1978). Some results concerning linear codes and $(k; 3)$-arcs in 3-dimensional Galois spaces, to appear.

[19] H. Karzel and K. Sörensen. Projective Ebenen mit einem pascalschen Oval, Abh. Math. Sem. Univ. Hamburg 36 (1971) 123–125.

[20] H. Karzel and K. Sörensen, Die lokalen Sätze von Pappus und Pascal, Mitt. Math. Gesellsch. Hamburg 10 (1971) 28–55.

[21] G. Korchmaros, Una generalizzazione del teorema di F. Buekenhout sulle ovali pascaliane, to appear.

[22] N. Krier, A Buekenhout oval which is not projective, Arch. Math. 28 (1977) 323–324.

[23] A.C. Mukhopadhyay, Lower bounds on $m_t(r, s)$, J. Combin. Theory 25 (A) (1978) 1–13.

[24] G. Pellegrino, Sulle calotte massime dello spazio $S_{4,3}$, Atti Accad. Sci. Lett. Arti Palermo 34 (4) (1976) 297–328.

[25] B. Qvist, Some remarks concerning curves of the second degree in a finite plane, Ann. Acad. Sci. Fenn. 134 (1952) 1–27.

[26] J.F. Rigby, Pascal ovals in projective planes, Can. J. Math. 21 (1969) 1462–1476.

[27] B. Segre, Ovals in a finite projective plane, Can. J. Math. 7 (1955) 414–416.

[28] B. Segre, Sui k-archi nei piani finiti di caratteristica due, Rev. Math. Pures Appl. 2 (1957) 283–294.
[29] B. Segre, On Galois geometries, in: Proc. Int. Congr. Math., Cambridge (1960) 488–499.
[30] B. Segre, Lectures on Modern Geometry (Cremonese, Roma, 1961).
[31] B. Segre, Introduction to Galois geometries, Memorie Lincei 7 (8) (1967) 133–236.
[32] B. Segre, Sugli insiemi finiti di punti di uno spazio grafico, Rend. Mat. 8 (1) (1975) 37–63.

Annals of Discrete Mathematics 6 (1980) 7–18

COMBINATORIAL PROBLEMS OF EXPERIMENTAL DESIGN II: FACTORIAL DESIGNS*

R.C. BOSE

Colorado State University, Fort Collins, CO 80523, U.S.A.

The designs used for statistically controlled experiments fall into two important classes: Incomplete block designs and factorial designs. The combinatorial problems arising in connection with incomplete block designs were surveyed in an earlier paper, Bose [13]. This is a companion paper which surveys the combinatorial problems of factorial designs.

1. Introduction

The principles of experimental design as we know it today were formulated by R.A. Fisher in his famous book, *Statistical Methods for Research Workers* [25] and in his paper on "The Arrangement of Field Experiments" [26]. They arose out of his own attempts and those of his precursors to increase the precision of field experiments. The object of the experimenter is to investigate the response of experimental units to a set of treatments. The experimental units are divided into sets called *blocks* which are relatively homogeneous. A certain set of treatments is applied to the units (*plots*) in a given block and the response observed. Statistical analysis is then used to estimate the effects of treatments. The three principles formulated by Fisher are *randomization*, *replication* and *local control.* Randomization dictates that once the set of treatments which are to be applied to the units of a given block have been chosen they should be applied randomly to the different units. The principle of replication requires that each treatment be applied to more than one experimental unit. In general, these units belong to different blocks. Due to the variability of the experimental material, the responses of the units to which the same treatment has been applied are not identical. Statistical analysis allows us to estimate the precision of the estimated treatment effects. If the number of treatments is large and experimental material is not homogeneous, precision can be increased by not accommodating all treatments in a given block. Thus blocks are incomplete. The problem of design is to select sets of treatments to be applied to the units of different blocks. The selection is subject to combinatorial constraints, since apart from the question of precision, it is desirable to have a design such that statistical analysis is as simple as possible, and such that its results are easily interpreted. There are two important classes of designs. In incomplete block designs (so called) the treatments are simple; for example, we may be interested in comparing the yields or some other characteristic of a large number of varieties

* Research supported by AFOSR Grant Number 77-3127.

of corn. On the other hand, treatments may be complex. Thus each treatment may consist of a number of factors, each at a certain level. For example, in studying effects of fertilizers, each treatment may contain nitrogen, potash and sulphate at different levels. Designs with complex treatments are called *Factorial Designs*. In this paper, we shall attempt a survey of combinatorial problems related to Factorial Designs. The combinational problems related to Incomplete Block Designs have been surveyed in an earlier paper, Bose [13].

2. Factorial designs

Suppose there are m factors; $A_1, A_2, \ldots, A_m$ in a factorial experiment and the factor A_i can be applied at any one of s_i levels. A treatment in which the factor A_i has been applied at the level x_i $(i = 1, 2, \ldots, m)$ may be denoted by $a_1^{x_1} a_2^{x_2} \cdots a_m^{x_m}$. We say that the experiment is of the type $s_1 \times s_2 \times \cdots \times s_m$. The number of different treatments is obviously $N = s_1 s_2 \cdots s_m$. If $s_1 = s_2 = \cdots = s_m$ we have a symmetric factorial experiment of the type s^m.

The effect of the treatment $a_1^{x_1} a_2^{x_2} \cdots a_m^{x_m}$ can be denoted by $t(x_1, x_2, \ldots, x_m)$. There are $N = s_1 s_2 \cdots s_m$ treatment effects. We consider all possible linear functions (with real coefficients) of the treatment effects. A typical linear function is

$$L = \sum C_{x_1 x_2, \ldots, x_m} t(x_1, x_2, \ldots, x_m) \tag{2.1}$$

where the summation is over all possible levels, $1 \leq x_i \leq s_i$, $i = 1, 2, \ldots, m$.

Orthogonality and independence of these linear functions are defined in the usual way. The mean of the treatment effects is given by

$$\mu = \frac{1}{N} \sum t(x_1, x_2, \ldots, x_m) \tag{2.2}$$

A linear function is said to be a contrast if it is orthogonal to the mean. Hence a necessary and sufficient condition for L given by (2.1) to be a contrast is

$$\sum C_{x_1 x_2 \cdots x_m} = 0. \tag{2.3}$$

Each independent linear function of the treatment effects can be regarded as carrying a *degree of freedom* (d.f.). One of these linear functions can be taken as the mean μ. Hence $N-1$ independent contrasts carry $N-1$ degrees of freedom. These degrees of freedom can be further subdivided into main effects and interactions as follows: A contrast is said to belong to the main effect A_i if the coefficients in the linear function constituting the contrast are independent of the levels of the factors other than A_i. Thus there are $s_i - 1$ independent contrasts which belong to the main effect A_i, and they carry $s_i - 1$ d.f.

A contrast is said to belong to the interaction of the factors A_i and A_j $(i \neq j)$, if it is orthogonal to any main effect, and the coefficients of the linear function constituting the contrast are independent of the levels of the factors other than A_i

and A_j. Thus there are $(s_i - 1)(s_j - 1)$ independent contrasts which belong to the interaction of A_i and A_j, and they carry $(s_i - 1)(s_j - 1)$ d.f.

In general, a contrast is said to belong to the interaction of the factors $A_{i_1}, A_{i_2}, \ldots, A_{i_r}$ if the linear function constituting the contrast is orthogonal to any main effect or to any k-factor interaction $k < r$, and the coefficients of the linear function constituting the contrast are independent of the levels of the factors other than $A_{i_1}, A_{i_2}, \ldots, A_{i_r}$. Thus there are $(s_{i_1} - 1)(s_{i_2} - 1) \cdots (s_{i_r} - 1)$ independent contrasts which belong to the interaction $A_{i_1}, A_{i_2}, \ldots, A_{i_r}$ and they carry as many d.f. Now

$$1 + \sum_{r=1}^{m} \binom{m}{r} (s_{i_1} - 1)(s_{i_2} - 1) \cdots (s_{i_r} - 1) = s_1 s_2 \cdots s_m = N.$$

Hence the mean, main effects and the various interactions carry the N d.f. which is the number of independent linear functions.

In a complete factorial experiment with r replications, there are r blocks, and each block has N plots (experimental units) to which the treatments are applied at random. The main effects and various interactions can then be estimated.

3. Confounding and fractional replication

In a complete factorial the block size tends to be large. Within block variability increases the error and reduces the efficiency. Yates [48, 49] developed the concept of confounding to overcome this difficulty. In many practical situations higher order interactions tend to be negligible. Yates method takes advantage of this fact to increase the efficiency of the experiment. We illustrate this concept by considering a symmetrical 2^3 experiment with three factors A_1, A_2, A_3 each at two levels 0 and 1.

In Yates' notation the treatment $a_1^{x_1} a_2^{x_2} \cdots a_m^{x_m}$ is regarded as a formal algebraic product. Since $a_i^0 = 1$ and $a_i^1 = a_i$, so for any treatment in which the level $x_i = 0$, the factor $a_i^{x_i}$ is dropped, and is replaced by a_i if $x_i = 1$. When all the levels are zero, the treatment is denoted by 1. Thus the eight treatments of a 2^3 experiment can be denoted by

$$1, a_1, a_2, a_3, a_2a_3, a_1a_3, a_1a_2, a_1a_2a_3.$$

The mean can then be written as the formal product

$$\mu = \tfrac{1}{8}(a_1 + 1)(a_2 + 1)(a_3 + 1) \tag{3.1}$$

since the linear function constituting the mean is obtained by formally developing the product on the right-hand side of (3.1).

The main effects and interactions can be written as

$$\tfrac{1}{4}(a_1 \pm 1)(a_2 \pm 1)(a_3 \pm 1) \tag{3.2}$$

where the +ve or the −ve sign is taken in $a_i \pm 1$ according as the factor A_i occurs at the level 0 or 1 in the effect. The numerical factor is conventionally taken as the reciprocal of the number of positive terms in any main effect or interaction. Thus the interaction between the three factors A_1, A_2, A_3, is written as

$$A_1A_2A_3 = \tfrac{1}{4}(a_1 - 1)(a_2 - 1)(a_3 - 1). \tag{3.3}$$

The treatments can be split in two groups $a_1a_2a_3$, a_1, a_2, a_3, and 1, a_2a_3, a_3a_1, a_1a_2. The treatments of the first (second) group carry a positive (negative) sign in the formal expansion of the right-hand side of (3.3).

If the treatments of different groups are assigned to different blocks as in Fig. 1 and we take the usual linear model

$$E\{y(a_1^{x_1}\, a_2^{x_2}\, a_3^{x_3})\} = t(x_1, x_2, x_3) + b_i + g \tag{3.4}$$

where $y(a_1^{x_1}\, a_2^{x_2}\, a_3^{x_3})$ denotes the observed response from the plot to which the treatment $a_1^{x_1}\, a_2^{x_2}\, a_3^{x_3}$ has been applied, b_i is the effect of block which contains this plot and g is the general effect, then we can estimate the main effects and two factor interaction, but the estimate of the three factor interaction is confounded with blocks since we can only estimate

$$\tfrac{1}{4}\{(a_1a_2a_3 + a_1 + a_2 + a_3) - (1 + a_1 + a_2 + a_3) + b_1 - b_2\} \tag{3.5}$$

where b_i is the effect of the ith block.

The general problem of confounding in the symmetrical case can be stated as follows:

Let $s = p^n$ where p is a prime. Consider a symmetric factorial experiment s^m. It is required to split the s^m treatments into s^{m-r} blocks, each of size s^r, so that as far as possible lower order interactions are unconfounded.

A general theory of confounding was developed by Bernard [7] for the 2^m factorial experiment. We can write the r factor interaction

$$A_{i_1}A_{i_2}\cdots A_{i_r} = \prod_{j=1}^{m} (a_j \pm 1)/2^{m-1} \tag{3.6}$$

where the negative sign is taken in $a_j \pm 1$ if j belongs to the set $\{i_1, i_2, \ldots, i_r\}$ and the positive sign is taken otherwise. Consider the Abelian group G generated by the symbols $A_1, A_2, \ldots, A_m$ with the relations

$$A_1^2 = A_2^2 = \cdots = A_m^2 = I,$$

where I is the identity element of the group. Bernard shows that any confounded arrangement corresponds to a subgroup of G. If we take any r independent

a_3	a_1
a_2	$a_1a_2a_3$

a_2a_3	1
a_2a_3	a_1a_2

Fig. 1

elements of G, then the subgroup generated by these elements consists of I (corresponding to the mean) and 2^{r-1} main effects or interactions all of which will be confounded. Thus in a 2^4 experiment in which a replication is to be split up into 2^2 blocks of size 2^2 if $A_1A_2A_3$ and $A_1A_2A_4$ are confounded, then A_3A_4 is also confounded. Thus it is easy to check that there exists no arrangement in which all the main effects and two factor interactions remain unconfounded.

The general theory of confounding was developed by Bose and Kishen [17], and by Fisher [27, 28]. The theory was further elaborated by Bose [8].

Consider an s^m symmetrical factorial experiment, $s = p^n$. In the Bose–Kishen theory, each level is identified with an element of the Galois field GF(p^n). Then the treatment $a_1^{x_1} a_2^{x_2} \cdots a_m^{x_m}$ can be represented by the point $(x_1, x_2, \ldots, x_m)$ of the finite affine geometry EG(m, p^n). This gives a (1, 1) correspondence between the set of treatments and points of the geometry. Consider the set of $(m-1)$-flats $\sum_\alpha$ with equations

$$c_1x_1 + c_2x_2 + \cdots + c_mx_m = \alpha \tag{3.7}$$

where α takes all values in GF(p^n). Then these s flats form a parallel pencil which may be denoted by

$$c_1x_1 + c_2x_2 + \cdots + c_mx_m = \alpha \tag{3.8}$$

or by the symbol $\{c_1, c_2, \ldots, c_m\}$. If P is any non-zero element of GF(p^n), then clearly the pencil $\{c_1, c_2, \ldots, c_m\}$ is the same as the pencil $\{\rho c_1, \rho c_2, \ldots, \rho c_m\}$.

We may denote by $(\sum_\alpha)$ the set of treatments corresponding to the $(m-1)$-flat $\sum_\alpha$. Then the s^m treatments are partitioned into s sets, corresponding to the flats of the pencil, and the $s-1$ independent contrasts between these sets carry $s-1$ d.f., which may be defined as the degrees of freedom carried by the pencil $\{c_1, c_2, \ldots, c_m\}$.

Bose and Kishen [17] showed that the d.f. carried by the pencil (3.8) belong to the interaction of the factors $A_{i_1}, A_{i_2}, \ldots, A_{i_r}$ if and only if the coefficients $c_{i_1}, c_{i_2}, \ldots, c_{i_r}$ in (3.8) are non-zero, and all the other coefficients are zero. Since there are $(s-1)^{r-1}$ parallel pencils for which this condition holds all the $(s-1)^r$ d.f. belonging to the interaction between the factors $A_{i_1}, A_{i_2}, \ldots, A_{i_r}$ are accounted for. They also show that the contrasts belonging to two distinct parallel pencils are mutually orthogonal. Thus the totality of $s^m - 1$ d.f. belonging to the contrasts are split up into $(s^m - 1)/(s-1)$ sets, with mutually orthogonal d.f.

If it is now required to split the s^m treatments into s^{m-r} blocks, this can be done by choosing any $k = m - r$ independent parallel pencils

$$\begin{aligned} L_1 &= a_{11}x_1 + a_{12}x_2 + \cdots + a_{1m}x_m = \text{const}, \\ L_2 &= a_{21}x_1 + a_{22}x_2 + \cdots + a_{2m}x_m = \text{const}, \\ &\cdot \\ &\cdot \\ &\cdot \\ L_k &= a_{k1}x_1 + a_{k2}x_2 + \cdots + a_{km}x_m = \text{const} \end{aligned} \tag{3.9}$$

where $L_1, L_2, \ldots, L_k$ are said to be independent if the corresponding coefficient vectors are independent, and dependent otherwise. Then the treatments belonging to any particular block are obtained by solving the equations

$$L_1 = c_1, \qquad L_2 = c_2, \quad \ldots, \quad L_k = c_k. \tag{3.10}$$

Choosing the values of $c_1, c_2, \ldots, c_k$ in all possible ways we get the s^{m-r} blocks. The degrees of freedom confounded belong to the pencils

$$\lambda_1 L_1 + \lambda_2 L_2 + \cdots + \lambda_k L_k = \text{const.} \tag{3.11}$$

Remembering that $(\lambda_1, \lambda_2, \ldots, \lambda_k) \neq (0, 0, \ldots, 0)$, and the pencil represented by (3.11) is unaltered by multiplication with non-zero ρ, we see that there are $(s^{m-r}-1)/(s-1)$ parallel pencils of the type (3.11). They carry the $s^{m-r}-1$ d.f. belonging to the contrasts between the s^{m-r} blocks. Thus to keep r-factor and all lower order interactions unconfounded, we must choose the pencils (3.9) in such a way that no pencil included in (3.11) has less than $r+1$ non-zero coefficients.

In Fisher's theory of confounding, [27, 28] the number of levels $s = p$ is a prime. His method is a generalization of Bernard's [7]. Let the m factors correspond to the symbols $A_1, A_2, \ldots, A_m$. Consider the group generated by these symbols, with the relations

$$A_1^p = A_2^p = A_m^p = I,$$

then the interaction components can be denoted by $A_1^{\lambda_1} A_2^{\lambda_2} \cdots A_m^{\lambda_m}$. Each subgroup with $k = m - r$ independent generators then leads to a design. The special case $p = 2$ then gives Bernard's results.

When $s = p$ is prime the Bose–Kishen and Fisher theories are equivalent. If $\lambda_{i_1}, \lambda_{i_2}, \ldots, \lambda_{i_r}$ are non-zero, then the component

$$A_{i_1}^{\lambda_{i_1}} A_{i_2}^{\lambda_{i_2}} \cdots A_{i_r}^{\lambda_{i_r}}$$

carries $s-1$ d.f. belonging to the interaction between the r factors $A_{i_1}, A_{i_2}, \ldots, A_{i_r}$. The equivalence becomes evident by noting that these degrees of freedom are carried by the pencil

$$\lambda_{i_1} x_{i_1} + \lambda_{i_2} x_{i_2} + \cdots + \lambda_{i_r} x_{i_r} = \text{const.}$$

If no main effects and two factor interactions are to be confounded then no pencil in (3.11) should have less than three non-zero coefficients or equivalently no element (except unity) of the subgroup of the design should contain less than three symbols. Fisher in this connection proved the following beautiful theorem.

Fisher's Theorem. *If no main effects and two factor interactions are to be confounded, then the maximum number of factors we can accommodate in a symmetric factorial experiment with s^{m-r} blocks, each of size s^r is*

$$m_2(r, s) = (s^r - 1)/(s-1). \tag{3.12}$$

4. Fractional replication

When s^m is large even one complete replication may be too expensive. If we assume that interactions between d or more factors are zero or negligible. Finney [24] showed that it is possible in certain cases to conduct an experiment with only s^{m-k} treatments (i.e., a $1/s^k$ fraction of all the treatments) so that the interactions of interest to the experimenter can be estimated. Each interaction component with $s-1$ d.f. is a member of an alias set consisting of s^k interaction components (each with $s-1$ d.f.) such that only the sum of s^k analogous contrasts, one belonging to each member of the alias set is estimable. Thus no alias set should contain more than one component which we desire to estimate, and all other components must belong to interactions assumed to be negligible or zero.

Finney's results can be elegantly expressed when viewed in the light of Bose–Kishen representation already discussed. This has been done by Kishen [29] and Bose [9]. The main result can be stated as follows:

In an s^m symmetrical factorial experiment, if we take a $1/s^k$ fraction defined by

$$L_1 = c_1,\ L_2 = c_2,\ \ldots,\ L_k = c_k \tag{4.1}$$

and further subdivide it into blocks by using the equations

$$L_{k+1} = \text{const.},\quad L_{k+2} = \text{const.},\quad \ldots\ L_{k+r} = \text{const.} \tag{4.2}$$

where each choice of the constants in (4.2) gives a block, then the degrees of freedom confounded with the block effects belong to pencils

$$\lambda_1 L_1 + \cdots + \lambda_k L_k + \cdots + \lambda_{k+r} L_{k+r} = \text{const.} \tag{4.3}$$

and the set of pencils

$$L + \mu_1 L_1 + \cdots + \mu_k L_k = \text{const.} \tag{4.4}$$

is the complete alias set of $L = \text{const.}$

If we want to estimate main effects and interactions under the assumption that all interactions between d or more factors are zero, then no pencil belonging to (4.3) should carry d.f. belonging to a main effect or two factor interaction, and for any $(\mu_1, \mu_2, \ldots, \mu_k) \neq (0, 0, \ldots, 0)$ the pencil (4.4) should carry d.f. belonging to an interaction between d or more factors.

5. The packing problem

We now return to the generalization of Fisher's theorem. This depends on the solution of the packing problem, formulated by Bose [8].

Let $M_t(r, s)$, $s = p^n$ denote the maximum number of columns, that we can choose in a r-rowed matrix M whose elements belong to $GF(p^n)$ and which has the property P_t that no t columns are dependent. Then the maximum number of factors we can accommodate in a symmetrical factorial experiment with s^{m-r} blocks each of size s^r, and such that no t-factor or lower order interaction is confounded is $M_t(r, s)$.

The coefficient vectors of the pencils chosen to construct the treatment sets should be orthogonal to the rows of M. Fisher's theorem follows at once by taking $t=2$, and noting that when $t \geqslant 2$, the columns of M can be identified with the points of the projective space PG$(r-1, s)$. Since the number of points in this space is $(s^r-1)/(s-1)$, Fisher's theorem follows.

If we want a $1/s^k$ fractional replication $k=m-r$ (without blocking), then we need a matrix with the property P_{2t} if no t-factor or lower order interaction is to be aliassed with a two factor or lower order interaction. A slight generalization of the packing problem suffices for the situation when blocking is wanted.

The packing problem is thus a fundamental combinatorial problem. The same problem arises in coding theory [16] and in the construction of combinatorial information retrieval systems [14]. It is far from solved in general.

When $t=3$, $m_3(r, s)$ equals the maximum number of points we can find in PG$(r-1, s)$ so that no three are collinear. In this case the following results are known [8, 32].

$$m_3(3, s) = s+2, \quad s=2^h,$$

$$m_3(3, s) = s+1, \quad s=p^h, \ p \text{ an odd prime},$$

$$m_3(4, s) = s^2+1, \quad s=p^h \neq 2, \ p \text{ a prime},$$

$$m_3(r, s) = 2^{r-1}, \quad s=2.$$

The value of $m_t(r, s)$ is known in a few other special cases. For example [10, 11, 12],

$$m_3(5, 3) = 20, \qquad m_4(5, 3) = 11, \qquad m_5(6, 3) = 12.$$

6. Construction and analysis of factorial designs

In a fractional design of resolution R, a p-factor interaction is not aliassed with any other effect containing less than $R-p$ factors, and is therefore estimable under the assumption that all interactions with $R-p$ or more factors are zero or negligible. The design is said to be orthogonal if the unbiased least square estimates are uncorrelated. Two important monographs containing plans for fractional factorial experiments of the type 2^m and 3^m were published by the National Bureau of Standards [30, 31]. Fractional factorial designs of the type 2^m and of resolutions 3, 4, 5 were studied by Box and Hunter [22].

So far we have considered only orthogonal symmetric designs. The study of unsymmetric designs started with Yates as early as [50], but serious work in this direction was not taken up for many years. Connor and Young [23] catalogue

$2^m \times 3^m$ fractional factorial designs, some orthogonal, and some in which correlation between estimates can be avoided by non-least square estimates. Addleman in a series of papers discussed the use and construction of both orthogonal and non-orthogonal fractions for symmetric as well as unsymmetric designs [1–6]. Bose and Srivastava give general methods of analysis for irregular factorial fractions [18, 19] and study a special type of balance viz. partial balance and use it to construct good factorial fractions [20, 21].

Much of the recent work on factorial designs is based on the concept of balanced arrays, which are a generalization of orthogonal arrays first introduced by Rao [33], and further properties of which were studied by Bose and Bush [15].

The concept of balanced arrays was first introduced by Chakravarti [51] who used them for construction of asymmetrical factorial designs. A balanced array (B-array) T of strength t, s symbols and size $(m \times N)$ is an $m \times N$ matrix T with elements belonging to a set S containing s symbols, such that for every $t \times N$ submatrix T_0 of T, and for every vector $\boldsymbol{v}$ of size $(t \times 1)$ with elements from S we have

$$\lambda(\boldsymbol{v}, T_0) = \lambda(P(\boldsymbol{v}), T_0) \quad \text{for every } P,$$

where $\lambda(\boldsymbol{v}, T_0)$ stands for the number of columns of T_0 which are identical with $\boldsymbol{v}$, and $P(v)$ is any vector obtained by permuting the elements of v. Thus P stands for a permutation.

The matrix T above is called an orthogonal array with the same parameters if for every $(t \times N)$ submatrix T_0 of T, and for every pair of vectors $\boldsymbol{v}$ and $\boldsymbol{v}^*$ of size $(t \times 1)$ each, we have

$$\lambda(\boldsymbol{v}, T_0) = \lambda(\boldsymbol{v}^*, T_0).$$

Clearly an orthogonal array is a special type of balanced array. It is an important combinatorial problem to study the conditions which the parameters must satisfy for the array to exist. Existence conditions for balanced arrays with two symbols were studied by Srivastava [34], and the general arrays with s levels were investigated by Srivastava and Chopra [45].

When there are only two levels, 0 and 1, then the columns of a balanced array with $S = \{0, 1\}$, provide a balanced fraction of 2^m factorial treatments. The variance covariance matrix Σ of the estimates of the interactions depends only on the order of the interactions and not on the factors contained in them. The information matrix is the inverse of Σ. To obtain an optimal design one should choose the balanced array such that for the corresponding design Σ^{-1} should be optimal with respect to some criterion. Optimal designs of the 2^m series have been extensively studied by Srivastava and her collaborators. They use the trace criterion and maximize Σ^{-1}; Srivastava and Anderson [39, 40], Srivastava and Chopra [42, 43, 44, 45]. Another possible criterion is to maximize det Σ, or to minimize the largest characteristic root of Σ. Srivastava and Anderson [41] compare the relative advantages and disadvantages of the three optimality criteria.

An excellent survey of recent work on optimal factorial designs has been given by Srivastava [38].

7. Search designs

Another important concept viz. that of search designs has been introduced by Srivastava. The application of this to the factorial case may be described as follows:

Suppose we want to estimate interactions up to a given order (say up to interactions between t or lesser number of factors) but allow for the possibility that a fixed number k of the remaining effects may be non-negligible. We then want to estimate up to t-factor interactions and search and estimate the k non-negligible interactions. Srivastava and her collaborators have given criteria for the existence of these designs, and also discussed methods of construction and analysis of these designs [35, 36, 37, 46, 47]. This work is still going on.

References

[1] S. Addleman, Irregular fractions of the 2^m factorial experiments, Technometrics 3 (1961) 479–496.

[2] S. Addleman, Augmenting factorial plans to accomodate additional two-level factors, Biometrics 18 (1962) 308–322.

[3] S. Addleman, Orthogonal main effect plans for asymmetrical factorial experiments, Technometrics 4 (1962) 21–46.

[4] S. Addleman, Symmetrical and asymmetrical fractional factorial plans, Technometrics 4 (1962) 47–58.

[5] S. Addleman, Techniques for constructing fractional replicate plans, J. Am. Statist. Assoc. 58 (1963) 45–71.

[6] S. Addleman, The construction of a 2^{17-9} resolution V plan in eight blocks of 32, Technometrics 7 (1965) 439–443.

[7] M.M. Bernard. An enumeration of the confounded arrangements in the $2\times2\times2\cdots$ factorial designs, Suppl. J. Roy. Statist. Soc. 3 (1936) 195–202.

[8] R.C. Bose, Mathematical theory of the symmetrical factorial design, Sankhya 8 (1947) 107–166.

[9] R. C. Bose. Mathematics of factorial designs, in: Proc. Int. Congr. Math., Cambridge, MA (1950) 543–548.

[10] R.C. Bose. On some connections between the design of experiments and information theory, Bull. Inst. Int. Statist. 38(4) (1961) 257–271.

[11] R.C. Bose, Some ternary error correcting codes and the corresponding fractionally replicated designs, Coll. Int. du C.N.R.S. le Plan d'Experiences, No. 110 (1963) 21–32.

[12] R.C. Bose, Error correcting, error detecting and error locating codes, in: Essays in Probability and Statistics, S.N. Roy Memorial Volume (Univ. N. C. Press, NC, 1970) 147–178.

[13] R.C. Bose, Combinatorial problems of experimental design I. Incomplete block designs, in: Proc. Symposia in Pure Math., Vol. 34: Relations between Combinatorics and Other Parts of Mathematics (Am. Math. Soc., Providence, RI, 1979) 47–68.

[14] R.C. Bose, C.T. Abraham and S.P. Ghosh, in: Proc. Conference on Combinatorial Mathematics and its Applications (Univ. N. C. Press, NC, 1969) 277–297.

[15] R.C. Bose and K.A. Bush, Orthogonal arrays of strength 2 and 3, Ann. Math. Statist. 23 (1952) 508–524.

[16] R.C. Bose and D.K. Ray-Chaudhari, Information and Control 3 (1960) 68–79.
[17] R.C. Bose and K. Kishen, On the problem of confounding in the general symmetrical factorial design, Sankhya 5 (1940) 21–36.
[18] R.C. Bose and J.N. Srivastava, Mathematical Theory of Factorial Designs. I Analysis; II Construction, Bull. Inst. Int. Statist. 40 (2c) (1964) 786–794.
[19] R.C. Bose and J.N. Srivastava, Analysis of irregular fractions, Sankhya, Ser. A 26 (1964) 177–144.
[20] R.C. Bose and J.N. Srivastava, Multidimensional partially balanced designs, Sankhya, Ser. A 26 (1964) 145–168.
[21] R.C. Bose and J.N. Srivastava, Ann. Inst. Statist. Math. 18 (1966) 57–73.
[22] G.E.P. Box and J.S. Hunter, The 2^{k-p} fractional factorial designs, Part I and Part II, Technometrics 3 (1961) 311–351 and 449–458.
[23] W.S. Connor and S. Young, Fractional factorial designs for experiments with factors at two and three levels, NBS Appl. Math. Ser. 58 (1961).
[24] D.J. Finney, The fractional replication of factorial arrangements, Ann. Eugen. 12 (1945) 291–301.
[25] R.A. Fisher, Statistical Methods for Research Workers (Oliver and Boyd, Edinburgh, 1925).
[26] R.A. Fisher, Arrangement of field experiments, J. Min. Agric. G.B. 33 (1926) 503–513.
[27] R.A. Fisher, The theory of confounding in factorial experiments in relation to the theory of groups, Ann. Eugen. 11 (1942) 341–353.
[28] R.A. Fisher, A system of confounding for factors with more than two alternatives, giving completely orthogonal cubes and higher powers, Ann. Eugen. 12 (1945) 283–290.
[29] K. Kishen, On fractional replication of the general symmetrical factorial design, J. Ind. Soc. Agric. Statist. 1 (1948) 91–106.
[30] National Bureau of Standards, Fractional factorial experiment designs for factors at two levels, Appl. Math. Series 48 (1957).
[31] National Bureau of Standards, Fractional factorial designs for experiments with factors at three levels, Appl. Math. Series 54 (1959).
[32] B. Quist, Some remarks concerning curves of the second degree on a finite plane, Ann. Acad. Sci. Finland 134 (1952) 1–27.
[33] C.R. Rao, Factorial experiments derivable from combinatorial arrangements of arrays, Suppl. J. Roy Statist. Soc. 9 (1947) 128–139.
[34] J.N. Srivastava, Some general existence conditions for balanced arrays of strength t and 2 symbols, J. Combin. Theory 13 (1972) 198–206.
[35] J.N. Srivastava, Designs for searching non-negligible effects, in: J.N. Srivastava, ed., A Survey of Stastical Designs and Linear Models (North Holland, Amsterdam, 1975) 507–519.
[36] J.N. Srivastava, Some further theory of search linear models, in: Contributions to Applied Statistics (Swiss Australian Region of Biometry Soc.) 249–256.
[37] J.N. Srivastava, Optimal search designs or designs optimal under bias-free optimality criteria, in: (S.S. Gupta and D.S. Moore, eds.) Statistical Decision Theory and Related Topics, Vol. 2 Academic Press, New York, 1977) 375–409.
[38] J.N. Srivastava, A review of some recent work on discrete optimal factorial designs for statisticians and experimenters, in: Developments in Statistics, Vol. 1 (Academic Press, New York, 1978) 267–329.
[39] J.N. Srivastava and D.A. Anderson, Optimal fractional factorial plans for main effects orthogonal two factor interactions 2^m series, J. Am. Statist. Assoc. 65 (1970) 828–843.
[40] J.N. Srivastava and D.A. Anderson, Factorial sub-assembly association scheme and multidimensional partially balanced designs, Ann. Math. Statist. 42 (1971) 1167–1181.
[41] J.N. Srivastava and D.A. Anderson, A comparison of the determinant, trace and largest root optimality criteria, Comm. Statist. 3 (1974) 933–940.
[42] J.N. Srivastava and D.V. Chopra, On the comparison of certain classes of balanced 2^8 fractional factorial designs of resolution V, with respect to the trace criterion, J. Ind. Soc. Agric. Statist. 23 (1971) 124–131.
[43] J.N. Srivastava and D.V. Chopra, On the characteristic roots of the information matrix of 2^m balanced factorial designs of resolution V with applications, Ann. Math. Statist. 42 (1971) 722–734.

[44] J.N. Srivastava and D.V. Chopra. Balanced optimal 2^m fractional factorial designs of resolution V, $m \leq 6$. Technometrics 13 (1971) 257–269.
[45] J.N. Srivastava and D.V. Chopra, Balanced arrays and orthogonal arrays, in: (J.N. Srivastava, ed.,) A Survey of Combinatorial Theory (North-Holland, Amsterdam, 1973) 411–428.
[46] J.N. Srivastava and S. Ghosh, Balanced 2^m factorial designs of resolution V which allow search and estimation of one extra unknown effect $4 \leq m \leq 8$, Comm. Statist. Theor. Methods A6(2) (1977) 141–166.
[47] J.N. Srivastava and D.M. Mallenby, Some studies on a new method of search linear models, to appear.
[48] F. Yates, The principles of orthogonality and confounding in replicated experiments, J. Agric. Sci. 23 (1933) 108–145.
[49] F. Yates, Complex experiments, Suppl. J. Roy Statist. Soc. 2 (1935) 181–223.
[50] F. Yates, The design and analysis of factorial experiments, Imperial Bureau of Soil Science, Technical Communications, No. 35 (1937).
[51] I.M. Chakravarti, Fractional replications in asymmetrical factorial designs and partially balanced arrays, Sankhya 17 (1956) 143–164.

Annals of Discrete Mathematics 6 (1980) 19–29

PARTIAL λ-GEOMETRIES OF SMALL NEXUS

Peter J. CAMERON

Merton College, Oxford OX1 4JD, England

David A. DRAKE

University of Florida, Gainesville, FL 32611, U.S.A.

In a partial λ-geometry, each two points are joined by 0 or λ blocks; each two blocks have 0 or λ points in common; the block size k is constant; and for each nonflag (p, G), there are precisely e blocks X with p in X such that $X \cap G$ is not empty. Generalized quadrangles are partial 1-geometries with nexus $e = 1$. If $\lambda = 2$, then $e \geq 3$; and the first author has determined that partial 2-geometries with nexus 3 exist precisely for the values $k = 3, 4, 8, 24$. We prove (1) that if $\lambda > 2$ and $k > e + 1$, then $e > 2\lambda$; (2) that $\lambda = 3$, $e = 7$ implies k is one of 7, 15, 21, 24 or 36. We call a partial λ-geometry extremal if $|G \cap H \cap K| > 1$ implies $|G \cap H \cap K| \geq \lambda$. There are no extremal partial λ-geometries with $e \leq \lambda^2 - \lambda$. Such a geometry with $e = \lambda^2 - \lambda + 1$ and $k > e$ is called a λ-quadrangle. We determine all λ-quadrangles with $\lambda > 2$. They are constructed from quadratic forms of Witt index 4 on finite 8-dimensional vector spaces.

0. Introduction

In Section 1, we observe that the block graph of a partial λ-geometry is strongly regular and apply the rationality-integrality conditions for strongly regular graphs to obtain a non-existence criterion for partial λ-geometries. In Section 2, we obtain a lower bound on the nexus of a partial λ-geometry. In Section 3, we obtain a better lower bound for the nexus e of the subclass of extremal partial λ-geometries; namely, $e > \lambda^2 - \lambda$. A proper extremal partial λ-geometry with $e = \lambda^2 - \lambda + 1$ is called a λ-quadrangle. In the remainder of Section 3, we determine the λ-quadrangles with $\lambda > 2$. (Those with $\lambda = 2$ have already been determined up to possible non-uniqueness when $k = 24$.)

To obviate excessive repetition of the word "finite", we now assert that all incidence structures considered in this paper are tacitly assumed to be finite. À la Dembowski, we write $[p_1, \ldots, p_n]$ to denote the number of blocks which contain the point set $\{p_1, \ldots, p_n\}$, $[G_1, \ldots, G_n]$ for the dual notion.

1. The block graph of a partial λ-geometry

Definition 1.1. For $\lambda > 0$, a *partial λ-geometry* (with *nexus* $e > 0$) is an incidence structure with b blocks and v points which satisfies:

(i) $[p, q] = 0$ or λ for each point pair (p, q) with $p \neq q$;

(ii) $[G, H]=0$ or λ for each block pair (G, H) with $G \neq H$;

(iii) for each non-incident point-block pair (p, G), there exist precisely e blocks X with $p \in X$ and $[X, G] \neq 0$;

(iv) $\lambda<[p]<b$ for every p, and $\lambda<[G]<v$ for every G; if $\lambda=1$, we also assume the existence of integers k, r such that $[G]=k$ and $[p]=r$ for all G, p.

Partial 1-geometries were first studied by Bose who called them simply "partial geometries". He proved the following result in the special case that $\lambda=1$ (see [2, p. 398]). The result with $\lambda \neq 1$ is due to the second author [6, Lemma 1.3].

Lemma 1.2. *Let $\mathcal{G}$ be a partial λ-geometry. Then there are integers k, r such that $[G]=k$ and $[p]=r$ for all G, p. Further,*

$$v=[k(r-1)(k-\lambda)/e\lambda]+k,.$$

$$b=[r(r-1)(k-\lambda)/e\lambda]+r.$$

If $\lambda \neq 1$, $r=k$; and therefore $b=v$.

If $e=k$, each pair of points in a partial λ-geometry $\mathcal{G}$ is joined, hence $\mathcal{G}$ is a (balanced incomplete) block design. If $e<k$, we shall call $\mathcal{G}$ a *proper* partial λ-geometry.

A valuable tool in our study is the notion of the *block graph* $\mathcal{B}$ of an incidence structure $\mathcal{G}$: $\mathcal{B}$ is obtained by using the blocks of $\mathcal{G}$ as vertices and taking two blocks to be adjacent if and only if they have a nonempty intersection. Recall that a *strongly regular graph* is a graph with the following properties: (1) each vertex is adjacent to the same number a of vertices; (2) $0<a<n-1$ where n denotes the total number of vertices; (3) the number of vertices adjacent to both of G and H is a constant c if G and H are themselves adjacent, otherwise is the constant d.

Proposition 1.3. *Let $\mathcal{G}$ be a proper partial λ-geometry with $\lambda \neq 1$. Then the block graph of $\mathcal{G}$ is strongly regular and has the following parameters:*

$$n=v=[k(k-1)(k-\lambda)/e\lambda]+k,$$

$$a=k(k-1)/\lambda,$$

$$c=[k\lambda-k-\lambda+e(k-\lambda)]/\lambda,$$

$$d=ek/\lambda.$$

Proof. The value for n is given by Lemma 1.2. To compute n, however, one need only fix a block G and count (in two ways) the double flags (x, y, Z) with x in G, y not in G. To compute d, one fixes a pair of non-intersecting blocks B and G and observes that the number of double flags (x, y, Z) with x in B and y in G is $ke\lambda=d\lambda^2$. Next, let B and G be fixed blocks with a non-trivial intersection; let f denote the number of blocks which have non-trivial intersection with B but are parallel to G. Counting the flags (x, Y) such that $x \in B - G$, $[Y, G] = 0$, one obtains $(k-\lambda)(k-e) = f\lambda$. Clearly, $c = a - 1 - f$. Now one easily computes a, hence c. Using the computed values for a and n and assuming that $a = n - 1$, one sees that k must be either e or λ, a contradiction. The following theorem yields the so-called "rationality conditions" or "integrality conditions" for strongly regular graphs. For a proof, one may consult Cameron and Van Lint [5, pp. 15, 16].

Theorem 1.4. *The adjacency matrix of a strongly regular graph has three eigenvalues whose multiplicities are* 1 *and*

$$\frac{1}{2} \cdot \left[n - 1 \pm \frac{(n-1)(d-c) - 2a}{((d-c)^2 + 4(a-d))^{1/2}} \right].$$

Proposition 1.5. *Let $\mathcal{G}$ be a proper partial λ-geometry with $\lambda \neq 1$. Then the following expression is an integer*:

$$\frac{k^2(k-1)(k-\lambda)}{e\lambda[k(\lambda+1) - \lambda(e+1)]}.$$

Proof. Apply Proposition 1.3 and Theorem 1.4, using the plus-sign in the expression in Theorem 1.4.

Proposition 1.6. *Let $\mathcal{G}$ be a proper partial λ-geometry with $\lambda \neq 1$. Then $p(k) \equiv k(\lambda+1) - \lambda(e+1)$ is a nonzero integer which evenly divides the integer $\lambda^3(e+1)^2(\lambda e - 1)(e - \lambda)$.*

Proof. By Proposition 1.5, $p(k)$ divides $k^2(k-1)(k-\lambda)$. The desired conclusion follows from the following facts: GCD$(p(k), k)$ divides $\lambda(e+1)$; GCD$(p(k), k-1)$ divides $(k-1)(\lambda+1) - p(k) = \lambda e - 1$; and GCD$(p(k), k-\lambda)$ divides $(k-\lambda)(\lambda+1) - p(k) = \lambda(e-\lambda)$.

Corollary 1.7. *For each pair (λ, e) with $\lambda > 1$, $e > 0$, there are only finitely many proper partial λ-geometries of nexus e.*

Proof. By Proposition 1.6, there are only finitely many possible k's for a given pair (λ, e); hence, by Proposition 1.3, only finitely many possible v's.

2. A lower bound for the nexus of a partial λ-geometry

Generalized quadrangles are partial 1-geometries of nexus 1. Thus there are many partial λ-geometries which satisfy $\lambda = e = 1$; however, $\lambda \neq 1$ clearly implies that $e > \lambda$. The first author has investigated the situation when $\lambda = 2$ and $e = 3$, obtaining the following

Theorem 2.1. (Cameron [4, Theorem 5.12]). *A partial 2-geometry with nexus 3 exists for precisely the values $k = 3$, 4, 8 and 24. With the possible exception of $k = 24$, the structure is determined by k.*

We shall devote Section 2 of this paper to proving

Theorem 2.2. *Assume the existence of a partial λ-geometry $\mathcal{G}$ with $\lambda > 2$ and $k > e + 1$. Then $e > 2\lambda$; and if λ is even, $e > 2\lambda + 1$.*

Remark 2.3. The requirement $k > e + 1$ is necessary: the existence of a Hadamard matrix of order $2\lambda > 3$ implies the existence of a partial λ-geometry with $k = 2\lambda$ and $e = 2\lambda - 1$ (see [6, Proposition 2.3 and Lemma 1.8]).

Proof of Theorem 2.2. To begin, let $\{G, H\}$ be a fixed pair of intersecting blocks of $\mathcal{G}$. Let c_i be the number of blocks Y such that $G \neq Y \neq H$, $|G \cap Y| > 0$, $|H \cap Y| > 0$ and $|G \cap H \cap Y| = i$, $0 \leq i \leq \lambda$. Then the formula for c in Proposition 1.3 yields the equation

$$\sum_{i=0}^{\lambda} \lambda c_i = k(e + \lambda - 1) - \lambda(e + 1). \tag{1}$$

Counting flags (x, Y) such that $x \in G \cap H$ and $H \neq Y \neq G$, one obtains

$$\sum_{i=0}^{\lambda} i c_i = \lambda(k - 2). \tag{2}$$

Counting double flags (x, y, Z) such that $x, y \in G \cap H$, $x \neq y$ and $G \neq Z \neq H$, one gets

$$\sum_{i=0}^{\lambda} i(i-1)c_i = \lambda(\lambda - 1)(\lambda - 2). \tag{3}$$

We now multiply (1) by $(j^2 + j)$ where j is some *integer*, (2) by $-2j\lambda$, and (3) by λ, and add the three resultant equations. The left-hand side of this linear combination is $\sum f(i)c_i$ where

$$f(i) = i^2\lambda - i(2j\lambda + \lambda) + (j^2\lambda + j\lambda).$$

Then $f'(i) = 2i\lambda - (2j\lambda + \lambda) = 0$ when $i = j + \frac{1}{2}$, and $f''(i) = 2\lambda > 0$; so $f(i)$ has an absolute minimum at $i = j + \frac{1}{2}$. Since $f(j) = f(j+1) = 0$, $f(i) \geq 0$ for all *integral* values of i. Since the c_i's are also all non-negative, the left-hand side of the linear combination of (1)–(3) is non-negative; and

$$0 \leq (j^2 + j)[k(\lambda + e - 1) - \lambda(e + 1)] - 2j\lambda(k\lambda - 2\lambda) + (\lambda^4 - 3\lambda^3 + 2\lambda^2).$$

then

$$kj[2\lambda^2-(j+1)(\lambda+e-1)]\leqslant\lambda^4-3\lambda^3+2\lambda^2+4j\lambda^2-\lambda(e+1)(j^2+j). \tag{4}$$

Then if $2\lambda^2-(j+1)(\lambda+e-1)>0$, we may replace k by $(e+1)$ in (4), obtaining a strict inequality. Restating, we have

$$(e+1)[2j\lambda^2-(j^2+j)(\lambda+e-1)]<\lambda^4-3\lambda^3+2\lambda^2+4j\lambda^2-\lambda ej^2-\lambda ej-\lambda j^2-\lambda j \tag{5}$$

provided that

$$\frac{2\lambda^2-\lambda-e+1}{\lambda+e-1}>j. \tag{6}$$

Now (5) simplifies to

$$0<(\lambda^4-3\lambda^3+2\lambda^2)+j(-2e\lambda^2+2\lambda^2+e^2-1)+j^2(e^2-1). \tag{7}$$

Consider the quantity

$$d=\lambda^2/(e+1). \tag{8}$$

We wish to substitute $[d]$, the largest integer $\leqslant d$, for j in (7). We may do this if we can show that $j=d$ satisfies (6); i.e., that

$$(2\lambda^2-\lambda-e+1)(e+1)>\lambda^2(\lambda+e-1). \tag{9}$$

This inequality is equivalent to

$$0>e^2+e(-\lambda^2+\lambda)+(\lambda^3-3\lambda^2+\lambda-1)\equiv h(e). \tag{10}$$

Now $h(e)=0$ if and only if $2e=\lambda^2-\lambda\pm(\lambda^4-6\lambda^3+13\lambda^2-4\lambda+4)^{1/2}$. Since $(\lambda^2-3\lambda+2)^2<\lambda^4-6\lambda^3+13\lambda^2-4\lambda+4$, $h(e)$ satisfies (10) when $2e$ lies between the values $\lambda^2-\lambda\pm(\lambda^2-3\lambda+2)$; i.e., when $\lambda-1\leqslant e\leqslant\lambda^2-2\lambda+1$. If $\lambda>3$, then inequality (10) holds when $\lambda-1\leqslant e\leqslant 2\lambda+1$.

Now denote the right-hand side of inequality (7) by $g(j)$, regarding e and $\lambda>3$ as fixed. Since $j=d$ satisfies (6), $0<g(j)$ for every integer $j\leqslant d$. Now

$$g'(j)=j(2e^2-2)-2e\lambda^2+2\lambda^2+e^2-1,$$

so $g'(j)=0$ if and only if

$$j=\frac{2e\lambda^2-2\lambda^2-e^2+1}{2e^2-2}=\frac{2d-1}{2}.$$

Since $g''(j)=2e^2-2>0$, $g(j)$ is a declining function to the left of $j=\frac{1}{2}(2d-1)$ and a rising function on the right.

We wish to prove that $g(d-1)=g(d)\leqslant 0$. It will follow that $g([d])\leqslant 0$, violating (7) and proving the non-existence of a partial λ-geometry with nexus e. One easily computes

$$g(d-1)=g(d)+2e\lambda^2-2\lambda^2-e^2+1+(1-2d)(e^2-1)=g(d).$$

Next, one computes

$$(e+1)g(d)/\lambda^2 = 2\lambda^2 - 3\lambda(e+1) + (e+1)^2$$

and concludes that $g(d) \leq 0$ if $\lambda - 2 < e < 2\lambda$. We have proved

Proposition 2.4. *There are no proper partial* λ*-geometries whose parameters satisfy* $\lambda > 3$, $k > e+1$ *and* $e < 2\lambda$.

Now, assume $\lambda = 3$. Then (6) is satisfied by $j = 1$ for all $e < 7$. Setting $\lambda = 3$ and $j = 1$ in (7) yields a contradiction if $\lambda < e < 7$. Next, assume the existence of a partial λ-geometry with parameter $e = 2\lambda$. Then (7) simplifies to

$$0 < (\lambda^4 - 3\lambda^3 + 2\lambda^2) + j(-4\lambda^3 + 6\lambda^2 - 1) + j^2(4\lambda^2 - 1). \tag{11}$$

The preceding inequality must hold when j satisfies (6); i.e., when

$$2\lambda^2 - 3\lambda + 1 > j(3\lambda - 1). \tag{12}$$

Assume that λ is an even integer. Then $j = \frac{1}{2}(\lambda - 2)$ satisfies (12), hence may be substitued into (11) to yield $0 < -\lambda^2 + 2\lambda$, a contradiction. If λ is odd and if $e = 2\lambda$, one observes that $j = \frac{1}{2}(\lambda - 1)$ satisfies (12); then one may substitute $j = \frac{1}{2}(\lambda - 1)$ into (11) to obtain the contradiction $0 < -\lambda^2 + 1$.

Lastly, assume that $e = 2\lambda + 1$, λ even. Then (7) becomes

$$0 < (\lambda^4 - 3\lambda^3 + 2\lambda^2) + j(-4\lambda^3 + 4\lambda^2 + 4\lambda) + j^2(4\lambda^2 + 4\lambda). \tag{13}$$

The preceding inequality must hold when j satisfies (6); i.e., when $2\lambda^2 - 3\lambda > 3j\lambda$. Then (13) must hold when $j = \frac{1}{2}(\lambda - 2)$; this requirement yields another contradiction and thus completes the proof of Theorem 2.2.

We conclude this section by considering the possible 3-geometries of nexus 7. Proposition 1.6 implies that $4k - 24$ divides $2^{10} \cdot 3^3 \cdot 5$ evenly. Then $4k - 24 = 2^a \cdot 3^b \cdot 5^c$ where $a \leq 10$, $b \leq 3$, $c \leq 1$. Since $4k - 24$ is a multiple of 4, $a \geq 2$.

By Proposition 1.3, the integer $d = ek/\lambda = \frac{7}{3}k$, so $3 \mid k$. Then $4k - 24$ is a multiple of 3, so $b \geq 1$. By Proposition 1.3, the integer $v - k = \frac{1}{21}k(k-1)(k-3)$, so $7 \mid [k(k-1)(k-3)]$. Then $k \equiv 0$, 1 or 3 (mod 7), and $k = 2^{a-2} \cdot 3^b \cdot 5^c + 6$, so

$$2^{a-2} \cdot 3^b \cdot 5^c \equiv 1,\ 2 \text{ or } 4 \pmod 7.$$

Thus $3^b \cdot 5^c \equiv 1$, 2 or 4 (mod 7), and hence $3^b \cdot 5^c$ is one of 9, 15, 135. We have proved that $k = 6 + \alpha \cdot 2^\beta$ where $\alpha = 9$, 15 or 135 and $0 \leq \beta \leq 8$.

Since $\lambda = 3$ and $e = 7$, the integer in Proposition 1.5 is

$$\begin{aligned} D_k &= \frac{k^4 - 4k^3 + 3k^2}{21(4k - 24)} \\ &= \frac{(k-6)^4 + 20(k-6)^3 + 147(k-6)^2 + 468(k-6) + 540}{84(k-6)} \\ &= \frac{\alpha^4 \cdot 2^{4\beta} + 20\alpha^3 \cdot 2^{3\beta} + 147\alpha^2 \cdot 2^{2\beta} + 468\alpha \cdot 2^\beta + 540}{21\alpha \cdot 2^{\beta+2}} \end{aligned}$$

If $\beta \geq 2$, $2\beta \geq \beta+2$; so $2^{\beta+2} \mid 540$, and hence $\beta \leq 1$. Then the only possible values of k are 15, 24, 21, 36, 141 and 276. One can see that D_k is not an integer for the two largest of these values, so we have proved

Proposition 2.5. *If $\mathscr{G}$ is a partial* 3-*geometry with nexus* 7, *then the block size in $\mathscr{G}$ must be* 7, 15, 21, 24 *or* 36.

3. λ-quadrangles

In this section, we define λ-quadrangles. For $\lambda = 2$, these are simply the proper partial λ-geometries of nexus 3 discussed in Theorem 2.1. We will completely determine the λ-quadrangles with $\lambda > 2$.

Definition 3.1. A partial λ-geometry will be called *extremal* if the condition $|G \cap H \cap K| > 1$ for blocks G, H, K implies that $|G \cap H \cap K| \geq \lambda$.

Proposition 3.2. *The inequality $e \geq \lambda^2 - \lambda + 1$ holds in every extremal partial λ-geometry $\mathscr{G}$. Equality holds if and only if the following criterion is satisfied: for each nonflag (p, G), the e blocks through p which intersect G induce a projective plane of order $\lambda - 1$ on the e points of G which are joined to p (degenerate unless $\lambda > 2$).*

Proof. Let (p, G) be a nonflag, S be the set of e points of G which are joined to p. Since $\mathscr{G}$ is extremal, the blocks through p induce a "partial plane" E on S; i.e., no two points of S lie in more than one common induced block. The number of pairs of points of S which are joined in E is $\frac{1}{2}e\lambda(\lambda-1)$ while the total number of pairs of points of S is $\frac{1}{2}e(e-1)$.

Definition 3.3. A *λ-quadrangle* is a proper extremal partial λ-geometry with $e = \lambda^2 - \lambda + 1$.

We describe next a family of λ-quadrangles. We refer to [1, 7, 8] for the geometric background: see in particular [1, Theorems 3.8 and 3.9 (Witt's Theorem)].

Example 3.4. Let V be an 8-dimensional vector space over the field GF(q). Let Q be a quadratic form of Witt index 4 on V. The totally singular subspaces of dimension 4 will be called *half spaces*. The half spaces can be divided into two families with the property that, for any two half spaces S and T, the codimension of $S \cap T$ in S is even if and only if S and T belong to the same family. (This is implicit in [8, 7.12 and 8.4.3]; we give a short direct proof in the Appendix.) Define an incidence structure $\mathscr{G}(q)$ as follows: the points of $\mathscr{G}(q)$ are the totally singular 1-dimensional subspaces of V; the blocks are the members of one family of half spaces; and incidence is the relation of containment.

Proposition 3.5. *$\mathcal{G}(q)$ is a $(q+1)$-quadrangle with $e=q^2+q+1$, $k=q^3+q^2+q+1$.*

Proof. Any block, with the points it contains, is a projective 3-space over GF(q); and two blocks meet in the empty set or a line. Given a totally singular 3-dimensional subspace $U=\langle v_1, v_2, v_3\rangle$, Witt's Theorem yields a basis $\{v_1, \ldots, v_4, w_1, \ldots, w_4\}$ for V consisting of singular vectors with $B(v_i, w_i)=1$, $B(v_i, v_j)=B(v_i, w_j)=B(w_i, w_j)=0$ for $i\neq j$ (where B is the bilinear form associated with Q). If follows that $\langle v_1, v_2, v_3, v_4\rangle$ and $\langle v_1, v_2, v_3, w_4\rangle$ are the only half spaces containing U and that exactly one of these is a block. Counting arguments (which we omit) now show that any point lies in q^3+q^2+q+1 blocks and that any two points on a singular line L lie in $q+1$ blocks whose intersection is L. Of course, two points on a hyperbolic line cannot lie in a block. Finally, if S is a half space and $p=\langle v\rangle$ a point not in S, the map $w \rightarrow B(v, w)$ is a nonzero linear form on S. A point incident with S is contained in the kernel of this map if and only if it is joined to p by a singular line. Then p is joined to q^2+q+1 points of S. It follows that the same number q^2+q+1 of blocks containing p meet S. This completes the proof.

Our main result is the converse to Proposition 3.5.

Theorem 3.6. *A λ-quadrangle with $\lambda \geqslant 3$ is isomorphic to $\mathcal{G}(q)$ where $\lambda=q+1$.*

Proof. We let $\mathcal{G}$ be a λ-quadrangle with $\lambda \geqslant 3$. Put $\lambda=q+1$, hence $e=q^2+q+1$. A non-empty intersection of two blocks will be called a *line.* If (p, G) is a non-flag, the set $\Pi(p, G)$ of points of G collinear with p will be called a *plane.* Proposition 3.2 shows that a plane, equipped with the lines it contains, is a projective plane (of order q). It is clear that any two points of a block lie in a unique line, and we prove

Lemma 3.7. *Three non-collinear points of a block lie in a unique plane.*

Proof. Let p, q, r be non-collinear points of G. There is a block $H\neq G$ containing p and q; r is joined to a point s of H outside G (indeed, to q^2 such points); and p, q, r lie in the plane $\Pi(s, G)$. This plane is the smallest linear subspace containing p, q and r, hence is uniquely determined by these points.

It follows from the classical axioms of projective geometry that a block, together with its points, lines and planes, is a projective space. We let $d-1$ denote its dimension, so that $k=(q^d-1)/(q-1)$.

Lemma 3.8. *d is a multiple of 4.*

Proof. Let G and H be disjoint blocks. Define a relation $\sim$ on G by the rule $p \sim q$ if either $p = q$ or there is a block containing p and q and meeting H. This relation is clearly reflexive and symmetric; we show it is transitive. Suppose p, q and r are distinct points with $p \sim q \sim r$. Let X and Y be blocks containing p, q and q, r, respectively, and meeting H. Then $X \cap H$ and $Y \cap H$ are lines of the plane $\Pi(q, H)$; so there is a point s in $X \cap Y \cap H$. Then p, q, r lie in the plane $\Pi(s, G)$; so there is a block containing p, r and s, whence $p \sim r$. Thus $\sim$ is an equivalence relation. Clearly every equivalence class contains the line through any two of its points, hence is a subspace of the projective space G. The number of lines through a point p lying in the equivalence class is equal to the number of lines in the plane $\Pi(p, H)$, that is, q^2+q+1; so the size of an equivalence class is q^3+q^2+q+1. Since G is partitioned into equivalence classes, $(q^4-1)/(q-1)$ divides $(q^d-1)/(q-1)$, whence 4 divides d.

Lemma 3.9. $d = 4$.

Proof. Proposition 1.6 asserts that $k(\lambda+1)-\lambda(e+1)$ divides- $\lambda^3(e+1)^2(\lambda e-1)(e-\lambda)$. Substituting the values of λ, e and k, we conclude that

$$D(q, d) \equiv (q+2)(q^{d-4}+q^{d-6}+\cdots+1)-1 \quad \text{divides}$$
$$F(q) \equiv (q+1)^2(q^2+q+2)^2(q^2+2q+2).$$

If $q>2$, each of $(q+1)^2$ and (q^2+q+2) is $<q^2+2q+2<2q^2$; so $F(q)<16q^8<q^{11}$, hence $d<14$ when $q>2$. Since $F(2)=2^7 \cdot 45<2^{13}$, $d \leqslant 14$ in general. By Lemma 3.8, d must be 4, 8 or 12.

Supposing that $d=12$, we denote $D(q, 12)$ by D. Use of the Euclidean algorithm shows that $(q+1, D)=(q+1, 4)$, $(q^2+q+2, D)=(q-45, 518)$ and $(q^2+2q+2, D)=(q+64, 397)$. Then D must divide $4^2 \cdot 518^2 \cdot 397 \equiv N$. Certainly $D=D(q, 12)$ is too large to divide N if $q \geqslant 300$. Then we may assume $q<300$; so $(q+64, 397)=1$, and D must divide $N/397=2^6 \cdot 7^2 \cdot 37^2 \equiv M$. If $q \geqslant 7$, then $D>7^9>M$, so $q \leqslant 5$. One easily eliminates the remaining possibilities for q, so $d \leqslant 8$. One eliminates the possibility that $d=8$ by a shorter calculation of the same type.

Lemma 3.10. *The geometry of points and lines is a polar space.*

Proof. We verify the axioms of Buekenhout and Shult [3]. It is clear that any line has at least three points and that no point is collinear with all others. Let L be a line and p a point not on L. We must show that p is collinear with one or all points of L. Let G be any block containing L. If $p \in G$, then p is collinear with every point of L; so suppose not. Then the points of L collinear with p are the members of $\Pi(p, G) \cap L$; this set is a point or a line, since G is a projective 3-space.

The rank of the polar space is four, since a block is a maximal totally singular subspace. Comparing the number of points of $\mathcal{G}$ (Lemma 1.2) with those in each of the six types of finite polar spaces of rank 4, we see that it is derived from a quadratic form of Witt index 4 on an 8-dimensional vector space; that is, it is of type $D_4(q)$ (see [8, p. 219]). Since the number of blocks is half the total number of half spaces and since two blocks meet in the empty set or a line, the blocks must comprise one family of half spaces. The theorem is proved.

Remark. Another way of making the identification is to observe that the only candidates for maximal totally singular subspaces containing a plane $\Pi = \Pi(p, G)$ are the unique block containing Π and union of the lines joining p to points of Π. Thus each plane lies in just two maximal subspaces. Now apply [8, 7.12 and 8.4.3].

Appendix

Let V be a vector space of dimension $2n$ over a commutative field, Q a quadratic form of Witt index n, B the associated bilinear form. As observed in the proof of Proposition 3.5, a totally singular subspace of dimension $n-1$ lies in exactly two half spaces. Form a graph whose vertices are half spaces, two vertices adjacent if their intersection has dimension $n-1$.

Let S be a vertex, $\{T_1, T_2\}$ an edge of the graph. Put $T_1 \cap T_2 = U$, and let $d = \dim(S \cap U)$. The functions $s \to B(s, u)$, $u \in U$, form an $(n-d-1)$-dimensional subspace of the dual space S'; its annihilator V is thus a $(d+1)$-dimensional subspace containing $S \cap U$. Then $\langle U, V\rangle$ is a half space containing U, whence $\langle U, V\rangle = T_1$ or T_2, say $\langle U, V\rangle = T_1$. Then

$$\dim(S \cap T_2) = \dim(S \cap T_1) - 1.$$

It follows that the distance function in the graph is given by the rule $d(S, T) = n - \dim(S \cap T)$, and that there are no circuits of odd length, that is, the graph is bipartite. Vertices in the same bipartite block lie at even distance. The claim made in Example 3.4 is proved.

Acknowledgment

The second author wishes to thank Mark Hale for helpful discussions on symplectic geometries.

References

[1] E. Artin, Geometric Algebra, Tracts in Pure and Applied Math. 3 (Interscience Publ., New York, 1957).
[2] R.C. Bose, Strongly regular graphs, partial geometries and partially balanced designs, Pacific J. Math. 13 (1963) 389–419.
[3] F. Buekenhout and E.E. Shult, On the foundations of polar geometry, Geometriae Dedicata 3 (1974) 155–170.
[4] P.J. Cameron, Parallelisms of Complete Designs, London Math. Soc., Lect. Note Series 23 (Cambridge Univ. Press, Cambridge, 1976).
[5] P.J. Cameron and J.H. van Lint, Graph Theory, Coding Theory and Block Designs, London Math. Soc., Lect. Note Series 19 (Cambridge Univ. Press, Cambridge, 1975).
[6] D.A. Drake, Partial λ-geometries and generalized Hadamard matrices over groups, to appear.
[7] B. Huppert, Geometric algebra, Lecture notes, Dept. Math., Univ. of Illinois at Chicago Circle, Chicago, IL (1970).
[8] J. Tits, Buildings of Spherical Type and Finite BN-Pairs, Lecture Notes in Math. 386 (Springer Verlag, Berlin, 1974).

Annals of Discrete Mathematics 6 (1980) 31–39

A CHARACTERIZATION OF GROUP-DIVISIBLE DESIGNS AND SOME RELATED RESULTS

Ching-Shui CHENG

University of California, Berkeley, CA. U.S.A.

Len J. GRAY

Union Carbide Corporation, Oak Ridge, TN, U.S.A.

1. Introduction

The optimality of some special types of regular graph designs (to be defined later) for the elimination of one-way heterogeneity has been proven by Cheng [1]. The main purpose of this paper is to show that these designs are group-divisible. The whole problem will be formulated and solved in the context of graph theory.

A *regular graph* G is a graph with no multiple edges and loops (i.e., there is at most one edge connecting two different vertices and any vertex is not joined with itself) such that each of its vertices is adjacent (or connected) to the same number of other vertices. This number is called the *degree* of G.

The *adjacency matrix* $T_G = (t_{ij})$ of a graph G with v vertices is the $v \times v$ matrix defined by

$$t_{ij} = \begin{cases} 1 & \text{if vertices } i \text{ and } j \text{ are connected,} \\ 0 & \text{otherwise.} \end{cases}$$

Clearly, if G is regular with degree t, then T_G is symmetric with $t_{ii} = 0$ for all i and $\sum_{j=1}^{v} t_{ij} = t$ for all i.

John and Mitchell [5] defined a *regular graph design* (RGD) to be an incomplete block design with v varieties and b blocks of size k satisfying the following conditions:

(a) Each variety appears in each block at most once.

(b) All the varieties are equally replicated.

(c) Let λ_{ij} be the number of times that varieties i and j appear together. Then there are only two possible values of λ_{ij}; namely λ or $\lambda + 1$.

Suppose d is an RGD with v varieties and b blocks of size k. Let N_d be the variety-block incidence matrix of d. Then all the diagonal elements of $N_d N_d'$ (the prime indicating transpose) are equal to r, the number of replications of each variety, and the off-diagonal elements are λ or $\lambda+1$. Since

$$r(k-1)=\sum_{j\neq i}\lambda_{ij} \quad \text{for all } i,$$

each row of N_dN_d' has

$$r(k-1)-\text{int}[r(k-1)/(v-1)]\cdot(v-1)$$

entries equal to $\lambda+1$, where int$[x]$ is the largest integer $\leq x$. Thus $N_dN_d'-(r-\lambda)I_v-\lambda J_v$ (where I_v is the $v\times v$ identity matrix and J_v is the $v\times v$ matrix consisting entirely of 1's) is the adjacency matrix of a regular graph with v vertices and degree

$$r(k-1)-\text{int}[r(k-1)/(v-1)]\cdot(v-1).$$

This is the reason why such designs are called regular graph designs.

More generally, if d is a design satisfying (a) and (b) such that there are only two distinct values of λ_{ij}, say $\lambda_{[1]}$ and $\lambda_{[2]}$ with $\lambda_{[1]}>\lambda_{[2]}$ (not necessarily of the form $\lambda+1$ and λ), then

$$(\lambda_{[1]}-\lambda_{[2]})^{-1}[N_dN_d'-(r-\lambda_{[2]})I_v-\lambda_{[2]}J_v]$$

is also the adjacency matrix of a regular graph. Thus, we call this kind of design a *regular graph design in the wide sense.* For convenience, we denote the corresponding regular graph of d (defined in the manner indicated above) by $G(d)$.

If any two adjacent vertices of a regular graph are joined to exactly p_{11}^1 other vertices, and any two nonadjacent vertices are joined to exactly p_{11}^2 other vertices, then the graph is said to be *strongly regular.* It can be shown that $G(d)$ is strongly regular if and only if d is a partially balanced incomplete block design (PBIBD) with two associate classes.

For any regular graph G with degree t, t is always an eigenvalue of T_G. Similarly, if d is an RGD in the wide sense, then rk is an eigenvalue of N_dN_d'. If the other eigenvalues of N_dN_d' are $\mu_1, \mu_2, \ldots$ and μ_{v-1}, then the eigenvalues of $T_{G(d)}$ besides the degree of $G(d)$ are

$$(\lambda_{[1]}-\lambda_{[2]})^{-1}[\mu_1-(r-\lambda_{[2]})], \ldots,$$

and

$$(\lambda_{[1]}-\lambda_{[2]})^{-1}[\mu_{v-1}-(r-\lambda_{[2]})].$$

Hereafter, $\mu_1, \mu_2, \ldots, \mu_{v-1}$ are called the *eigenvalues of d.* Note that one multiplicity of the trivial eigenvalue rk is excluded for convenience. Similarly, if G is a regular graph with v vertices and degree t, and the eigenvalues of T_G are $t, \mu_1', \ldots, \mu_{v-1}'$, then $\mu_1', \ldots, \mu_{v-1}'$ are called the *eigenvalues of G.* Again, one multiplicity of t is excluded for convenience.

The complementary graph of G is defined to be the graph $\bar{G}$ with adjacency matrix $J_v - I_v - T_G$, i.e., any two vertices are adjacent in $\bar{G}$ if and only if they are not adjacent in G. Clearly, if G is a regular graph with degree t and eigenvalues $\mu_1, \ldots, \mu_{v-1}$, then $\bar{G}$ is also regular with degree $v-1-t$ and eigenvalues $-1-\mu_1, \ldots, -1-\mu_{v-1}$.

A *group-divisible design* is an RGD in the wide sense such that the v varieties can be divided into several groups of equal size such that any two varieties in the same group appear together in the same number of blocks and those from different groups also appear together in the same number of blocks. Clearly, a group-divisible design is a PBIBD with two associate classes. Let $G_{l,n}$ be the disjoint union of l complete graphs (a complete graph is one in which every two vertices are adjacent) each with n vertices. Then d is group-divisible with l groups of size n if and only if $G(d)$ is $G_{l,n}$ or $\bar{G}_{l,n}$. Note that $\bar{G}_{l,n}$ is the complete l-partite graph $K_{n_1,n_2,\ldots,n_l}$ where $n_1 = n_2 = \cdots = n_l = n$. It is easy to see that $G_{l,n}$ only has two distinct eigenvalues $\mu_1 > \mu_2$, with $\mu_2 = -1$ and the multiplicity of μ_1 is $l-1$. Similarly, $\bar{G}_{l,n}$ has two distinct eigenvalues $\mu'_1 > \mu'_2$, with $\mu'_1 = 0$ and the multiplicity of μ'_2 is $l-1$. Group-divisible designs also have the same eigenvalue structures.

Roughly speaking, the main purpose of this paper is to characterize some group-divisible designs by the eigenvalue structure of $N_d N'_d$. This consideration arises from a problem in the theory of optimal experimental designs. For fixed parameter values v, b, and k, it is well-known (see, e.g., [6, 7]) that if there is a balanced incomplete block design, then it is optimal for the elimination of one-way heterogeneity. Cheng [1] considered the problem of finding an optimal design when there is no BIBD, and proved the optimality of some RGD's with two distinct eigenvalues such that one of them is simple (i.e., the multiplicity = 1). By the observation in the last paragraph, a group-divisible design with 2 groups is such an example. Cheng has conjectured that this special structure of eigenvalues implies group-divisibility (with 2 groups), and this turns out to be true (Corollary 2.3 and Corollary 2.4). Its proof is the main subject of the present paper. We find it easier to present the proof in the context of graph theory. A partial result in the case that neither of the two eigenvalues are simple is also proved. As a byproduct, we obtain a necessary condition for the existence of a strongly regular graph (or a PBIBD with 2 associate classes), which is discussed in Section 3.

It is known that a regular graph with two distinct eigenvalues is strongly regular. (See, e.g., [3]). Hence an RGD in the wide sense with two distinct eigenvalues is a PBIBD with two associate classes. Using a different method, we obtained another proof which is also presented in this paper as an Appendix.

For convenience, some conventions are made. For a regular graph G, we use $\mu_{\max}$($\mu_{\min}$, respectively) to denote the maximum (minimum, respectively) of the $v-1$ eigenvalues of G. If G only has two distinct eigenvalues (i.e., G is strongly regular), then we write $\mu_1 = \mu_{\max}$ and $\mu_2 = \mu_{\min}$. All these notations will also be used for RGD's in the wide sense, with similar meanings.

2. Main results

Theorem 2.1. *Let G be a regular graph. Then* $\mu_{\min} \leqslant -1$, *and* $\mu_{\min} = -1$ *if and only if G is the disjoint union of several complete graphs.*

Proof. This actually has been proved implicitly by Takeuchi [8], and we therefore omit the proof. This result is also well-known to the graph theorists.

By considering the complementary graphs, we have

Theorem 2.2. *Let G be a regular graph. Then* $\mu_{\max} \geqslant 0$, *and* $\mu_{\max} = 0$ *if and only if G is a complete l-partite graph for some l.*

Theorem 2.3. *Let G be a regular graph with v vertices and degree t. Then G is the disjoint union of two complete graphs if and only if G has two distinct eigenvalues* $\mu_1 > \mu_2$ *and* μ_2 *has multiplicity* $v-2$.

Proof. We only have to show that the assumed eigenvalue structure implies group-divisibility. By Theorem 2.1, it suffices to show $\mu_2 = -1$. Using the fact that the eigenvectors of μ_2 form a $(v-2)$-dimensional space, we will show that μ_2 must be one of five possible values, all except -1 being easily eliminated.

Since G has only two distinct eigenvalues, G is strongly regular and hence $\mu_1\mu_2$ and $\mu_1 + \mu_2$ are integers. (See, e.g., [3] or the Appendix of the present paper.)

For each i, $1 \leqslant i \leqslant v$, let $f_i = (\delta_{i1}, \delta_{i2}, \ldots, \delta_{iv})'$, i.e., the standard coordinate vectors. Conceive of $\{f_1, f_2, \ldots, f_v\}$ as the v vertices of G. Then f_i and f_j are adjacent if and only if $(T_G f_i, f_j) = 1$. Without loss of generality, we may assume that $f_2, f_3, \ldots, f_{t+1}$ are the t vertices connecting to f_1.

Since the case $t = 1$ is trivial, we may assume that $(T_G f_1, f_2) = (T_G f_1, f_3) = 1$. To prove that $\mu_2 = -1$, we first demonstrate that a linear combination of f_1, f_2, and f_3 is an eigenvector with eigenvalue μ_2. From the well-known relation [(4,) p. 19]

$$\dim(E_j \cap E_k) = \dim(E_j) + \dim(E_k) - \dim(E_j + E_k),$$

It follows that the intersection of a three-dimensional subspace and a $(v-2)$-dimensional subspace of a v-dimensional space is at least one-dimensional. By choosing the $v-2$ dimensional subspace of eigenvectors with eigenvalue μ_2 and the three-dimensional space spanned by $\{f_1, f_2, f_3\}$, we conclude that there exist a_1, a_2, and a_3, not all zero, such that $g = a_1 f_1 + a_2 f_2 + a_3 f_3$ is an eigenvector with eigenvalue μ_2. Since $(T_G f_1, f_2) = (T_G f_1, f_3) = 1$, if we take the inner product of both sides of the equation

$$T_G g = \sum_{i=1}^{3} a_i T_G f_i = \mu_2 \sum_{i=1}^{3} a_i f_i = \mu_2 g$$

with f_j, $j=1, 2, 3$, then we arrive at two possible sets of simultaneous homogeneous equations for the a_i's (depending upon whether $(T_G f_2, f_3)$ equals zero or one); namely

$$\begin{bmatrix} -\mu_2 & 1 & 1 \\ 1 & -\mu_2 & (T_G f_3, f_2) \\ 1 & (T_G f_2, f_3) & -\mu_2 \end{bmatrix} \begin{bmatrix} a_1 \\ a_2 \\ a_3 \end{bmatrix} = \begin{bmatrix} 0 \\ 0 \\ 0 \end{bmatrix}.$$

For a non-trivial solution, the determinant of the coefficient matrix must equal zero, and thus we find that μ_2 must equal $-1, 2, 0$, or $\pm\sqrt{2}$. The values 2, 0, and $\sqrt{2}$ are eliminated because $\mu_2 \leqslant -1$, and it thus remains to show that $\mu_2 = -\sqrt{2}$ is impossible. If $\mu_2 = -\sqrt{2}$, we use the fact that $t + \mu_1 + (v-2)\mu_2 = \operatorname{tr} T_G = 0$ to obtain $\mu_1 = -t + (v-2)\sqrt{2}$. Then $\mu_1\mu_2 = \sqrt{2}\,t - 2(v-2)$ is not an integer. Thus, $\mu_2 = -1$ and the proof is completed.

Considering the complementary graphs, we have

Theorem 2.4. *Let G be a regular graph with v vertices and degree t. Then G is a complete bipartite graph if and only if G has two distinct eigenvalues $\mu_1 > \mu_2$, and μ_1 has multiplicity $v-2$.*

In the above theorems, if neither of μ_1 and μ_2 are simple eigenvalues, then G is not necessarily the disjoint union of several complete graphs or the complementary graph of that. However, we can prove the following partial results:

Theorem 2.5. *Let G be a regular graph with v vertices and degree t. Suppose G has two distinct eigenvalues $\mu_1 > \mu_2$. Let m be the multiplicity of μ_1. Then $t \geqslant v/(m+1) - 1$ and $m \geqslant v/(t+1) - 1$. The equalities hold if and only if G is the disjoint union of $m+1$ complete graphs.*

Proof. Since $\operatorname{tr} T_G = 0$ and $\operatorname{tr}(T_G^2) = vt$, we have

$$m\mu_1 + (v-1-m)\mu_2 = -t$$

and

$$m\mu_1^2 + (v-1-m)\mu_2^2 = vt - t^2.$$

Solving these equations, we obtain

$$\mu_1 = (v-1)^{-1}\{-t + m^{-1}[mvt(v-1-m)(v-1-t)]^{\frac{1}{2}}\} \tag{2.1}$$

and

$$\mu_2 = (v-1)^{-1}\{-t - (v-1-m)^{-1}[mvt(v-1-m)(v-1-t)]^{\frac{1}{2}}\}. \tag{2.2}$$

By Theorem 2.1, $\mu_2 \leqslant -1$, i.e.,

$$-t - (v-1-m)^{-1}[mvt(v-1-m)(v-1-t)]^{\frac{1}{2}} \leqslant 1 - v.$$

Therefore,

$$v-1-t \leqslant (v-1-m)^{-1}[mvt(v-1-m)(v-1-t)]^{\frac{1}{2}}.$$

It follows that

$$(v-1-t)^2 \leqslant (v-1-m)^{-1} mvt(v-1-t),$$

i.e.,

$$v-1-t \leqslant mvt(v-1-m)^{-1}.$$

After simplification, we obtain

$$t \geqslant v/(m+1)-1.$$

This is equivalent to

$$m \geqslant v/(t+1)-1.$$

From the above argument, equality holds $\Leftrightarrow \mu_2=-1 \Leftrightarrow G$ is the disjoint union of several complete graphs (Theorem 2.1). The number of graphs can be easily determined to be $m+1$.

Again, we have the following

Theorem 2.6. *Let G be a regular graph with v vertices and degree t. Suppose G has two distinct eigenvalues $\mu_1 > \mu_2$. Let m be the multiplicity of μ_1. Then $t \leqslant v - v/(v-m)$ and $m \leqslant v - v/(v-t)$. The equalities hold if and only if G is the complete $(v-m)$-partite graph $K_{n_1,n_2,\ldots,n_{v-m}}$, where $n_1 = n_2 = \cdots = n_{v-m} = v/(v-m)$.*

All these results can be translated into the context of experimental designs as follows:

Corollary 2.1. *Let d be an RGD in the wide sense. Then $\mu_{\min} \leqslant r - \lambda_{[1]}$ and $\mu_{\min} = r - \lambda_{[1]}$ if and only if d is a group-divisible design such that any pair of varieties from the same group appear together in $\lambda_{[1]}$ blocks, and those from different groups appear together in $\lambda_{[2]}$ blocks.*

Corollary 2.2. *Let d be an RGD in the wide sense. Then $\mu_{\max} \geqslant r - \lambda_{[2]}$, and $\mu_{\max} = r - \lambda_{[2]}$ if and only if d is a group-divisible design such that any pair of varieties from the same group appear together in $\lambda_{[2]}$ blocks, and those from different groups appear together in $\lambda_{[1]}$ blocks.*

Corollary 2.3. *Let d be an RGD in the wide sense with v varieties. Then d is a group-divisible design with 2 groups such that any pair of varieties from the same group appear together in $\lambda_{[1]}$ blocks and those from different groups appear together in $\lambda_{[2]}$ blocks if and only if d has two distinct eigenvalues $\mu_1 > \mu_2$, and μ_2 has multiplicity $v-2$.*

Corollary 2.4. *Let d be an RGD in the wide sense with v varieties. Then d is a group-divisible design with 2 groups such that any pair of varieties from the same group appear together in* $\lambda_{[2]}$ *blocks and those from different groups appear together in* $\lambda_{[1]}$ *blocks if and only if d has two distinct eigenvalues* $\mu_1 > \mu_2$, *and* μ_1 *has multiplicity* $v-2$.

To save space, the results corresponding to Theorem 2.5 and Theorem 2.6 are omitted.

Remark. From Corollary 2.3 and Corollary 2.4, it follows that some of the optimal designs obtained in Cheng [1] do not exist. But even if they do not exist, the results in Cheng [1] can still be used to establish lower bounds on the efficiencies of a regular graph design under various criteria, as is treated in Cheng [2].

3. A necessary condition for the existence of strongly regular graphs

Let G be a strongly regular graph with degree t which is neither a void nor a complete graph. Then G has only two distinct eigenvalues $\mu_1 > \mu_2$ and $\mu_1 + \mu_2$ is an integer. Let m be the multiplicity of μ_1. By (2.1) and (2.2),

$$\mu_1 + \mu_2 = (v-1)^{-1}\{-2t + (v-1-2m)m^{-1}(v-1-m)^{-1}[vmt(v-1-m)(v-1-t)]^{\frac{1}{2}}\}.$$

This is an integer only if $m = \frac{1}{2}(v-1)$ or $vmt(v-1-m)(v-1-t)$ is a perfect square. This gives a necessary condition for the existence of a strongly regular graph in terms of the number of vertices, degree, and the multiplicity of the bigger eigenvalue.

Since both of $\mu_1 + \mu_2$ and $\mu_1\mu_2$ are integers, it can be easily seen that either both of μ_1 and μ_2 are integers or they are conjugate to each other. Therefore the condition that $vmt(v-1-m)(v-1-t)$ is a perfect square is also a necessary and sufficient condition for μ_1 and μ_2 to be integers. This is equivalent to the well-known condition that $(p_{12}^2 - p_{12}^1)^2 + 2(p_{12}^1 + p_{12}^2) + 1$ is a perfect square.

Acknowledgments

We would like to thank D.G. Wilson and T.J. Mitchell for helpful conversations. The first author was supported by the National Science Foundation. By acceptance of this article, the publisher or recipient acknowledges the U.S. Government's right to retain a nonexclusive royalty-free license in and to any copyright covering the article.

Appendix

We prove that a regular graph with two distinct eigenvalues is strongly regular. Let G be a regular graph with degree t and two distinct eigenvalues μ_1 and μ_2. It suffices to show that any nonadjacent vertices are joined to p_{11}^2 (independent of the pair of vertices) other vertices; for adjacent vertices, the result follows from the complementary graph. We only have to consider the case where t is simple.

Define $\{f_1, f_2, \ldots, f_v\}$ as in the proof of Theorem 2.3. We now proceed to use the Lanczos tridiagonalization procedure on T_G, with $(f_1 - f_2)/\sqrt{2}$ as the starting vector. [Given a symmetric matrix T, the Lanczos algorithm generates a new basis under which the matrix is tridiagonal. If basis vectors $g_1, \ldots, g_k$ have been generated, then

$$g_{k+1} = (Tg_k - \alpha_k g_k - \beta_{k-1} g_{k-1})/\|Tg_k - \alpha_k g_k - \beta_{k-1} g_{k-1}\|,$$

where α_k and β_{k-1} are chosen such that g_{k+1} is orthogonal to g_k and g_{k-1}.]

Recall that a symmetric tridiagonal matrix with nonzero off diagonals has distinct eigenvalues; thus, any starting vector can generate at most a 3×3 matrix. However, since $(f_1 - f_2)/\sqrt{2} = g_1$ is orthogonal to the eigenvector of T_G with eigenvalue t $(=(1/\sqrt{v}\sum_{i=1}^{v} f_i)$, we must generate only a 2×2 matrix (whose eige values are μ_1 and μ_2). Furthermore, $T_G g_1$ is orthogonal to g_1 (f_1 and f_2 are not connected), and thus $g_2 = T_G g_1/\|T_G g_1\|$, and the matrix for T_G with the basis $\{g_1, g_2\}$ looks like

$$\begin{bmatrix} 0 & \beta \\ \beta & \alpha \end{bmatrix}.$$

However, since the eigenvalues are μ_1 and μ_2, we must have $\beta = (-\mu_1\mu_2)^{\frac{1}{2}}$ and $\alpha = \mu_1 + \mu_2$. But note that

$$\beta = \|T_G g_1\| = \frac{1}{\sqrt{2}}(2t - 2c)^{\frac{1}{2}},$$

where c = number of vertices that are connected to both f_1 and f_2. Thus,

$$\frac{1}{\sqrt{2}}(2t-2c)^{\frac{1}{2}} = (-\mu_1\mu_2)^{\frac{1}{2}},$$

$$t - c = -\mu_1\mu_2, \qquad c = t + \mu_1\mu_2,$$

and thus c is a function of $\mu_1\mu_2$ and t, and not f_1 and f_2.

By considering the complementary graphs, one can show that any two adjacent vertices are joined to $\mu_1\mu_2 + \mu_1 + \mu_2 + t$ other vertices.

Note added in proof

It was brought to our attention that our main results followed from a stronger result of P. Delsarte, J.M. Goethals, and J.J. Seidel, (Spherical codes and designs, Geometriae Dedicata (1977) 363–388). The methods are different.

References

[1] C.S. Cheng, Optimality of certain asymmetrical experimental designs, Ann. Statist. 6 (1978) 1239–1261.

[2] C.S. Cheng, A note on (M.S.)-optimality, Commun. Statist. A7 (1978) 1327–1338.

[3] M. Doob, Graphs with a small number of eigenvalues, Ann. New York Acad. Sci. 175 (1970) 104–110.

[4] P.R. Halmos, Finite Dimensional Vector Spaces (Springer-Verlag, New York, 1974).

[5] J.A. John and T.J. Mitchell, Optimal incomplete block designs, J. R. Statist. Soc. B 39 (1977) 39–43.

[6] J. Kiefer, On the nonrandomized optimality and randomized nonoptimality of symmetrical designs, Ann. Math. Statist. 29 (1958) 675–699.

[7] J. Kiefer, Construction and optimality of generalized Youden designs, in: J.N. Srivastava, ed., A Survey of Statistical Designs and Linear Models (North-Holland, Amsterdam, 1975) 333–353.

[8] K. Takeuchi, On the optimality of certain types of PBIB designs, Rep. Statist. Appl. Res., JUSE 8 (1961) 140–145.

Annals of Discrete Mathematics 6 (1980) 41–55

ON PERMUTATION CLIQUES

M. DEZA
C.N.R.S., Paris, France

The symmetric group S_n is a metric space with distance $d(a, b)=|E(a^{-1}b)|$ where $E(c)$ is the set of points moved by $c\in S_n$. Let L be a given subset of $\{1, 2, \ldots, n\}$, a *permutation clique* $A=A(L, n)$ is any subset $A\subseteq S_n$ with $d(a, b)\in L$ whenever $a, b\in A$, $a\neq b$. We give a framework of new and known information on some special $A=A(L, n)$: maximal, largest, largest subgroups of S_n, subscheme of Hamming metric scheme, permutation geometry and some other problems related to this metric space. Some links with classical problems of classification of permutation groups and with extremal problems on finite sets are given.

1. Introduction

We will use the following notations. Let n be a given integer, $n\geq 2$; $N=\{1, 2, \ldots, n\}$; S_n is the symmetric group of all permutations of N.

We will consider each $a\in S_n$ as an n-vector $a=(a_1, a_2, \ldots, a_n)$ over N where $a(i)=a_i$; denote $E(a)=\{i\in N: a_i\neq i\}$; $d(a, b)=|E(a^{-1}b)|$ for $a, b\in S_n$ (it is the *distance* on S_n; $d(a, b)=|\{i\in N: a_i\neq b_i\}|$, i.e. it is the Hamming distance on S_n considered as a subset of a set of all n-vectors over N; $d(ac, bc)=d(a, b)=d(ca, cb)$).

Let $L=\{l_1, l_2, \ldots, l_r\}$ be a given set of integers, $1\leq l_1<l_2<\cdots<l_r\leq n$, $r\geq 1$; denote $\bar{L}=N-L$. Let $A\subseteq S_n$; denote

$$A=A(L, n) \quad \text{if } d(a, b)\in L \text{ for any } a, b\in A, \quad a\neq b;$$

we will call $A(L, n)$ an *(L, n)-clique*; below, we will study the following special (L, n)-cliques.

Let $A=A(L, n)$; we call it:

(1) $A^*(L, n)$ if A is a subgroup of S_n;

(2) *$A(L, n)$-scheme* if A is a metric association scheme where we define $a, b\in A$ is *i-associated* whenever $d(a, b)=l_i$;

(3) *maximal* $A(L, n)$ if $\{a\}\cup A\neq A(L, n)$ for any $a\in S_n-A$;

(4) *largest* $A(L, n)$ if $A'=A(L, n)$ implies $|A'|\leq|A|$; *largest* $A^*(L, n)$ if $A'=A^*(L, n)$ implies $|A'|\leq|A|$;

(5) *sharp* $A(L, n)$ if $|A(L, n)|=\prod_{i\in L} i$.

(6) $A(\leq t, n)$ if $L=\{1, 2, \ldots, t\}$; $A(\geq t, n)$ if $L=\{t, t+1, \ldots, n\}$; $A(t, n)$ if $L=\{t\}$.

Denote $\langle 1\rangle$ as the *identity* permutation; the subset $A\subset S_n$ is *unitary* iff $\langle 1\rangle\in A$. Of course, for any $A=A(L, n)$ and $b\in S_n$ we have $B=\{ab: a\in A\}=A(L, n)$ with

$|B| = |A|$; B has any of the properties (2)–(7) iff A has that property. In the case $b^{-1} \in A$ the set B will be unitary.

The distance $d(a, b)$ was introduced in [16]. Some other distances on S_n (for example, two in [6, p. 97, Vol. 2], four in [15] and many others in [5]) have interesting links with statistics, number theory, calculation of permanents etc., but we will consider below only $d(a, b)$.

The metric space (S_n, d) gives a good common point of view for many different objects. For example:

(a) $A(t, n)$ is an orthogonal (r, λ)-design $(r = n, \lambda = n - t)$ or E.P.A. (i.e. equidistant permutation array) in other terms;

(b) sharp $A(\geqslant t, n)$ is sharply $(n-t+1)$-transitive subset of S_n or (in terms of German geometers [17]) Minkowski $(n-t-1)$-structure of order $t+1$;

(c) some classification theorems for permutation groups are equivalent to classification of some sharp $A^*(L, n)$;

(d) some $A(\geqslant t, n)$ have good properties as codes (for example, for majority type decoding [4]); the coding aspect of $A(L, n)$ is considered in [1].

My essential motivation, however, was to start a large project of extremal theory on subsets of S_n by analogy with an extremal theory on systems of finite sets. For example, we have easily $\max |A^*(\leqslant t, n)| = t!$, i.e. an analog to Erdos–Ko–Rado's theorem for $A^*(\leqslant t, n)$. (We conjecture $\max |A(\leqslant t, n)| = t!$ for $n \geqslant n_0(n-t)$; Frankl (personal communication) gave $\max |A \leqslant t, n)| \leqslant (t + \log n)!$ for any $n, t > \frac{1}{2}n$).

Imitating my teacher Paul Erdös, I tried, during the last 2 years, to push this project with my mathematician friends around the world. The survey of bounds on largest $A(L, n)$ with specified L is given in [9, 11], and on maximal $A(\leqslant t, n)$ is studied in [14]. Below we give information on some $A(L, n)$ with arbitrary L.

In Section 2, we will give, via $\bar{L} = N - L$, a characterization of maximal $A(L, n)$ and some bounds for its cardinality. In Section 3 we shall give (via permutation representation of S_n) a bijection between the set of all $A(L, n)$ and the set of all systems B of n-subsets of a given n^2-set such that $L = \{\frac{1}{2}|C \cup D - C \cap D| : C, D \in B\} - \{0\}$, and the system B is 2-resolvable. In Section 4, we will give an application of Q^v (i.e. the set of all v-vectors over $Q = \{0, 1, \ldots, q-1\}$) into S_{vq}, permitting us to give examples of $A(L, n)$-schemes. We give some other examples also. Sections 3, 4 are slight developments of [10, 12], respectively.

In Section 5, we introduce a semilattice such that S_n is its upper fiber; it gives a permutation analog of the packing and covering problem on sets.

In Section 6, permutation analog of a matroid is introduced.

In Section 7 is given a link with group theory. The proofs are usually omitted.

2. Duality

Proposition 2.1. $|A(L, n)| \cdot |A(\bar{L}, n)| \leqslant n!$

Proposition 2.2 (generalization). *Let* $A = A(L, n)$, *such that either* $A = A^*(L, n)$ *or* A *is a* $A(L, n)$*-scheme. Let* $L = L' \cup L''$ *where* $L' \cap L'' = \emptyset \neq L', L''$. *Let* $B', B'' \subset A$ *and* $B' = A(L', n)$, $B'' = A(L'', n)$. *Then*

$$|B'| \cdot |B''| \leqslant |A|.$$

Proposition 2.1 is the case $A = S_n$ of Proposition 2.2. The case "A is an $A(L, n)$-scheme" is a special case of the inequality proved by Delsarte for any association scheme. In the case "$A = A^*(L, n)$" of Proposition 2.2 (actually, we need only $a, b \in A \Rightarrow a \cdot b \in A$, i.e. A is a semigroup), we have $a'b' \neq a''b''$ (for any $a', a'' \in B'$; $b', b'' \in B''$, $a' \neq a''$) because otherwise $b' = (a')^{-1}a''b''$, $b'(b'')^{-1} = (a')^{-1}a''$, $|E(b' \cdot (b'')^{-1})| = |E((a')^{-1} \cdot a'')|$, $d(b', b'') = d(a', a'')$, $L' \cap L'' \neq \phi$, a contradiction. So $|B'B''| = |B'| \cdot |B''|$, and Proposition 2.2 follows from $B'B'' \subseteq A$.

Proposition 2.3. *Let* $A = A(L, n)$. *Then* A *is a maximal* $A(L, n)$ *iff*

$$\bigcup_{a \in A} S^a_{\bar{L},n} = S_n$$

where

$$S^a_{\bar{L},n} = \{b \in S_n : d(a, b) \in \bar{L}\} \cup \{a\}.$$

In fact, from the definition of $S^a_{\bar{L},n}$, it follows that $A \cup \{c\} = A(L, n)$ iff $c \in S_n \setminus (\bigcup_{a \in A} S^a_{\bar{L},n})$.

We remark that, for any $A = A(L, n)$, $B \subset A$, $L \supset L' \neq \phi$, $B = A(L', n)$, we have that B is $A(L', n)$ *maximal in* A iff

$$A \cap \left(\bigcup_{a \in B} S^a_{\bar{L}-L',n} \right) = A.$$

We remark also that (for any $a \in S_n$)

$$|S^a_{\bar{L},n}| = 1 + \sum_{i \in L} \binom{n}{i} D_i,$$

where D_i is the number of *derangements* of degree i ($b \in S_i$ is a derangement iff $|E(b)| = i$; we know that

$$D_i = i! \sum_{0 \leqslant j \leqslant i} (-1)^j (1/j!) \sim i!/e$$

and $D_i = 0, 1, 2, 9, 44, 265, \ldots$ for $i = 1, 2, 3, 4, 5, 6, \ldots$). For $L = \{1, 2, \ldots, t\}$, the number $|S^a_{L,n}|$ is just the volume of a sphere of radius t in the metric space (S_n, d); denote it $|S_{t,n}|$.

Corollary 2.4. *Let* A *be a maximal* $A(L, n)$. *Then*

$$n!/|S^a_{\bar{L},n}| \leqslant |A| \leqslant n!/|A(\bar{L}, n)|$$

for any $A(\bar{L}, n)$.

In the special case of $L=\{t+1, t+2, \ldots, n\}$ we have, for any $A=A(>t, n)$ maximal, the following inequality

$$n!/|S_{t,n}| \leqslant |A| \leqslant n!/|S_{[t/2],n}|.$$

The right side of this inequality is an analog to Rao-Hamming's (sphere-packing) upper bound for largest error correcting codes; C. Landauer studied (in [21]) the case of equality. The left side is an analog to the Gilbert lower bound, but we have no lower bound for largest $A^*(L, n)$ (i.e. an analog to Varshamov's modification of Gilbert's bound for subgroups) because S_n is not commutative. But for any *abelian* $A=A^*(L, n)$, $B \subset A$, $L \supset L' \neq \phi$, $B=A^*(L', n)$ *maximal* (as $A^*(L', n)$) *in* A we have

$$|B| \geqslant |A|/\max_{a \in B} |A \cap S^a_{L-L',n}|.$$

The equality in Proposition 2.1 corresponds to a "factorization" of S_n by 2 dual cliques. In the case of 2 dual sharp cliques, we know only that

(a) for $L=\{t\}$ a sharp $A^*(L, n)$ exists; there exists sharp $A(\bar{L}, n)$ if and only if $t=n$; for $t=n$ $(S_n)_\alpha$ is a sharp $A^*(\bar{L}, n)$;

(b) for $L=\{1, 2, \ldots, n-t\}$ a sharp $A^*(L, n)$ exists; a sharp $A(\bar{L}, n)$ (or sharp $A^*(\bar{L}, n)$) exists iff there exists a sharply t-transitive subset (or subgroup) of S_n;

(c) $\max |A(\geqslant 5, 6)| = 18$, $\max |A(<5, 6)| = 4!$

Some information on $A=A(L, n)$ or maximal $A(L, n)$ can be obtained from the set $E(A)=\{E(a) : a \in A\}$ using the evident inclusion

$$E(a) \cup E(b) \supseteq E(a^{-1}b) \supseteq E(a) \Delta E(b).$$

For example, in the case $|E(a) \cap E(b)| \leqslant 1$ $\forall a, b \in A$ (i.e. $E(A)$ is finite linear space) we have

$$d(a, b) = |E(a) \cup E(b)| \quad \forall a, b \in A.$$

Proposition 2.5. *Let* A *be a complete* A (L, n), *i.e.* $E(a) \in E(A) \Rightarrow a \in A$. *Then* $|E_1 \cup E_2| \in L$ $\forall E_1, E_2 \in E(A)$ *and* A *is a maximal* $A(L, n)$ *if*

$$B \subseteq \{1, \ldots, n\}, |B| \neq 1, B \notin E(A) \Rightarrow |B \cup E_1| \notin L \quad \text{for some } E_1 \in E(A).$$

In [14], the author considered *complete* maximal $A(\leqslant t, n)$; they correspond (via $a \to E(a)$) to upper fibres of hereditary families $\{E_i\}$ such that $E_i \subseteq \{1, \ldots, n\}$, $|E_i| \neq 1$, $|E_i \cup E_j| \leqslant t$ and $\{E_i\}$ is maximal (non extendable). Similar results can be given a link of $d(a, b)$ with $|E(a) \Delta E(b)|$, i.e. with Hamming distance on $E(A)$.

We remark also that for any set $\alpha^0, a^1, \ldots, a^s \in S_n$, we have

$$\{i \in \{1, \ldots, n\} : \alpha^0_i = a^1_i = \cdots = a^s_i\} = \bigcap_{j=1}^{s} \{i \in \{1, \ldots, n\} : a^0_i = a^j_i\}$$
$$= \{1, \ldots, n\} - \bigcup_{j=1}^{s} E((a^0)^{-1} a^j).$$

3. A characterization of permutation cliques

For any permutation $a \in S_n$ we denote by $m(a)$ the following binary n^2-vector

$$m(a) = (m_{11}, m_{12}, \ldots, m_{1n}, m_{21}, \ldots, m_{2n}, \ldots, m_{n1}, \ldots, m_{nn}),$$

where $m_{ij} = 1$ if $a_i = j$ and $m_{ij} = 0$ otherwise, i.e. $m(a)$ is just a "linearized" permutation matrix $((m_{ij}))_n^n$ representing a in a *natural* (*permutation*) representation of S_n (the trace of this matrix equals $n - |E(a)|$; it is the number of fixed points of a and value $\theta(a)$ on a of permutation character θ). Replacing double indexing with i, j by indexing with t, we obtain

$$m(a) = (m_1, m_2, \ldots, m_t, \ldots, m_{n^2})$$

where

$$m_{ij} = m_t \ (t = (i-1)n + j) \quad \text{and} \quad m_t = m_{ij} (t \equiv j \ (\text{mod}\ n),\ i = (t-j)/n).$$

Denote

$$\tilde{a} = \{t \in \{1, 2, \ldots, n^2\} : m_t = 1\}.$$

It is easy to see that $\tilde{a}$ is an n-subset of the n^2-set $\{1, 2, \ldots, n^2\}$ and

$$\tfrac{1}{2} d(a, b) = \sum_{t=1}^{n^2} (m_t(a) \oplus m_t(b))/\text{mod } 2) \tag{1}$$

i.e. equal to the Hamming distance of $m(a)$, $m(b)$; $\frac{1}{2}d(a, b) = |\tilde{a} \,\Delta\, \tilde{b}|$ where $\tilde{a} \,\Delta\, \tilde{b} = \tilde{a} \cup \tilde{b} - \tilde{a} \cap \tilde{b}$ is the symmetric difference of $\tilde{a}$, $\tilde{b}$;

$$n - d(a, b) = (m(a) \cdot m(b)), \tag{2}$$

i.e. equal to the scalar product of $m(a)$, $m(b)$; $n - d(a, b) = |\tilde{a} \cap \tilde{b}|$.

Denote by $[n^2]^n$ the set of all n-subsets of the n^2-set $\{1, 2, \ldots, n^2\}$. Let $B \subseteq [n^2]^n$; suppose that $R = \{R_1, R_2, \ldots, R_n\}$ is a partition of $\{1, 2, \ldots, n^2\}$ such that for any i, $1 \leqslant i \leqslant n$, the sets $\{b \in B : t \in b\}$, $t \in R_i$, form a partition of B. In the case of the existence of such a partition R we will call B *resolvable* (or 1-*resolvable*), and we will call $R = \{R_1, R_2, \ldots, R_n\}$ the *resolution* of B with *classes* $R_1, R_2, \ldots, R_n$. Given an integer $f \geqslant 0$, we will call a subset $B \subset [n^2]^n$ $(f+1)$-*resolvable* iff there exist $f+1$ resolutions $R^{(0)}, R^{(1)}, \ldots, R^{(f)}$ of B such that $|R_{i'}^{(i)} \cap R_{j'}^{(j)}| \leqslant 1$ for any $0 \leqslant i < j \leqslant f$ and $1 \leqslant i', j' \leqslant n$. The referee of this paper remarked that condition $|R_i| = n$, $1 \leqslant i \leqslant n$, follows from $(f+1)$-resolvability only for $f \geqslant 1$.

Proposition 3.1. *Let $B \subseteq [n^2]^n$. Then*

(a) *B is resolvable with resolution $\{R_1, R_2, \ldots, R_n\}$, $|R_i| = n$, iff $B = \{\tilde{a} : a \in A\}$ for some set of n-vectors over N;*

(b) *B is 2-resolvable iff $B = \{\tilde{a} : a \in A\}$ for some set $A \subseteq S_n$;*

(c) *B is $(f+1)$-resolvable (f integer, $f \geqslant 0$) iff $B = \{\tilde{a} : a \in A^{(1)}\}$ for some orthogonal system $A^{(1)}, \ldots, A^{(f)}$ of subsets of S_n.*

Here we call *orthogonal* any system $A^{(1)}, \ldots, A^{(f)}$ of equicardinal (with $|A^{(i)}| = m$) ordered (with $A^{(i)} = \{a^{(i)}_{(1)}, \ldots, a^{(i)}_{(m)}\}$) subsets of S_n such that

$$d(a^{(i)}_{(i')}, a^{(i)}_{(j')}) = d(a^{(j)}_{(i')}, a^{(j)}_{(j')}) \tag{a}$$

for any $1 \leq i, j \leq f$ and $1 \leq i' < j' \leq m$ (i.e. any two $A^{(i)}$, $A^{(j)}$ are isometric),

$$(a^{(g)}_{(i)j}, a^{(g')}_{(i)j}) = (a^{(g)}_{(i')j'}, a^{(g')}_{(i')j'}) \tag{b}$$

iff $j = j'$ and

$$a^{(1)}_{(i)j} = a^{(1)}_{(i')j}, \qquad a^{(2)}_{(i)j} = a^{(2)}_{(i')j}, \ldots, a^{(f)}_{(i)j} = a^{(f)}_{(i')j}.$$

So, $(f+1)$-resolvable subsets of $[n^2]^n$, consisting of m *disjoint* n-subsets of $\{1, 2, \ldots, n^2\}$, correspond (via c) to a system of f pairwise orthogonal Latin rectangles $(A^{(i)})$ with m rows and n columns; in particular, for $m = f+1 = n$, it corresponds to a complete set of Latin squares.

Proposition 3.2. *Let* $B \subseteq [n^2]^n$, $|B| > 1$, B *is* $(f+1)$*-resolvable. Then*

$$2(f+1) \leq \max_{a,b \in B} |a \,\Delta\, b| \leq 2n$$

with both equalities if B *consists of* n *disjoint sets and there exists* PG(2, n), *i.e. a projective plane of order* n.

The orthogonal system $A^{(1)}, \ldots, A^{(f)}$ of ordered subsets of S_n can be visualized as a 3-dimensional matrix of size $n \times m \times f$ over N; we can consider ordered subsets of S_n (following the terminology of Cheema and Motzkin) as *multipermutations.*

From the mapping $a \to m(a)$, and the corresponding bound for n-subsets of a given n^2-set follows

Proposition 3.3. *Given* $A = A(L, n)$, *we have* g.c.d.$\{i \in L\} \nmid n \Rightarrow |A| \leq n^2$.

4. Examples of an $A(L, n)$-scheme

The set of all binary n^2-vectors with n ones (and, in general, the set Q^v of all v-vectors over $Q = \{0, 1, \ldots, n-1\}$) is a *metric association scheme* if one defines $x = (x_1, \ldots, x_v)$, $y = (y_1, \ldots, y_v)$ to be i-associates whenever their Hamming distance

$$d(x, y) = |\{j \in \{1, 2, \ldots, v\} : x_j \neq y_j\}| = i$$

(see e.g. [7]). It is called the *Hamming scheme.* Some subsets of a Hamming scheme are also schemes with the same definitions of distance and i-association (for example, the set of all binary v-vectors, each with the same number of ones). Now we will give an example of such subschemes of a Hamming scheme consisting only of permutations of $\{0, 1, \ldots, q-1\}$.

Let $n = n_1 n_2$ where n_1, n_2 are integers >1, i.e. n is not a prime. For any n_2-vector $x = (x_1, \ldots, x_{n_2})$ over $\{0, 1, \ldots, n_1 - 1\}$, we define the following permutation of

$$\{1, \ldots, n\} = \{i + jn_2 : i \in \{1, \ldots, n_2\},\ j \in \{0, 1, \ldots, n_i - 1\}:$$

$$l(x) = \prod_{i=1}^{n_2} (c_0^i, \ldots, c_{n_1-1}^i) \quad \text{where } c_j^i = i + n_2((x_i \oplus j)/\text{mod } n_1)$$

i.e. $l(x)$ is a product of some disjoint cycles of length n_2 whenever $x_i \neq 0$. For $n_1 = 2$ (i.e. for n even) $l(x)$ is some involution because it is a product of disjoint transpositions.

Proposition 4.1. *Let x, y be two n_2-vectors over $\{0, 1, \ldots, n_1 - 1\}$. Then*
(a) $l((x \oplus y)/\text{mod } n_1) = l(x)l(y)$;
(b) $n_1 d(x, y) = d(l(x), l(y))$.

Corollary 4.2. *Let B be a set of n_2-vectors over $\{0, 1, \ldots, n_1 - 1\}$. Then*
(a) *B is a subgroup (whose addition is defined componentwise and modulo n_1) of the group of all n_2-vectors over $\{0, 1, \ldots, n_1 - 1\}$ iff $\{l(x) : x \in B\}$ is a subgroup of S_n (the symmetric group under usual composition of permutations);*
(b) *B is a subscheme of the Hamming scheme of all n_2-vectors over $\{0, 1, \ldots, n_1 - 1\}$ iff $\{l(x) : x \in B\}$ is a subscheme of the Hamming scheme of all n-vectors over $\{1, 2, \ldots, n\}$.*

So, for example, the set of all involutions of S_n of even degrees n contains a subgroup of order $2^{n/2}$, which is a metric association scheme (with $1 + \frac{1}{2}n$ classes) with respect to our Hamming distance on permutations.

Corollary 4.3. *Let $R = ((r_{ij}))_n^m$ be the distance matrix which is isometric to some ordered set of m binary n-vectors with Hamming distance. Then $2R$ is isometric to some ordered set of m involutions of S_{2n} (with Hamming distance).*

Another example of an $A(L, n)$-scheme is provided for the case $L = \{l_1, l_2\}$ by an $A = A(L, n)$ which is such that the *graph* defined on A (for $a, b \in A$ the edge (a, b) exists iff $d(a, b) = l_2$) is strongly regular. In this case, $A(L, n)$ will be a scheme with $|L| + 1 = 3$ classes. Any complete multipartite graph is strongly regular. For example, a sharp $A(\{n, n-1\}, n)$ corresponds to a complete n-partite graph on $n(n-1)$ vertices; a sharp $A(\{n, n-1\}, n)$ exists iff there exists PG(2, n).

5. Subpermutations; packing, covering

Let us fix some element $\alpha = -\infty$ ("joker") *exterior to* $N = \{1, 2, \ldots, n\}$. Denote by $(N \cup \{\alpha\})^v$ the set of all v-vectors $a = (a_1, \ldots, a_v)$ over $N \cup \{\alpha\}$. This set is a partially ordered set with order $a \leqslant b$ iff $a_i \neq b_i$, $1 \leqslant i \leqslant v \Rightarrow a_i = \alpha$. Define the *height* $\|a\|$ of an element a as $|\{i\in[1,v]: a_i\neq\alpha\})$. The smallest element is $0=(\alpha,\alpha,\ldots,\alpha)$ having height 0. The largest height is v, but the largest element does not exist for $|N|>1$.

Let $v=n$; denote by $\mathcal{P}$ the subset of $(N\cup\{\alpha\})^v$ which consists only of all n-vectors $a=(a_1,\ldots,a_n)$ such that $a_i=a_j$, $1\leq i\neq j\leq n\Rightarrow a_i=a_j=\alpha$. Any $a\in\mathcal{P}$ we will call *subpermutation* (of height $\|a\|$); the set of all subpermutations of largest height n is just our symmetric group S_n.

Let $A\subseteq S_n$, $1\leq t\leq n-1$; we call it

t-packing $P(t,n)$ if for any $a\in\mathcal{P}$ with $\|a\|=t$ there exists *at most* one $b\in A$ with $a<b$;

t-covering $C(t,n)$ if for any $a\in\mathcal{P}$ with $\|a\|=t$ there exists *at least* one $b\in A$ with $a<b$;

sharply t-transitive set $T(t,n)$ if for any $a\in\mathcal{P}$ with $\|a\|=t$ there exists *exactly* one $b\in A$ with $a<b$.

Proposition 5.1. *Let* $A\subseteq S_n$, $1\leq t\leq n-1$. *Then:*

(a) $A=P(t,n)$ *iff* $A=A(>n-t,n)$;

(b) $A=T(t,n)$ *iff* A *is a sharp* $A(>n-t,n)$;

(c) *in the case* $\langle 1\rangle\in A$

$$A=C(t,n)\quad \textit{iff } A \textit{ is } t\textit{-transitive subset of } S_n;$$

(d) $|P(t,n)|\leq n!/(n-t)!\leq|C(t,n)|$ *and each of these inequalities become equality iff there exists a sharp* A $(>n-t,n)$.

So we have a direct analogy with the situation for finite sets and, in particular, $T(t,n)$ is a permutation analog to the Steiner system $S(t,k,v)$ (i.e. to t-design). We remark that both of them are special cases of q-ary T-*designs* introduced by Delsarte [7]. C. Landauer remarked that $42\leq\min|C(2,6)|\leq60$.

Proposition 5.2. (a) $A=P(t,n)$ *iff* $\{\tilde{a}: a\in A\}$ *is a* 2-*resolvable packing of* t-*subsets of an* n^2-*set by its* n-*subsets;*

(b) *Let* B *be a* 1-*resolvable covering of* t-*subsets*, $1\leq t\leq n-1$, *of an* n^2-*set by its* n-*subsets; then it is the smallest covering of* 1-*subsets, consisting of* n *disjoint* n-*subsets and* $\{a\in S_n:\tilde{a}\in B\}=T(1,n)$.

So a unique $A=C(t,n)$ such that $\{\tilde{a}: a\in A\}$ is a t-*covering of* t-subsets, is a Latin square $T(1,n)$.

Denote $A=\bar{P}(t,n)$ (*maximal* t-packing) if $A\cup\{a\}\neq P(t,n)$ $\forall a\in S_n-A$; denote $A=\underline{C}(t,n)$ (*minimal* t-covering) if $A-\{a\}\neq C(t,n)$ $\forall a\in A$. We have $\min|\bar{P}(1,n)|=n$ (and also smallest maximal packing of 1-subsets of an n^2-set consists of n disjoint subsets). Largest minimal covering of 1-subsets of an n^2-set consists of n^2-n+1 n-subsets containing a given $(n-1)$-set. We have $\max|\underline{C}(1,n)|=n(n-2)$ (see [3]).

We have $\min|\bar{P}(2, 3)| = 6$ and, of course, $\min|\bar{P}(n-1, n)| = n!$. But $\min|\bar{P}(2, n)| \leq n$ for any even n, because any cyclic Latin square has no transversals. For $t > 2$ we can use the lower bound of Proposition 2.4. $\bar{P}(t, n)$ is just maximal $A(n-t, n)$; so

$$\min|\bar{P}(t, n)| \geq n!/|S_{n-t,n}|.$$

We remind that

$$\exists T(2, n) \Leftrightarrow \exists S(2, n, n^2),$$

because $S(2, n, n^2)$ is just a set of lines of $AG(2, n)$, and

$$\exists T(2, n) \Leftrightarrow \exists \text{ sharp } A(\{n-1, n\}, n) \Leftrightarrow \exists PG(2, n) \Leftrightarrow \exists AG(2, n).$$

For a given subset $B \subset S_n$ we call *B-sorting* any $A \subset S_n$ such that $b \in B$, $a \in A$ implies $b \not< a$. A. Nozaki communicated to me that the problem of finding the largest B-sorting (for the case of B consisting of all subpermutations a of height 4 such that

$$a_{i_1}, a_{i_2}, a_{i_3}, a_{i_4} \neq \alpha, \quad i_1 < i_2 < i_3 < i_4 \Rightarrow a_{i_1} > a_{i_2} > a_{i_3} > a_{i_4})$$

is equivalent to some problem in computer science (best sorting of permutations).

We call a given t-covering $A \subset S_n$ *c-uniform* if for any $a \in \mathcal{P}$ with $\|a\| = t$, there exist exactly c elements $b_1, \ldots, b_c \in A$ with $a < b_1, \ldots, a < b_c$. It is easy to see that any c-uniform t-covering A has $|A| = c \cdot n!/(n-t)!$ and that any t-transitive group G is a c-uniform t-covering with $c = |G| \cdot (n-t)!/n!$. K. Lieberherr (private communication) raised the following problem: to find an infinite sequence $\{A(n)\}$, $n \to \infty$, of c-uniform t-coverings $A(n)$, such that c is bounded by a polynom of n. Of course, either sharply 1-transitive (or 2, or 3) groups are c-uniform with $c = 1$ for either any n (or $n = p^a$ or $n = p^a + 1$); so we consider only $t > 3$ and also we prefer c as small as possible. This problem comes from computer science also (design of fast algorithms for construction of interpretations of conjunctive normal forms).

6. Sharp $A(L, n)$'s: permutation geometries $PG(L, n)$

We come back to the set (considered in the beginning of Section 5) $(\{\alpha\} \cup N)^v$ of all v-vectors over $N \cup \{\alpha\}$. Let Q be any given subset of $(\{\alpha\} \cup N)^v$. Let $a, b \in Q$ and $a \leq b$; we denote by $[a, b]$ the set $\{c \in Q : a \leq c \leq b\}$ and call it a *segment*. For any $c \in Q$ define $\hat{c} = \{i \in [1, v] : c_i \neq \alpha\}$; so $\|c\| = |\hat{c}|$ and $c \leq d \Leftrightarrow \hat{c} \subseteq \hat{d}$ (for any $c, d \in Q$). Denote by $c \wedge d$ (and call *meet* of c, d) the vector $(l_1, \ldots, l_v)$ where $l_i = c_i$ if $c_i = d_i$ and $l_i = \alpha$ otherwise. Now we specify c, d to be elements of some segment $[a, b]$. Denote by $c \vee d$ (and call *join* of c, d) the vector $(l_1, \ldots, l_v)$ where $l_i = \alpha$ whenever $c_i = d_i = \alpha$ and $l_i = b_i$ otherwise. It is possible because $c \leq b$, $d \leq b$ implies that all $c_i = \alpha, b_i$ and all $d_i = \alpha, b_i$. In some cases (and, in particular, for $Q = (N \cup \{\alpha\})^v$), Q is a lower semilattice under the operation $c \wedge d$

and $\|c\wedge d\|=|\hat{c}\cap\hat{d}|$. In some cases (and, in particular, for $Q=(N\cup\{\alpha\})^v$) the segment $[a, b]$ is an upper semilattice under the operation $c\vee d$ and $\|c\vee d\|=|\hat{c}\cup\hat{d}|$.

We can consider the number $\frac{1}{2}(\|c\|+\|d\|-2\|c\wedge d\|)$ as a modified Hamming "distance". It is not a metric; it was introduced for the case $|N|=2$ (actually, for $N=\{0, 1\}$) by Graham and Pollak for addressing of loop switching in some data communication system. For the case $\|c\|=\|d\|=v$ it is just the usual Hamming distance $v-\|c\wedge d\|$ on N^v. It will be zero iff the join $c\wedge d$ exists in $(N\cup\{\alpha\})^v$, i.e. iff $c\leq b$, $d\leq b$ for some $b\in(N\cup\{\alpha\})^v$.

We come back to the set L given at the beginning of Section 1 but for the case $v\neq n$ (so, only, in this Section 6) we will consider L as a subset of $\{1, 2, \ldots, v\}$. Let $l'_i=vl_{r+1-i}$, $L'=\{l'_i : l_i\in L\}$. Suppose $0, 1\in L'$. Denote by $D(L, n, v)$ any Q such that

(a) Q contains all elements of height ≤ 1 and at least one element of height v;

(b) $a\in Q$, $\|a\|\neq v \Rightarrow \|a\|\in L'$;

(c) for any $a\in Q$ with $\|a\|=l'_i\in L'$ and any element b of height 1 there exists exactly one $c\in Q$ with $\|c\|=l'_{i+1}$ (we denote here and below $l'_{r+1}=v$), $a<c$, $b\leq c$ whenever the following condition holds:

$$\{i\in[1, v] : b_i\neq\alpha\}\not\subseteq\{i\in[1, v] : a_i\neq\alpha\}. \tag{i}$$

Suppose now that $v=n$, $Q=Q\cap\mathcal{P}$, i.e. Q contains only subpermutations from $(N\cup\{a\})^n$. Suppose that Q has the properties (a), (b), (c), but condition (i) in (c) is replaced by the stronger condition

$$\begin{aligned}&\{i\in[1, v] : b_i\neq\alpha\}\not\subseteq\{i\in[1, v] : a_i\neq\alpha\},\\ &\{b_i : i\in[1, v], b_i\neq\alpha\}\not\subseteq\{a_i : i\in[1, v], a_i\neq\alpha\}.\end{aligned} \tag{ii}$$

Then we call Q a *permutation geometry* or for short PG(L, n). It will be considered in detail in [2].

In the case $N=\{1\}$ the condition (i) takes the form

$$\{i\in[1, v] : b_i=1\}\not\subset\{i\in[1, v] : a_i=1\},$$

and Q will be the $D(L, n, v)$ iff $\{\hat{c} : c\in Q\}$ is the lattice of all flats of *simple perfect matroid-designs* on $\{1, \ldots, v\}$; we will call it PMD for short (a survey on PMD's is given in [13]).

Proposition 6.1. *Let* $|N|>1$ *and let* Q *be either a* $D(L, n, v)$ *or a* PG(L, n). *Then*

(a) *for any segment* $[a, b]$ *of* Q *the set* $\{\hat{c} : c\in Q, a\leq c\leq b\}$ *is the set of all flats of simple* PMD's on $\hat{b}$.

(b) $(l_r-l_{r-1})\,|(l_{r-1}-l_{r-2})|\cdots|(l_2-l_1)|\,l_1$.

In fact, (b) follows from (a) and the necessary conditions

$$(l'_2-l'_1)\,|(l'_3-l'_2)|\cdots|l'_r-l'_{r-1})|\,(l'_{r+1}-l_r)$$

(given by Edmonds–Murty–Young) for the existence of PMD.

Suppose now that Q is either a $D(L, n, v)$ or a $\mathrm{PG}(L, n)$. Let $t(i, j, k) = |\{c \in Q : a \leqslant c \leqslant b, \|c\| = l'_j\}|$, where $a, b \in Q$ are given and $a \leqslant b$, $\|a\| = l'_i$, $\|b\| = l'_k$, $1 \leqslant i \leqslant j < k \leqslant r+1$ (remind that $l'_{r+1} = v$). From (a) of Proposition 6.1, it follows that

$$t(i, j, k) = \prod_{s=i}^{j-1} \left[\frac{l'_k - l'_s}{l'_j - l'_s}\right].$$

Let $t(i, j) = |\{c \in Q : a \leqslant c,\ \|c\| = l'_j\}|$, where $a \in Q$ is given and $\|a\| = l'_i$, $1 \leqslant i \leqslant j \leqslant r+1$.

Proposition 6.2. (a) *Let Q be a $D(L, n, v)$, then*

$$t(i, j) = |N|^{j-i} \prod_{s=i}^{j-1} \left[\frac{v - l'_s}{l'_j - l'_s}\right]$$

and Q contains $t(1, r+1) = |N|^r$ elements of height v;

(b) *Let $v = n$ and Q be a $\mathrm{PG}(L, n)$, then*

$$t(i, j) = \prod_{s=i}^{j-1} \left[\frac{(n - l'_s)^2}{(l'_j - l'_s)}\right],$$

and Q contains

$$t(1, r+1) = \prod_{s=1}^{r} (n - l'_s) = \prod_{i \in L} i$$

elements of height n.

In fact, in both cases (a), (b), we have

$$t(i, j) = \prod_{s=1}^{j-1} \frac{t(s, s+1)}{t(i, s, s+1)}.$$

It is easy to check that

$$\prod_{s=1}^{j-1} t(i, s, s+1) = \prod_{s=1}^{j-1} \left[\frac{l'_j - l'_s}{l'_{s+1} - l'_s}\right].$$

We have

$$t(s, s+1) = \frac{(v - l'_s)}{(l'_{s+1} - l'_s)} |N|$$

for the case (a), and

$$t(s, s+1) = \frac{(v - l'_s)(v - l'_s)}{(l'_{s+1} - l'_s)} = \frac{(n - l_s)^2}{l'_{s+1} - l'_s}$$

for the case (b). Proposition 6.2 follows.

Corollary 6.3. *Let $Q \subseteq \{N \cup \alpha\}^n$, and suppose that Q is a $\mathrm{PG}(L, n)$. Then the set of* *all its elements of height n is a sharp* $A(L, n)$ *and*

$$(l_r - l_{r-1})\,|(l_{r-1} - l_{r-2})|\cdots|(l_2 - l_1)|\,l_1.$$

Proposition 6.4. *Any sharp* $A = A(>n-t, n)$ *is a set of all elements of height n of a* $\mathrm{PG}(L, n)$, Q *with* $L = \{n-t+1, \ldots, n\}$.

In fact we can take

$$Q = \{c \in \mathscr{P} : \text{either } c \in A \text{ or } |\hat{c}| \leq t-1\}.$$

For any $c \in A$ the set $\{\hat{d} \in Q : d \leq c\}$ will be PMD (actually, a truncation of a boolean algebra, i.e. trivoid in terminology of PMD's). So we have to prove the condition (c) in the definition of $\mathrm{PG}(L, n)$ only for a with $\|a\| = t-1 = n - l_r = l_r'$. But this condition for our case just consists of saying that for any subpermutation a of height t there exists exactly one $c \in A$ with $a < c$. In other words, A is a sharply t-transitive subset of S_n. From Section 5, we know that A is a sharp $A(>n-t, n)$ iff A is a sharply t-transitive subset of S_n.

Actually, we have to call a $\mathrm{PG}(L, n)$ *simple* because of the restriction $l_1' = 0$, $l_2' = 1$ (i.e. $l_{r-1} = n-1$, $l_r = n$). But it is easy to extend the above definition of a $\mathrm{PG}(L, n)$ to more general L. From Proposition 6.4 follows

Corollary 6.5. *Let* $L = \{t_1, t_1+1, \ldots, t_2\}$, $1 \leq t_1 \leq t_2 \leq n$. *Then*

(a) *any sharp* $A(L, n)$ *having* $n - t_2$ *trivial orbits (i.e. fixing some* $n - t_2$ *points of* N*) is the set of all elements of height n of some* $\mathrm{PG}(L, n)$;

(b) *any sharp* $A^*(L, n)$ *has the same property if* $t_1 < t_2$.

In fact, (b) is a special case of (a) as described in Section 7. Any sharp $A^*(t, n)$ corresponds to $\mathrm{PG}(L, n)$ ($L = \{t\}$) for $t = n, n-1$ or for $t = n-2$, n odd, but for any even n there exists a sharp $A^*(n-2, n)$ which does not correspond to a $\mathrm{PG}(L, n)$.

We remark that conditions (i) and (ii) in the definitions of $D(L, n, v)$ and $\mathrm{PG}(L, n)$, correspond to 1- and 2-resolvability considered in Section 3. In both cases we replace in the axiom for flats of matroid ("for any i-flat and a point exterior to it, there exists exactly one $(i+1)$-flat containing this i-flat and point") the word "exterior" by "strongly exterior" specified by either (i) or (ii). It suggests the construction of other such things (for example, for $(f+1)$-resolvability).

Perhaps also, it will be interesting to study the sets Q, such that the sets $\{\hat{c} : c \in Q, a \leq c \leq b\}$ are not PMD's but either matroid-designs or matroids.

In order to get more simularity with matroids one can use, for example, following concepts of orthogonality and parallelism on subpermutations $(p, q, t, \ldots) - p \perp q$ iff $p \wedge q = 0$, $p \vee q \in S_n$; $p \| q$ iff $p \perp t$, $q \perp t$ for some t so, $p \| q$ is an equivalence and $p \| q$ iff $\hat{p} = \hat{q}$ and $p = sq$ for some permutation s of $\hat{p}$).

Perhaps (because of (a) of Proposition 6.1) it will be useful to consider the set of all elements of height n of a PG(L, n) as a "good" set of automorphisms of corresponding PMD's. The existence of a PG(L, n) (especially of subgroups) and the existence of a PMD which is an extension of this PMD can be related.

$A(L, n)$-schemes coming from PG(L, n) can be obtained via "regularity" of its semilattice considered in [7].

7. Sharp $A^*(L, n)$

Let $G = A^*(L, n)$, i.e. G is a subgroup of S_n. Denote $f(a) = n - |E(a)|$, the number of fixed points of $a \in G$; it is the number of 1-cycles in a. The set $\{n - i : i \in L\} = \{f(a) : a \in G,\ a \neq \langle 1 \rangle\}$ is called the *type* of the group G. Sometimes one is interested only in $\{f(a) : a \in G,\ a^2 = \langle 1 \rangle\}$, i.e. only in involutions.

Proposition 7.1 (Bannai–Deza's conjecture proved by Kiyota [22]).

$$|G| \text{ divides } \prod_{i \in L} i.$$

In particular, $|G| \leq \prod_{i \in L} i$, i.e. any sharp $A^*(L, n)$ is a largest $A^*(L, n)$.

Proposition 7.2 ([23]). *Let X_t be the Bell number, i.e. $X_0 = 1 = X_1$, $X_{j+1} = \sum_{i=1}^{j} \binom{j}{i} X_{j-i}$. Then*

$$|G| \leq \sum_{a \in G} (f(a))^t / X_t = \left(n^t + \sum_{i \in L} (n-i)^t c_i) / X_t\right)$$

(here $c_i = |\{a \in G : f(a) = n - i\}|$) with equality iff G is t-transitive.

It is well-known also that G has $r = (\sum_{a \in G} f(a))/|G|$ orbits. It proves directly that $|A^*(t, n)| \leq n - t$ $(r = (r + (|G| - 1)t)/|G|$, $|G| = (n-t)/(r-t) \leq n - t)$ and $A^*(\leq n - t, n) \leq (n-t)!$ $(r \geq (n + (|G| - 1)t)/|G| > t$, $|G| = n_1!\, n_2! \cdots n_r!$ where $n = n_1 + \cdots + n_r$ is a partition of n by lengths of orbits; so the maximum of G corresponds to $n = 1 + \cdots + 1 + (n - t)$, i.e. to $|G| \leq (n-t)!)$ but the bound for an arbitrary L was proved via theory of characters.

In the problem of upper bound for $|G|$ (in the absence of sharp $A^*(L, n)$) we can use, for $|A^*(\geq t, n)|$ some characterization theorems. The open questions for small $n - t$ are to find largest:

(1) $A^*(\geq n - 1, n)$, i.e. Frobenius group, for $n \neq p^a$;

(2) $A^*(\geq n - 2, n)$ which is not 2-transitive, etc.

From now on we suppose G to be a sharp $A^*(L, n)$. We have a non-extremal (but typical for group theory) classification problem (for some L, for special G, either for n sufficiently small for counting or for a sufficiently large n). The problem of classification of sharp $A^*(L, n)$ (which I proposed to Bannai in Tokyo, 1977) was finally solved by his students [20] for the following cases:

(a) $L = \{t_1 + 1, t_1 + 2, \ldots, t_1 + t_2\}$, $1 \leq t_2 \leq n - t_1$; here G has $n - t_1 - t_2$ fixed points and sharply t_2-transitive on the remaining elements of N (so we can apply Jordan's theorem that only sharply t-transitive groups with $t>3$ are $M_{11}, M_{12}, A_{t+2}, S_{t+1}, S_t$);

(b) $L=\{t, t+2\}$, here $t=4, 6, 8, 14$;

(c) $L=\{t, t+3\}$, here $t=6, 9, 15, 24, 27$. The case $L=\{n-2, n\}$ done in [25] is a case of a group of rank 3 with orbits 1, 1, $|G_a|$ of G_a. As a transitive extension of sharp $A^*(\{n-2, n\}, n)$, [25] gives $n=7, 9, 15$ only. (In general, the transitive extension of G is a sharp $(A^*(L\cup\{n+1\}, n+1)$ and we can use it for induction.)

So the degree of a sharp group of type $\{0, 1, 3\}$ can be only 7, 9, 15. We remark that a perfect matroid-design with flat sizes $\{0, 1, 3\}$ exists, for hyperplane-sizes 7, 9 and $n=13, 15, 19, 21, \ldots$ are next candidates [13].

Also in [19] a sharp $A^*(t, n)=G$ was classified for either $n-t\leq 3$ or G abelian.

As a final remark, we say that it will be interesting to see the relations between possibilities on the structure of $A=A(L, n)$:

(a) A is the set of all elements of height n of some PG(L, n); denote it by $A=\tilde{A}(L, n)$,

(b) $A=A^*(L, n)$, i.e. A is a subgroup of S_n,

and the possibilities on its cardinality:

(1) A is sharp, write $A=\hat{A}(L, n)$,

(2) A is largest, write $A=\bar{\bar{A}}(L, n)$,

(3) A is maximal, write $A=\bar{A}(L, n)$.

For $L=\{n-t+1, \ldots, n\}$, (1)$\Rightarrow$(2), (1)$\Rightarrow$(a) ("sharp" is just sharply t-transitive set), but (1)$\not\Rightarrow$(b) in general (for $t=1$ it is a Latin square which is not a group, for $t=2$ it comes from PG$(2, n)$ which is not over near-field, for $t=3$ C. Pedrini (1966) constructed it). (But for all these counterexamples, sharp $A^*(>n-t, n)$ exists as well.) On the other hand, for $L=\{t\}$, $t\neq n$, (1) and (b)$\not\Rightarrow$(a) (at least for $t=n-2$, any even n), and (1) and (b)$\not\Rightarrow$(2). A pair of nonisomorphic sharp $A^*(\{n-q, n\}, n)$ (and, moreover, exactly one of them is a permutation geometry) exists, for example, for $n=q^2$, $q=p^a$ (see [2]) and for $n=6$, $q=2$ (see [25, Theorem 1]).

In [2] the concept of permutation geometry PG(L, n) will be done in detail; in particular, a characterization of sharp $A^*(L, n)$'s which are simple PG(L, n) will be given via some Jordan groups considered by W. Kantor. Also sharp $A^*(L, n)$ is t-transitive iff $n-t'+1\in L$ for all $t'\leq t$.

References

[1] I. Blake, G. Cohen and M. Deza, Coding with permutations, Information and Control, 43 (1) (1979) 1–19.

[2] P. Cameron and M. Deza, On permutation geometries, J. London Math. Soc., 20 (3) (1979) 373–386.

[3] U. Celmins and E.T.H. Wang, Transitive sets of permutations, to appear.

[4] G. Cohen and M. Deza, Décodage des codes de permutations, Proc. du Colloque Int. du C.N.R.S., No 276, Théorie de l'information, Cachan (1977) 203–207.

[5] G. Cohen and M. Deza, Some metrical problems on S_n, Proc of France–Canada Meeting, Montreal (1979), to appear in Annals of Discrete Math. 8–9.
[6] L. Comtet, Analyse combinatoire. (Presses univ. de France, Paris, 1970).
[7] P. Delsarte, Association schemes and t-designs in regular semi-lattices, J. Combin. Theory 20 (A) (1976) 230–243.
[8] M. Deza, Matrices dont deux lignes quelconques coïncident dans un nombre donné de positions communes, J. Combin. Theory 20 (A) (1976) 306–318.
[9] M. Deza and P. Frankl, On the maximum number of permutations with given maximal or minimal distance, J. Combin. Theory 22 (A) (1977) 352–360.
[10] M. Deza, R. Mullin and S. Vanstone, Orthogonal systems, Aequationes Math. 17 (1978) 322–330.
[11] M. Deza and S. Vanstone, Bounds for permutation arrays, J. Statist. Planning and Inference 2(1978) 197–209.
[12] M. Deza and S. Foldes, Some remarks on combinatorial metric spaces and association schemes, SEA Bull Math. 2 (1978) 26–28.
[13] M. Deza. On perfect matroid–designs. Proc. of Symposium on Construction and Analysis of Designs, Kyoto Univ. (1977) 98–108.
[14] M. Deza. On maximal permutation anticodes, Proc. of 10 S.E. Conference on Combinatorics, Boca Raton, (1979) 381–392.
[15] P. Diaconis and R.L. Graham, Spearman's footrule is a measure of disarray, J. R. Statist. Soc. Ser. B 39–2 (1977) 262–268.
[16] M.K. Farahat, The symmetric group as metric space, J. London Math. Soc. 35 (1960) 215–220.
[17] W. Heise and H. Karzel, Laguerre und Minkowski-m-structuren, Rend. Ist. Mat. Univ. Trieste 4 (1972) 139–147.
[18] W. Heise, On sharply transitive sets of permutations, J. Geometry 7 (1976) 9.
[19] N. Iwahori, On a property of a finite group. Part 1, J. Fac. Sci. Univ. Tokyo X (1964) 47–64. N. Iwahori and T. Kondo. On a property of a finite group. Part 2, J. Fac. Sci. Univ. Tokyo XI (1965) 113–147.
[20] T. Ito and M. Kiyota, Sharp permutation groups, to appear.
[21] C. A. Landauer, Perfect packing theorems for groups, Notices of AMS, Oct. 1978, A–629.
[22] M. Kiyota, An inequality for finite permutation groups, J. Combin.Theory, 27 (A) (1979) 119.
[23] R. Merris and S. Pierce, The Bell numbers and 2-fold transitivity, J. Combin.Theory 12 (A) (1971) 155–157.
[24] W.H. Mills, An application of linear programming to permutation groups, Pacific J. Math. 13 (1963) 197–213.
[25] T. Tsuzuku, Transitive extensions of certain permutation groups of rank 3, Nagoya Math. J. 31 (1967) 31–36.

Annals of Discrete Mathematics 6 (1980) 57–76

SOME PROPERTIES OF PERFECT MATROID DESIGNS

M. DEZA

C.N.R.S., 3, rue Duras, Paris, 75008, France

N.M. SINGHI

School of mathematics, T.I.F.R., Colaba, Bombay 400005, India

Structure of perfect matroid designs is studied. Various known results and some new results on them are given. A list of small possible parameters of perfect matroid designs of rank 4 is given. Some unsolved problems are suggested.

1. Statements of problems and results

A *design* $D=(X, \beta)$ is a pair, where X is a finite set and $\beta \subseteq P(X)$, the set of all subsets of X. Elements of X are called the *treatments*, while those of β are called the *blocks*.

We define a *matroid* by the "hyperplane axioms". A matroid is a design $M=(X, \beta)$ such that

(i) $H_1, H_2 \in \beta, \quad H_1 \neq H_2 \Rightarrow H_1 \not\subset H_2$.

(ii) $H_1, H_2 \in \beta$, $H_1 \neq H_2$, $x \in X$, $x \notin H_1 \cup H_2 \Rightarrow$ there exists unique $H_3 \in \beta$ such that $H_1 \cap H_2 \cup \{x\} \subseteq H_3$.

Blocks of a matroid are called *hyperplanes*. For various definitions and results connected with matroids, see [26]. Subsets of X, which are intersections of hyperplanes are called *flats* of a matroid. Each subset $Y \subseteq X$ has a well-defined rank. If F is a flat of rank i and $x \in X \setminus F$, then, there is a unique flat of rank $(i+1)$ which contains $F \cup \{x\}$. Rank of X is said to be the rank of matroid.

Let M be a matroid of rank r. We will denote by M_i the matroid $M_i=(X, \beta_i)$, where β_i is the set of all i-flats (i.e., flats of rank i) of M, $1 \leqslant i \leqslant r$. M_i is called $(r-i)$th *truncation* of M. We will use the usual geometric terminology and call 1-flats points, 2-flats lines, etc.

A *perfect matroid design* (PMD) is a matroid of rank r, such that, for any integer i, the cardinality of all i-flats is the same number α_i, $0 \leqslant i \leqslant r$. For simplicity we consider only *simple* PMDs, i.e., PMD for which $\alpha_0=0$ and $\alpha_1=1$. We do not lose much generality by this assumption (see Section 2). We will write $\alpha_2=l$, $\alpha_{r-1}=k$ and $\alpha_r=|X|=v$. The set $\alpha=\{\alpha_0=0,\ \alpha_1=1,\ \alpha_2=l,\ \alpha_3, \ldots, \alpha_{r-1}=k,$

*The paper was written when the authors were visiting the Department of Combinatorics and Optimization, University of Waterloo, Waterloo, Ont.

$\alpha_r = v\}$ will be called the set of parameters for a PMD. We define the set $L = \alpha \setminus \{v, k\}$. PMD were first studied by Murty [19] and Murty *et al.* [20]. See also [10] for a short summary of various results and problems connected with PMDs. The following classical designs are examples of PMDs:

(i) Let $P_m(n, q) = (X, \beta)$, $0 < m < n$. X = set of points of an n-dimensional projective space PG(n, q) over a finite field GF(q); β = set of all m-dimensional subspaces of PG(n, q). $P_m(n, q)$ is a PMD of rank $m+2$ with

$$\alpha_i = \frac{q^i - 1}{q - 1}, \quad 0 \leqslant i \leqslant m+1 \quad \text{and} \quad \alpha_{m+2} = \frac{q^{n+1} - 1}{q - 1}.$$

(ii) Let $A_m(n, q) = (X, \beta)$, $0 < m < n$. X = set of points of an n-dimensional affine space AG(n, q) over a field GF(q); β = set of m-dimensional subspaces of AG(n, q). $A_m(n, q)$ is a PMD of rank $m+2$, with $\alpha_i = q^i$, $0 \leqslant i \leqslant m+1$ and $\alpha_{m+2} = q^n$.

(iii) $D = (X, \beta)$ be $S(t, k, v) = S_\lambda(t, k, v)$, i.e., a t-design, with $\lambda = 1$ (see Section 2 for definition of $S_\lambda(t, k, v)$). D is a PMD of rank $t+1$, with $\alpha_i = i$, $0 \leqslant i \leqslant t-1$, $\alpha_t = k$, $\alpha_{t+1} = v$.

(iv) A particular case of (iii) is the design $D = (X, \beta)$, $\beta = P_k(X)$, the set of k-subsets of X. D is a PMD of rank $k+1$, with $\alpha_i = i$, $0 \leqslant i \leqslant k$, $\alpha_{k+1} = v$. Such a PMD is called a *trivoid*. Clearly a trivoid is a $(v-k)$-truncation of a Boolean algebra.

(v) An affine-triple system (ATS(m)) of order m is a PMD of rank 4 with $\alpha_0 = 0$, $\alpha_1 = 1$, $\alpha_2 = 3$, $\alpha_3 = 9$, $\alpha_4 = 3^m$, $m \geqslant 4$. Affine triple systems were first studied by Hall [13] These systems have a closed relationship with commutative Moufang loops. Zassenhauss [27] was first to construct such a loop of order 81, which gives an affine triple system of order 4. Hall [13] independently constructed the system of order 4 and proved that it is unique. Beneteau [1] has shown recently that ATS(5) is unique ATSs have also been studied by Young [25]. Kantor [8] gives an interesting review of various results on these systems. We note that parameters of these systems are the same as those $A_2(m, 3)$, however, these systems are not isomorphic to $A_2(m, 3)$.

So far the above five examples are the only known PMDs. In Section 6 we will give a list of possible set of small parameters of PMDs of rank 4. We note that any PMD whose parameters are the same as those of $P_m(n, q)$ $m \geqslant 2$, $n \geqslant 3$, or those of $A_m(n, q)$ over GF(q), $q > 3$, $m \geqslant 2$, $n \geqslant 3$ are necessarily isomorphic to $P_m(n, q)$ or $A_m(n, q)$. This follows from a result of Tierlink [24] (see also [3, 21, 23]). Similarly any PMD with parameters of $S(t, k, v)$ is indeed $S(t, k, v)$. In view of above five examples, it seems interesting to develop the language of PMD. We believe that there are PMDs different from the above examples. In this paper we study PMDs in general. We ask some interesting questions connected with them; we answer them partially. A few unsolved problems are suggested.

Let L be a set of integers, k and v be any given integers. A design $D = (X, \beta)$ is said to be an *$A(L, k, v)$-system* [8,9] if

(i) $|B_1 \cap B_2| \in L$ for all $B_1, B_2 \in \beta$, $B_1 \neq B_2$.
(ii) $|B| = k$ for all $B \in \beta$.

Any PMD is an $A(L, k, v)$ where $L = \alpha \setminus \{v, k\}$. Any $A(L, k, v)$-system $D = (x, \beta)$ is said to be *maximal* if $K \subseteq X$, $|K| = k$, $|K \cap B| \in L$ for all $B \in \beta$ implies that $K \in \beta$. Any PMD is said to be *maximal* if it is maximal as an $A(L, k, v)$. The following result [8, 9] gives an upper bound on the number of blocks in an $A(L, k, v)$ and characterizes PMDs with large v. In fact it shows that PMDs with large v are $A(L, k, v)$'s with a maximum number of blocks.

Theorem 1a. *There exists a polynomial $v_0(x)$ such that for any $A(L, k, v)$, with b blocks and $v > v_0(k)$*

$$b \leq \prod_{X \in L} \frac{v - X}{k - X},$$

equality holds if and only if $A(L, k, v)$ is a PMD.

Conjecture on maximality. *Every simple PMD is maximal.*

Theorem 1a implies that any PMD with $v > v_0(k)$ is maximal. In Section 3 we will prove

Theorem 3a. *Any simple PMD of rank ≥ 3, with*

$$v \leq k^2/\alpha_{r-2}$$

is maximal.

Theorem 3b. *Any simple PMD of rank 4 is maximal, unless $l > 2$ and*

$$\frac{(k-l)^2}{l-2} \leq v - k \leq \frac{(k-l)^2}{l-2} - \frac{(k-l)^2}{2(l-2)(l-1)^2}[-l + (4l-3)^{1/2}(l-2)].$$

Corollary 3c. *Any PMD of rank 4, with $l \leq 3$ is maximal.*

We remark that the examples (iii), $S(t, k, v)$'s are maximal. This follows from the well-known Johnson–Schönheim upper bound for codes and packings. In fact, the above results imply that all known PMDs (i)–(v) are maximal.

A matroid M is said to be *complete* if given any flat F there exist two hyperplanes H_1 and H_2 such that $F = H_1 \cap H_2$. In Section 4 we study the problem of completeness of PMDs. We prove that completeness depends only on parameters.

Theorem 4c. *Any PMD $M=(X, \beta)$ with parameters $\alpha=\{\alpha_0, \alpha_1, \ldots, \alpha_r\}$ is complete if and only if numbers $c(i, r-1)>0$ for all $0 \leqslant i \leqslant r-1$.*

(See Theorem 4a below for the definition and value of $c(i, r-1)$.)

In the process of proving Theorem 4c, we also obtain some necessary conditions for the existence of PMDs.

Theorem 4a. *Let M be a PMD of rank r, $\alpha=\{\alpha_0, \alpha_1, \ldots, \alpha_{r-1}, \alpha_r\}$ is the set of parameters for M. Let F be any i-flat of M and H be any u-flat of M, $F \subseteq H$. Then the number of u-flats H_1 of M, such that $F=H \cap H_1$ is given by $c(i, u)$, where*

$$c(i, u)=\sum_{i \leqslant w \leqslant u} t(w, u, r) \sum_{\boldsymbol{a} \in S(i, w)} (-1)^h \prod_{g=1}^{h} t(a_g, a_{g+1}, u).$$

In the summation

$$S(i, w)=\{\boldsymbol{a}=(a_0, a_1, \ldots, a_h) \mid (i=a_0<a_1<a_2<\cdots<a_h)=w\}$$

is the set of all chains of integers, starting at i and ending at w, h is the length of chain $\boldsymbol{a}$ and $t(i, j, k)$ are defined in Section 2.

Corollary 4b. *A necessary condition for the existince of a PMD of rank r, with parameters $\alpha=\{\alpha_0, \alpha_1, \ldots, \alpha_r\}$ is that $c(i, u) \geqslant 0$ for all $0 \leqslant i<u \leqslant r$.*

The problem of completeness for t-designs has been completely studied by Gross [12]. He has proved

Theorem 1b. *Any $S(t, k, v)$ is not complete if and only if (t, k, v) is one of the following:*

(i) $(2, n+1, n^2+n+1)$;
(ii) $(3, 6, 22)$;
(iii) $(4, 7, 23)$;
(iv) $(t, t+1, 2t+3)$ *and* $t+3$ *is a prime number*;
(v) $(5, 8, 24)$;
(vi) $(3, 4, 8)$;
(vii) $(3, 12, 122)$ *if it exists*;
(viii) $(t, t+1, 2t+2)$ *and* $t+2$ *is a prime number.*

The expressions $c(i, u)$ described in Corollary 4b and Theorem 4c are messy and hence difficult to apply. A simple Gross type formula [12, Lemma 6] for general PMDs would be very much desirable. In Section 4 we will also show that all large PMDs are complete.

Theorem 4d. *There exists a polynomial $v_1(x)$ such that any PMD M with $v>v_1(k)$ is complete.*

The following result follows easily from the properties of vector-spaces and the fact that completeness depends only on parameters.

Theorem 1c.

(i) *$P_m(n, q)$ is complete if and only if $n \geq 2m+1$.*
(ii) *$A_m(n, q)$ is complete if and only if $n \geq 2m$.*
(iii) *Affine triple systems are always complete.*

Thus for all known PMDs the problem of completeness is solved. A coline of a PMD is always the intersection of two hyperplanes. In Section 4 we show that any PMD in which a coplane is not the intersection of two hyperplanes is an extension of a projective plane. We do not have any interesting observation for flats of smaller rank. For an interesting characterization of $P_{d-1}(d, q)$ or $A_{d-1}(d, q)$ as matroids which are extensions of symmetric BIBDs, see [28]. A PMD $M=(X, \beta)$ is said to be a *PMD-scheme* if β is a subscheme of a Johnson association scheme (see Section 5 for definitions). PMD-schemes are discussed briefly in Section 5.

Section 6 is devoted to the study of PMDs of rank 4, essentially from the point of view of existence. Existence of a PMD M implies existence of many BIBDs (see Section 2). Hence from the known conditions for BIBDs we get many necessary conditions on the parameters of a PMD. A list of small possible parameters of rank 4 PMD is given in Section 6. We also discuss some general results.

Finally we state the following result, which is an immediate consequence of Theorem 1a.

Theorem 1d.

(i) *If $D=(X, \beta)$ is PMD of rank r and with parameters $\alpha = \{\alpha_0, \alpha_1, \ldots, \alpha_{r-1}=k, \alpha_r=v\}$, then D is also BIBD (v, b, r, k, λ) such that*
(a) $b=\prod(v-\alpha_i)/(k-\alpha_i)$, $0 \leq i \leq r-2$.
(b) $|B_1 \cap B_2| \in \alpha$ for all $B_1, B_2 \in \beta$.

(ii) *There exists a polynomial $v_0(x)$ such that if D is any BIBD (v, b, r, k, λ) with $v > v_0(k)$ and satisfying conditions* (a) *and* (b) *for a given set α, then D is a PMD with parameters α.*

2. Preliminaries

We first define a $S_\lambda(t, k, v)$. A design $D(X, \beta)$ is called an $S_\lambda(t, k, v)$ (also called a t-design) if

(i) $|X|=v$;
(ii) $|B|=k$ for all $B \in \beta$;
(iii) Every t-subset of X is in exactly λ blocks.

$S_\lambda(t, k, v)$ with $t=2$, is also called BIBD (b, v, r, k, λ) where b is the number of blocks in D and r is the number of blocks containing any given $x \in X$. BIBDs and $S_\lambda(t, k, v)$s have been widely studied. see for example [20, 26] for various properties of BIBDs or $S_\lambda(t, k, v)$s.

Now let $M=(X, \beta)$ be a PMD of rank r and parameters L. Let F be any i-flat of M, K be an m-flat of M, $F \subseteq K$. Let us denote by $t(i, j, m)$ the number of j-flats of M containing F and which are contained in K. Then $t(i, j, m)$ is given by (see [20])

$$t(i, j, m) = \prod_{g=i}^{j-1} \frac{(\alpha_m - \alpha_g)}{(\alpha_j - \alpha_g)}. \tag{2.1}$$

The $(r-i)$th truncation M_i defined in Section 2 is a BIBD $(b_1, v_1, r_1, k_1, \lambda_1)$ with $v_1 = v$, $b_1 = t(0, i, r)$, $r_1 = t(1, i, r)$, $k_1 = t(0, 1, i)$, $\lambda_1 = t(2, i, r)$.

Given any matroid M one can define in a natural way a "projective" matroid (see [26]) $P(M) = (X_1, T)$. X_1 is the set of all 1-flats of M. T is the set of subsets $H \subseteq X_1$ such that $\bigcup_{h \in H} h$ is a hyperlane H_1 of M and $h \in H$ if and only if $h \subseteq H_1$. If M is a PMD of rank r and parameters α, then $P(M)$ is also PMD of rank r and parameters

$$\alpha' = \{\alpha'_i \mid 0 \leqslant i \leqslant r\}, \quad \text{where} \quad \alpha'_i = \frac{\alpha_i - \alpha_0}{\alpha_1 - \alpha_0}, \quad 1 \leqslant i \leqslant r$$

Let $M=(X, \beta)$ be any simple matroid. Let $K \subseteq X$, $M_K = (K, \beta_K)$, where $\beta_K = \{H \cap K \mid H \in \beta\}$ is a matroid, called the *induced matroid* on K. The rank of M_K is the rank of K in M. If M is PMD with parameters α and K is i-flat of M, then M_K is a PMD of rank i. Flats of M_K are flats of M contained in K.

Let M be any matroid. Let A be any $i-$flat of M. let $M_1^A = (X, \beta')$ be the design, where β' is the set of all hyperplanes of M containing A. Then M_1^A is a matroid. We will denote by M^A the matroid $P(M_1^A)$. If M is a PMD of rank r and parameters α, then M^A is a PMD of rank $r-1$ and parameters

$$\alpha' = \{\alpha'_l = (\alpha_{l+i} - \alpha_i)/(\alpha_{i+1} - \alpha_i) \mid 0 \leqslant l \leqslant r-1\}.$$

Thus the existance of a PMD implies the existance of many other PMDs and hence many BIBDs. This gives many necessary conditions on the parameters of PMDs. In Section 6 we use these conditions for the case of rank 4 PMDs. A matroid M of rank r will be called an *extension* of order i of matroid N of rank $r-1$ if $N \simeq M^F$ for some flat F of M of rank i.

3. Maximality

Proof of Theorem 3a. The technique used to prove the result is essentially the same as that in [22]. In fact the result is implicitly contained in the theorem proved in [22], however, we give a short proof here. Let us assume that $M=(x, \beta)$ is a

PMD of rank r, which is not maximal. Hence there is a set $K \subseteq X$, $|K| = k$, such that for all $H \in \beta$, $|K \cap H| = \alpha_i$, $0 \leqslant i < r-1$ but $K \notin \beta$. In particular K satisfies

$$|K \cap H| \leqslant \alpha_{r-2} \quad \text{for all } H \in \beta. \tag{3.1}$$

Now M is a BIBD (b, v, r', k, λ) with parameters

$$b = \frac{v(v-1)(v-\alpha_2)\cdots(v-\alpha_{r-2})}{k(k-1)(k-\alpha_2)\cdots(k-\alpha_{r-2})}, \tag{3.2}$$

$$\lambda = \frac{(v-\alpha_2)\cdots(v-\alpha_{r-2})}{(k-\alpha_2)\cdots(k-\alpha_{r-2})}, \tag{3.3}$$

$$r' = \frac{(v-1)\cdots(v-\alpha_{r-2})}{(k-1)\cdots(k-\alpha_{r-2})}. \tag{3.4}$$

We will show that the hypothesis on M implies $\alpha_{r-2}v > k^2$. Let $\beta = \{H_1, H_2, \ldots, H_b\}$ and let $x_i = |K \cap H_i|$, $i = 1, \ldots, b$. Using Fisher equations for K in BIBD M we have

$$\sum_{i=1}^{b} x_i = r'|K| = r'k \tag{3.5}$$

and

$$\sum_{i=1}^{b} x_i(x_i - 1) = \lambda|K|(|K|-1) = \lambda k(k-1). \tag{3.6}$$

Using (3.1) we have $x_i \leqslant \alpha_{r-2}$ for all i. Hence using (3.5) we get

$$k \cdot \frac{kb}{v} = k\frac{(v-1)(v-\alpha_2)\cdots(v-\alpha_{r-2})}{(k-1)(k-\alpha_2)\cdots(k-\alpha_{r-2})} \leqslant \alpha_{r-2}b$$

and so

$$\alpha_{r-2}v \geqslant k^2. \tag{3.7}$$

Now suppose $\alpha_{r-2}v = k^2$. Then we must have equality in (3.7) and hence $x_i = \alpha_{r-2}$ for all i. Using (3.6) and (3.4) we have

$$\alpha_{r-2}(\alpha_{r-2}-1)\frac{v(v-1)\cdots(v-\alpha_{r-2})}{k(k-1)\cdots(k-\alpha_{r-2})} = k(k-1)\frac{(v-\alpha_2)\cdots(v-\alpha_{r-2})}{(k-\alpha_2)\cdots(k-\alpha_{r-2})}. \tag{3.8}$$

Simplifying, using (3.8), we get

$$v + \alpha_{r-2} = 2k.$$

Hence

$$\frac{v}{k} - 1 = \frac{k-\alpha_{r-2}}{k} < \frac{k-\alpha_{r-2}}{\alpha_{r-2}} = \frac{k}{\alpha_{r-2}} - 1 \quad \text{or} \quad v - \alpha_{r-2} < k^2.$$

This contradicts the hypothesis and completes the proof of Theorem 3a.

Corollary 3.9 *The examples* (i), (ii) *and* (iv) *are maximal* (*see* [22]).

We now assume that $M=(X,\beta)$ is a PMD of rank 4, which is not maximal. Hence there is a set $K\subseteq X$, $K\notin\beta$, $|K|=k$, such that $|H\cap K|=\alpha_i$, $0\leqslant i\leqslant r-1$. If x and y are two points of M, we will denote by xy the line containing points x and y of M.

Lemma 3.10. $|K\cap xy|=l-(k-l)^2/(v-k)$ *for all* $x, y\in K$ (*note that* $l=\alpha_2$).

Proof. Let $x, y\in K$ and $|K\cap xy|=p$. Let H be any hyperplane of M, $x, y\in H$. Now $|H\cap K|\geqslant 2$, hence $|H\cap K|=l$. Also, if $z\in K$, $z\notin K\cap xy$, then clearly there is a unique hyperplane of M, containing x, y, z. Hence counting occurrences of such z in hyperplanes, we have

$$k-p=\lambda(l-p),$$

where $\lambda=(v-l)/(k-l)$ is the number of hyperplanes containing x and y. Thus

$$k-p=\frac{v-l}{k-l}(l-p).$$

Hence $p=l-((k-l)^2/(v-k))$.

Corollary 3.11. *If M is simple PMD of rank* 4, *which is not maximal, then*

(i) $(k-l)^2\equiv 0 \mod(v-k)$;
(ii) $(l-2)(v-k)\geqslant(k-l)^2$ *and* $l\neq 2$.

Proof. (i) is immediate from Lemma 3.10. (ii) follows from the fact that $p\geqslant 2$ and $k\neq l$.

Again consider the PMD M as described above. We can also assume $l\neq 2$ in view of the above corollary. From the proof of Lemma 3.10, it is clear that the induced matroid M_K is a PMD of rank 4, with parameters $\alpha'=\{0, 1, p, l, k\}$. Now let H_1 be a hyperplane of M_K. Consider the induced matroid $(M_K)_{H_1}$ of M_K. $(M_K)_{H_1}$ is clearly a BIBD with parameters $(l, l(l-1)/p(p-1), (l-1)/(p-1), p, 1)$. Now using Fisher's inequality we have $l-1\geqslant p(p-1)$. Substituting $p=l-((k-l)^2/(v-k))$, we get

$$l-1\geqslant\left(l-\frac{(k-l)^2}{v-k}\right)^2-l+\frac{(k-l)^2}{v-k}.$$

Simplifying we get

$$(l-1)^2(v-k)^2-(2l-1)(k-l)^2(v-k)+(k-l)^4\leqslant 0. \tag{3.12}$$

The left-hand side of above equation is a quadratic in $(v-k)$. The two roots of the quadratic are

$$\frac{1}{2(l-1)^2}[(2l-1)(k-l)^2 \pm ((2l-1)^2(k-l)^2 - 4(k-l)^4(l-1)^2)^{1/2}]$$
$$= \frac{(k-l)^2}{2(l-1)^2}[2l-1 \pm (4l-3)^{1/2}]$$
$$= \frac{(k-l)^2}{l-2} + \frac{1}{2(l-2)(l-1)^2}[-l \pm (4l-3)^{1/2}(l-2)].$$

Hence using corollary 3.11(ii) and (3.12) we have

$$\frac{(k-l)^2}{l-2} \leqslant v-k \leqslant \frac{(k-l)^2}{l-2} + \frac{(k-l)^2}{2(l-2)(l-1)^2}[-l+(4l-3)^{1/2}(l-2)].$$

Thus we have proved Theorem 3.6.

4. Completeness

Let M be a PMD of rank r, with parameters $\alpha = \{\alpha_0 = 0, \alpha_1 = 1, \alpha_2 = l, \ldots, \alpha_{r-1} = k, \alpha_r = v\}$. Let F be any i-flat of M. Let H be any u-flat of M, $F \subseteq H$. We will denote by $c(i, u)$ the number of u-flats H_1 of M, such that $H_1 \cap H = F$. We will show that $c(i, u)$ is independent of choice of F and H. Let us define for a given u function

$$\alpha_u(i, j) = t(i, j, u).$$

Clearly

$$\alpha_u(i, j) = 1, \quad 0 \leqslant i \leqslant u,$$
$$\alpha_u(i, j) = 0 \quad \text{unless } 0 \leqslant i \leqslant j \leqslant u.$$

Let $P_u = \{x \mid x \text{ is an integer } 0 \leqslant x \leqslant u\}$. P_u is a partially ordered set under the relation $\leqslant$. Let $I(P_u, R)$ denote the incidence algebra of partially ordered set P_u with coefficients in the field of real numbers (see [11, p. 81] or [2, p. 270]). The product in $I(P_u, R)$ is given by convolution $*$. $[f * g(i, j) = \sum_{i \leqslant u \leqslant j} f(i, u) g(u, i)]$. Now for given integers $0 \leqslant i \leqslant j$,

$$S(i, j) = \{\boldsymbol{a} = (a_0, a_1, \ldots, a_h) \mid a_0 = i < a_1 a_2 \cdots a_h = j\}$$

be the set of all chains of integers with starting point i and final point j.

For $\boldsymbol{a} = (a_0, a_1, \ldots, a_t) \in S(i, j)$ define

$$\alpha_u(\boldsymbol{a}) = (-1)^t \prod_{1<i<t} \alpha_u(a_i, a_{i+1})$$

and

$$\beta_u(i, j) = \sum_{\boldsymbol{a} \in S(i, j)} \alpha_u(\boldsymbol{a}).$$

Clearly

$$\beta_u(i, i) = 1, \quad 0 \leq i \leq u$$

$$\beta_u(i, j) = 0, \quad \text{unless } 0 \leq i \leq j \leq u.$$

Thus $\beta_u \in I(P_u, R)$.

Lemma 4.1. *$\alpha_u * \beta_u(i, j) = \delta_{ij}$, $0 \leq i \leq j \leq u$, i.e., α_u and β_u are inverses of each other in $I(P_u, R)$.*

Proof. $\alpha_u(i, i) = 1$, $0 \leq i \leq l$.

Hence α_u has an inverse in $I(P_u, R)$ (see [11, pp. 81–83] or [2]). α_u^{-1} is defined by

$$\alpha_u^{-1}(i, i) = \frac{1}{\alpha_u(i, i)} = 1,$$

$$\alpha_u^{-1}(i, j) = -\sum_{i \leq w < j} \alpha_u^{-1}(i, w)\alpha_u(w, j). \tag{4.2}$$

We will show by induction on $j - i$, that $\alpha_u^{-1} = \beta_u$. Let i be given and

$$\alpha_u^{-1}(i, i) = 1 = \beta_u(i, i), \quad 0 \leq i \leq u.$$

Suppose $\alpha_u^{-1}(i, j_1) = \beta_u(i, j_1)$ for all $0 \leq i \leq j_1 < j \leq u$. We will show that

$$\alpha_u^{-1}(i, j) = \beta_u(i, j).$$

Using the induction hypothesis and (4.2) we have

$$\alpha_u^{-1}(i, j) = -\sum_{i \leq w < j} \beta_u(i, w)\alpha_u(w, j), \tag{4.3}$$

$$\beta_u(i, w) = \sum_{\boldsymbol{a} \in S(i, w)} \alpha_u(\boldsymbol{a}).$$

Now for $i < j$

$$S(i, j) = \bigcup_{w=i}^{j-1} \{\boldsymbol{a} = (a_0, a_1, \ldots, a_{t-1}, a_t) \mid a_t = j, (a_0, a_1, \ldots, a_{t-1}) \in S(i, w)\}.$$

Hence clearly, the sum on the right-hand side of (4.3) can be written as

$$\sum_{\boldsymbol{a} \in S(i, j)} \alpha_u(\boldsymbol{a}) = \beta_u(i, j).$$

This completes the proof.

Lemma 4.4. *$c(i, u)$ is independent of choice of F and H and is given by*

$$c(i, u) = t(i, u, r) - \sum_{i < w \leq u} t(i, w, u)c(w, u). \tag{4.5}$$

Proof. We prove by induction on $u - i$. The result is clearly true for $i = u$. Assume that the lemma is true for all F of rank $i+1$, $F \subseteq H$. Let the given F be of rank i. There are $t(i, w, u)$ flats $F_1 \subseteq H$, such that $F_1 \supseteq F$ and rank of F_1 is w. For each of these F_1 There exist $c(w, u)$ u-flats H_1 such that $H_1 \cap H = F_1$. There are $t(i, u, r)$ flats H_1 of rank u, in all, containing F. Hence we have

$$c(i, u) = t(i, u, r) - \sum_{i<w\leq u} t(i, w, u)c(w, u).$$

This completes the proof.

Proof of Theorem 4a. If $0 \leq i \leq j \leq u$ define $f(i, j) = c(i, j)$, $g(i, j) = t(i, j, r)$ and $g(i, j) = f(i, j) = 0$ otherwise. Then clearly $f, g \in I(P_u, R)$.

(4.5) may be written as

$$g(i, u) = \sum_{i\leq w\leq u} \alpha_u(i, w)f(w, u).$$

Hence using the fact that $\alpha_u * \beta_u = \delta_{ij}$ we have

$$f(i, u) = \sum_{i\leq w\leq u} \beta_u(i, w)g(w, u),$$

i.e.,

$$c(i, u) = \sum_{i\leq w\leq u} \beta_u(i, w)t(w, u, r). \tag{4.6}$$

Substituting values for $\beta_u(i, w)$, we get

$$c(i,u) = \sum_{i\leq w\leq u} t(w, u, r) \sum_{\boldsymbol{a}\in S(i, w)} (-1)^h \prod_{g=1}^{h} t(a_g, a_{g+1}, u)$$

where h is the length of chain $\boldsymbol{a}$. This completes the proof of Theorem 4a.

The Corollary 4b is immediate. Now a PMD M is complete if and only if $c(i, r-1) > 0$, hence we have also proved Theorem 4c.

The expressions for $c(i, u)$s given in Theorem 4a is messy, and hence difficult to apply.

The following trivial lemma is useful to check the completeness for PMD.

Lemma 4.7. *Let $M = (X, \beta)$ be a PMD of rank r. Let F be any i-flat of M. Then in M $H_1 \cap H_2 \neq F$ for all $H_1, H_2 \in \beta$ if and only if in M^F $H'_1 \cap H'_2 \neq \emptyset$ for all hyperplanes H'_1 and H'_2 of M^F.*

Corollary 4.8. *In a PMD M of rank r, $c(r-3, r-1) = 0$ if and only if M is an extension of order $r-2$ of a projective plane.*

Proof. Let F be any $r-3$ flat of M. M^F is clearly a BIBD with $\lambda = 1$. Using the lemma we have $|B_1 \cap B_2| = 1$ for all blocks B_1, B_2 of M^F. Hence M^F is a projective plane.

The following result of Gross [12] is immediate from the corollary and Cameron's characterizations [4] of Steiner systems which are extensions of projective plane.

Theorem 4.9. *In a Steiner system* $S(t, k, u)$, $t>2$, $|B_1 \cap B_2| \neq t-2$ *for any two blocks of* $S(t, k, v)$ *if and only if* (t, k, v) *is one of the following.*

(i) (3, 6, 22),
(ii) (4, 7, 23),
(iii) (5, 8, 24),
(iv) (3, 4, 8),
(v) (3, 12, 112), *if it exists.*

We now give a proof of Theorem 4d.

Proof of Theorem 4d. Suppose $M=(X, \beta)$ is a PMD of rank r. Let F be any i-flat of M, $i<r-1$, such that $H_1 \cap H_2 \neq F$ for any $H_1, H_2 \in \beta$. We will show that $v \leqslant v_1(k)$ for some polynomial $v_1(x)$. Consider the PMD M^F of rank $r-i$. From the hypothesis, it is clear that M^F is $A(L_1, k_1, v_1)$ with

$$L_1=\left\{\beta_g=\frac{\alpha_g-\alpha_i}{\alpha_{i+1}-\alpha_i}\,\middle|\, i+1 \leqslant g \leqslant r-2\right\}, \qquad k_1=\frac{k-\alpha_i}{\alpha_{i+1}-\alpha_i}, \qquad v_1=\frac{v-\alpha_i}{\alpha_{i+1}-\alpha_i}.$$

Note that β_g is the size of (g−u)-flats in M^F. Now let $b_1=$ the number of blocks in M^F.

$$b_1=\prod_{t=i}^{r-2}\frac{(v_1-\beta_t)}{(k_1-\beta_t)}, \quad \text{where } \beta_i=0. \tag{4.10}$$

Also using Theorem 1a we can find a polynomial $v_1'(x)$ such that if $v>v_1'(k)$, then

$$b_1 \leqslant \prod_{g=i+1}^{r-2}\frac{(v_1-\beta_g)}{(k_1-\beta_g)}. \tag{4.11}$$

Using (4.10) and (4.11) we can find a polynomial $v_1''(x)$ such that $v \leqslant v_1''(k)$. Thus $v \leqslant v_1(k)$, where $v_1(x)=\max\{v_1'(x), v_1''(x)\}$. This completes the proof.

5. PMD-schemes

An *association scheme* with n classes on a set X (see, for example, [5, 6, 14]) is a partition of the set of 2-subsets of X into classes $T_1, \ldots, T_n$ such that

(i) the number a_i of $y \in X$ with $\{x, y\} \in T_i$ for given $x \in X$ depends only on i;
(ii) the number b_{ijk} of $z \in X$ with $\{x, z\} \in T_i$, $\{z, y\} \in T_j$ for given $\{x, y\} \in T_k$ depends only on i, j, k.

A *Johnson scheme* is an association scheme with k classes of all k-subsets of given v-set, such that $\{x, y\} \in T_i$ if and only if $|x \cap y| = k - i$. We will call a PMD $M = (X, \beta)$ with $\alpha_{r-1} = k$, $\alpha_r = v$ *PMD-scheme* if it is subscheme of Johnson scheme.

Any PMD of rank 3 is BIBD with $\lambda = 1$ and so it is a PMD-scheme (see [5]). From now on we consider only PMD of rank $r > 3$.

Any $P_i(d, q)$, $2 \leq i \leq d-1$, is PMD-scheme (This follows from [6, Theorems 4, 7] and from [14]).

Only $A_i(d, q)$, $3 \leq i \leq d-1$, which is a PMD-scheme is $A_{d-1}(d, q)$ (see [14, Theorem 5.2]).

Let $M = (X, \beta)$ be a Steiner system $S(t, k, v)$. Such a PMD will be a PMD-scheme if $|\{|H_1 \cap H_2| \mid H_1, H_2 \in \beta, H_1 \neq H_2\}| \leq \frac{1}{2}t + 1$ (see Cameron–Delsarte theorem [5]). Suppose now that $M = S(3, k, v)$ is PMD-scheme; then Cameron [5] proved that $v \leq 2 + \frac{1}{2}k(k-1)(k-2)$. Only known $S(3, k, v)$-scheme are Möbius (inversive) planes $I(q) = S(3, q+1 = k, q^2 + 1 = v)$ with 3 classes and $S(3, 4, 8)$, $S(3, 6, 22)$ and $S(3, 12, 112)$ (if it exists) with 2 classes. These 3 Steiner systems and also $S(4, 5, 11)$, $S(4, 7, 23)$, $S(5, 8, 24)$ are the only Steiner systems with $t > 2$ for which $|\{|H_1 \cap H_2| \mid H_1, H_2 \in \beta, H_1 \neq H_2\}| \leq \frac{1}{2}t + 1$ (see [12]) and so Cameron-Delsarte theorem [5] is applicable.

6. PMD of rank 4

Let M be a simple PMD of rank 4 with parameters $\alpha = \{0, 1, l, k, v\}$. Let $t = (k-1)/(l-1)$, $u = kt/l$, $\lambda = (v-l)/(k-l)$.

Let us consider the parameters of four following BIBD $(b_i, v_i, r_i, k_i, \lambda_i)$, $1 \leq i \leq 4$,

$$D_1 = S(2, l, k),$$

$$D_2 = S(2, (k-1)/(l-1), (v-1)/(l-1),$$

$$D_3 = S(2, l, v),$$

$$D_4 = S_\lambda(2, k, v).$$

Lemma 6.1.

(1) $$b_1 = \frac{k(k-1)}{l(l-1)} = t^2 - \frac{t(t-1)}{l} = u,$$

$$(v_1, r_1, k_1, \lambda_1) = \left(k - tl - t + 1, \frac{k-1}{l-1} = t, l, 1\right);$$

(2) $$b_2 = \frac{(v-1)(v-l)}{(k-1)(k-l)} = \lambda^2 - \frac{\lambda(\lambda-1)}{t},$$

$$(v_2, r_2, k_2, \lambda_2) = \left(\frac{v-1}{l-1} = \lambda t - \lambda + 1, \frac{v-l}{k-l} = \lambda, \frac{k-1}{l-1} = t, 1\right);$$

(3) $$b_3 = \frac{v(v-1)}{l(l-1)} = ub_2 - (\lambda - 1)v_2$$

$$(v_3, r_3, k_3, \lambda_3) = (v, v_2, l, 1);$$

(4) $$b_4 = \frac{v(v-1)(v-l)}{k(k-1)(k-l)} = \lambda b_2 - \frac{\lambda(\lambda-1)v_2}{u},$$

$$(v_4, r_4, k_4, \lambda_4) = (v, b_2, k, \lambda).$$

In fact, the lemma follows from $k = tl - t + 1$, $v = \lambda k - \lambda l + l$ (from definitions of t, λ) and $r_i = \lambda_i(v_i - 1)/(k_i - 1)$, $b_i = v_i r_i / k_i$ (well-known properties of BIBD). For example,

$$b_3 = \frac{v}{l} r_3 = \frac{\lambda k - \lambda l + l}{l} r_3 = (1-\lambda)r_3 + r_3 \frac{k}{l}\lambda = (1-\lambda)r_3 + r_3 \frac{\lambda u}{t}$$

$$= (1-\lambda)r_3 + \frac{\lambda t - \lambda + 1}{t}\lambda u = \lambda^2 u - (\lambda - 1)r_3 - u\frac{\lambda(\lambda-1)}{t} = ub_2 - (\lambda-1)v_2.$$

$$b_4 = \frac{v}{k} r_4 = \frac{\lambda k - \lambda l + 1}{k} r_4 = \lambda r_4 - \frac{l(\lambda-1)}{k} r_4 = \lambda r_4 - \frac{t(\lambda-1)}{u} r_4$$

$$= \lambda r_4 - \frac{(\lambda-1)(\lambda^2 t - \lambda^2 + \lambda)}{u} = \lambda r_4 - \frac{\lambda(\lambda-1)(\lambda t - \lambda + 1)}{u} = \lambda b_2 - \frac{\lambda(\lambda-1)v_2}{u}.$$

Let $(b_i, v_i, r_i, k_i, \lambda_i)$ be an ordered set of five positive integers such that $b_i = v_i r_i / k_i$, $r_i = \lambda(v_i - 1)/(k_i - 1)$, $r_i \geqslant k_i$; then we say that this *five-tuple* is *admissible*. The existence of a BIBD $(b_i, v_i, r_i, k_i, \lambda_i)$ will imply that $(b_i, v_i, r_i, k_i, \lambda_i)$ is an admissible five-tuple (from well-known divisibility conditions and Fisher inequality for BIBD).

Let us call *triple* of integers (l, k, v), $0 < l < k < v$, *admissible*, if all $(b_i, v_i, r_i, k_i, \lambda_i)$, $1 \leqslant i \leqslant 4$, are admissible five-tuples and if all numbers

$$\frac{k-1}{l-1}, \quad \frac{k(k-1)}{(l-1)}, \quad \frac{v-l}{k-l}, \quad \frac{(v-1)(v-l)}{(k-1)(k-l)} \quad \text{and} \quad \frac{v(v-1)(v-l)}{k(k-1)(k-1)}$$

are integers.

Lemma 6.2. *The triple (l, k, v) of integers, $0 < l < k < v$, is admissible if and only if*

$$t, \quad \lambda, \quad \frac{t(t-1)}{l}, \quad \frac{\lambda(\lambda-1)}{t}, \quad \frac{\lambda(\lambda-1)(\lambda t - \lambda + 1)}{u}$$

are integers and $l \leqslant t \leqslant \lambda$.

Proof. t, λ are integers if and only if $(l-1) \mid (k-l) \mid (v-k)$. Let t, λ be integers. $(b_1, v_1, r_1, k_1, \lambda_2)$ is admissible if and only if u is an integer, $l \leqslant t$, i.e., if and only if $t(t-1)/l$ is integer, $l \leqslant t$.

$(b_2, v_2, r_2, k_2, \lambda_2)$ is admissible if and only if $\lambda(\lambda-1)/t$ is integer, $t \leqslant \lambda$.

$(b_3, v_3, r_2, k_3, \lambda_3)$ is admissible if both $(b_i, v_i, r_i, k_i, \lambda_i)$, $i = 1, 2$ are admissible.

$(b_4, v_4, r_4, k_4, \lambda_4)$ is admissible if and only if $\lambda(\lambda-1)/t$, u, $\lambda(\lambda-1)(\lambda t-\lambda+1)/u$ are integers and $b_2 \geqslant k$.

$$b_2 \geqslant k = \lambda^2(t-1)+\lambda-kt \geqslant 0 \Leftrightarrow \lambda \geqslant \lambda'$$

where

$$\lambda' = -\frac{1}{2(t-1)}+\left(\frac{1}{4(t-1)^2}+\frac{kt}{t-1}\right)^{1/2}.$$

But

$$l \leqslant t \Leftrightarrow k \leqslant t^2-t+1 \Rightarrow \lambda' \leqslant -\frac{1}{2(t-1)}+\left(\frac{1}{2(t-1)}+\frac{t}{t-1}+t^2\right)^{1/2}=t.$$

Also $l=t$ if $b_2=k$.

The numbers mentioned in the statement of the lemma are integers if and only if t, u, λ, b_2, b_4 are integers.

Let $0<l<k<v$ be given integers such that $t=(k-1)/(l-1) \leqslant 15$ is an integer. Below we give a list of possible PMDs for $l=2$, $t-1$, t and all possible parameters (l, k) for the remaining case $2<l<t-1$. For $l=2$ the PMD is $S(3, k, v)$. The existence of $S(2, k-1, v-1)$ is necessary for the existence of $S(3, k, v)$. $S(3, k, v)$ exists for $k=4$ if and only if $v \equiv 2, 4 \pmod 6$, $v \geqslant 8$ (Hanani [15]). It also exists for $k-1=q$ (a power of prime), $v=q^2+1$ (i.e., inversive plane IP(q)). Also there exists $S(3, 6, 22)$ and $S(3, 4, 10)$ related to Mathieu groups. Now for $l=t>2$ the PMD is $P_2(d, l-1=q)$ where $v=(q^{d+1}-1)/(q-1)$, it exists if and only if q is a power of a prime, since $d \geqslant 3$. For $l=t-1>2$ the PMD is $A_2(d, l)$ where $v=l^d$; it exists if and only if l is a power of prime, as $d \geqslant 3$.

For $2<l<t-1 \leqslant 14$ we give in Table 1 all possible (l, k) such that $l \mid k(k-1)$; for all of these (l, k), except $(l, k)^*$ BIBD $D_1=S(2, l, k)$ is known to exist.

Table 1

$t=(k-1)/(l-1)$	$(l, k)=(2, t+1)$	$(l, k)=(t, t^2-t+1)$	$(l, k)=(t-1, (t-1)^2)$	All other possible (l, k)s
2	$S(3, 3, v)$	$S(3, 3, v)$	—	—
3	$S(3, 4, v)$	$P_2(d, 2)$	$S(3, 4, v)$	—
4	$S(3, 5, v)$	$P_2(d, 3)$	$A_2(\lg_3 v, 3)$, ATSs	—
5	$S(3, 6, v)$	$P_2(d, 4)$	$A_2(\lg_4 v, 4)$	—
6	$S(3, 7, v)$	$P_2(d, 5)$	$A_2(\lg_5 v, 5)$	(3, 13)
7	$S(3, 8, v)$	—	—	(3, 15)
8	$S(3, 9, v)$	$P_2(d, 7)$	$A_2(\lg_7 v, 7)$	(4, 25)
9	$S(3, 10, v)$	$P_2(d, 8)$	$A_2(\lg_8 v, 8)$	(3, 19), (4,28), (6, 46)*
10	$S(3, 11, v)$	$P_2(d, 9)$	$A_2(\lg_9 v, 9)$	(3, 21), (5, 41), (6, 51)*
11	$S(3, 12, v)$	—	—	(5, 45)
12	$S(3, 13, v)$	$P_2(d, 11)$	$A_2(\lg_{11} v, 11)$	(3, 25), (4, 37), (6, 61)*
13	$S(3, 14, v)$	—	—	(3, 27), (4, 40), (6, 66)*
14	$S(3, 15, v)$	$P_2(d, 13)$	$A_2(\lg_{13} v, 13)$	(7,85)*
15	$S(3, 16, v)$	—	—	(3.31), (5, 61), (6, 76)* (7, 91), (10, 136)*

— means "does not exist".

Lemma 6.3. *All admissible (l, k, v) for $2<l<t-1\leqslant 8$ are;*

(1) $(l, k)=(3,13)$ $\lambda\equiv 0, 1, 13, 18, 27, 31 \pmod{39}$,
$(t, u)=(6, 26)$ $v\equiv 3, 13, 133, 183, 273, 313 \pmod{390}$,
$v=10\lambda+3>k$ $\frac{v-1}{l-1}\equiv 1, 6, 66, 91, 136, 156 \pmod{195}$;

(2) $(l, k)=(3, 15)$ $\lambda\equiv 0, 1, 14, 15, 21, 29 \pmod{35}$,
$(t, u)=(7, 35)$ $v\equiv 3, 15, 171, 183, 255, 351 \pmod{420}$,
$v=12\lambda+3>k$ $\frac{v-1}{l-1}\equiv 1, 7, 85, 91, 127, 175 \pmod{210}$;

(3) $(l, k)=(4, 25)$ $\lambda\equiv 0, 1, 25, 32, 57, 176 \pmod{200}$,
$(t, u)=(8, 50)$ $v\equiv 4, 25, 529, 676, 1201, 3700 \pmod{4200}$,
$v=21\lambda+4>k$ $\frac{v-1}{l-1}\equiv 1, 8, 176, 225, 400, 1233 \pmod{1400}$;

(4) $(l, k)=(3, 19)$ $\lambda\equiv 0, 1, 19, 45, 64, 153 \pmod{171}$,
$(t, u)=(9, 57)$ $v\equiv 3, 19, 307, 723, 1027, 2451 \pmod{2736}$,
$v=16\lambda+3>k$ $\frac{v-1}{l-1}\equiv 1, 9, 153, 361, 513, 1225 \pmod{1368}$;

(5) $(l, k)=(4, 28)$ $\lambda\equiv 0, 1, 27, 28, 36, 55 \pmod{63}$,
$(t, u)=(9, 63)$ $v\equiv 4, 28, 652, 676, 868, 1324 \pmod{1512}$,
$v=24\lambda+4>k$ $\frac{v-1}{l-1}\equiv 1, 9, 217, 225, 289, 441 \pmod{504}$;

(6) $(l, k)=(6, 46)$ $\lambda\equiv 0, 1, 46, 135, 162, 181 \pmod{207}$,
$(t, u)=(9, 69)$ $v\equiv 6, 46, 1846, 5406, 6786, 7246 \pmod{8280}$,
$v=40\lambda+6>k$ $\frac{v-1}{l-1}\equiv 1, 9, 369, 1081, 1297, 1449 \pmod{1656}$.

Proof. We will use the conditions $t\,|\,(\lambda(\lambda-1)$, and $u\,|\,\lambda(\lambda-1)(\lambda t-\lambda+1)$ of Lemma 6.2 and following observations:

(a) if t is a power of prime, then $t\,|\,\lambda(\lambda-1)$ if and only if $\lambda\equiv 0, 1 \pmod{t}$;
(b) if $\frac{1}{2}t$ is a power of odd prime, then $t\,|\,\lambda(\lambda-1)$ if and only if $\lambda\equiv 0, 1 \pmod{\frac{1}{2}t}$.
Case $(l, k)=(3,13)$: $t\,|\,\lambda(\lambda-1)\Leftrightarrow\lambda\equiv 0, 1 \pmod{3}$, because $\frac{1}{2}t=3$ is prime.

$$u\,|\,\lambda(\lambda-1)(t\lambda-\lambda+1)\Leftrightarrow 26\,|\,\lambda(\lambda-1)(5\lambda+1-26)\Leftrightarrow 2\cdot 13\,|\,5\cdot\lambda(\lambda-1)(\lambda-5)$$
$$\Leftrightarrow 13\,|\,\lambda(\lambda-1)(\lambda-5),$$

i.e., $\lambda\equiv 0, 1, 5 \pmod{13}$. So

$$\lambda\equiv 0 \pmod{3}\Leftrightarrow\lambda\equiv 0, 27, 18 \pmod{39},$$
$$\lambda\equiv 1 \pmod{3}\Leftrightarrow\lambda\equiv 13, 1, 31 \pmod{39}.$$

Case $(l, k)=(3, 15)$: $t\,|\,\lambda(\lambda-1)\Leftrightarrow\lambda\equiv 0, 1 \pmod{7}$, because $t=7$ is prime.

$$u\,|\,\lambda(\lambda-1)(t\lambda-\lambda+1)\Leftrightarrow 35\,|\,\lambda(\lambda-1)(6\lambda+1+35)\Leftrightarrow 35\,|\,\lambda(\lambda-1)(\lambda+6).$$

In the subcase $\lambda \equiv 0 \pmod 7$ we have $\lambda = 7x$,

$$35 \mid 7x(7x-1)(7x+6) \Leftrightarrow 5 \mid x(7x-1)(7x+6) \Leftrightarrow 5 \mid x(7x-1+15)(7x+6+15)$$
$$\Leftrightarrow 5 \mid 7 \cdot 7x(x+2)(x+3) \Leftrightarrow 5 \mid x(x-3)(x-2),$$

i.e., $x \equiv 0, 2, 3 \pmod 5$ and $\lambda \equiv 0, 14, 21 \pmod{35}$.

In the subcase $\lambda \equiv 1 \pmod 7$ we have

$$\lambda = 7x+1,\ 35 \mid (7x+1)7x(7x+7) \Leftrightarrow 5 \mid x(x+1)(7x+1) \Leftrightarrow 5 \mid x(x+1-5)(7x+1-15)$$
$$\Leftrightarrow 5 \mid x(x-2)(x-4),$$

i.e., $x \equiv 0, 2, 4 \pmod 5$ and $\lambda = 1, 15, 29 \pmod{35}$.

The other cases are proved, using similar techniques.

Let (l, k) be integers $2 \geqslant l \leqslant t = (k-1)/(l-1)$ and let t, $t(t-1)/l$ be integers. Denote by $v_0 = (k-l)k + l$. Let $\alpha = \{0, 1, l, k, v_0\}$ be the parameters of a PMD M, which may or may not exist.

Proposition 6.4. (a) *(l, k, v_0) is admissible;*

(b) *for $l = 2$ PMD M exists if and only if $S(3, k, (k-1)^2+1)$ exists; so a PMD M exists (it is inversive plane* IP *(q)) if $k - 1 = q$ is a power of a prime;*

(c) *for $l = t$, M must be $P_2(4, q)$;*

(d) *for $2 < l < t-1 \leqslant 8$, $v_0 = \min v$, such that (l, k, v) is admissible, except for the cases $(l, k) = (3, 15), (4, 28)$;*

(e) *for $2 < l < t - l \leqslant 7$, $v = v_0$ D_1, D_3 exists always, D_2 exists for $(l, k) = (3, 15)$,* (3, 19).

In fact, (a) follows from Lemma 6.2. (b), (c) we can see from cases $l = 2, t-1, t$. (d), (c) follow from Lemma 6.3 and known lists of BIBDs.

Let $v'_0 = (k-l)(k-1)+l$. It is easy to see that (l, k, v'_0) is admissible if and only if $k \mid 2l(2l-1)$; thus for $l = 3, 4$ the only admissible (l, k, v'_0) are $(3, 15, 171)$ and $(4, 28, 652)$. $171 = \min v$ such that $(3, 15, v)$ is admissible, $652 = \min v$ such that $(4, 28, v)$ is admissible, any (l, k, v) with $v'_0 < v < v_0$ is not admissible.

Denote by $m(l, k)$ the number $\frac{1}{2}$l.c.m.(t, u), if both $\frac{1}{2}t$, $\frac{1}{2}u$ are odd integers and the number l.c.m.(t, u) otherwise.

Proposition 6.5. *For any $\lambda \equiv 0, 1 \pmod{m(l, k)}$, $v > k$, the triple (l, k, v) is admissible.*

Proof. Let at least one of the numbers $\frac{1}{2}t$, $\frac{1}{2}u$ be not an odd integer. We have $\lambda \equiv 0, 1$ (mod(l.c.m.(t, u))). Hence t, u both divide $\lambda(\lambda-1)$ and (l, k, v) is admissible by Lemma 6.2. Let both $\frac{1}{2}t, \frac{1}{2}u$ be odd integers. Then l.c.m.$\frac{1}{2}(t, u) - 1 = m(l, k) - 1$ is even. We have $\lambda \equiv 0, 1$ (mod(l.c.m.$\frac{1}{2}(t, u)$)). Hence $\lambda \equiv 0, 1$

(mod(l.c.m.($\frac{1}{2}t, \frac{1}{2}u$))): so either $\frac{1}{2}t, \frac{1}{2}u$ both divide odd λ (and then t, u both divide $\lambda(\lambda-1)$) or $\frac{1}{2}t, \frac{1}{2}u$ both divide odd $\lambda - 1$ (and then t, u both divide $\lambda(\lambda-1)$). So (l, k, v) is admissible via Lemma 6.2.

We have for $2<1<t-1\leq 8$ only one possible case ($(l, k) = (3, 13)$) when both $\frac{1}{2}t, \frac{1}{2}u$ are odd integers. From Lemma 6.3 it is clear that for each (l, k), $2<l<t-1\leq 8$, all v's, such that (l, k, v) is admissible, have 6 possible values modulo $m(l, k)$. Perhaps, any v from admissible (l, k, v) for a given (l, k) also has 6 possible values modulo $m(l, k)$, including $0, 1, k$ and sometimes $\pm(k-1)$ (as it happens for $(l, k) = (3, 15), (4, 28)$).

We collect a few simple properties of PMD of rank 4 with $\alpha = \{0, 1, l, k, v\}$, in the following:

Proposition 6.6. (i) *If a PMD of rank 4, M exists then BIBDs D_1, D_2, D_3, D_4 exist.*

(ii) *D_1 is a symmetric BIBD, if and only if M is $P_2(d, q)$, $d\geq 3$ (see* [21]).

(iii) *D_2 is a symmetric BIBD, if M is $P_2(3, q)$ or $A_2(3, q)$ or $S(3, 6, 22)$. (if* rank *$M\geq 5$, D_2 is a symmetric BIBD iff M is $P_{d-1}(d, q)$ or $A_{d-1}(d, q)$, see* [28].)

(iv) *D_3 is never symmetric BIBD.*

(v) *D_4 is a symmetric BIBD if and only if M is $P_2(3, q)$. (Dembowski and Wagner's Theorem* [7], *see also* [23].)

Using Theorem 4c and some messy computing we can give the following information on completeness of PMD $M = (X, \beta)$ of rank 4 with $\alpha = \{0, 1, l, k, v\}$.

Proposition 6.7. (i) *A point $x \in X$ such that $H_1 \cap H_2 \neq \{x\}$ for any $H_1, H_2 \in \beta$ exists if and only if $\lambda = t$ (i.e., if and only if D_2 is a projective plane).*

(ii) *$H_1 \cap H_2 \neq \emptyset$ for all $H_1, H_2 \in \beta$ if and only if*

$$g = (k-1)(k\lambda - k + l) + l\left(\lambda^2 - \frac{\lambda(\lambda-1)}{t}\right)\left(\frac{v}{k} - l\right) = 0;$$

in particular, for $\lambda = k$ (i.e., for $v = k^2 - kl + l = v_0$) $g \geq 0$ with $g = 0$ if and only if $M = P_2(4, q)$.

(iii) *Any M with parameters (l, k, v) given in Lemma 6.3 is complete and maximal if it exists.*

Here we have for $(l, k) = (2, k), (3, 7), (3, 9), (4, 13), (4,16)$ all possible v's such that a PMD with $\alpha = \{0, 1, l, k, v)$ exists. For others (l, k) we give first few v's such that (l, k, v) is admissible and, moreover, BIBDs D_1, D_2 exist ([16]); By $\underline{v}$ we denote such v's from this list that the BIBD D_2 is known to exist (see, for example, [18]). So now we need, for example, $D_4 = S_{18}(2, 13, 183)$, $S_{15}(2, 15, 183)$,

Table 2. PMD of rank 4 with small α

α_0,	α_1,	$\alpha_2 = l$;	$\alpha_3 = k$	Possible $\alpha_4 = v$
0	1	2	k	any v such that $S(3, k, v)$ exists
0	1	3	7	$2^n - 1$, $n \geq 4$ (truncated PG $(n-1, 2)$)
0	1	3	9	3^n, $n \geq 3$ (truncated AG $(n, 3)$ and ATSs)
0	1	3	13	133, 183, 273, 313, 393, 403, . . . , 663, . . .
0	1	3	15	171, 183, 225, 351, . . . , 675, 77, 843, . . .
0	1	3	19	307, 723, 1027, 2451, 2739, 2755, 3043, 3459, . . .
0	1	4	13	$\frac{1}{2}(3^n - 1)$, $n \geq 4$ (truncated PG$(n-1, 3)$)
0	1	4	16	4^n, $n \geq 3$ (truncated AG $(n, 4)$)
0	1	4	25	529, 676, 1201, 3700, 4204, 4225, . . . , 8404, . . .
0	1	4	28	652, 676, 868, 1324, 1516, 1540, 2164, 2188. . . .

$S_{19}(2, 19, 307)$, $S_{31}(2, 13, 313)$ with only 0, 1, 3 as sizes of block-intersections for $(l, k, v) = (3, 13, 183)$, $(3, 15, 183)$, $(3, 19, 307)$, $(3, 13, 313)$.

References

[1] L. Beneteau, Topics about 3-Moufang loops and Hall triple systems, to appear.

[2] C. Berge, Principles of Combinatorics. (Acad. Press, New York, 1971).

[3] F. Buekenhout, Characterization des espaces affine basée sur la notion de droit, Math. Z. 3 (1969) 367–371.

[4] P.J. Cameron, Extending symmetric designs, J. Combin. Theory 14 (A) (1973) 215–220.

[5] P.J. Cameron, Two remarks on Steiner systems, Geometriae Dedicata 4 (1975) 403–418.

[6] P. Delsarte, Association schemes and t-designs in regular semilattices, J. Combin. theory 20 (A) (1976). 230–243.

[7] P. Dembowski and A. Wagner, Some characterizations of finite projective spaces, Arch. Math. (Basel) II (1960). 465–469.

[8] M. Deza, P. Erdös and P. Frankl, Intersection properties of the systems of finite sets, Proc. London math. soc. 36 (3) (1978). 369–384.

[9] M. Deza, Pavage généralisé parfait comme generalisation de matroïde-configurations et de simples t-configutations, in: Coll. int. CNRS. No. 260–Problèmes combinatoires, Paris-Orsay (1978). 97–100.

[10] M. Deza, On perfect matroid designs, in; Proc. Symposium on Construction and Analysis and Designs, Kyoto (1977) 98–108.

[11] P. Doubilet, G.C. Rota and R. Stanley, On the foundations of combinatorial theory (VI): The idea of generating functions, in: Proc. of Sixth Berkeley Symposium of Math. Stat. and Probability, Vol. 2 (Univ. of California Press, Berkeley, C. A. (1972) 267–318.

[12] B. M. Gross, Intersection triangles and block intersection numbers of steiner systems, Math. Z. 139 (1974) 87–104.

[13] M. Hall, Automorphisms of Steiner triple systems, IBM J. Res. Develop. 4 (1960) 460–472.

[14] N. Hamada and F. Tamari, Duals of balanced incomplete blocks designs derived from an affine geometry, Ann. Statist. 3 (1975) 926–938.

[15] H. Hanani, On some tactical configurations, Can. J. Math. 15 (1963) 702–722.

[16] H. Hanani, The existance and construction of balanced incomplete block designs, Ann. Math. Statist. 32 (1961) 361–386.

[17] W.M. Kantor, Math. Reviews 50 (1975), Review No. 142.

[18] W.H. Mills, The construction of balanced incomplete block designs with $\lambda = 1$, in: Proc. of Seventh Manitoba Conference, to appear.

[19] U.S.R. Murty, Equicardinal matroids and finite geometries, in: Calgary Int. Conference on Comb. Structures and Appl. (Gordon and Breach, new York, 1970) 289–291.
[20] U.S.R. Murty, H.P. Young and J. Edmonds, Equicardinal on Comb. Math. and Appl., Chapel Hill, N.C. (1970) 498–547.
[21] D.K. Ray-Chaudhuri and N.M. Singhi, A characterization of line-hyperplane design of projective space and some extremal theorems for matroid-designs, in: H. Zassenhaus, ed., Number Theory and Algebra (Acad. Press, New York, 1977) 289-301.
[22] B.L. Rothschild and N.M. Singhi, Characterizing k-flats in geometric designs. J. Combin. Theory 20 (A) (1976) 398–403.
[23] N.M. Singhi, Characterizion of finite projective spaces, Math. Student, to appear.
[24] L. Teirlinck, On linear spaces in which every plane is projective or affine, Geometriae Dedicata 4 (1975) 39–44.
[25] H.P. Young, Affine triple systems and matroid-designs, Math. Z. 132 (1973) 343–359.
[26] D.J.A. Welsh, Matroid Theory (Acad. Press, London, New York, 1976).
[27] H. Zassenhaus, in: G. Bol, Math. Ann. 114 (1937) 414–431; Zbl. 16, 226.
[28] W.M. Kantor, Characterizations of projective and affine spaces, Can. J. Math. 21 (1969) 64–75.

Annals of Discrete Mathematics 6 (1980) 77–87

A COMBINATORIAL INTERPRETATION OF THE SEIDEL GENERATION OF GENOCCHI NUMBERS*

Duminique DUMONT

Université de Strasbourg, Departement de Mathématiques, 67084 Strasbourg, France

Gérard VIENNOT

University of California, San Diego, Department of Mathematics, La Jolla, CA 92093, U.S.A.

1. Introduction

The *Genocchi numbers* G_{2n} are integers defined from the *Bernoulli numbers* by the relation:

$$G_{2n} = 2(2^{2n} - 1)B_{2n} \quad (n \geq 1). \tag{1}$$

They are related to the *tangent numbers* T_{2n-1} (or *Euler numbers* of the second kind) by the relation:

$$2^{2n-2}G_{2n} = nT_{2n-1} \quad (n \geq 1). \tag{2}$$

The first values of these sequences are given in the Table 1. The exponential generating functions of these numbers (more exactly of $(-1)^n G_{2n}$ and $(-1)^{n+1}B_{2n}$) are respectively:

$$\frac{2t}{e^t+1} = t + \sum_{n\geq 1} (-1)^n G_{2n} \frac{t^{2n}}{(2n)!},$$

$$\frac{t}{e^t-1} = 1 - \tfrac{1}{2}t + \sum_{n\geq 1} (-1)^{n+1} B_{2n} \frac{t^{2n}}{(2n)!},$$

$$\tan t = \sum_{n\geq 1} T_{2n-1} \frac{t^{2n-1}}{(2n-1)!}.$$

The relations (1) and (2) are easily derived from these expressions.

Dumont gave in [4] the first combinatorial interpretation of the Genocchi numbers. It is related to a generation of these numbers conjectured in 1970 by Gandhi [8] and proved by Carlitz [2] and Riordan and Stein [12] by means of analytic calculus.

Here we give another interpretation related to the Seidel definition of the Bernoulli and Genocchi numbers [13]. Though being very natural and simple, this definition has been somewhat forgotten and is restated in Section 4. The following

*Work partially supported by NSF-CNRS Exchange Visitor Program No. g–05–0252.

Table 1

n	1	2	3	4	5	6	7
B_{2n}	$\frac{1}{6}$	$\frac{1}{30}$	$\frac{1}{42}$	$\frac{1}{30}$	$\frac{5}{66}$	$\frac{691}{2730}$	$\frac{7}{6}$
G_{2n}	1	1	3	17	155	2973	38227
T_{2n-1}	1	2	16	272	7936	353792	22368256

identity can be deduced (where $G_0=0$ and $n \geqslant 1$ fixed)

$$\sum_{0 \leqslant k \leqslant n/2} (-1)^k \binom{n}{2k} G_{2n-2k} = 0. \tag{3}$$

This identity (called Seidel identity by Nielsen [11, pp 186–187]) leads to the combinatorial interpretation with "*alternating pistols*" in Section 2 and, in Section 3, with a subclass of the well-known *alternating permutations* enumerated by the tangent numbers.

This combinatorial interpretation enables us to give a very convenient way to calculate the Genocchi (and hence Bernoulli) numbers, by constructing a tableau of positive integers with summations two by two of positive integers. These integers appear in Seidel's definition and, roughly speaking, are equivalent to that definition. In Sections 4 and 5 we explain this fact by calculus and give some additional results, beyond the original paper of Seidel.

In Section 6 we give a geometric correspondance between the interpretation [4] of the Gandhi generation and the permutations of Section 3 interpreting the Seidel generation. This correspondance is briefly recalled as being a consequence of some techniques involving permutations introduced by Françon and Viennot in [7] (where one can find its complete definition).

2. Alternating pistols

Notations. In this paper $[m]$ denote the set of integers $[1, m]=\{1, 2, \ldots, m\}$, and $|E|$ the cardinality of every finite set E. The map $\pi: \mathbb{N} \to \mathbb{N}$ is defined by the following condition

$$\pi(1)=1 \quad \text{and} \quad \forall i \geqslant 1, \quad \pi(2i)=\pi(2i-1), \qquad \pi(2i+1)=\pi(2i)+1. \tag{4}$$

Definition 1. Let $m \geqslant 1$. A *pistol* on $[m]$ is a map $h:[m]\to[m]$ with $h(i) \leqslant \pi(i)$ for every $i \geqslant 1$.

Definition 2. A pistol h on $[m]$ is said to be *alternating* if the following condition holds:

$$\begin{aligned} \forall i \in [m-1], \quad & h(i) \geqslant h(i+1) \quad \text{if } i \text{ is odd}, \\ & h(i) \leqslant h(i+1) \quad \text{if } i \text{ is even}. \end{aligned} \tag{5}$$

For example, the pistol $h = h(1) \cdots h(8) = 11213332$ is displayed in Fig. 1.

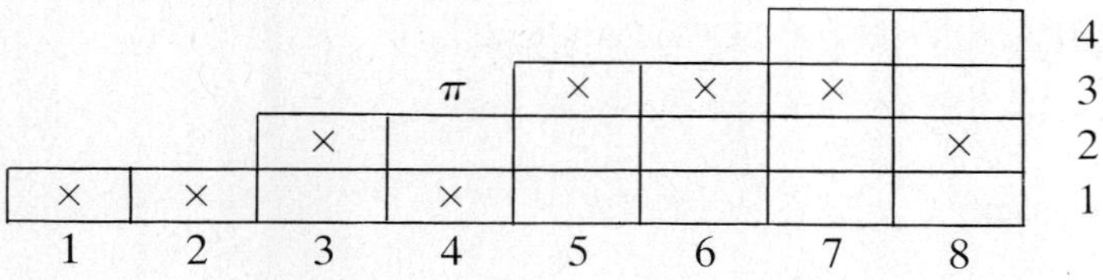

Fig. 1.

We denote $\mathcal{P}_m$ (resp. $\mathcal{AP}_m$) the set of pistols (resp. alternating pistols) on $[m]$.

The next definition is introduced for two reasons: we need it in Section 3, and the proof of Proposition 3 below will be more clear.

Let $\mathcal{AP}'_m$ be the set of pistols h of $\mathcal{P}_m$ satisfying the following condition:

$$\forall i \in [m-1], \quad h(i) \geqslant h(i+1) \quad \text{if } i \text{ is odd},$$
$$h(i) < h(i+1) \quad \text{if } i \text{ is even}. \tag{5'}$$

For every pistol $h \in \mathcal{P}_m$, we define the map $\alpha(h) = h' : [m+1] \to [m+1]$ by the condition:

$$h'(1) = 1 \quad \text{and} \quad \text{for every } i, \quad 2 \leqslant 2i \leqslant m,$$
$$h'(2i) = i + 1 - h(2i-1),$$
$$h'(2i+1) = i + 2 - h(2i). \tag{6}$$

For example, the pistol $h = 11213332$ gives the pistol $h' = 112131224$ displayed in Fig. 2.

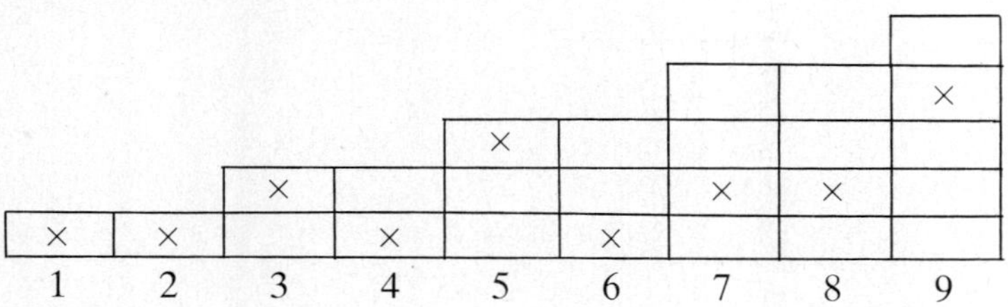

Fig. 2.

It is easily checked that the map α is a bijection from $\mathcal{P}_m$ onto the set of pistols $h \in \mathcal{P}_{m+1}$ such that $h(i) \geqslant 2$ for every odd index $i \in [3, m]$. This bijection α maps $\mathcal{AP}_m$ onto $\mathcal{AP}'_{m+1}$.

Proposition 3. *For every $n \geqslant 1$, the number of alternating pistols on $[2n]$ is the Genocchi number G_{2n+2}, or:*

$$\forall i \geqslant 1, \quad |\mathcal{AP}_{2n}| = |\mathcal{AP}'_{2n+1}| = G_{2n+2}.$$

Denote $a_{2n} = |\mathcal{AP}'_{2n+1}|$, and for $p \in [n]$, define a_{2n}^{2p} the number of pistols

$h \in \mathscr{P}_{2n+1}$ such that the following holds:

$$\forall i \in [2n+1], \quad h(i) \geqslant h(i+1) \quad \text{if } i \text{ is odd and } i \in [2p],$$
$$h(i) < h(i+1) \quad \text{in the opposite case.} \tag{$5''_p$}$$

It is easy to prove the relation:

$$\text{For } 2 \leqslant p \leqslant n, \quad a_{2n}^{2p} = \binom{n+1}{2n-2p+2} a_{2p-2} - a_{2n}^{2p-2}. \tag{7}$$

A recursive application of (7) shows that the sequence $b_{2n} = a_{2n-2} = a_{2n-2}^{2n-2}$ $(n \geqslant 2)$ with $b_2 = 1$, satisfies the relation (3). This relation gives b_{2n} or G_{2n} as an expression involving the a_{2i} (resp. G_{2i}) for $1 \leqslant i < n$. The initial conditions $b_2 = G_2 = 1$ are the same, hence the two sequences are identical. Thus $a_{2n} = G_{2n+2}$, and the proposition results using the bijection α.

We give another way to enumerate these numbers. Define the tableau of integers h_m^k by $h_1^1 = 1$ and the following induction:

$$\text{for } m = 2n \geqslant 2, \quad h_m^n = h_{m-1}^n \quad \text{and} \quad \text{for } k \in [n-1], \quad h_m^k = h_m^{k+1} + h_{m-1}^k, \tag{8}$$

$$\text{for } m = 2n-1 \geqslant 3, \quad h_m^1 = h_{m-1}^1 \quad \text{and} \quad \text{for } k \in [2, n-1],$$
$$h_m^k = h_m^{k-1} + h_m^k, \qquad h_m^n = h_m^{n-1}. \tag{8'}$$

The first values are given in Table 2.

Table 2

								155	155	5
						17	17	155	310	4
				3	3	17	34	138	448	3
		1	1	3	6	14	48	104	552	2
1	1	1	2	2	8	8	56	56	608	1
1	2	3	4	5	6	7	8	9	10	k \ m

Define $\mathscr{AP}_m^k$ to be the set of alternating pistols h with $h(m) = k$. It is easy to prove from (5) that the numbers $|\mathscr{AP}_m^k|$ satisfy the recurrences (8) and (8′), with same initial conditions. Thus, from Proposition 3, we can state.

Corollary 4. *The Genocchi number G_{2n+2} is the sum of the $2n$th column of the tableau (h_n^k) defined by (8) and (8′):*

$$G_{2n+2} = \sum_{k \in [n]} h_n^k \qquad (n \geqslant 1).$$

3. Combinatorial interpretation of Genocchi numbers with alternating permutations

A permutation σ of $[2n+1]$ is said to be *alternating* if:

$$\forall i \in [n], \quad \sigma(2i-1) \geqslant \sigma(2i) \leqslant \sigma(2i+1). \tag{9}$$

Since André [1], it is known that the number of such permutations is the tangent number T_{2n+1} defined in Section 1.

Let $m \geqslant 1$ and σ a permutation of $[m]$. The *inversion table* (also called *Lehmer code*) is the map $l(\sigma) = f:[m] \to [0, m-1]$ defined by:

$$\forall i \in [m], \quad f(i) \quad \text{is the number of indices } j \tag{10}$$
$$\text{such that } 1 \leqslant j < i \quad \text{and} \quad \sigma(j) < \sigma(i).$$

The map f is a *subexceedant function* on $[m]$, that is a map $f:[m] \to [0, m-1]$ such that $f(i) < i$ for every $i \in [m]$. It is well-known (see for example Knuth [9, p. 12]) that the map $l: \sigma \to f$ is a bijection between the set $\mathfrak{S}_m$ of permutations on $[m]$ and the set $\mathfrak{F}_m$ of subexceedant functions on $[m]$. Furthermore, the following relation holds:

$$\forall i \in [n], \qquad \sigma(i) > \sigma(i+1) \qquad \text{if} \qquad f(i) \geqslant f(i+1). \tag{11}$$

Hence, the set of alternating permutations on $m = 2n+1$ is in bijection by l with the set of subexceedant functions on $[m]$ satisfying (5′).

Now to every pistol h of $\mathcal{P}_{2n+1}$ we associate the map $\phi(h) = f:[2n+1] \to [0, 2n]$ by the following

$$\forall i \in [2n+1], \quad f(i) = 2(h(i) - 1). \tag{12}$$

Condition (4) implies $f \in \mathfrak{F}_{2n+1}$. The correspondance ϕ is a bijection between $\mathcal{P}_{2n+1}$ and subexceedant functions of $\mathfrak{F}_{2n+1}$ with even values. The pistol $h \in \mathcal{P}_{2n+1}$ satisfies (5′) iff $\phi(h)$ satisfies also (5′). Hence, from above, the map $l^{-1} \circ \phi \circ \alpha$, defined by (6), (11), (12), is a bijection between the set $\mathcal{AP}_{2n+1}$ of alternating pistols, and the set of alternating permutations with inversion table having only even values.

For example, with alternating pistol $h = 11213332$ of Section 2, we have successively $h' = \alpha(h) = 112131224$, $h'' = \phi(h') = 002040226$, $\alpha = l^{-1}(h'') = 658291437$.

Proposition 3 becomes here:

Proposition 5. *For every $n \geqslant 0$, the Genocchi number G_{2n+2} is the number of alternating permutations σ of $[2n+1]$ having an inversion table $l(\sigma)$ (defined by* (10)*) with even values.*

This proposition is essential for Section 6.

4. Seidel matrices

We show now the relation between the Seidel identity (3), the tableau defined by (8), (8′) and the Seidel definition of Genocchi numbers.

Definition 6. A *Seidel matrix* is an infinite matrice $A=(a_n^k)_{n,k\geq 0}$, with elements in $\mathbb{Q}$, where the sequence of elements of the first row is given $a^0=(a_0^0, a_1^0, \ldots, a_n^0, \ldots)$, the others elements being defined by the induction:

$$\forall n\geq 0, \quad \forall k\geq 1, \qquad a_n^k = a_n^{k-1} + a_{n+1}^{k-1}.$$

Following Seidel, we call the first row the *initial sequence* and the first column $a_0=(a_0^0, a_0^1, \ldots)$ the *terminal sequence.* The following relations give the expression of every element of A from the initial sequence or from the terminal sequence.

$$\forall k, \qquad n\geq 0, \qquad a_n^k = \sum_{i=0}^{k} \binom{k}{i} a_{n+i}^0, \tag{14}$$

$$\forall k, \qquad n\geq 0, \qquad a_n^k = \sum_{i=0}^{n} (-1)^i \binom{n}{i} a_0^{k+n-i}. \tag{14′}$$

These relations are easily deduced by induction from (13) or the equivalent relation:

$$\forall k\geq 0, \qquad \forall n\geq 1, \qquad a_n^k = a_{n-1}^{k+1} - a_{n-1}^k. \tag{13′}$$

Let $a(t)$ the exponential generating function of the initial sequence:

$$a(t) = \sum_{n\geq 0} a_n^0 \frac{t^n}{n!}.$$

Proposition 7. *The generating function f of the Seidel matrix* $A=(a_n^k)$ *is:*

$$\sum_{n,k\geq 0} a_n^k \frac{t^n}{n!}\frac{u^k}{k!} = \mathrm{e}^u a(t+u). \tag{15}$$

From (14), the (exponential) generating function of the kth row is:

$$\sum_{n\geq 0} a_n^k \frac{t^n}{n!} = \sum_{i=0}^{k} \binom{k}{i} a^{(i)}(t) \tag{16}$$

where $a^{(i)}(t)$ denote the ith derivative of $a(t)$. With Leibnitz formula, we deduce

$$\mathrm{e}^t \Big(\sum_{n\geq 0} a_n^k \frac{t^n}{n!} \Big) = [\mathrm{e}^t a(t)]^{(k)}. \tag{17}$$

Then, applying Taylor formula:

$$\mathrm{e}^t \Big(\sum_{n,k\geq 0} a_n^k \frac{t^n}{n!}\frac{u^k}{k!} \Big) = \mathrm{e}^{t+u} a(t+u).$$

By setting $t=0$ in (15), we obtain the proposition of Seidel:

Corollary 8 (Seidel). *In a Seidel matrix $A=(a_n^k)$, the generating function of the terminal sequence is the product by the exponential of the generating function of the initial sequence*:

$$\sum_{k\geq 0} a_0^k \frac{u^k}{k!} = e^u a(u).$$

Hence, the transformation mapping the initial sequence onto the terminal sequence has no invariant. But, with Seidel, we can ask for "almost invariant" sequence in the following sense:

$$b_0^0=1, \qquad b_0^1=-b_1^0, \quad \text{and} \quad \forall i\geq 2, \qquad b_0^i=b_i^0. \tag{18}$$

There exist a unique Seidel matrix $b=(b_n^k)$ satisfying (18). The initial sequence can be calculated: with (14) and (18), and form Corollary 8, its (exponential) generating function is $b(t)=t/(e^t-1)$. Thus this initial sequence is the sequence of Bernoulli numbers:

$$\forall n\geq 1, \quad b_{2n}^0=(-1)^{n+1}B_{2n}, \qquad b_0^0=1, \quad b_1^0=-\tfrac{1}{2}, \quad b_{2n+1}^0=0.$$

Similarly, we can ask for some "almost anti-invariant" of the correspondence defined by the Seidel process, that is:

$$g_0^0=0, \qquad g_1^0=g_0^1=1, \quad \text{and} \quad \forall i\geq 2, \qquad g_0^i=-g_i^0. \tag{19}$$

Again, relation (14) with $n=0$, defines by induction a unique initial sequence g_n^0 satisfying (19) and a unique Seidel matrix denoted by $G=(g_n^k)$. From Corollary 8, its (exponential) generating function is $g(t)=2t/(e^t+1)$. The initial sequence is the sequence of Genocchi numbers:

$$\forall i\geq 1, \qquad g_{2n}^0=(-1)^n G_{2n}, \qquad g_0^0=0, \qquad g_1^0=1, \qquad g_{2n+1}^0=0.$$

The relations (18) and (19) are, what we call in this paper the Seidel definitions of Bernoulli and Genocchi numbers.

Remark 9. If one takes as initial sequence the tangent numbers $(-1)^n T_{2n-1}$ (with $t_0^0=1$), with exponential generating function $2/(e^{2t}+1)$, the terminal sequence is the sequence of *secant numbers* $(-1)^n E_{2n}$ (or Euler numbers of the first kind), with exponential generating function $1/(\operatorname{ch} t)$.

5. The Seidel matrix of Genocchi numbers

The first values of the Seidel matrix $G=(g_n^k)$ defined above by (19) are given in Table 3.

Table 3

0	1	−1	0	1	0	−3	0	17	0	−155
1	0	−1	1	1	−3	−3	17	17	−155	
1	−1	0	2	−2	−6	14	34	−138		
0	−1	2	0	−8	8	48	−104			
−1	1	2	−8	0	56	−56				
0	3	−6	−8	56	0					
3	−3	−14	48	56						
0	−17	34	104							
−17	17	138								
0	155									
155										

The reader recognizes the integers of Table 2. The following proposition makes explicit the signs and symetries of the integers g_n^k.

Proposition 10. *In the Seidel matrix* $G=(g_m^k)$ *of Genocchi numbers defined by* (19), *the following properties hold*:

$$\forall m,\ k \geqslant 0, \quad g_m^k = (-1)^{m+k-1} g_k^m, \tag{20}$$

$$\forall m \geqslant k \geqslant 0, \quad \textit{let} \quad m+k=2n \quad \textit{or} \quad 2n+1, \quad \textit{then if } g_m^k \text{ is not equal to } 0, \textit{ its sign is } (-1)^n. \tag{21}$$

From (19) and the fact that the generating function $g(t)-t=t(1-e^t/(1+e^t)$ is an even function, identity (20) holds for $k=0$.

The general case is proved by induction, using (14) and (14′) for $a_m^k = g_m^k$.

The relation (21) is proved by induction on n. From (19), it is true for $n=0$. The reader will easily prove by induction from (14′) and (14), that the sign of g_m^k changes when replacing "$m+k=2n-1$" by "$m+k=2n$", and that the sign does not change when replacing "$m+k=2n$" by "$m+k=2n+1$".

A corollary of (20) is the relation:

$$\forall n \geqslant 0, \qquad g_n^n = 0. \tag{22}$$

Taking the expression of g_n^n from (14) and using $g_{2n+1}^0 = 0$, the relation (22) is nothing but the Seidel identity (3). Lucas [10, p. 252], write also (22) or (3), under symbolic notations, by:

$$G^n(G+1)^n \equiv 0 \quad \text{where} \quad G^i \equiv (-1)^i Gi. \tag{23}$$

With these notations, the relations (14′) and (19) give the identity:

$$(G+1)^n + G^n \equiv 1. \tag{24}$$

Relations (13) and (21) show that the upper triangular matrix $(g_m^k)_{m \geqslant k \geqslant 1}$ is, under an appropriate bijection converting diagonals into rows, and by cancelling the signs given by (21), the triangle of integers h_m^k defined by (8) and (8′).

Remark 11. The analogue of this transformation for the Seidel matrix of Euler numbers (Remark 9) gives the tableau of Entringer [5], that is the distribution of alternating permutations (for m odd or even) according to their last letters. We give in Table 4 the first values of this Seidel matrix.

Table 4

1	−1	0	2	0	−16	0	272
0	−1	2	2	−16	−16	272	
−1	1	4	−14	−32	256		
0	5	−10	−46	224			
5	−5	−56	178				
0	−61	122					
−61	61						
0							

6. Geometric relation with the Gandhi generation of Genocchi numbers.

Gandhi [8] defined, in a slightly different way, the following polynomials:

$$Q_0(t)=1 \quad \text{and} \quad \forall i \geqslant 0, \quad Q_{i+1}(t)=t^2 Q_i(t+1)-(t-1)^2 Q_i(t) \tag{25}$$

and conjectured that

$$\forall n \geqslant 0, \quad Q_n(1)=G_{2n+2}. \tag{26}$$

This was proved by Carlitz [2] and Riordan and Stein [12]. Dumont gave in [4] the following geometric interpretation of this generation.

Define $\mathcal{SP}_{2n}$ the set of *surjective pistols* on $(2n]$, that is pistols $h \in \mathcal{P}_{2n}$ such that the map h, considered as a map $[2n] \rightarrow [n]$, is a surjection. Then G_{2n+2} is the number of such pistols. In fact the cofficients of the polynomials $Q_{n-1}(t+1)$ give the distribution of these pistols according to the number of $i \in [2n]$ with $h(i)=1$.

In [4], the first author gave also a bijection between $\mathcal{SP}_{2n}$ and the set of permutations $\sigma \in \mathfrak{S}_{2n+1}$ satisfying the relation below:

$$\forall i \in [2n], \quad \sigma(i)<\sigma(i+1) \quad \textit{iff } \sigma(i) \textit{ is even}. \tag{27}$$

We briefly recall some techniques of Françon and Viennot [7].

Define a contraction γ to be a map $[n] \rightarrow [n]$ satisfying $\gamma(1)=\gamma(n)=1$ and $|\gamma(i+1)-\gamma(i)| \leqslant 1$ for every $i \in [n-1]$. A *colored contraction* γ_c is a contraction with possibility of coloring the *levels* (that is $\gamma(i+1)=\gamma(i)$) with two colors: blue and red. Let $\mathfrak{I}_n$ the set of pairs (γ_c, f) where γ_c is a colored contraction and f a map $[n] \rightarrow [n]$ with $f(i) \leqslant \gamma(i)$ for every $i \in [n]$. The set $\mathfrak{I}_n$ has $n!$ elements.

This fact is proved in [7] by defining a bijection $\theta: \mathfrak{I}_n \rightarrow \mathfrak{S}_n$ having the following property: for every $\sigma=\theta(\gamma_c, f)\in\mathfrak{S}_n$, and with convention $\sigma(0)=\sigma(n+1)=0$, $\forall i\in[n-1]$,

$$\begin{aligned}
&\gamma(i)>\gamma(i+1) \quad \text{iff} \quad \sigma(i-1)<\sigma(i)>\sigma(i+1),\\
&\gamma(i)<\gamma(i+1) \quad \text{iff} \quad \sigma(i-1)>\sigma(i)<\sigma(i+1),\\
&\gamma(i)=\gamma(i+1) \quad \text{with red,} \quad \text{iff} \quad \sigma(i-1)<\sigma(i)<\sigma(i+1),\\
&\gamma(i)=\gamma(i+1) \quad \text{with blue,} \quad \text{iff} \quad \sigma(i-1)>\sigma(i)>\sigma(i+1).
\end{aligned} \tag{28}$$

From (28), the alternating permutations on $[2n+1]$ correspond to the pair (γ_c, f) where γ has no levels (and thus no color), called also *strict contractions* (and enumerated by the Catalan numbers $C_n=(1/(n+1))\binom{2n}{n}$). It is shown in [7] that the alternating permutations of Proposition 5 correspond to the pairs (γ, f) where γ is a strict contraction and f a map with only odd values.

Furthermore, relation (28) shows that the permutations satisfying (27) correspond to the pairs (γ_c, f), where γ_c is an *alternating contraction*, that is a contraction satisfying condition (5), and with colors given by the rule:

$$\text{the level } \gamma(i)=\gamma(i+1) \text{ is} \begin{cases} \text{red} & \text{for } i \text{ even},\\ \text{blue} & \text{for } i \text{ odd}.\end{cases} \tag{29}$$

This means in fact that the pair is defined by (γ, f) and we don't care about colors.

Thus the problem to find a geometric relation between the Gandhi and the Seidel generation of Genocchi numbers remains to find a bijection between the two sets of pairs (γ, f) defined above. A very simple one is given in [7]. Let (γ, f) a pair of $\mathfrak{I}_{2n+1}$ with γ alternating contraction. Define $\rho(\gamma, f)=(\gamma', f')$ by the following relation:

$$\forall i\in[2n+1], \qquad \gamma'(i)=2\gamma(i)-1 \quad f'(i)=2f(i)-1. \tag{30}$$

Then $(\gamma', f')\in\mathfrak{I}_{2n+1}$ with γ' strict contraction and f' function with only odd values.

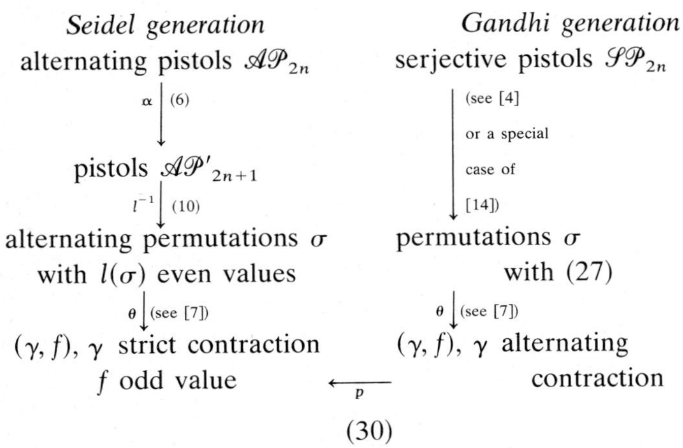

Fig. 3.

The map ρ is the desired bijection. The following diagram recalls bijections used to go from alternating pistols to surjective pistols (see Fig. 3).

A straightforward bijection between alternating and surjective pistols would be welcome. A simplification can be derived from [14] for the bijection between surjective pistols and permutations (25).

Note. In our lecture in Fort Collins, we give also more details about combinatorial interpretations of Euler numbers. They will be found in [15] where the second author gives, after the interpretation of André [1] (alternating permutations) and of Foata and Schützenberger [6] (André permutations), a third one (Jacobi permutations). This one leads to a combinatorial interpretation of Jacobi elliptic functions *sn*, *cn* and *dn*. The Seidel generation can also be interpreted as a subclass of Jacobi permutations.

References

[1] D. André, Sur les permutations alternées, J. Math. Pures Appl. 7 (1881) 167–184.
[2] L. Carlitz, A conjecture concerning Genocchi numbers, K. Norske Vidensk. Selsk. Sk. 9 (1972) 1–4.
[3] L. Comtet, Advanced Combinatorics (D. Reidel Publ. Comp., Dordrecht, Boston, 1974).
[4]. D. Dumont, Interprétations combinatoires des nombres de Genocchi, Duke Math. J. 41 (2) (1974) 305–318.
[5] R.C. Entringer, A combinatorial interpretation of the Euler and Bernoulli numbers, Nieuw Arch. Wisk. 14 (1966) 241–246.
[6] D. Foata and M.P. Schützenberger, Nombres d'Euler et permutations alternantes, in: J.N. Srivastava et al., eds., A survey of Combinatorial Theory, Symp. Colorado State Univ, Fort Collins (1971) (North-Holland, Amsterdam, 1973).
[7] J. Françon and G. Viennot, Permutations selon les pics, creux, doubles montées et doubles descentes, nombres d'Euler et nombres de Genocchi, submitted for publication.
[8] J.M. Gandhi, A conjectured representation of Genocchi numbers, Am. Math. Monthly (1970) 505–506.
[9] D. Knuth, The Art of Computer Programming, Vol 3 (Addison-Wesley, Reading, MA, (1974).
[10] A. Lucas, Théorie des nombres, Vol 1 (Gauthiers-Villars, Paris, 1891).
[11] N. Nielsen, Traité élémentaire des nombres de Bernoulli (Gauthiers Villars, Paris, 1923).
[12] J. Riordan and P.R. Stein, Proof of a conjecture on Genocchi numbers, Discrete Math. 5 (1973) 381–388.
[13] L. Seidel, Über eine einfache Entshehungsweise der Bernoullischen Zahlen und einiger verwandten Reihen, Sitzungsberichte Münchener Akad, Math. Phys. Classe (1877) 157–187.
[14] G. Viennot, Finite difference calculus and permutations with prescribed set of rises, submitted for publication.
[15] G. Viennot, Une interpretation combinatoire des coefficients des développements en série entière des fonctions elliptiques de Jacobi, to appear in J. Combinatorial Theory (A) (1980).

Annals of Discrete Mathematics 6 (1980) 89–115

A SURVEY OF PROBLEMS IN COMBINATORIAL NUMBER THEORY

Paul ERDÖS

Hungarian Academy of Science, Budapest, Hungary

During my long life I published and stated many conjectures and wrote several papers which consisted entirely of stating old and new problems. In this paper I hope to survey some of them — not to make the paper too long I restrict myself to problems on the borderline of combinatorics and number theory fields which are — as stated in a previous paper — closest to my heart or rather than to my brain. Again, to avoid excessive length, I restrict myself as much as possible to problems which either were raised by my collaborators and myself, or we worked on them. I do not want to imply that these problems are more interesting or important than the ones I neglect, but I am likely to know more about my own problems than the reader and finally, despite my great age, my memory and mind are still more or less intact. Thus I remember how and why I or we came to consider these problems, and what are the connections with other questions.

First, I list some of my papers in which relevant problems are stated.

(i) *Problems and results on combinatorial number theory* I, II, II′ *and* III. Paper I was given at the first conference at Fort Collins (in honor of Professor Bose) also edited by Srivastava: A survey of combinatorial theory (North Holland Amsterdam, 1973) 117–138; the pair II and II′ is due to my mistake: II′ should have been III, II is in Journée Arithmetique de Bordeaux, Astérisque 24–25 (Juin 1974), 295–210, and II′ is in J. Indian Math. Soc. 40 (1976) 285–298. Paper III is in Number theory day, held at Rockerfeller Univ., edited by M. Natanson, Lecture Notes in Mathematics 626 (Springer Verlag, Berlin) 43–72.

(ii) *Some older papers of mine:* Quelques problèmes de la théorie des nombres, Monographies de l'Enseignement Mathématique No. 6, Univ. de Geneva (1963) 81–135. Graham and I will soon publish a long paper in the same journal which brings this paper up to date and give also many new problems.

P. Erdös, Some unsolved problems, Michigan Math. J. 4 (1957) 291–300 and Publ. Math. Inst. Hung. Acad. Sci. 6 (1961) 2216–254. The paper in Michigan Math. Journal was my first paper on unsolved problems.

P. Erdös, Extremal problems in number theory, Proc. Symp. in Pure Math. VIII, Theory of Numbers (Amer. Math. Soc., Providence, RI, 1965) 181–189. Several of the results stated in this paper were improved and extended in various papers by S.L.R. Choi.

P. Erdös, Some recent advances and current problems in number theory,

Lectures on Modern Mathematics, edited by L. Saaty, (Wiley, New York, 1965) III. 196–244.

P. Erdös, Some extremal problems in combinatorial number theory, Math. essays dedicated to A.J. Macyntyre, edited by L. Shankar, (Ohio Univ. Press, Athens, OH, 1970) 123–133.

P. Erdös, Some problems in number theory, Computers in number theory, conference held in Oxford 1969 (Academic Press, London, 1971) 406–414.

P. Erdös, Remarks on number theory IV and V. Extremal problems in number theory I and II. Mat. Lapok 13 (1962) 28–38; 17 (1966) 135–166 (in Hungarian).

I divide the present paper into several sections, one long one is for problems which do not seem to fit into any of the classifications. I try as much as possible to indicate what happened to the problems stated in previous papers. So as not to make the paper too long I do not give proofs and I discuss some of the old problems only if there is some recent progress or if I find them very attractive and feel that perhaps they have been neglected.

1. Van der Waerden's and Szemerédi's theorem

In this Section I discuss problems connected with Van der Waerden's and Szemerédi's theorem. These questions have been discussed adequately in my previous papers. Thus I give details only if there is some important new development.

According to Alfred Brauer, Schur conjectured more than 50 years ago that if we divide the integers into two classes at least one of them contains arbitrarily long arithmetic progressions. Van der Waerden proved this conjecture in the following stronger form: There is an $f(n)$ so that if we divide the integers $1 \leq t \leq f(n)$ into two classes, at least one of them contains an arithmetic progression of n terms. There is no satisfactory upper bound for $f(n)$ — in view of the recent surprising results of Paris and his colleagues on a modification of Ramsey's theorem. I am no longer absolutely certain that Van der Waerden's original bound can be very much improved. The best lower bounds are due to Berlekamp and Lovász and myself: $f(p) > cp2^p$ if p is a prime and $f(n) > c2^n$ for all n. I am of course sure that

$$\lim_{n=\infty} f(n)/2^n = \infty. \tag{1}$$

(1) will perhaps be not too difficult to establish, but I offer 500 dollars for the proof or disproof of

$$\lim_{n=\infty} f(n)^{1/n} = \infty. \tag{2}$$

(2) seems very difficult. In view of the difficulty in dealing with $f(n)$ I modified the problem as follows: Let $\frac{1}{2}n < u \leq n$. Let $f(u, n)$ be the smallest integer such that if

we divide the integers $1 \leq t \leq f(u, n)$ into two classes there always is an arithmetic progression of n terms in which one of the classes has at least u terms. the probability method easily gives $f(u; n) > (1 + c(\varepsilon))^n$ for $u > \frac{1}{2}n(1+\varepsilon)$. J. Spencer determined $f(n+1; 2n)$. Otherwise nothing is known. Clearly very many interesting problems remain here; e.g. determine or estimate $f(n+k; 2n)$ for fixed k if n tends to infinity.

Analogously to the Ramsey numbers we can define the Van der Waerden numbers as follows: $f_{u,v}$ is the smallest integer so that if we divide the integers $1 \leq t \leq f_{u,v}$ into two classes either class I contains an arithmetic progression of u terms or class II contains a progression of v terms. Very little is known about these Van der Waerden numbers — in particular it is not known if $f_{3,v}$ tends to infinity polynomially or faster. Another analogy with Ramsey numbers is this: Denote by $f_{(4,3),v}$ the smallest integer so that if we divide the integers $1 \leq t' \leq f_{(4,3),v}$ into two classes then either the first class contains three numbers of an arithmetic progression of four terms or the second contains an arithmetic progression of v terms. Clearly many related questions can be asked whose formulation can be left to the reader, unfortunately so far there are practically no non-trivial bounds for any of these problems.

We can extend these functions to more than two variables — in fact this was already done in the original paper of Van der Waerden. Here we only state one problem: Let $g(l)$ be the smallest integer so that if we divide the integers $1 \leq t \leq g(l)$ into l classes at least one of them contains an arithmetic progression of 3 terms. The bound $g(l) < \exp \exp cl$ follows from Roth's result but as far as I know it has not yet been shown that $g(l)$ tends to infinity faster than polynomially.

In view of the difficulty of obtaining a satisfactory upper bound for $f(n)$ Turán and I conjectured more than 45 years ago that every sequence of positive upper density contains arbitrarily long arithmetic progressions. More precisely: Let $r_k(n)$ be the smallest integer l so that if

$$1 \leq a_1 < \cdots < a_l \leq n \tag{3}$$

then the a's contain an arithmetic progression of k terms. Our conjecture with Turán states

$$r_k(n) = o(n) \tag{4}$$

We first believed $r_k(n) < n^{1-\varepsilon_k}$, but Salem and Spencer showed $r_3(n) > n^{1-\varepsilon}$ for every $\varepsilon > 0$ and $n > n_0(\varepsilon)$. I realised the real difficulty of (4) only after this result of Salem and Spencer. Behrend and Roth proved

$$n \exp(-c_1(\log n)^{\frac{1}{2}}) < r_3(n) < \frac{c_2 n}{\log \log n} \tag{5}$$

Yudin states without proof (Abstracts of Number Theory meeting in Vilnius 1974) that he can get $r_3(n) < cn/\log n$.

Finally Szemerédi proved (4); his proof is a masterpiece of combinatorial reasoning. I offered 1000 dollars for the proof of (4).

Recently, Fürstenberg proved (4) by using methods of ergodic theory and topological dynamics. This proof was recently simplified by Katz-Nelson, Ornstein and Varadhan. At this moment it is impossible to decide on the importance of this invasion of ergodic methods into combinatorial number theory. It is conceivable that it will be like the application of analysis to number theory, but perhaps it is too early to form an opinion — anyway the future will soon decide.

I conjectured long ago that if

$$\sum_i 1/a_i = \infty \tag{6}$$

then the a_i's contain arbitrarily long arithmetic progressions. If true this of course implies that there are arbitrarily long arithmetic progressions all whose terms are primes. I offer 3000 dollars for a proof or disproof of (6). In fact it would be very desirable to obtain an asymptotic formula for $r_k(n)$. I would be very pleased to have reasonably good upper and lower bounds for it.

Szemerédi remarks that we can not even prove that $r_k(n)/r_{k+1}(n) \to 0$ as $n \to \infty$. As far as I know this has not been proved for $k = 3$.

There is an interesting finite version of the 3000 dollar conjecture: Put

$$A_k = \max \sum_i 1/a_i$$

where the maximum is taken over all the sequences which contain no arithmetic progressions of k terms. It is not clear that $A_k < \infty$, I can not even prove $A_3 < \infty$. Gerver recently proved $A_k \geq (1+o(1)k \log k$. For further related questions see III.

Here I state only one more related problem which seems interesting: Is it true that to every $\varepsilon > 0$ and k there is an $n_0 = n_0(\varepsilon, k)$ so that if $n_0 < a_1 < a_2 \cdots$ is a sequence of integers which do not contain an arithmetic progression of k terms then $\sum_i 1/a_i < \varepsilon$?

J. Spencer observed that there is an $h(a)$ which increases very slowly (like the inverse function of the Van der Waerden function) so that we can divide the integers into three classes so that for every a all arithmetic progressions with first term a and all whose terms belong to the same class have length less than $h(a)$. He asked what happens for two classes? I can divide the integers into two classes so that all arithmetic progressions all whose terms are in the same class and whose first term is a are shorter than $c_1 a^{1-c_2}$, very likely this remains true for a^{ε}, unfortunately I have no non trivial lower bound.

A problem in measure theory very recently led Mauldin and myself to ask: Is it true that to every $c > 0$ there is a $t = t(c)$ so that if $1 \leq a_1 < \cdots < a_k \leq n$, $k > cn$ is a sequence of integers and $1 \leq b_1 < \cdots < b_t \leq n$ is an arbitrary set of t integers then the a's always contain an arithmetic progression of three terms and difference $b_i - b_j$. Probably the same result holds with arithmetic progressions of k terms.

Let $l(d)$ be an increasing function. More than 25 years ago Cohen asked: Divide the integers into two classes. Is there for some an arithmetic progression of

$l(d)$ terms and difference d? I showed that if $l(d) > cd$, the answer is negative and Petruska and Szemerédi proved that if $l(d) > cd^{\frac{1}{2}}$ the answer is also negative. They expect a negative answer for $l(d) > d^{\varepsilon}$ and think that their method of proof might settle this question. Unfortunately no lower bound for $l(d)$ is in sight.

Denote by $A(n; k)$ the largest integer so that if we divide the integers $1 \leq t \leq n$ into two classes there are at least $A(n; k)$ k-term arithmetic progressions all whose terms are in the same class. It is easy to see that

$$c_k n^2 < A(n; k) < (1 + o(1)) \frac{n^2}{2(k-1)2^{k-1}}. \tag{7}$$

The lower bound follows from Van der Waerden's theorem and the upper bound from the probability method. Perhaps one can get an asymptotic formula in (7). For Ramsey's theorem, A. Goodman and others obtained related and in some cases sharp results.

Denote by $f_k(n, \alpha)$ the largest integer so that every set of αn integers not exceeding n contains at least $f_k(n, \alpha)$ arithmetic progressions of k terms. The inequality $f_k(n, \alpha) > c(\alpha, k)n^2$ follows from Szemerédi's theorem (for $k = 3$ this was shown by Varnavides, this was before Szemerédi). It would be interesting to have an asymptotic formula for $f_k(n, \alpha)$.

Is it true that if we divide the integers into two classes then there always is a three term arithmetic progression all whose elements are in the same class and whose difference is larger than its first term? If true this is best possible. To see this put in the first class the integers $3^{2k} \leq t < 3^{2k+1}$, $t = 1, 2, \ldots$ and in the second class the other integers. Clearly none of the classes contains a four term arithmetic progression whose difference is larger than its first term.

Van der Waerden's theorem can be formulated in terms of hypergraphs as follows: Consider the hypergraph whose vertices are the integers and whose edges are the k term arithmetic progressions. This hypergraph has chromatic number infinity.

Does this remain true if we restrict ourselves to arithmetic progressions whose difference is a prime number or all whose terms are primes? Clearly many further questions could be posed.

On final question which was already stated in I: Let $h(n; k, l)$ $(l > k)$ be the smallest integer so that if the sequence $\{a_1, \ldots, a_n\}$ contains $h(n; k, l)$ arithmetic progressions of k terms then it contains at least one progression of l terms. Estimate $h(n; k, l)$ as well as possible. In particular prove

$$h(n; k, l) = o(n^2), \tag{8}$$

for every k and l, (8) is open for $k = 3$, $l = 4$. Suppose $\{a_1, \ldots, a_n\}$ contains cn^2 progressions of three terms. Perhaps it must then contain a progression of length $\varepsilon \log n$, $\varepsilon = \varepsilon(c)$, but as I just stated we can not even prove that it contains an arithmetic progression of four terms. Below some papers in this general field are listed.

E.R. Berlekamp, A construction for partitions which avoid long arithmetic progressions, Canad. Math. Bull. 11 (1968) 409–419.

E. Szemerédi, On sets of integers containing no k elements in arithmetic progression, Acta Arith. 27 (1975), 299–315. For further literature and history of the problem see I, II, II′ and III and the paper of Szemerédi. Here I only make a historical remark which can not entirely be documented. E. Rothe in 1944 told me that his wife Dr. Hildegard Ille was given the problem of estimating $r_k(n)$ by I. Schur sometime in the 1930's. Thus perhaps Schur conjectured $r_k(n)=o(n)$ before Turán and myself.

J. Spencer, Bull. Canad. Math. Soc. 16 (1973) 464.

H. Fürstenberg, Ergodic behaviour of diagonal measures and a theorem of Szemerédi on arithmetic progressions, J. Analyse Math. 31 (1977) 204–265. See also the preprint H. Fürstenberg and B. Weiss Topological dynamics and combinatorial number theory and the lecture of Jean Paul Thouvenot, La démonstration de Fürstenberg du théorème de Szemerédi sur les progressions arithmétiques, Sem. Bourbaki, Vol. 1977–78 (Février 1978) 518, 1–11.

J.L. Gerver, The sum of the reciprocals of a set of integers which no arithmetic progression of k terms, Proc. Amer. Math. Soc. 62 (1977) 211–214.

A.W. Goodman, On sets of acquaintences and strangers at any party, Amer. Math. Monthly 69 (1962) 114–120.

P. Varnavides, On certain sets of positive density, J. London Math. Soc. 34 (1959) 358–360.

G.J. Simmons and H.L. Abbott, How many three-term arithmetic progressions can there be if there are no longer ones? Amer. Math. Monthly 84 (1977) 633–635.

J. Paris, Independence result for Peano arithmetic using inner models, to appear; see also J. Paris and L. Harrington, A mathematical incompleteness in Peano arithmetic, Handbook of Mathematical logic, edited by J. Barwise, Studies in Logic and Foundation of Math., Vol. 90, (North Holland, Amsterdam, 1977) 1133–1142.

H. Fürstenberg and Y. Katz-Nelson, An ergodic theorem for commuting transformations, J. Analyse Math. 34 (1978) 275–291. See also the forthcoming book of Fürstenberg, Recurrence in Ergodic Theory and Combinatorial Number Theory (Princeton Univ. Press).

2. Covering congruences and related questions

These problems are very adequately discussed in the papers quoted in the introduction and in my forthcoming paper with R.L. Graham. Thus I make this section short.

A system of congruences

$$a_i (\text{mod } n_i), \quad n_1 < \cdots < n_k \tag{1}$$

is called a covering if every integer satisfies at least one of the congruences (1). The basic unsolved problem which I formulated in 1934 states: Can n_1 be chosen arbitrarily — in particular can it be arbitrarily large? I offer 500 dollars for a proof or disproof. The current record is still $n_1=20$ due to S.R.L. Choi. Another question: Is it true that for every $d>1$ there is a system (1) with $(n_i, d)=1$?

A set of integers $1<n_1<\cdots<n_k$ is called a covering set if they can be the moduli of the system (1). Such a covering set is called irreducible if no subset of it is a covering. It is easy to see that there are only a finite number of irreducible covering sets of size k. How large is their number and how small (resp. large) can n_k be for an irreducible covering set $1\leqslant n_1<\cdots<n_k$? Also how many irreducible covering systems $1\leqslant n_1<\cdots<n_l\leqslant x$ (l is variable) are there? Let $1<n_1<\cdots< n_k\leqslant x$ be an irreducible covering system. Determine or estimate $\max\sum 1/n_i$.

Are there infinitely many integers n so that the set of divisors $d\geqslant 2$ of n form an irreducible covering system? $n=12$ is such an n.

I conjectured that for every C there is an integer N with $\sigma(N)/N>C$ ($\sigma(N)$ is the sum of the divisors of N) so that the divisors of N do not form a covering set. J. Haight recently proved this conjecture, his paper will soon appear in Mathematica. The following extremal problem can now be posed: Put

$$f(x)=\max_{m<x}\sigma(m)/m$$

where the maximum is to be taken over all the m for which the divisors of m do not form a covering system. By Haight's theorem $f(x)$ tends to infinity as $x\to\infty$. Is it true that $f(x)=o(\log\log x)$. In other words does $f(x)$ tend to infinity much slower than $\max\sigma(m)/m$.

Some more extremal problems: Determine or estimate the maximum number j_x of covering systems $\{n_1^{(i)}<n_2^{(i)}<\cdots<n_{k_i}^{(i)}\}$ where all the n's are distinct and less than x. We do not even know that j_x tends to infinity with x since this would imply that there is a covering system for which n_1 is arbitrarily large. I expect that j_x tends to infinity very slowly.

Determine or estimate

$$\max\sum 1/m_i$$

where the maximum is taken over all the sequences $\{m_i\}$ which do not form a covering system and all terms of which are less than x. This maximum is perhaps greater than $\log x-C$; if not then again there is a covering system for which n_1 is as large as we please.

Romanoff proved in 1934 that the density of integers of the form 2^k+p is positive. In a letter he then asked me if there are infinitely many odd numbers not of this form. Using covering congruences (see my paper in Summa Bras. Math.). I proved that there are arithmetic progressions of odd numbers $u \pmod v$ and a finite set of primes $P=\{p_1,\ldots,p_k\}$ so that for every $n\equiv u \pmod v$ every number $n-2^l$ is divisible by one of the primes P. Consider all the arithmetic progressions no term of which is of the form 2^k+p — they probably are all included in the progressions determined by covering congruences. Is it true that almost all odd integers not contained in any of these progressions are of the form 2^k+p?

If j_x tends to infinity — or what is the same there are systems (1) with n_1 as large as we please then for every r there is an arithmetic progression no term of which is the form 2^k+Q_r where Q_r has at most r prime factors. Surely every large n is of the form 2^k+Q_r, $r<\log\log n$, but I have not been able to prove this.

Are there infinitely many odd integers n for which $n-2^u$, $1\leq 2^u\leq n$ is never squarefree? In fact is there any such an integer?

One more question which can be formulated in terms of covering congruences. Is there a finite set $p_1, \ldots, p_k$ of primes and an infinite sequence of integers $\{n_i\}$ so that all positive numbers of the form $n-2^u-2^v$ are multiples of the least one of the p_i $(i=1,\ldots,k)$? I conjecture with some trepidation that the answer is negative. With even more trepidation, I conjecture that almost all odd numbers are of the form $p+2^u+2^v$ and that for some r every integer is of the form $p+2^{u_1}+\cdots+2^{u_r}$. A theorem of Gallagher asserts that for $r>r_0(\varepsilon)$ the lower density of the integers of this form is $>1-\varepsilon$; Crocker proved that there are infinitely many odd integers not of the form $p+2^u+2^v$. Covering congruences can clearly be generalised for groups and many interesting problems can be posed, Schinzel used covering congruences to study the irreducibility of polynomials, he and Selfridge made interesting observations on the existence of covering congruences.

A system of congruences

$$a_i \pmod{n_i}, \quad n_1<\cdots<n_k \tag{2}$$

is called disjoint if there is no integer which satisfies two of the congruences (2). It would be of some interest to find necessary and sufficient conditions on the $\{n_i\}$ which would imply that a disjoint system (2) exists for suitable a_i. Stein and I asked: Determine or estimate max $k=f(x)$ where the maximum is extended over all the disjoint systems (2). Szemerédi and I proved that

$$x\exp(-(\log x)^{\frac{1}{2}+\varepsilon})<f(x)<\frac{x}{(\log x)^c}.$$

The lower bound we obtained with the help of Stein. We believe that the lower bound is closer to the true order of magnitude of $f(x)$.

One of the few theorems on covering congruences was proved by L. Mirsky and D. Newman. A disjoint system can never be covering, or (2) implies $\sum 1/n_i<1$. Put

$$\varepsilon_m=\max\sum 1/n_i$$

where the maximum is taken over all disjoint systems for which $n_1>m$. Determine or estimate ε_m as well as possible. I could not even decide whether $\varepsilon_m\to 0$ as $m\to\infty$.

Some further work is contained in the following papers:

P. Erdös and E. Szemerédi, On a problem of Erdös and Stein, Acta Arithmetica 15 (1968) 85–90.

A. Schinzel, Reducibility of polynomials and covering systems of congruences, Acta Arithmetica 13 (1967) 91–101.

For a further extremal problem see: B.B. Crittenden and C.L. Van der Eynden, Any k arithmetic progressions covering the first 2^k integers cover all integers, Proc. Amer. Math. Soc. 24 (1970) 475–481.

P. Erdös, On systems of congruences (in Hungarian), Mat. Lapok 3 (1952) 122–128; On the integers of the form $p+2^k$ and some related problems, Summa Basil. Math. 11 (1950) 113–123.

S.L.G. Choi, Covering the set of integers by congruence classes of distinct moduli, Math. Comp. 25 (1971) 885–895.

R. Crocker, On the sum of a prime and of two powers of two, Pacific J. Math. 36 (1971) 103–107.

J.A. Haight, Covering systems of congruences. A negative result, Mathematika 26 (1979) 53–61.

3. Some problems in additive number theory

Here are some of my oldest problems and conjectures. Unfortunately not much progress has been made with these questions in the last few years.

Let $1 \leqslant a_1 < \cdots < a_k \leqslant x$ be a sequence of integers. Assume that the sums $\sum_{i=1}^k \varepsilon_i a_1$, $\varepsilon_i = 0$ or 1 are all different. Put max $k = F(x)$. I conjectured more than 45 years ago that

$$F(x) = \frac{\log x}{\log 2} + O(1) \tag{1}$$

A simple counting argument gives

$$F(x) < \frac{\log x}{\log 2} + \frac{\log\log x}{\log 2} + O(1)$$

and L. Moser and I using second moments improved this to

$$F(x) < \frac{\log x}{\log 2} + \frac{\log\log x}{2\log 2} + O(1) \tag{2}$$

As far as I know (2) has never been improved.

I asked: Is it true that $f(2^k) \geqslant k+2$ for $k > k_0$? (Both this question and (1) were posed independently also by L. Moser.) I offer 500 dollars for a proof or disproof of (1).

Conway and Guy proved $F(2^l) \geqslant k+2$ for $k > k_0$ and it has been conjectured that for $k > k_0 F(2^k) = k+2$. I have no opinion.

In 1932 or 1933, S. Sidon defined a sequence $1 \leq a_1 < a_2 < \cdots$ to be a B_r sequence if the sums $\sum \varepsilon_i a_i$, $\sum \varepsilon_i \leq r$ are all distinct ($\varepsilon_i = 0$ or 1). In other words the sums taken r (or fewer) at a time should be all different. He wanted to estimate the slowest possible growth of a B_r sequence. He was led to these problems from his well known work on lacunary trigonometric series.

Also he asked the following very fruitful question: Let $1 \leq a_1 < a_2 < \cdots$ be an infinite sequence of integers. Denote by $f_2(n)$ the number of solutions of $n = a_i + a_j$. Is there a sequence A for which $f_2(n) > 0$ for $n > n_0$ but $f_2(n)m^{-\varepsilon} \to 0$ for all $\varepsilon > 0$?

Sidon mentioned these problems to me when we first met in 1932 or 1933. By the greedy algorithm I easily constructed a B_2 sequence satisfying $a_k < ck^3$. We both conjectured that this is very far from the truth and probably there are B_2 sequences with $a_k < k^{2+\varepsilon}$ for every $\varepsilon > 0$ if $k > k_0(\varepsilon)$.

We are still very far from being able to settle this question. Using the probabilistic method Rényi and I proved the existence of a sequence with $a_k < k^{2+\varepsilon}$ and $f_2(n) < C_\varepsilon$. The problem whether a B_2 sequence exists with $a_k = o(k^3)$ is still open and I offer a hundred dollars for a proof or disproof. (*Added in proof.* Ajtai, Komlos and Szemerédi proved this conjecture; their proof will appear in Europ. Comb. J.)

Using the probability method I proved the existence of a sequence $A = \{a_1 < \cdots\}$ satisfying

$$c_1 \log n < f_2(n) < c_2 \log n \tag{3}$$

which answers affirmatively Sidon's question on whether $f_2(n)n^{-\varepsilon} \to 0$. Is (3) the best possible? Turán and I conjectured that if $f_2(n) > 0$ for all $n > n_0$, then $f_2(n)$ can not be bounded. I offer 500 dollars for a proof or disproof. Perhaps $f_2(n) > 0$ for all n already implies that $f_2(n) > c \log n$ for infinitely many n. Is there a sequence A for which

$$f_2(n)/\log n \to 1? \tag{4}$$

The probability method easily gives that if $h(n) \to \infty$ monotonically, then there is a sequence A satisfying $f_2(n)(h(n) \log n)^{-1} \to 1$. I expect that there is no sequence satisfying (4).

I proved that for a B_2 sequence

$$\limsup a_k/k^2 \log k > 0. \tag{5}$$

Perhaps in (5) the lim sup is in fact infinity. On the other hand perhaps there is a B_2 sequence satisfying (for all k)

$$a_k < c_1 k^2 (\log k)^{c_2}. \tag{6}$$

I offer a thousand dollars for clearing up the problems raised by (5) and (6).

These problems change character completely if we restrict ourselves to finite sequences. Denote by $F_r(n)$ the largest integer l for which there is a sequence $1 \leq a_1 < \cdots < a_l \leq n$ so that all the sums $\sum \varepsilon_i a_i$, $\sum \varepsilon_i \leq r$, $\varepsilon_i = 0$ or 1 are different.

Turán and I conjectured

$$F_2(n) = n^{\frac{1}{2}} + O(1). \tag{7}$$

We only could prove $F_2(n) < n^{\frac{1}{2}} + cn^{\frac{1}{4}}$, $F_2(n) < n^{\frac{1}{2}} + n^{\frac{1}{4}} + 1$ was proved by Lindstrom. I offered (and offer) 500 dollars for a proof or disproof of (7). Bose and Chowla showed that $F_r(n) \geq (1 + o(1))n^{1/r}$ follows by using finite geometries. Bose called attention to the fact that the proof of

$$F_r(n) < (1 + \varepsilon)n^{1/r} \tag{8}$$

presents great difficulties for $r > 2$. Our proof with Turán for $r = 2$ does not work and at the moment this attractive problem seems intractable. Perhaps $F_r(n) = n^{1/r} + O(1)$ holds for every r.

I was not able to prove that if $1 \leq a_1 < a_2 < \cdots$ is an infinite B_3 sequence, then

$$\limsup a_k/k^3 = \infty. \tag{9}$$

The same problem arises for every $r > 3$.

A further generalisation: Let $a_k < ck^r$ for every k. Is it true that

$$\limsup f_r(n) = \infty?$$

This is open even for $r = 2$ and perhaps here the real difficulty occurs already for $r = 2$.

Rényi and I proved by the probabilistic method that there is a sequence $a_k < ck^r$ for which

$$\sum_{n=1}^{x} f_r(n)^2 < Cx. \tag{10}$$

Probably (10) holds even for a basis of order r. In other words there is a sequence A satisfying (10) and $f_r(n) > 0$ for every n.

I proved that there is an infinite B_2 sequence for which

$$\limsup A(n)n^{\frac{1}{2}} \geq \tfrac{1}{2},$$

where

$$A(n) = \sum_{a_i < n} 1.$$

this has been improved to $1/\sqrt{2}$ by Krickeberg and I conjectured that it can be improved to 1, which of course would be best possible.

This would follow from one of my favourite recent conjectures: Let $1 \leq a_1 < \cdots < a_k$ be a finite B_2 sequence. Prove that it can be imbedded into a perfect difference set, i.e. there is a prime p and a set of $p+1$ residues $u_1, \ldots, u_{p+1} \bmod p^2 + p + 1$ so that all the differences $u_i - u_j$ are incongruent mod $p^2 + p + 1$ and the a's all occur amongst the u's. I offer a thousand dollars for a proof or disproof of this conjecture.

Recently Nathanson and I published several papers on bases and we will soon (I hope) publish a survey paper on this subject — here I only state one of our most attractive problems: Is there an infinite sequence $a_1 < a_2 < \cdots$ of integers so that

$f_2(n)>0$ for all $n>n_0$ but if we omit any a_i, then the number of $n<x$ with $f_2(n)=0$ is $>c_ix$ for all (or perhaps only infinitely many) x?

A recent conjecture of D. Newman and myself states as follows: There is a sequence $a_1<a_2<\cdots$ of integers for which $f_2(n)$ is bounded but which is not the union of a finite number of B_2 sequences. (*Added May 1980;* I proved this conjecture in 1979 and my proof will soon appear in the first issue of Europ. Combinatorial J.)

Let $g(n)>0$ be a non-decreasing function of n. I conjecture that the lower density of the integers n for which $f_2(n)=g(n)$ is 0. The upper density can be positive but I believe it is bounded away from 1.

It is easy to construct a sequence of integers $1\leqslant a_1<\cdots$ so that every integer n can be uniquely written in the form a_i-a_j. It is easy to see that

$$\limsup_{n\to\infty} a_i^{(n)}/n=\infty \quad \text{where } n=a_i^{(n)}-a_j^{(n)}. \tag{11}$$

I can not determine how fast (11) must increase.

In general the problems here can at present be attacked only by the probability method and where these methods do not apply we have not been able to make much progress.

Many of the problems stated here are discussed in the excellent book of Halberstam and Roth Sequences, (Oxford Univ. Press, Oxford, 1966). The book has very extensive references.

For many problems and results on bases and related questions see A. Stöhr, Gelöste and ungelöste Fragen über Basen der natürlichen Zahlenreihe I and II, J. Reine u. Angew. Math. 194 (1955) 40–64, 110–140.

B. Lindström, An inequality for B_2 sequences, J. Combinatorial Theory, 6 (1969) 211–212.

P. Erdös, Problems and results in additive number theory, Colloque sur la Théorie des Nombres, Bruxelles, George Thone, Liège, Manon and Cie, Paris 1959, 127–137.

In 1959, 1963 and 1972 there were meetings on number theory at the University of Colorado in Boulder, many interesting problems can be found in the conference reports which were published but unfortunately are not easily available. Also see the book H.H. Ostmann, Additive Zahlentheorie, Ergebnisse der Math, Heft 7 und 11 contains a review of the older litterature and also H. Rohrbach, Ein Beitrag zur additiven Zahlentheorie, Math. Zeitschrift 42 (1937) 1–30 and Einige neuere Untersuchungen über die Dichte in der additiven Zahlentheorie, Jahresbericht D.M.V. 48 (1938) 199–236.

4. Solutions of equations in dense sets of integers or real numbers

A sequence of real numbers is called primitive if no one divides any other. A sequence of real numbers $1<a_1<\cdots$ is called primitive if for every i, j and k

$$|ka_i-a_j|\geqslant 1. \tag{1}$$

If the a's are integers (1) means that no a divides any other. In 1934 Besicovitch made the surprising discovery that there is a primitive sequence of positive upper density. Behrend and I proved that the lower density of a primitive sequence must be 0. Behrend proved that if $1<a_1<\cdots<a_k\leq x$ is a primitive sequence, then

$$\sum_{i=1}^{k} \frac{1}{a_i} < c \log x (\log\log x)^{-\frac{1}{2}}. \tag{2}$$

Szemerédi, Sárközi and I proved that for an infinite primitive sequence

$$\sum_{a_i<x} \frac{1}{a_i} = o(\log x (\log\log x)^{-\frac{1}{2}}). \tag{3}$$

(2) and (3) are both best possible.

Perhaps (2) and (3) hold if A is a sequence of real numbers satisfying (1). In fact I can not even prove that if (1) is satisfied, then $\limsup a_n/n=\infty$.

The only result in this direction is an unpublished one of J. Haight. He proved that if the $\{a_i\}$ are rationally independent and satisfy (1), then $a_n/n\to\infty$, or in other words $[a_n]$ is a sequence of density 0. Observe that by the example of Besicovitch this does not have to hold if the a's are integers.

A question of W. Schmidt states: Is there a set S of real numbers of infinite measure so that x/y, $x\in S$, $y\in S$ is never an integer?

J. Haight and E. Szemerédi (independently) constructed such a set. Denote by $m(S, x)$ the measure of the intersection of S with $(0, x)$. How fast can $m(S, x)$ tend to infinity? It is easy to see that $m(S, x)=o(x)$ but probably very much more is true. Immediately two questions can be posed: Let $F(x)\to\infty$; is there a set S so that x/y $x\in S$, $y\in S$ is never an integer and $m(S, x)>F(x)$ for all $x>x_0$, or only for a sequence $x_m\to\infty$? The answer to these two questions will no doubt be quite different.

A conjecture of Sárközi, Szemerédi and myself which has perhaps been neglected states: To every $\varepsilon>0$ there is a k so that if $k<a_1<a_2<\cdots$ is any primitive sequence then

$$\sum_{i=1}^{\infty} \frac{1}{a_i \log a_i} < 1+\varepsilon. \tag{4}$$

The following question seems difficult and perhaps has no reasonable solution: Let $b_1<b_2<\cdots$ be an infinite sequence. What is the necessary and sufficient condition that there should exist a primitive sequence $\{a_n\}$ satisfying $a_n<Cb_n$ for every n? Perhaps there is more chance to get an answer for the following question: Let $n_1<n_2<\cdots$. What is the necessary and sufficient condition that there should exist a primitive sequence $\{a_k\}$ satisfying $A(2^{n_i})>c2^{n_i}$ for every i and an absolute constant c?

Now we discuss some multiplicative extremal problems. Let $1<a_1<\cdots<a_k\leq x$ be a sequence of integers. Assume that the products a_ia_j are all distinct. I

proved

$$\pi(x)+c_1x^{\frac{3}{4}}(\log x)^{-\frac{3}{2}}<\max k<\pi(x)+c_2x^{\frac{3}{4}}(\log x)^{-\frac{3}{2}}. \tag{5}$$

where $\pi(x)$ is the number of primes $\leqslant x$.

(5) is perhaps unexpectedly accurate I am sure that there is an absolute constant c so that

$$\max k=\pi(x)+(c+o(1))x^{\frac{3}{4}}(\log x)^{-\frac{3}{2}}. \tag{6}$$

I do not at the moment see how to prove (6) and I do not believe that there is a simple explicit formula for max k.

Let $F(x)$ be the largest l for which there is a sequence of real numbers $1\leqslant a_1<\cdots<a_l\leqslant x$ for which

$$|a_ia_j-a_ra_s|\geqslant 1 \tag{7}$$

for all choices of the indices i, j, r, s. I was sure that $l=o(x)$ and in fact I expected that an estimation like (6) will hold here too. In fact it turned out that I completely misjudged the situation. Ralph Alexander constructed a sequence satisfying (7) and $l>cx$. Here is the outline of Alexander's construction. Put $x=4e^{N+1}$. By a theorem of Turán and myself there are integers

$$1<y_1<\cdots<y_t<e^{2N},\quad t>(1-\varepsilon)e^N \tag{8}$$

so that the sums y_i+y_j, $1\leqslant i<j\leqslant l$ are all distinct. By removing at most half of the y's we obtain a subsequence

$$1\leqslant y_1'<\cdots<y_l'<e^{2N},\qquad l\geqslant\tfrac{1}{2}t,\quad y_{i+1}-y_i'>\tfrac{1}{4}e^N \tag{8'}$$

Put

$$x_i=\exp(N+y_i'/e^{2N}). \tag{9}$$

From (9) we obtain by a simple computation that

$$|x_i-x_j|>\tfrac{1}{4}\quad\text{and}\quad|x_ix_j-x_rx_p|>1.$$

Put finally $a_i=4x_i$. (8), (8′) and (9) show that (7) and $l>cx$ are satisfied. After this surprising result I am no longer so sure that (1) implies $\limsup a_n/n=\infty$ as I used to be. The following problem might be still of some interest: Let $1\leqslant a_1<\cdots<a_{t_n}\leqslant n$ satisfy (7). Determine of estimate max t_n. Perhaps $\lim t_n/n=C$. Determine C, this perhaps is not a hopeless task. Finally is there an infinite sequence $1\leqslant a_1<\cdots$ satisfying (7) and $A(x)>cx(\log x)^{-\frac{1}{2}}$ for every x? If true this is clearly best possible.

Some more problems. Let $1\leqslant a_1<\cdots<a_{t_n}\leqslant n$ be again a sequence of integers. Assume that all the products $\prod_i a_i^{\alpha_i}$ ($\alpha_i\geqslant 0$ integer) are distinct. Then it is easy to see that $\max t_n=\pi(n)$. Assume next that the products $\prod a_i^{\varepsilon_i}$, $\varepsilon_i=0$ or 1 are all distinct. I suspect that then

$$\max t_n=\pi(n)+\pi(n^{\frac{1}{2}})+o(n^{\frac{1}{2}}/\log n)$$

I could only prove that $\max t_n < \pi(n) + Cn^{\frac{1}{2}}(\log n)^{-1}$. Perhaps the reader will forgive a very old man to tell how Pósa and I conjectured in 1963 how large t_n can be. In 1962 or 1963 I talked to high school students at the Bolyai Math. Soc. I stated that perhaps

$$\max t_n = \pi(n) + \sum_{k=1}^{\infty} \pi(n^{1/2^k}),$$

but then immediately noticed that this is wrong. Not wanting to state more nonsense I talked immediately about something else. Before the end of my talk I formulated in my mind the following conjecture:

$$\max t_n = \pi(n) + \pi(n^{\frac{1}{2}}) + \pi(n^{\frac{1}{4}}) + \pi(n^{\frac{1}{7}}) + \cdots \tag{10}$$

where in the sum (10) $\pi(n^{1/k})$ occurs if and only if $F(k) > F(k-1)$ where $F(k)$ is defined in the first problem of III (it is the largest l so that there is a sequence $1 \leq a_1 < \cdots < a_l \leq k$, all the sums $\sum_{i=1}^{l} \varepsilon_i a_i$ are different)

$$\max t_n \geq \sum \pi(n^{1/a_k}) \tag{11}$$

is of course easy — it is not clear if the opposite inequality is also true. After the lecture I talked to Pósa (who was then about 14) and he formulated the same conjecture during my talk and noticed of course that $\sum \pi(n^{1/2^k})$ is wrong. We do not know at present if (10) is true.

The following nice conjecture is due to Beurling. Let $1 < P_1 < \cdots$ be a sequence of real numbers. $b_1 < b_2 < \cdots$ is the set of real numbers of the form $\prod P_i^{\alpha_i}$ where the α_i are non-negative integers. Assume that

$$B(x) = \sum_{b_i < x} 1 = x + o(\log x). \tag{12}$$

Then the P's are the primes. It is not hard to see that if (12) is true it is best possible.

In this connection H.N. Shapiro asked the following question: Assume that the numbers $\prod_i P_i^{\alpha_i}$ differ by at least one. Is it then true that

$$\sum_{P_i \leq x} 1 \leq \pi(x), \tag{13}$$

The equality occurring only if the P's are the primes?[1]

To end this section I state the following interesting measure theoretic conjecture of John Haight: Let E be a set of a positive measure in $(0, \infty)$. Consider the

[1] I just notice (1978.IX.17) that I stated nearly the same conjecture in my paper "Some applications of graph theory to number theory". The many Facets of Graph Theory, Lecture Notes in Mathematics 110 (Springer Verlag, Berlin) 77–82, see p. 82. I did not conjecture though that equality holds only if the P's are the primes. In view of this I should perhaps with regret modify what I said about my memory and mind. When Shapiro told me this problem I felt foolish that I did not think of it myself.

set $E' = \bigcup_{r=1}^{\infty} rE$. In other words $z \in E'$ if for some integer r, $z/r \in E$. Is it true that for almost all x there is an $M(x)$ so that for every integer $n > M(x)$, $nx \in E'$?

For primitive sequences see Halberstam–Roth, "Sequences" and P. Erdös, A. Sárközy and E. Szemerédi, On the divisibility properties of sequences of integers, Coll. Math. Soc. J. Bolyai, Vol. 2. Number theory (North Holland Amsterdam, 1968) 35–49, this paper has extensive references.

J.A. Haight, A linear set of infinite measure with no two points having integral ratio, Mathematika 17 (1970) 133–138.

E. Szemerédi, On a problem of W. Schmidt, Studia Sci. Math. Hungar. 6 (1971) 287–288.

About Beurling primes see e.g. H.G. Diamond, When do Beurling generalised integers have a density, J. Reine Angew. Math. 295 (1977) 22–29. This paper contains many references to older results.

P. Erdös, On some applications of graph theory to number theoretic problems, Publ. Ramanujan Inst. 1 (1969) 131–136.

5. Some problems on infinite subsets

Graham and Rothschild conjectured that if the integers are split into two classes there always is an infinite sequence $1 < a_1 < \cdots$ so that all the sums $\sum \varepsilon_i a_i$, $\varepsilon_i = 0$ or 1 are in the same class. This beautiful conjecture was proved by Hindman whose proof was greatly simplified by Baumgartner. Later a different and perhaps the simplest proof was given by Glaser.

Some time ago, I thought of the following fascinating problem: Divide the integers into two classes. Is it true that there always is an infinite sequence $a_1 < \cdots$ so that all the multilinear expressions formed from the a's are all in the same class. One would perhaps guess that the answer must be "no" but no counterexample is in sight.

The following much weaker conjecture is also open: Divide the integers into two classes. Is there an infinite sequence $a_1 < \cdots$ so that all the sums $a_i + a_j$ and the products $a_i a_j$ are in the same class? A further complication arises if we also insist that $a_1, a_2, \ldots$ should also belong to the same class. In fact very little is known. Graham proved that if we divide the integers $1 \leq t \leq 252$ there are always 4 distinct numbers x, y, $x+y$, xy all in the same class. The number 252 is the best possible. Hindman showed that if we require $x > 1$, $y > 1$ then the same result holds for $t \leq 990$ and 990 is the best possible. As far as I know this is all that is known. (Recently Hindman found some very interesting counterexamples.)

Ewing conjectured that there always is an infinite sequence $a_1 < \cdots$ where all the sums $a_i + a_j$ ($i = j$ permitted) are in the same class. It is rather annoying that this simple and interesting question is still open. Hindman has certain preliminary results. Among others he proved that this conjecture fails certainly for three classes. In fact Hindman observes that one of his sequences can have density 0 and in fact

$$A_1(x) = \sum_{a_i < x} 1 < Cx^{\frac{1}{2}} \tag{1}$$

It is not yet clear if (1) is the best possible.

I thought of strengthening Hindman's theorem in the same way as Szemerédi strengthened Van der Waerden's theorem: Let A be a sequence of positive density. Is there an infinite subsequence $a_1 < \cdots$ and an integer t so that all the integers a_i, $a_i + a_j + t$ belong to the same class?

I first hoped that Hindman's theorem on all subsums $\sum \varepsilon_i a_i$ can be generalised for sets of positive density, but the following example of E. Straus seems to give a counterexample to all such attempts: Let $p_1 < \cdots$ be a set of primes tending to infinity fast, and let $\sum \varepsilon_i < \infty$. Consider the set of integers $a_1 < a_2 < \cdots$ so that $a_u \not\equiv \alpha \pmod{p_i}$ where $|\alpha| \leq \varepsilon_i p_i$. The density of this sequence is $\prod_i (1 - \varepsilon_i) > 0$, and it is not difficult to see that this sequence furnishes the required counterexample.

More than 10 years ago, Graham and I conjectured that if we split the integers into k classes, then

$$1 = \sum \frac{1}{x_i}, \quad x_1 < \cdots \quad \text{(finite sum)} \tag{2}$$

is always solvable with all the x_i in the same class. In the language of hypergraphs the conjecture states: The chromatic number of the non-uniform hypergraph whose vertices are the integers and whose edges are the sets satisfying (2) is infinite. A finite form of this problem states that if $\sum_{a_i < n} 1/a_i$ is large enough, then one can select amongst the a's a sequence satisfying (2). "Large enough" as a function of n? We really have no idea what "large enough" should mean. It could be $o((\log\log n)^\alpha)$ or $\varepsilon \log n$.

Silverman and I conjectured that if we split the integers into k classes, then there are always two integers in the same class whose sum is an rth power (in particular a square). It would be of interest to characterise the sequences for which this conjecture holds. We also conjectured that if $1 \leq a_1 < \cdots < a_k \leq n$ is such that $a_i + a_j$ is never a square, then $k \leq (1 + o(1))n/3$, $n/3$ is given by the integers $\equiv 1 \pmod 3$. The exact determination of max k is perhaps not hopeless, but we certainly have not succeeded.

If we assume that $a_k - a_i$ is never a square, then the situation is radically different. Fürstenberg and Sárközi independently of each other proved that here $k = o(m)$. Sárközi obtains explicit bounds for k but no doubt they are very far from being best possible. Many further conjectures could be formulated but we must leave these to the reader.

Glazer's proof is given in the excellent survey paper of W.W. Comfort, Ultrafilters: Some old and some new results, Bull. Amer. Math. Soc. 83 (1977) 417–455, see 449–452.

N. Hindman, Finite sums from sequences within cells of a partition of N, J. Combinatorial theory 17 (1974) 1–11; Baumgartner, A short proof of Hindman's theorem, 17 (1974) 384–386.

A. Sárközy, On difference sets of integers I, II and III. II will appear in Annales Univ. Sci. Budapest, I and III, Acta Math. Acad. Sci. Hungar. 31 (1978) 125–149 and 355–386.

N. Hindman, Ultrafilters and combinatorial number theory, Number Theory Carbondale, 1979, Lecture Notes in Math. 751 (Springer-Verlag, Berlin) 119–184. This paper has a very extensive list of references.

6. Some problems on sieve methods

During my long life I often applied Brun's method but unfortunately I contributed next to nothing to the improvement of the method or its generalisations. All I can do is to state a few problems which are directly or indirectly connected with sieve methods.

1. Let $f(x)$ be the smallest integer so that there is a set of residues

$$a_p \pmod{p}, \quad p<f(x) \tag{1}$$

so that every integer $n<x$ satisfies one of the congruence (1). In particular must $f(x)$ be significantly larger than $x^{\frac{1}{2}}$? Also it would be very useful to estimate $F(x)$ if we only require that the number of integers $n \leqslant x$ not satisfying any of the congruences (1) is $o(x/\log x)$ (where $p<F(x)$). Is $F(x)$ significantly smaller than $f(x)$? These problems can of course be extended if more than one residue is omitted.

It is not clear who first formulated this problem — probably many of us did it independently. I offer max(1000 dollars, $\frac{1}{2}$ my total savings) for clearing up of this problem. It is clear that many important problems could be attacked if we would know a little more.

Let ε_x be the largest number for which there is a system of congruences

$$a_p \pmod{p}, \quad x^{\varepsilon_x}<p<x \tag{1'}$$

so that every integer $n<x$ satisfies at least one of the consequences (1'). It is not difficult to prove that

$$\varepsilon_x \geqslant \frac{c \log\log\log x}{\log\log x},$$

but perhaps ε_x is much larger.

Ruzsa and I made the following surprising conjecture. There is a constant C for which there is a set of primes $p_i<x$, $\sum 1/p_i<C$ and a system of congruences

$$a_i \pmod{p_i} \tag{1''}$$

so that every integer $n<x$ satisfies at least one of the congruences (1''). If we are right, then very likely $\varepsilon_x>c$ for some absolute constant c. (See our forthcoming paper in the Journal of Number Theory.)

2. Is it true that to every ε and η there is a k so that the density of integers n for which

$$\max_{1 \leq i \leq k} P(n+i) > n^{1-\varepsilon}$$

is greater than $1-\eta$, where $P(m)$ denotes the greatest prime factor of m. I can only do this for $\varepsilon = \frac{1}{2}$.

3. Is it true that to every $\varepsilon > 0$, $\eta > 0$ there is a $k = k_0(\varepsilon, \eta)$ so that $p_1 < \cdots < p_k < n^{1-\varepsilon}$ and $a_i^{(j)}$, $1 \leq i < (p_j - 1)/2$, $j = 1, \ldots, k$ are any set of $(p_j - 1)/2$ distinct residues mod p_j, $j = 1, \ldots, k$. Then the number of integers $m \leq n$ for which

$$m \not\equiv a_i^{(j)} \pmod{p_j}, \quad 1 \leq i \leq (p_j - 1)/2, \quad j = 1, \ldots, k$$

is less than εn? This follows easily from the large sieve if $p_k < n^{\frac{1}{2}}$ but the general case seems intractable at present.

4. Denote by $V(n)$ the number of distinct prime factors of n I conjecture that for infinitely many

$$\max_{m<n} m + V(m) \leq n \tag{2}$$

(2) seems unattackable. Replace $[m + V(m)]$ by $[m + \varepsilon V(m)]$ where $\varepsilon > 0$ is sufficiently small. It seems to me that even this weakened form of (2) is just beyond what we can do at the present time.

Selfridge and I considered the following much more difficult problem

$$\max_{m<n} m + d(m) \leq n + 2 \tag{3}$$

where $d(m)$ is the number of divisors of m.

$n = 24$ satisfies (3). We convinced ourselves that if (3) has a solution $n > 24$, then n must be enormously big–far beyond the limits of our tables and computations.

A related problem can be formulated as follows. Put

$$f(n, t) = \max_{1 \leq i \leq t} V(n+i)$$

Trivially $f(n, t) \geq (1 + o(1)) \log t (\log \log t)^{-1}$, but I expect that usually very much more will be true. More precisely put

$$\max_{t} f(n, t) \log \log t \, (\log t)^{-1} = F(n). \tag{4}$$

Then $F(n) \to \infty$ as $n \to \infty$. I am very far from being able to prove this.

4. Are there infinitely many integers n for which $n - i \not\equiv 0 \pmod{p^2}$ for $i = 0, 1, \ldots, p-1$ and for all $p^2 < n$? A further question is: are there infinitely many integers n for which $n - i \not\equiv 0 \pmod{k^2}$ for every $i = 0, 1, \ldots, k-1$ and $k^2 < n$?

The density of integers in both of these classes is 0, but I have no good estimation for these numbers. Some preliminary calculations of Selfridge seem to indicate that the number of these integers not exceeding x is $(c+o(1))x^{\alpha}$ for some $1>\alpha>0$ and c, but we are very far from being able to prove this.

5. Elliott considered the following problem: Let $0\leq a_1<\cdots<a_k$ be a sequence of integers which does not contain a complete set of residues mod p (p runs through the set of all primes). Clearly only the primes $p\leq k$ have to be considered. Elliott investigates the estimation of min $a_k=A(k)$. He proves

$$(1+o(1))k\log k\leq A(k)\leq(2+o(1)k\log k. \tag{5}$$

The lower bound in (5) is due to Davenport. Probably the lower bound is the correct one, this seems very hard and is of course connected with the problem stated in 1. The exact determination of a_k is probably hopeless.

It would be of interest to determine A_k for small values of k. Also determine or estimate

$$B_k=\min(a_1+\cdots+a_k)/k. \tag{6}$$

There is no reason to assume that the minimum in (5) and (6) is given by the same sequence.

One can define a sequence $\{a_1,\ldots,a_k\}$ by the greedy algorithm as follows: Assume $a_1,\ldots,a_{i-1}$ is already defined let a_i be the smallest integer greater than a_{i-1} so that $a_1,\ldots,a_{i-1},a_i$ does not form a complete set of residues (mod p). Estimate this a_k as well as possible. There is no reason to assume that this sequence gives the minimum in (5) or (6).

6. Let $f(x)$ be the largest integer for which there is a system of congruences

$$a_p \pmod p \tag{7}$$

(where p runs through the primes not exceeding x) so that every integer $n<x$ should satisfy at least $f(x)$ of the congruences (1). I can not even prove that $f(x)\geq 2$ for $x>x_0$. On the other hand perhaps $f(x)$ tends to infinity together with x.

Denote by $F(x)$ the largest integer for which there is a system of congruences

$$a_n \pmod n \tag{8}$$

(where n runs through the integers not exceeding x) so that every integer $t\leq x$ satisfies at least $F(x)$ of the congruences (8). Now it is a simple exercise to prove that $F(x)\to\infty$ as $x\to\infty$. The only problem is to determine or estimate how fast $F(x)$ tends to infinity.

Denote by $\pi(x)$ the number of primes not exceeding x. $\pi(x)=(1+o(1))x/\log x$ is the prime number theorem. The number of primes in short intervals is very difficult to estimate. I conjectured

$$\pi(x)-\pi(y)<c\frac{x-y}{\log x}\quad\text{for every } y<x-(\log x)^C \tag{9}$$

where C is a sufficiently large absolute constant.

The proof or disproof of (9) is probably hopeless at present.

P.D.T.A. Elliott, On sequences of integers, Quarterly J. Math. 16 (1965) 35–45.

About sieve methods in general see Halberstam–Roth, "Sequences" and the more recent Cambridge Tract of Hooley and the book of Halberstam–Richert on sieve methods.

7. Miscellaneous problems

This section is necessarily incomplete. I have to concentrate mainly on problems stated in previous papers. I discuss problems only if they were in my opinion neglected or if some progress has been made.

Riddell defines $g_k(n)$ as the largest integer so that among any n real numbers one can always find $g_k(n)$ of them which do not contain an arithmetic progression of k terms. Clearly $g_k(n) \leqslant r_k(n)$ and Riddell showed that $g_k(n)$ can be less than $r_k(n)$, $g_3(5)=3$, $r_3(5)=4$ (from the set 1, 3, 4, 5, 7) one can select only 3 integers not containing an arithmetic progression). Riddell also shows $g_3(14)=7<r_3(14)=8$. Perhaps for large n, $g_k(n)=r_k(n)$.

Riddell also investigated the following problem: Let $1 \leqslant a_1 < \cdots < a_n$ be any set of n integers (or real numbers). Denote by $F_2^*(n)$ the minimum of the largest r for which $a_{i_1} < \cdots < a_{i_r}$ is a B_2 sequence. The minimum is to be taken for all the sequences $\{a_1, \ldots, a_n\}$. I conjectured $F_2^*(n)=(1+o(1))n^{\frac{1}{2}}$ and perhaps $F_2^*(n)=F(n)$ for all large n. Komlós, Sulyok and Szemerédi proved a general theorem on the solution linear equations which implies $g_k(n)>cr_k(n)$ and $F_2^*(n)>cF_2(n)$. They proved that apart from a multiplicative constant the worst set for all such problems is to take the integers $1, 2, \ldots, n$. This remarkable theorem still leaves many questions unanswered. Very likely c can be chosen to be $1-\varepsilon$ for every $\varepsilon>0$ if $n>n_0(\varepsilon)$ and perhaps $g_k(n)=r_k(n)$, $F_2^*(n)=F_2(n)$ for $n>n_0$. If this conjecture seems too optimistic perhaps one should only expect $g_k(n)>r_k(n)-C$, $F_2^*(n)>F_2(n)-C$. (Recall the functions $r_k(n)$ and $F_k(n)$ defined in Sections 1 and 3 respectively.)

Also, what happens for non linear equations? Here are some of the questions I have in mind: Let $A_n=\{a_1, \ldots, a_n\}$ be a sequence of n integers. Denote by $g_k(A_n)$ the largest integer l for which there is a subsequence $\{a_{i_1}, \ldots, a_{i_l}\}$ so that $\{a_{i_j}^k\}$ is a B_2 sequence. Put $\min g_k(A_n)=G(n, k)$ where the minimum is to be taken over all the sequences A_n. Is it true that $G(n, k)$ is attained if A_n is $1, \ldots, n$? Is it true that $G(n, 2)>n^{1-\varepsilon}$ for every $\varepsilon>0$ if $n>n_0(\varepsilon)$? $G(n, k)>c_k n$ for $k \geqslant 3$? Perhaps these conjectures are completely wrongheaded.

We can ask questions for infinite sequences which are perhaps both interesting and fruitful: Is there an infinite sequence $a_1<\cdots$, $a_k<k^{1+\varepsilon}$, so that $\{a_i^2\}$ is a B_2 sequence? Further is there a sequence $a_t<Ct$ so that the sums $a_i^3+a_j^3$ are all distinct?

Riddell denotes by $P(n, k)$ the largest integer so that amongst n points in k-dimensional space one can always find $P(n, k)$ points which do not contain an isosceles triangle. It is not hard to prove that $P(n, k) > n^{\varepsilon_k}$, $\varepsilon_k \to 0$ as $k \to \infty$. Perhaps $P(n, 2) < n^{1-c}$. Here again perhaps the worst set of points is if they are lattice points in a sphere of as small radius as possible. By the way, $P(n, 1)$ reduces to having no three points in an arithmetic progression. Instead of no three points forming an isosceles triangle we could require that all the distances be distinct. We can leave the exact formulation of the problems to the reader.

One of my oldest unsolved problems states as follows: Let $f(n)$ be $+1$ or -1 for every n. Is it true that to every c there is a d and an m so that

$$\left|\sum_{k=1}^{m} f(kd)\right| > c? \tag{1}$$

I offer 500 dollars for a proof or disproof of (1). Perhaps (1) can be strengthened as follows: For suitable d and m, is it true that

$$\left|\sum_{\substack{k=1\\ md\le n}}^{m} f(kd)\right| > c \log n? \tag{1'}$$

It is not difficult to prove that (1′), if true, is the best possible.

Non-averaging sets. A set of integers $a_1 < a_2 < \cdots < a_k \le n$ is called not averaging by E. Straus if no $\underset{\sim}{a}$ is the arithmetic mean of other a's. Straus asked for an estimation of max $k = A(n)$. The best bounds are at present

$$c_1 n^{\frac{1}{10}} < A(n) < n^{\frac{2}{3}} + \varepsilon. \tag{2}$$

The lower bound is due to Abbott and the upper to Straus and myself. It would be of interest to prove that $\lim \log A(n)(\log n)^{-1} = \alpha$ exists and to determine α.

Graham conjectured: Let $1 \le a_1 < \cdots < a_n$ be n integers. Then

$$\max_{i,j} a_j/(a_i, a_j) \ge n.$$

Szemerédi proved this recently. The proof is not yet published.

I conjectured that if $1 \le a_1 < \cdots < a_k \le n$, $\sum 1/a_i < C$, then the number of integers $\le n$ not divisible by any of the a's is $> n^{1-\varepsilon}$ for every $\varepsilon > 0$ and C if $n > n_0(\varepsilon, C)$. It turned out here that my intuition completely misled me since Ruzsa proved that to every $\varepsilon > 0$ there is a $C = C(\varepsilon)$ so that there is a sequence A with $\sum_{i=1}^{k} 1/a_i < C$ so that the number of $m \le n$, with $m \not\equiv 0 \pmod{a_i}$ for $i = 1, \ldots, k$, is $< n^{\varepsilon}$, the exact dependence of ε from C is far from being known. Ruzsa proved that to every $\varepsilon > 0$ there is a $C = C(\varepsilon)$ so that there is a sequence A more than $x^{1-\varepsilon}$ numbers $m < x$, $m \not\equiv 0 \pmod{a_i}$, $i = 1, \ldots, k$. This is not yet cleared up.

the following extremal problem seems to be interesting. Let

$$\sum_i 1/a_i \le 1$$

be any sequence of integers. Denote by $A'(x)$ the number of integers not exceeding x and not divisible by any of the a's. Determine or estimate min $A'(x)$ where the minimum is to be taken over all sequences satisfying (3). I expect that is of the order of magnitude $x/(\log x)^\alpha$. It follows from a result of Schinzel and Szekeres that $A'(x) < cx/(\log x)^\alpha$ for some $\alpha > 0$.

Here also my intuition was wrong. In 1940 I conjectured that if $1 \leq a_1 < \cdots < a_k \leq x$ is a sequence of integers so that the least common multiple of any two a's is greater than x, then $A'(x) > cx$. Szekeres soon proved me wrong and in fact here we have

$$c_2 x/(\log x)^{\beta_1} < A'(x) < c_1 x/(\log x)^{\beta_2}$$

It would be of interest if one could give explicitely the sequence which solves the extremal problem (3), but I very much doubt if this is possible. Ruzsa remarks that trivially $A'(x) > cx/\log x$.

The prime k-tuple conjecture of Hardy and Littlewood states that if $a_1 < \cdots < a_k$ such that the a's do not form a complete set of residues mod p for every p, then there are infinitely many integers n so that all the numbers $n + a_i$, $i = 1, \ldots, k$ are primes. This is of course quite unattackable at present.

On the other hand it is a simple exercise to prove that if $a_1 < \cdots < a_k$ does not form a complete set of residues mod p^2 for every p, then there are infinitely many integers n for which $n + a_i$, $i = 1, 2, \ldots, k$ are all squarefree.

There are difficulties in formulating a reasonable conjecture for infinite sequences. Can the following conjecture be true: Let a_k tend to infinity sufficiently fast, and assume that there is an n so that all the $n + a_k$ are primes. Are there infinitely many such values of n? Is there any hope of proving this for squarefree numbers instead of primes? Or are there values of n for which say $n + 2^{2^k}$ is always a prime? Always squarefree, or infinitely often a prime, or infinitely often squarefree? Unless I overlook a trivial way of getting a counterexample these questions are quite hopeless.

Perhaps the following generalisation for infinite sequences is possible: Let $1 \leq a_1 < \cdots$ be a sequence which does not contain a complete set of residues mod p for every p, then there are infinitely many values of n for which all the integers $n + a_k$, $a_k < n$ are primes. Perhaps the following modification is slightly less hopeless: Let $1 \leq a_1 < \cdots$ be a sequence which does not contain a complete set of residues mod p^2 for every p. Then there are infinitely many values of n for which all the integers $n + a_k$, $a_k < n$ are squarefree.

These conjectures just occurred to me, and perhaps some further condition on the thinness of the sequence $A = \{a_k\}$ may be needed. If true they are of course hopeless for the primes and it seems to be hopeless for the square free numbers too. On the other hand if the squares of primes are replaced by a sequence which increases fast enough the conjecture becomes easy. Let $n_1 < n_2 < \cdots$, $(n_i, n_j) = 1$ tend to infinity sufficiently fast and let $a_1 < a_2 < \cdots$ be a sequence of integers which does not contain a complete set of residues (mod n_i) for every i. Then there are infinitely many integers x so that

$$x+a_i \not\equiv 0 \pmod{n_j}$$

for every n_j and $a_i < x$.

Let me call attention to the following beautiful conjecture of Ostman which is now about 25 years old: Prove that there do not exist two infinite sequences A and B so that the sequence $A+B$ differs from the set of primes in only a finite number of elements. Hornfeck proved that the sets A and B must both be infinite.

Here is a problem which I did not state quite correctly in III: Let $1 \leq a_1 < \cdots < a_k \leq n \cdots$ be a sequence of integers satisfying $(a_i, a_j) = 1$. Is it true that there exists an absolute constant C for which (p runs through the primes)

$$\sum \frac{1}{n-u_k} < C + \sum_{p<n} \frac{1}{p}.$$

I state some old problems which are perhaps not hopeless but which have been neglected: Denote by $f(k)$ the minimum number of terms of the square of a polynomial of k terms. Sharpening a result of Rédei and Rényi, I proved that $f(k) < k^{1-c}$. Rényi and I conjectured that $f(k) \to \infty$ as k tends to infinity. The value of $f(k)$ may depend what kind of coefficients we permit (integers, rationals, real or complex numbers). We believe that $f(k) \to \infty$ holds in all cases.

Let $a_1 < \cdots < a_k \leq n$ be any sequence of integers, $b_1 < b_2 < \cdots$ is the sequence of integers no one of which is the multiple of any of the a's. Put $B(x) = \sum_{b_i < x} 1$. Is it true that for every $m \geq n$

$$\frac{B(m)}{m} < 2\frac{B(n)}{n}? \tag{1}$$

It is easy to see that if (1) is true it is best possible. The a's consist only of a_1, $n = 2a_1 - 1$, $m = 2a_1$.

Roth conjectured that there is an absolute constant c so that to every k there exists an $n_0 = n_0(k)$ which has the following property: Let $n > n_0$, split the integers not exceeding n into k classes $\{a_i^{(j)}\}$, $j = 1, \ldots, k$. Then the number of distinct integers $m \leq n$ which for some j, $j = 1, 2, \ldots, k$ can be written in the form $a_{i_1}^{(j)} + a_{i_2}^{(j)}$ is greater than cn.

Let $a_1 < \cdots < a_n$ be a set of integers. Consider the sums $a_i + a_j$ and products $a_i a_j$. I conjectured that there are more than $n^{2-\varepsilon}$ distinct numbers amongst them. This if true seems very difficult. It is possible that Szemerédi and I will be able to prove that the number of distinct terms in this sequence is $> n^{1+c}$ for some $c > 0$.

Graham conjectured: Let $1 \leq a_1 < \cdots < a_p$ be p not necessarily distinct residues mod p. Assume that $\sum_{i=1}^{p} \varepsilon_i a_i \equiv 0 \pmod{p}$, $\varepsilon_i = 0$ or 1, and not all $\varepsilon_i = 0$ implies $\sum_{i=1}^{p} \varepsilon_i = r$. Does it then follow that there are at most two distinct residues mod p?

Szemerédi and I proved this if $p > C$ i.e. for sufficiently large p.

The following old problem of mine still seems to be open: Let $1 \leq a_1 < \cdots < a_t \leq n$ be such that there are at most r a's which are pairwise relatively prime.

Then we obtain max t by taking the multiples of the first r primes not exceeding n. Graham and I have the following related problem: Let $1 \leq a_1 < a_2 < \cdots < a_k = n$, $(a_i, a_j) \neq 1$. What is the maximum value of k? A reasonable guess seems to be that either $\max k = n/p(n)$, where $p(n)$ is the least prime factor of n, or it is the number of integers of the form $2t$, $t \leq \frac{1}{2}n$, $(t, n) \neq 1$. This problem is stated with many confusing misprints in I (p. 123).

Straus and I conjectured that for $k > k_0$ there always is an $i = i(k)$ so that $p_k^2 < p_{k+i}p_{k-i}$. Pomerance recently proved that this conjecture is completely wrongheaded and that if $a_n^{1/n} \to 1$, $a_n/n \to \infty$, then there are infinitely many values of k so that for every $i < k$, we have

$$a_k^2 > a_{k+1}a_{k-i}. \tag{2}$$

This was in fact conjectured by Selfridge who always disbelieved our conjecture.

Pomerance and I tried unsuccessfully to prove that if the a's are the primes then the density of the integers k for which (2) is satisfied for every $i < k$ is 0.

Sárközi and I proved that if $a_1 < a_2 < \cdots$ is an infinite sequence of integers where no a divides sum of two greater a's, then this sequence has density 0. We could not prove that $\sum 1/a_i < \infty$. Also the following finite problem seems interesting: Let $1 \leq a_1 < \cdots < a_k \leq x$ be such that no a_i divides the sum of two greater a's. Then $k \leq [\frac{1}{3}x] + 1$. Equality holds if $X = 3n$ and the a's are: $2n, 2n+1, \ldots, 3n$. The conjecture $k \leq [\frac{1}{3}x] + 1$ is still open.

To finish let me state four more problems: Is it true that almost all numbers n have two divisors d_1 and d_2 satisfying $d_1 < d_2 < 2d_1$. This is one of my oldest conjectures and I offer 300 dollars for a proof or disproof. I only proved that the density of these integers exists. I claimed once that I have a proof that the density is 1. I unfortunately have to withdraw this claim.

The second question is only a few days old. Divide the integers into two classes $n_1 < \cdots$ and $m_1 < m_2 < \cdots$. Denote by $N_1 < N_2 < \cdots$ and respectively $M_1 < M_2$ the integers which are distinct sums of the n's respectively the m's. It is easy to see that either the (N_i) or the (M_j) must have upper density 1 (both can of course have lower density 0). It is not clear to me at present how large

$$\limsup \frac{1}{\log x} \max\left(\sum_{N_1 < x} \frac{1}{N_i}, \sum_{M_i < x} \frac{1}{M_i}\right)$$

must be. It is easy to see that it can be less than 1, but I expect it to be greater than $\frac{1}{2}$

$$\left(\limsup\left(\frac{1}{\log x} \sum_{a_i < x} 1/a_i\right)\right.$$

is called the upper logarithmic density of the sequence $A\Big)$. Here I mention a theorem and problem of Davenport and myself on the existence of logarithmic density. Let $a_1 < a_2 < \cdots$ be any sequence of integers, $b_1 < b_2 < \cdots$ is the sequence of integers not divisible by any of the a's. Then the b's have logarithmic density. Now to our problem: To each a_i make correspond an arbitrary set of residues $\{x_j^{(i)}\}$, $j=1,2,\ldots,t_i$. Let $b_1<b_2<\cdots$ be the sequence of integers for which

$$b_u \not\equiv x_j^{(i)} \pmod{a_i}, \quad j=1,2,\ldots,t_i$$

always holds if $b_u>a_i$. Is it true that the b's have a logarithmic density? Perhaps this question is not very difficult as far as I know it has not been attacked really seriously.

A sequence of integers $A=\{a_1<a_2<\cdots\}$ is said to have property P if

$$\prod_i a_i = \prod_j a_j$$

(where each product runs over a subset of the a's) is possible only if the number of factors on both sides of the equation is the same. Clearly the integers $\equiv 2 \pmod 4$ have property P. I asked: is there a sequence with property P having density $>1-\varepsilon$? Is there a finite sequence $1<a_1<\cdots<a_k\leq n$ having property P and satisfying $k>(1-\varepsilon)n$? Ruzsa showed that the answer to both questions is negative. Ruzsa in fact shows that if A is an infinite sequence with property P then its upper density is less than 1/e. This is best possible. Also if $1\leq a_1<\cdots<a_k\leq x$ has property P, then $A(x)<(1-c)x$ for an absolute constant $c>0$. The best value of c is not known.

Finally let me tell of a problem where I somewhat made a fool of myself. Herzog and Stewart studied visible lattice points i.e. lattice points $\{u, v\}$, satisfying $(u, v)=1$. One joins two visible lattice points if they are neighbours i.e. if they differ in only one coordinate and there by ± 1. Herzog and Stewart prove that there is only one infinite component and they conjecture that (a, p) $a\not\equiv 0 \pmod p$, p prime, always belongs to the infinite component. Last year when I gave a talk at Michigan State University I asked: Is there an infinite path through visible points no coordinate of which is 1. I though that the answer will be not too hard, and affirmative, and follishly offered 25 dollars for a proof. In the evening Stewart gave the simple proof: $\{p_k, p_{k+1}\}$ can be joined to $\{p_{k+1}, p_{k+2}\}$ by Tchebicheff's theorem which gives the path in question. Perhaps I should have asked: Is there a path going to infinity which avoids points both coordinates of which are primes and also points one coordinate of which is 1. We could further demand that the path is monotone i.e. every step increases the distance from the origin. Is there a monotone path where we change direction after a bounded number of steps?

The following result surely holds, but there are some technical difficulties in giving a rigorous proof: To every ε there is a k so that for all but εn^2 lattice points $\{a, b\}$, $0\leq a$, $b\leq n$ there is a sequence of visible lattice points $\{u_k, v_k\}$, $k=1,2,\ldots,(u_0=v_0=1)$, where $u_0<u_1<\cdots$, $v_0<v_1<\cdots$ and $0\leq u_{k+1}-u_k<C$, $0\leq v_{k+1}-v_k<C$. In other words infinity can be reached with jumps of bounded length.

The following beautiful conjecture is due to Gordon and Motzkin: Is there a sequence of distinct Gaussian primes $P_1, P_2, \ldots,$ for which $|P_{k+1} - P_k| < C$ for some absolute constant C. The answer is almost certainly negative.

J. Komlós, M. Sulyok and E. Szemerédi, Linear problems in combinatorial number theory, Acta Math. Acad. Sci. Hungar. 26 (1975) 113–121.

H.L. Abbott, On a conjecture of Erdös and Strauss on non averaging sets of integers, Proc. 5th British Combinatorial Conference, Aberdeen (1975) 1–4.

P. Erdös and E. Szemerédi, On a problem of Graham, Publicationes Math. 23 (1976) 123–127.

P. Erdös and A. Sárközy, On the divisibility properties of sequences of integers, Proc. London Math. Soc. 21 (1970) 97–101.

P. Erdös, On the number of terms of the square of a polynomial, Nieun Arch. Wiskunde (1949) 63–65.

W. Verdenius, On the number of terms of the square and cube of polynomials Indig. Math. II (1949) 459–465.

Finally I would like to call attention to a forthcoming monograph of R.L. Graham and myself entitled "Old and new Problems and Results in Combinatorial Number Theory" which will soon appear in L'Enseignement Math. See further the forthcoming book of R.L. Graham, B. Rothschild and J. Spencer, Ramsey Theory (Wiley, New York, 1980).

Annals of Discrete Mathematics 6 (1980) 117–123

RESIDUALLY-COMPLETE GRAPHS

Paul ERDÖS

Hungarian Academy of Science and the University of Colorado

Frank HARARY

University of Michigan in Ann Arbor, Ann Arbor, MI, U.S.A.

Maria KLAWE

Oakland University in Rochester, Rochester, MI, U.S.A.

If G is a graph such that the deletion from G of the points in each closed neighborhood results in the complete graph K_n, then we say that G is K_n-residual. Similarly, if the removal of m consecutive closed neighborhoods yields K_n, then G is called m-K_n-residual. We determine the minimum order of the m-K_n-residual graphs for all m and n. The minimum order of the connected K_n-residual graphs is found and all the extremal graphs are specified.

1. Introduction

A graph G is said to be F-residual if for every point u in G, the graph obtained by removing the closed neighborhood of u from G is isomorphic to F. We inductively define multiply-F-residual graphs by saying that G is *m-F-residual* if the removal of the closed neighborhood of any point of G results in an $(m-1)$-F-residual graph, where of course a 1-F-residual graph is simply an F-residual graph.

We are concerned with residually-complete graphs, i.e., graphs which are m-K_n-residual for some m and n. It is easy to see that there exists such a graph for any m and n, since $(m+1)K_n$ is clearly such a graph. Actually we show that there exist infinitely many connected m-K_n-residual graphs for any m and n.

It is natural to ask what is the minimum number of points that an m-K_n-residual graph must contain. We easily prove that this number is $(m+1)n$ and that the only m-K_n-residual graph with this number of points is $(m+1)K_n$. The same question for connected m-K_n-residual graphs is more interesting. We are able to show that a connected K_n-residual graph must have at least $2n+2$ points if $n \neq 2$. Furthermore, the cartesian product $K_{n+1} \times K_2$ is the only such graph with $2n+2$ points for $n \neq 2, 3, 4$. We complete the result by determining all connected K_n-residual graphs of minimal order for $n = 2, 3, 4$.

Although we have not obtained the minimum number of points for a connected m-K_n-residual graph, we include some canonical examples which might be expected to have smallest order when n is large.

In general the notation follows that of [1]. In particular $p(G)$ is the number of points in a graph G, $N(u)$ is the neighborhood of a point u consisting of all points adjacent to u, $N^*(u)$ is the closed neighborhood of u. Also, for any real x, the symbol $\lceil x \rceil$ denotes the ceiling of x defined as the smallest integer $n \geqslant x$.

2. Residually-complete graphs of minimum order

We begin this section with a simple observation which will turn out to be extremely useful.

Remark 1. If G is F-residual, then for any point u in G, the degree $d(u) = p(G) - p(F) - 1$. Hence every F-residual graph is regular, though this is generally not true for multiply-F-residual graphs (see Example 3).

Theorem 1. *Every m-K_n-residual graph has at least $(m+1)n$ points, and $(m+1)K_n$ is the only m-K_n-residual graph with $(m+1)n$ points.*

Proof. Let G be K_n-residual, and u, v nonadjacent points in G. Then $H_1 = G - N^*(u)$ and $H_2 = G - N^*(v)$ are disjoint copies of K_n contained in G, so $p(G) \geqslant 2n$. If $p(G) = 2n$, then $G = H_1 \cup H_2$ so all that remains to be shown is that there are no lines between H_1 and H_2, which is clear since G is $(n-1)$-regular by Remark 1.

Using induction on m, the rest of the theorem can easily be proved by similar arguments.

Theorem 2. *Every connected K_n-residual graph has at least $2n+2$ points if $n \neq 2$.*

The proof of this theorem requires a few preliminary results. We begin with the following definition.

For two points u, v in G, we say u is *K_n-adjacent* to v if there exists a copy of K_n in G which contains both u and v.

Lemma 2a. *Let G be a K_n-residual graph with $p(G) < 2n + \lceil \frac{1}{2}n \rceil$, and let u, v, w be points in G such that u is K_n-adjacent to v and v is K_n-adjacent to w. Then u is adjacent to w, in fact, u is K_n-adjacent to w.*

Proof. Let H_1 and H_2 be copies of K_n contained in G with $u, v \in H_1$ and $v, w \in H_2$. Suppose u is not adjacent to w. Then $w \in H_3 = G - N^*(u)$ which is another copy of K_n in G. Clearly $H_1 \cap H_3 = \emptyset$ since $H_1 \subset N^*(u)$. Thus $p(H_2 - H_3) \geqslant p(H_2 \cap H_1)$ and we see that $p(H_1 - H_2) + p(H_2 - H_3) \geqslant p(H_1) = n$. This shows that $\max\{p(H_1 - H_2), p(H_2 - H_3)\} \geqslant \lceil \frac{1}{2}n \rceil$. Now consider the degrees of v and w. We have

$$d(v) \geqslant p(H_2) - 1 + p(H_1 - H_2) = n - 1 + p(H_1 - H_2)$$

and

$$d(w) \geq p(H_3) - 1 + p(H_2 - H_3) = n - 1 + p(H_2 - H_3).$$

Hence there exists a point y in G with $d(y) \geq n - 1 + \lceil \frac{1}{2}n \rceil$, showing that

$$p(G) \geq n + (n - 1 + \lceil \tfrac{1}{2}n \rceil) + 1 = 2n + \lceil \tfrac{1}{2}n \rceil$$

by Remark 1, which contradicts the hypothesis $p(G) < 2n + \lceil \frac{1}{2}n \rceil$. Thus we see that u is adjacent to w. By repeating this argument, it is clear that u is adjacent to every point in H_2, and hence u is K_n-adjacent to w.

Remark 2. If G is a K_n-residual graph with $p(G) < 2n + \lceil \frac{1}{2}n \rceil$, then for any two nondisjoint copies H_1 and H_2 of K_n contained in G, we have $H_1 \cup H_2 \cong K_s$ where $s = p(H_1 \cup H_2)$.

Proof. Choose $v \in H_1 \cap H_2$, and let u, w be any two points in $H_1 \cup H_2$. Clearly u is K_n-adjacent to v and v is K_n-adjacent to w, so by Lemma 2a, u and w are adjacent.

Lemma 2b. *If G is a connected K_n-residual graph with $p(G) < 2n + \lceil \frac{1}{2}n \rceil$, then G contains a copy of K_{n+1}.*

Proof. Since G is connected and K_n-residual, by Theorem 1 we have $p(G) \geq 2n + 1$. Choose some copy of K_n in G, denoted by H_1, and let u be a point in H_1. Since $p(G) \geq 2n + 1$, we have $d(u) \geq n$ and thus we can find $v \in N^*(u) - H_1$. If $\langle H_1 \cup \{v\} \rangle \cong K_{n+1}$ we are done, so assume there exists $w \in H_1 - N^*(v)$. let $H_2 = G - N^*(v)$. Now H_1 and H_2 are nondisjoint copies of K_n in G, so $\langle H_1 \cup H_2 \rangle \cong K_s$ where $s = p(H_1 \cup H_2) \geq n + 1$ since $u \in H_1 - H_2$.

We are now ready to prove Theorem 2. Let G be a connected K_n-residual graph. The case where $n = 1$ is obvious since neither of the connected graphs of order 3, P_3 and K_3, is K_1-residual. Thus we assume $n \geq 3$. If $p(G) \geq 2n + \lceil \frac{1}{2}n \rceil$ we are done since $\lceil \frac{1}{2}n \rceil \geq 2$. If $p(G) < 2n + \lceil \frac{1}{2}n \rceil$, then G contains a copy of K_{n+1} which we denote by H. Since G is connected and $G - H \neq \emptyset$, we must have $d(u) \geq n + 1$ for some point u in H, and thus

$$p(G) \geq n + (n + 1) + 1 = 2n + 2$$

by Remark 1.

The next result determines the connected K_n-residual graphs of minimum order. It is interesting to note that for $n \neq 3, 4$ the graph is unique.

Theorem 3. *If $n \neq 2$, then $K_{n+1} \times K_2$ is a connected K_n-residual graph of minimum order, and except for $n = 3$ and $n = 4$, it is the only such graph. For each of the cases*

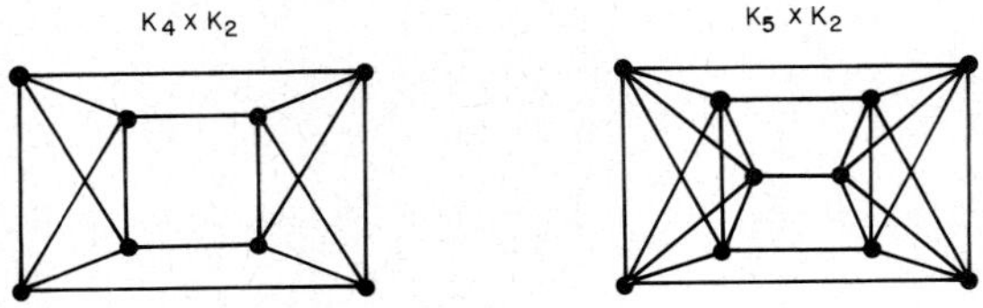

Fig. 1. Two examples of $K_{n+1} \times K_2$.

$n=3$ and $n=4$ there is exactly one other such graph. Finally, C_5 is the only connected K_2-residual graph of minimum order.

The graphs $K_4 \times K_2$ and $K_5 \times K_2$ are shown in Fig. 1 while the other smallest connected K_n-residual graphs for $n=3$ and 4 are given in Figs. 2 and 3.

Proof. It is easy to verify that $K_{n+1} \times K_2$ is a connected K_n-residual graph for any n. Since $p(K_{n+1} \times K_2) = 2n+2$, Theorem 2 shows that $K_{n+1} \times K_2$ has minimum order for $n \neq 2$. Suppose $n \geqslant 5$ and that G is a connected K_n-residual graph with $p(G) = 2n+2$. Then $p(G) < 2n + \lceil \frac{1}{2}n \rceil$ so G contains a copy of K_{n+1}, which we denote by $L = \langle x_1, \ldots, x_{n+1} \rangle$. Since $d(x_i) = n+1$, it follows that $N^*(x_i) - L = \{y_i\}$. Also $G = \bigcup_{i=1}^{n} N^*(x_i)$ since otherwise we would have $L \subset G - N^*(u)$ for some point u in G. This shows that $G - L = \langle y_1, \ldots, y_{n+1} \rangle$ and since $p(G-L) = n+1$ we find that the y_i's are distinct. Moreover, for $i \neq j$ we see that $y_i, y_j \in G - N^*(x_k)$ for any $k \neq i, j$ and hence y_i and y_j are adjacent, showing that $G - L \cong K_{n+1}$. Clearly $G \cong K_{n+1} \times K_2$.

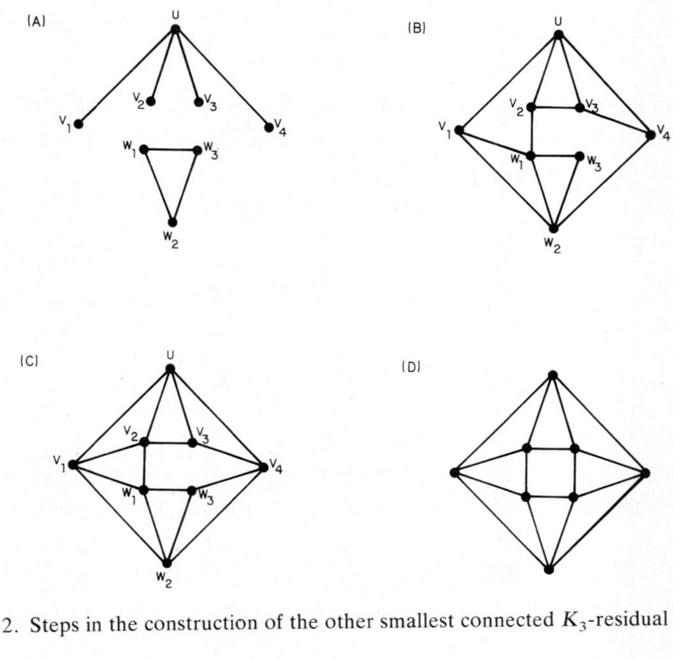

Fig. 2. Steps in the construction of the other smallest connected K_3-residual graph.

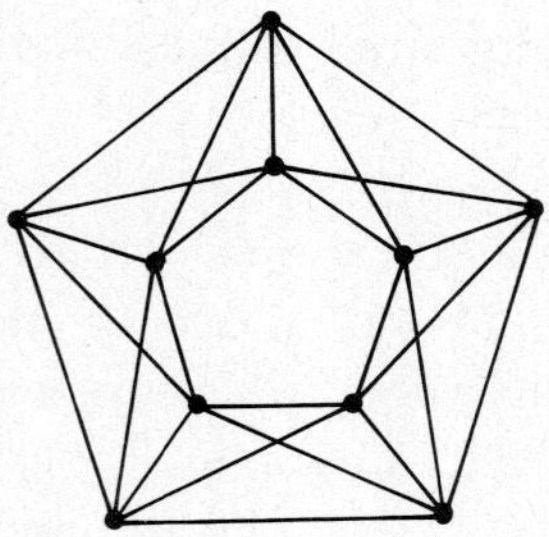

Fig. 3. The other smallest connected K_4-residual graph.

We now prove the remainder of the theorem involving the small cases $n \leq 4$. For $n = 1$, $K_2 \times K_2 = C_4$ is the only regular connected graph of degree 2 on 4 points, and similarly for $n = 2$, C_5 is the only regular connected graph of degree 2 on 5 points.

For $n = 3$, suppose G is a connected K_3-residual graph with $p(G) = 8$. If G contains a copy of K_4, then the same proof as for $n \geq 5$ will show that $G \cong K_4 \times K_2$. Thus we may assume that G does not contain a copy of K_4. Let

$$V(G) = \{u, v_1, v_2, v_3, v_4, w_1, w_2, w_3\}$$

where $N(u) = \{v_1, v_2, v_3, v_4\}$ and $\langle W \rangle = \langle w_1, w_2, w_3 \rangle \cong K_3$ (see Fig. 2a). Since $K_4 \not\subset G$ we see that $W \not\subset N(v_i)$ for any i, and for the same reason $N(w_i) \cap N(u) \neq N(w_j) \cap N(u)$ if $i \neq j$. Thus for each pair of distinct i and j we have

$$p(N(w_i) \cap N(u) \cap N(w_j)) = 1.$$

By symmetry we may assume that $N(w_1) = \{v_1, v_2, w_2, w_3\}$ and $N(w_2) = \{v_1, v_4, w_1, w_3\}$. These imply $\langle u, v_3, v_4 \rangle \cong K_3$ and $\langle u, v_2, v_3 \rangle \cong K_3$. In particular v_3 is adjacent to v_4 and v_2 is adjacent to v_3 (see Fig. 2b). Since $W \not\subset N(v_1)$, v_1 is not adjacent to w_3, hence either v_2 or v_4 is adjacent to w_3, and by symmetry we may assume v_4 is adjacent to w_3. Thus $N(v_4) = \{u, v_3, w_2, w_3\}$ so $\langle v_1, v_2, w_1 \rangle \cong K_3$ and in particular v_1 is adjacent to v_2 (see Fig. 2c). Finally, since $N(v_1) = \{u, v_2, w_1, w_2\}$ we have $\langle v_3, v_4, w_3 \rangle \cong K_3$ so v_3 is adjacent to w_3. Now every point in G has degree 4, so the construction is finished (see Fig. 2d).

For $n = 4$, suppose G is a connected K_4-residual graph with $p(G) = 10$. If G contains a copy of K_5, then as before one finds $G \cong K_5 \times K_2$. If G does not contain a copy of K_5, then similar arguments as for the case $n = 3$ will construct the graph shown in Fig. 3.

3. Multiply-K_n-residual graphs

In this section we first note that for any m and n there are infinitely many connected m-K_n-residual graphs, then exhibit some canonical examples, and close with some conjectures on the minimum number of points in a connected m-K_n-residual graph.

Remark 3. For any choice of positive integers m and n, there are infinitely many connected m-K_n-residual graphs.

Proof. Observe that if G_1 and G_2 are disjoint m-K_n-residual graphs, then their join G_1+G_2 (as in [1, p. 21]) is a connected m-K_n-residual graph. Since $(m+1)K_n$ is an m-K_n-residual graph, we can repeatedly use the above technique to construct an infinite collection of graphs.

It is easy to see that G is K_n-residual if and only if $\bar{G}$ is n-regular and contains no triangles. This observation of R.W. Robinson verifies Remark 3 at once for $m=1$.

Example 1. The join $(m+1)K_n+(m+1)K_n$ is a connected m-K_n-residual graph with $2n(m+1)$ points.

Example 2. The cartesian product $K_{n+m}\times K_{m+1}$ is a connected m-K_n-residual graph with $(n+m)(m+1)$ points. This is easily proved by induction on m since we have already noted that $K_{n+1}\times K_2$ is K_n-residual.

Notice that for $n=m$, the graphs of Examples 1 and 2 have the same order although they are not isomorphic unless $n=1$.

Example 3. For each $m\geqslant 1$, the graph G_m defined by

$$V(G_m)=\{u_0,\ldots,u_{m+1},\ v_1,\ldots,v_m,\ w_0,\ldots,w_{m-1}\}$$

and

$$E(G_m)=\{u_iu_{i+1},\ u_iw_i,\ u_iv_{i-1},\ v_iw_i,\ v_iw_{i-1}\}$$

can be shown to be a connected m-K_2-residual graph with $3m+2$ points. The graphs G_m for $m=1,2,3,4$ are shown in Fig. 4, as well as another connected 3-K_2-residual with 11 points. Notice that the graph G_m is not regular unless $m=1$.

4. Unsolved problems and conjectures

We have only determined the minimum order of the connected K_n-residual graphs. The question is open for m-K_n-residual graphs when $m\geqslant 2$.

Conjecture 1. *If $n\neq 2$, then every connected m-K_n-residual graph has at least* $\min\{2n(m+1),(n+m)(m+1)\}$ *points.*

Every connected m-K_2-residual graph has at least $3m+2$ points.

Note that this quantity agrees with that of Theorem 2 for $m=1$ when $n\neq 2$, and with Theorem 3 when $n=2$.

We believe that there will be an analogous uniqueness result for $m\geqslant 2$.

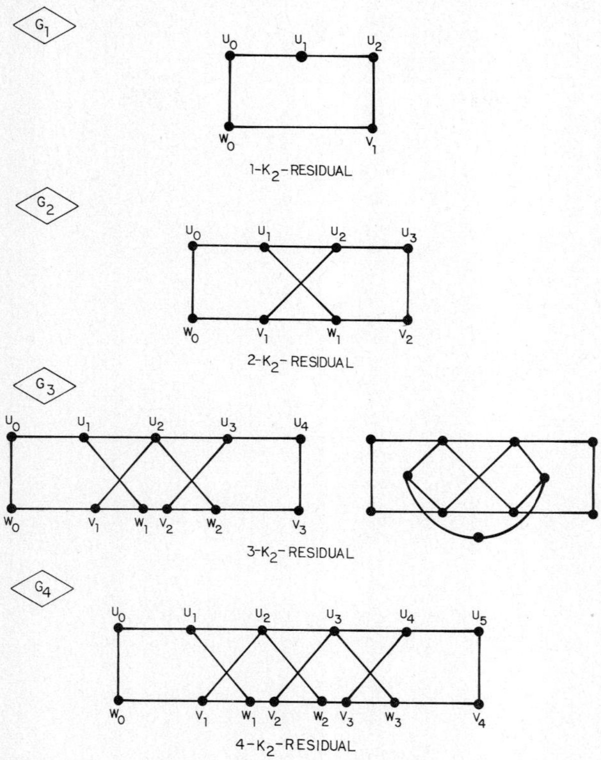

Fig. 4. Multiply-K_2-residual graphs of small order.

Conjecture 2. *For n large, there is a unique smallest connected m-K_n-residual graph.*

The *link* of a point u of a graph G, written $L(u)$, is the subgraph $\langle N(u)\rangle$ induced by the neighborhood of u. A graph G has *constant link* if for all $u, v \in V(G)$, $L(u) \cong L(v)$. Clearly G is K_n-residual if and only if its complement $\bar{G}$ has constant link $\bar{K}_n$.

In general, then, G is an F-residual graph if and only if $\bar{G}$ has constant link $\bar{F}$. In later communications we propose to investigate F-residual graphs for $F \neq K_n$, in order to determine the minimum order among such graphs, and to specify the corresponding extremal graphs.

References

[1] F. Harary, Graph Theory. (Addison-Wesley, Reading, MA, 1969).

Annals of Discrete Mathematics 6 (1980) 125–135

A COMBINATORIAL PROOF OF JACOBI'S IDENTITY*

Dominique FOATA

Département de Mathématique, Université de Strasbourg, 7, rue René-Descartes, 67084 Strasbourg, France

The Jacobi identity $\det\exp A = \exp \operatorname{Tr} A$ is proved by using combinatorial methods: partitional complex and free monoid factorizations.

1. Introduction.

Jacobi's identity

$$\det\exp A = \exp \operatorname{Tr} A \tag{1.1}$$

holds for every square matrix A with complex entries with $\operatorname{Tr} A$ standing for the trace of A. (See [8, pp. 152–156] for a classical Linear Algebra proof.) As the identity has been applied to several problems in Enumeration [4, 6], the question of doing the converse has been raised: use combinatorial techniques to derive Jacobi's identity. This is the purpose of this paper.

Let n be a fixed positive integer and $B = (b(i,j))$ $(1 \leqslant i, j \leqslant n)$ be a square matrix with n^2 commuting variables $b(i,j)$ $(1 \leqslant i, j \leqslant n)$. Using combinatorial techniques (essentially, partitional complex and free monoid factorizations) we show that the identity

$$\frac{1}{\det(I-B)} = \exp \operatorname{Tr} \log \frac{1}{I-B} \tag{1.2}$$

holds in the $\mathbb{C}$-algebra of formal power series with the n^2 variables $b(i,j)$ $(1 \leqslant i, j \leqslant n)$. From (1.2) it is straightforward to derive (1.1) with A a matrix whose entries are formal power series with constant terms equal to zero.

There are three steps in the proof of (1.2). First, the ratio $1/\det(I-B)$ is shown (see Section 2) to be the generating function for the so-called *color words* by a certain monomial-valued function β_{dec}, i.e.

$$\frac{1}{\det(I-B)} = 1 + \sum_{m \geqslant 1} \sum \{\beta_{\mathrm{dec}}(w) : w \in X^m\}. \tag{1.3}$$

* Supported in part by N.S.F. Grant MCS 77-02113 (Department of Mathematics, University of California, San Diego, La Jolla, CA 92093, U.S.A.).

Second, $\exp \operatorname{Tr} \log(1/I-B)$ is proved (see Section 3) to be the exponential generating function for the so-called *colored permutations* by another monomial-valued function β, i.e.

$$\exp \operatorname{Tr} \log \frac{1}{I-B} = 1 + \sum_{m \geqslant 1} \frac{1}{m!} \sum \{\beta(p) : p \in P_m\}. \tag{1.4}$$

The third (and crucial) step consists of constructing a bijection $(w, \sigma) \to p$ of $X^m \times \mathfrak{S}_m$ (with $\mathfrak{S}_m$ the symmetric group of degree m) onto P_m with the property that

$$\beta_{\text{dec}}(w) = \beta(p). \tag{1.5}$$

This is done in Section 4. Clearly, (1.3), (1.4) and the existence of the above bijection imply (1.2).

The following notation will be used throughout. If $\alpha(y)$ is a monomial for every element y of a finite set Y, then $\alpha\{Y\}$, will stand for the polynomial

$$\alpha\{Y\} = \sum \{\alpha(y) : y \in Y\}.$$

2. Decreasing words

Let X designate the totally ordered set $[n] = \{1 < 2 < \cdots < n\}$. Its elements are called *colors*, and X itself is referred to as the *color set*. For each $m \geqslant 1$ any word $w = x_1 x_2 \cdots x_m$ of length m with letters taken from X will be called a *color word* and X^m will denote the set of all color words of length m. When the m letters of the word $w = x_1 x_2 \cdots x_m$ are rearranged in non-decreasing order, we obtain a word $\bar{w} = \bar{x}_1 \bar{x}_2 \cdots \bar{x}_m$ called the *non-decreasing rearrangement* of w. Define $\alpha(w)$ to be the monomial

$$\alpha(w) = b(\bar{x}_1, x_1) b(\bar{x}_2, x_2) \cdots b(\bar{x}_m, x_m)$$

in the variables $b(i, j)$. Then the identity

$$\frac{1}{\det(I-B)} = 1 + \sum_{m \geqslant 1} \alpha\{X^m\} \tag{2.1}$$

is essentially the MacMahon's Master theorem identity ([7, pp. 93–98]; see also the nine proofs given by Cartier [1], or [2, (Chapter 5]). To transform (2.1) into (1.3) that with the above notations may be rewritten

$$\frac{1}{\det(I-B)} = 1 + \sum_{m \geqslant 1} \beta_{\text{dec}}\{X^m\},$$

it remains to define β_{dec} and show that

$$\alpha\{X^m\} = \beta_{\text{dec}}\{X^m\}.$$

A color word $w = x_1 x_2 \cdots x_m$ is said to be (initially) *minorized* if its first letter is strictly less than all its other letters. The *decreasing factorization* of a word w is a sequence $(d_1, d_2, \ldots, d_s)$ of minorized words the first letters of which are in non-increasing order ($Fd_1 \geq Fd_2 \cdots \geq Fd_s$; the symbol Fd_i standing for the first letter of d_i) and $w = d_1 d_2 \cdots d_s$. Clearly, every word has one and only one decreasing factorization.

For each color word $w = x_1 x_2 \cdots x_m$ of positive length let

$$\beta(w) = b(x_1, x_2) b(x_2, x_3) \cdots b(x_{m-1}, x_m) b(x_m, x_1), \tag{2.2}$$

Furthermore, if $(d_1, d_2, \ldots, d_s)$ is the decreasing factorization of w, let

$$\beta_{\text{dec}}(w) = \beta(d_1)\beta(d_2) \cdots \beta(d_s). \tag{2.3}$$

When w is the empty word, let $\beta(w) = \beta_{\text{dec}}(w) = 1$.

For instance, the decreasing factorization of the following word w of length $m = 20$ is shown with vertical bars drawn after every minorized factor

$$w = 6 \mid 23 \mid 243 \mid 13 \mid 1 \mid 13 \mid 1 \mid 13 \mid 1 \mid 13 \mid 1 \mid 13 \mid. \tag{2.4}$$

Accordingly

$$\beta_{\text{dec}}(w) = b(6,6) \cdot b(2,3)b(3,2) \cdot b(2,4)b(4,3)b(3,2) \cdot b(1,3)^5 \cdot b(3,1)^5 \cdot b(1,1)^4.$$

Finally, the identity

$$\alpha\{X^m\} = \beta_{\text{dec}}\{X^m\} \quad (m \geq 1)$$

is a consequence of the next theorem that can be found in [2, p. 51].

Theorem 2.1. *For each $m \geq 1$ there exists a bijection Φ of X^m onto itself with the following property: for each color word $w = x_1 x_2 \cdots x_m$ the color word $\Phi(w) = w' = x'_1 x'_2 \cdots x'_m$ is a rearrangement of w and*

$$\alpha(w') = \beta_{\text{dec}}(w).$$

This completes the proof of (1.3).

3. Colored Permutations

To establish (1.4) first expand $\operatorname{Tr}\log(1/I - B)$ to obtain

$$\operatorname{Tr}\log\frac{1}{I-B} = \operatorname{Tr}\sum_{m\geq 1}\frac{B^m}{m}.$$

As the (i, j)-entry in the matrix B^m is

$$\sum_{x_2, \ldots, x_m} b(i, x_2) b(x_2, x_3) \cdots b(x_m, j),$$

the trace of B^m is also equal to

$$\begin{aligned}\operatorname{Tr} B^m &= \sum_i \sum_{x_2,\ldots,x_m} b(i, x_2)b(x_2, x_3)\cdots b(x_m, i)\\ &= \sum \{\beta(w) : w \in X^m\}\end{aligned}$$

(see the definition of β in (2.2)). Thus

$$\operatorname{Tr} \log \frac{1}{I-B} = \sum_{m\geq 1} \frac{1}{m} \beta\{X^m\}. \tag{3.1}$$

The next step is to transform the right-side of (3.1) into an *exponential* generating function. This is achieved with the introduction of colored permutations and cycles. A *colored m-permutation* is nothing but a permutation graph with m vertices $1, 2, \ldots, m$ that has the further property that each arc or loop is colored. This means that an element of X is assigned to each arc or loop. For instance, the graph in Fig. 1 is a colored m-permutation with $m = 20$ and $X = \{1, 2, \ldots, 6\}$. In (1.4) the symbol P_m stands for the set of colored m-permutations and β enumerates the colors *by adjacency*. More precisely, let n_{ij} be the number of vertices of a colored m-permutation p that are ends of an arc colored i and beginnings of an arc colored j. Then

$$\beta(p) = \prod_{i,j} b(i, j)^{n_{ij}}. \tag{3.2}$$

For instance, with the above example

$$\beta(p) = b(6, 6)b(2, 3)b(3, 2)^2b(2, 4)b(4, 3)b(1, 3)^5b(3, 1)^5b(1, 1)^4. \tag{3.3}$$

Next a *colored m-cycle* is simply defined as a *connected* colored m-permutation. The set of all colored m-cycles will be denoted by C_m. An alternate definition can be given as follows. A two-row matrix

$$\binom{w}{\sigma} = \binom{x_1x_2\cdots x_m}{v_1v_2\cdots v_m} \tag{3.4}$$

with $w = x_1x_2\cdots x_m$ a color word of length m and $\sigma = v_1v_2\cdots v_m$ a permutation of $1\,2\cdots m$ is called a *colored m-biword*. Two colored m-biwords $\binom{w}{\sigma}$ and $\binom{w'}{\sigma'}$ are (cyclically) *equivalent* if they only differ by a cyclic rearrangement of their columns. Each equivalence class is then a *colored m-cycle*. Each colored m-cycle will be represented as a circular biword

$$c = \begin{bmatrix} x_1x_2\cdots x_m \\ v_1v_2\cdots v_m \end{bmatrix}. \tag{3.5}$$

The graphical representation of c can be obtained by drawing a graph with m vertices labeled $1, 2, \ldots, m$ and joining the vertex labeled v_i (resp. v_m) with the vertex labeled v_{i+1} (resp. v_1) by an arc colored x_i (resp. x_m) for every $i = 1, 2, \ldots, m-1$. Furthermore, when applied to the cycle c, the function $\beta(c)$

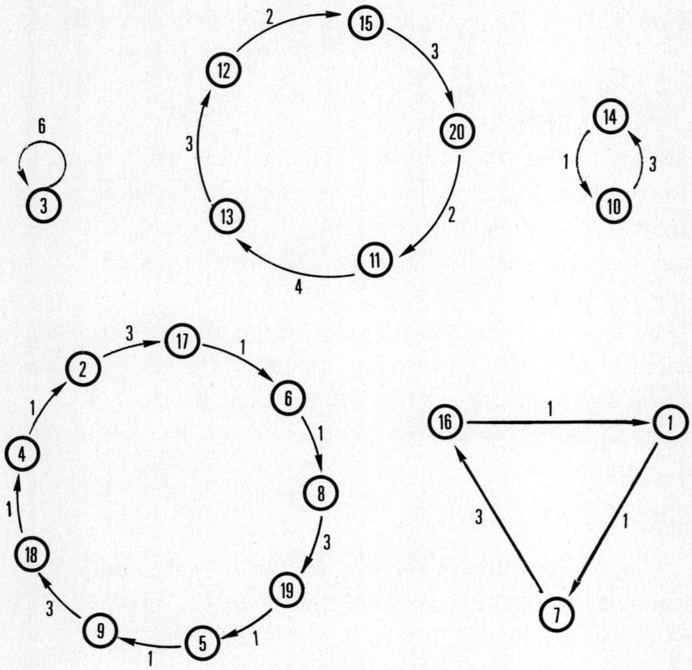

Fig. 1.

(given in (3.2)) simply reads

$$\begin{aligned}\beta(c) &= b(x_1, x_2)b(x_2, x_3)\cdots b(x_{m-1}, x_m)b(x_m, x_1) \\ &= \beta(w)\end{aligned} \tag{3.6}$$

with $w = x_1x_2\cdots x_m$.

Lemma 3.1. *For each* $m \geqslant 1$

$$\beta\{C_m\} = (m-1)!\ \beta\{X^m\}. \tag{3.7}$$

Proof. Let $\binom{m}{\sigma}$ be a colored m-biword. As σ is a permutation, the m cyclic rearrangements of $\binom{w}{\sigma}$ (including itself) are all distinct. Thus each colored m-cycle has exactly m representatives. Hence

$$\begin{aligned} m\beta\{C_m\} &= \sum\{\beta(w) : w \in X^m, \sigma \in \mathfrak{S}_m\} \quad \text{(from (3.6))} \\ &= m!\sum\{\beta(w) : w \in X^m\} \\ &= m!\ \beta\{X^m\}. \end{aligned}$$

Thus (3.1) and (3.7) imply

$$\operatorname{Tr}\log\frac{1}{I-B} = \sum_{m\geqslant 1}\frac{1}{m!}\beta\{C_m\}. \tag{3.8}$$

The proof of (1.4) will be completed, if the following identity

$$1+\sum_{m\geq 1}\frac{1}{m!}\beta\{P_m\}=\exp\sum_{m\geq 1}\frac{1}{m!}\beta\{C_m\} \tag{3.9}$$

can be established. But this can be achieved by using partitional complex techniques as follows. Let $I=\{i_1<i_2<\cdots<i_m\}$ be a finite subset of positive integers. When in (3.4) (resp. in (3.5)) $v_1v_2\cdots v_m$ stands for a permutation of $i_1i_2\cdots i_m$, the two-row matrix $\binom{w}{\sigma}$ (resp. $\left[{w \atop \sigma}\right]$) will be called a *colored I-biword* (resp. a *colored I-cycle*).

Let $c'=\left[{w \atop \sigma}\right]$ be a colored I-cycle and let φ be the unique increasing bijection of I onto $[m]$. Then, c' may be expressed as a pair (c, I) with c the colored m-cycle obtained from c' by replacing each label vertex v by $\varphi(v)$. Let p be a colored m-permutation. It is an unordered collection of its connected components $c'_1, c'_2, \ldots, c'_r$, namely

$$\{(c_1, I_1), (c_2, I_2), \ldots, (c_r, I_r)\} \tag{3.10}$$

with $\{I_1, I_2, \ldots, I_m\}$ a partition of $[m]$ and for $1\leq i\leq r$ the component c_i a colored $|I_i|$-cycle. Moreover, from the very definition of β ((3.2) and (3.6)) the following *multiplicative* property holds

$$\beta(p)=\beta(c_1)\beta(c_2)\cdots\beta(c_r). \tag{3.11}$$

The exponential formula (3.9) is a consequence of (3.10) and (3.11). More explicitly, (3.10) says that P_m is the so-called *partition complex* of degree m of the set $\bigcup_{m\geq 1} C_m$ (see e.g. [5, p. 51]). For each $p=\{(c_1, I_1), (c_2, I_2), \ldots, (c_r, I_r)\}$ let $\psi(p)=c_1c_2\cdots c_r$ be regarded as a commutative monomial in the variables c_i's. The abstract exponential formula (see e.g. [5, p. 51]) reads

$$1+\sum_{m\geq 1}\frac{1}{m!}\sum\{\psi(p):p\in P_m\}=\exp\sum_{m\geq 1}\frac{1}{m!}\sum\{c:c\in C_m\}. \tag{3.12}$$

As β is multiplicative (property (3.11)) and the degree of the monomial $\beta(p)$ is m whenever p belongs to P_m, the map β can be extended to a continuous homomorphism. Therefore, the image under β of (3.12) yields (3.9).

This completes the proof of (1.4).

4. The standard factorization

Now comes the crucial step: combine identities (1.3) and (1.4), that is, prove the combinatorial theorem described as step (iii) in the introduction.

First, a few basic notions are recalled. A color word is *primitive* if it is not equal to the power w_1^k of another word w_1 with $k\geq 2$. It is *standard* if it is less (with respect to the lexicographic order) than each of its cyclic rearrangements.

For instance, $w=23243$ is less than each of its cyclic rearrangements 32432, 24323, 43232 and 32324. It is then standard.

A standard word is always primitive. Each word w is the power of a unique primitive word denoted by $\sqrt{w}$ and referred to as the *root* of w. As before, Fw will designate the first letter of w.

The notion of standard word was introduced by Chen et al. [3] in the study of free differential calculus. Then, Schützenberger [9] developed a theory of free monoid factorizations, an example of which is the standard factorization described in Proposition 4.1 below. Finally, Viennot [10] made a systematic study of these factorizations and built up a unified theory of the basic commutator calculus. The proof of the next proposition can be found in Viennot ([10, p. 40, Lemma 1.11]).

Proposition 4.1. *Each non-empty color word w admits a unique factorization $(u_1, u_2, \ldots, u_t)$, called its lexicographic standard factorization, having the following three properties*

(i) $u_1, u_2, \ldots, u_t$ *are standard*;
(ii) $u_1 \geqslant u_2 \geqslant \cdots \geqslant u_t$ *(with respect to the lexicographic order)*;
(iii) $w = u_1 u_2 \cdots u_t$.

For instance, the standard factorization of 62324313113113113113 reads

$$(6, 23243, 13, 113, 113, 113, 113). \tag{4.1}$$

The notion of standard factorization is now extended to biwords. A colored I-biword

$$\binom{w}{\sigma} = \binom{x_1 x_2 \cdots x_m}{v_1 v_2 \cdots v_m}$$

is *bistandard*, if

(i) $w = u^k$ $(k \geqslant 1)$ and u is standard:
(ii) if $k \geqslant 2$ and $m = kl$, then

$$v_1 < \min\{v_{l+1}, v_{2l+1}, \ldots, v_{(k-1)l+1}\}.$$

For instance, the biword

$$\binom{w}{\sigma} = \begin{pmatrix} 1 & 1 & 3 & 1 & 1 & 3 & 1 & 1 & 3 \\ 16 & 6 & 8 & 19 & 5 & 9 & 17 & 4 & 2 \end{pmatrix}$$

is bistandard. With the above notations $u = 113$ is standard. Further $m = 9$, $k = 3$, $l = 3$ and $v_1 = 16 < \min\{v_4 = 19,\ v_7 = 17\}$.

Now a sequence

$$\binom{w_1}{\sigma_1}, \binom{w_2}{\sigma_2}, \ldots, \binom{w_q}{\sigma_q}$$

of colored biwords is said to be a *bistandard factorization* of a colored m-biword

$\binom{w}{\sigma}$ if the following properties hold

(i) each biword $\binom{w_i}{\sigma_i}$ is bistandard;

(ii) for $1 \leqslant i \leqslant q-1$ either $\sqrt{w_i} > \sqrt{w_{i+1}}$, or $\sqrt{w_i} = \sqrt{w_{i+1}}$ and $F\sigma_i > F\sigma_{i+1}$;

(iii) $\binom{w}{\sigma} = \binom{w_1}{\sigma_1}\binom{w_2}{\sigma_2} \cdots \binom{w_q}{\sigma_q}$.

In the next example it can be verified that the factorization of $\binom{w}{\sigma}$ determined by the vertical bars is a bistandard factorization.

$$\binom{w}{\sigma} = \left(\begin{array}{c|ccccc|cc|ccccccccc|ccc} 6 & 1 & 3 & 2 & 4 & 3 & 1 & 3 & 1 & 1 & 3 & 1 & 1 & 3 & 1 & 1 & 3 & 1 & 1 & 3 \\ 3 & 12 & 15 & 20 & 11 & 13 & 14 & 10 & 17 & 6 & 8 & 19 & 5 & 9 & 18 & 4 & 2 & 10 & 1 & 7 \end{array}\right). \tag{4.2}$$

The procedure described next will yield a bistandard factorization of each colored biword.

Procedure. (i) let $\binom{w}{\sigma}$ be a colored m-biword and $(u_1, u_2, \ldots, u_t)$ be the standard factorization of w;

(ii) cut the biword $\binom{w}{\sigma}$ at the end of each factor $u_1, u_2, \ldots, u_t$; this determines a factorization $(\tau_1, \tau_2, \ldots, \tau_t)$ of the second row σ of the biword, namely

$$\begin{array}{c|c|c|c|} u_1 & u_2 & \cdots & u_t \\ \tau_1 & \tau_2 & \cdots & \tau_t \end{array}. \tag{6.3}$$

(iii) For each $j = 1, 2, \ldots, t-1$ remove the jth vertical bar (standing between u_j and u_{j+1}) whenever the following two conditions hold:

(a) $u_{i-1} > u_i = u_{i+1} = \cdots = u_{j-1} = u_j$ for some i with $1 \leqslant i \leqslant j$ (the inequality $u_{i-1} > u_i$ being discarded when $i = 1$);

(b) $F\tau_{j+1}$ is *not* the minimum of the set $\{F\tau_i, F\tau_{i+1}, \ldots, F\tau_j, F\tau_{j+1}\}$.

(iv) The remaining vertical bars determine a bistandard factorization of $\binom{w}{\sigma}$.

Take again example (4.2) and disregard the vertical bars. By (4.1) its (4.3)-factorization reads

$$\begin{array}{c|ccccc|cc|ccc|ccc|ccc|ccc|} u_1 & & & u_2 & & & u_3 & & & u_4 & & & u_5 & & & u_6 & & & u_7 & \\ 6 & 2 & 3 & 2 & 4 & 3 & 1 & 3 & 1 & 1 & 3 & 1 & 1 & 3 & 1 & 1 & 3 & 1 & 1 & 3 \\ 3 & 12 & 15 & 20 & 11 & 13 & 14 & 10 & 17 & 6 & 8 & 19 & 5 & 9 & 18 & 4 & 2 & 16 & 1 & 7 \\ \tau_1 & & & \tau_2 & & & \tau_3 & & & \tau_4 & & & \tau_5 & & & \tau_6 & & & \tau_7 & \end{array}.$$

Then apply step (iii) of the above procedure. The first bar remains for $u_1 = 6 > 23243 = u_2$. Also the second and the third ones. Next $u_3 > u_4 = u_5$ and $17 = F\tau_4 < F\tau_5 = 19$. Thus the 4th bar is removed. Also, as $u_3 > u_4 = u_5 = u_6$ and $17 = F\tau_4 < F\tau_6 = 18$, the 5th bar is removed. But as $16 = F\tau_7$ is minimum among $F\tau_4$, $F\tau_5$, $F\tau_6$, $F\tau_7$, the 6th bar is not removed. This yields the factorization shown in (4.2).

Proposition 4.2. *Each colored m-biword* $\binom{w}{\sigma}$ *has a unique bistandard factorization, that can be obtained by applying to* $\binom{w}{\sigma}$ *the above procedure.*

Proof. Let $(u_1, u_2, \ldots, u_t)$ be the standard factorization of w. Each bistandard factorization

$$\binom{w_1}{\sigma_1}, \binom{w_2}{\sigma_2}, \ldots, \binom{w_q}{\sigma_q}$$

of $\binom{w}{\sigma}$ is such that $w_1 = u_1'^{k_1}$, $w_2 = u_2'^{k_2}, \ldots, w_q = u_q'^{k_q}$ with $u_1', u_2', \ldots, u_q'$ standard and $u_1' \geqslant u_2' \geqslant \cdots \geqslant u_q'$. As the standard factorization of w is unique, the sequence $(u_1', \ldots, u_1', u_2', \ldots, u_2', \ldots, u_q', \ldots, u_q')$ where each word u_i' is repeated k_i times, must be equal to $(u_1, u_2, \ldots, u_t)$. thus each w_i is a product of *equal* successive factors u_j.

Suppose now that $\binom{w}{\sigma}$ has two different bistandard factorizations

$$\binom{w_1}{\sigma_1}, \binom{w_2}{\sigma_2}, \ldots, \binom{w_q}{\sigma_q} \quad \text{and} \quad \binom{w_q'}{\sigma_q'}, \binom{w_2'}{\sigma_q'}, \ldots, \binom{w_{q'}'}{\sigma_{\sigma'}}.$$

Let i be the smallest integer with $\binom{w_i}{\sigma_i} \neq \binom{w_i'}{\sigma_i'}$. Assume that w_i is shorter than w_i'. Then $w_i = u^{p_i}$ and $w_i' = u^{p_i'}$ with u standard and $p_i < p_i'$. Also $w_{i+1} = u^{p_{i+1}}$ for some $p_{i+1} \geqslant 1$ and the same u because $p_i < p_i'$. Therefore, $F\sigma_i > F\sigma_{i+1}$. Also $F\sigma_i = F\sigma_i' < F\sigma_{i+1}$, because $\binom{w_i'}{\sigma_i'}$ is bistandard. This is a contradiction. Thus there exists at most one bistandard factorization.

Finally, it is straightforward to verify that the above procedure does yield a bistandard factorization.

Corollary 4.3. *Each colored I-cycle contains one and only one bistandard word.*

Proof. Let

$$\binom{w}{\sigma} = \binom{x_1 x_2 \cdots x_m}{v_1 v_2 \cdots v_m}$$

be a colored I-biword and u' be the root of w. Then there exist u' primitive and positive integers, k, l with $w = u'^k$ and $m = kl$. The rearrangement class of u' contains, by definition, a unique standard word u. Thus $\binom{w}{\sigma}$ is equal to the juxtaposition product

$$\binom{u_0'}{\sigma_0'}\binom{u_1}{\sigma_1}\binom{u_2}{\sigma_2} \cdots \binom{u_{k-1}}{\sigma_{k-1}}\binom{u_k'}{\sigma_k'}$$

where $u_1 = u_2 = \cdots = u_{k-1} = u_k' u_0' = u$. Let h be the integer defined by

$$F\sigma_h = \min\{F\sigma_1, F\sigma_2, \ldots, F\sigma_{k-1}, F\sigma_k'\}. \tag{4.4}$$

Then the product

$$\binom{u_h}{\sigma_h} \cdots \binom{u_{k-1}}{\sigma_{k-1}}\binom{u_k' u_0'}{\sigma_k' \sigma_0'}\binom{u_1}{\sigma_1} \cdots \binom{u_{h-1}}{\sigma_{h-1}}$$

is bistandard. Because of property (4.4) it is the only one in the cyclic rearrangement class.

Let $\binom{w}{\sigma}$ be a bistandard biword. Then $\binom{w}{\sigma} \to \left[{w \atop \sigma}\right]$ is a bijection onto the set of colored cycles. The inverse bijection will be denoted by

$$c \to \mathrm{BIS}(c).$$

The main theorem of this paper can now be stated.

Theorem 4.4. *Let* $\binom{w}{\sigma}$ *be a colored m-biword with bistandard factorization*

$$\binom{w_1}{\sigma_1}, \binom{w_2}{\sigma_2}, \ldots, \binom{w_q}{\sigma_q}.$$

Then, the mapping that sends $\binom{w}{\sigma}$ to the unordered collection of colored cycles

$$p = \left\{\left[{w_1 \atop \sigma_1}\right], \left[{w_2 \atop \sigma_2}\right], \ldots, \left[{w_q \atop \sigma_q}\right]\right\}.$$

is a bijection of $X^m \times \mathfrak{S}_m$ *onto* P_m *with the property that*

$$\beta_{\mathrm{dec}}(w) = \beta(p).$$

Proof. The bijectivity property follows from Proposition 4.2 and its corollary. Only the last property is to be verified. The function β_{dec} was defined in (2.3). Let $(d_1, d_2, \ldots, d_s)$ be the decreasing factorization of w. then

$$\beta_{\mathrm{dec}}(w) = \beta(d_1)\beta(d_2) \cdots \beta(d_s).$$

Also, by (2.2) and (3.11)

$$\beta(p) = \beta(w_1)\beta(w_2) \cdots \beta(w_q).$$

Accordingly, it suffices to verify that each factor w_i of the bistandard factorization is a product $d_j d_{j+1} \cdots d_{j+r}$ $(1 \leqslant j \leqslant j+r \leqslant s)$ and

$$\beta(w_i) = \beta(d_j)\beta(d_{j+1}) \cdots \beta(d_{j+r}).$$

But each w_i is a power of a standard word u, say, $w_i = u^k$, and $\beta(w_i) = \beta(u)^k$. Let $(u_1, u_2, \ldots, u_t)$ be the standard factorization of w. It remains to verify that

$$\beta(d_1)\beta(d_2) \cdots \beta(d_s) = \beta(u_1)\beta(u_2) \cdots \beta(u_t).$$

As $u_1 \geqslant u_2 \geqslant \cdots \geqslant u_t$, then $Fu_1 \geqslant Fu_2 \geqslant \cdots \geqslant Fu_t$. Hence, each u_i is a product of factors of the decreasing factorization. Write

$$u_i = d_j d_{j+1} \cdots d_{j+r}.$$

Then $Fu_i = Fd_j \geqslant Fd_{j+1} \geqslant \cdots \geqslant Fd_{j+r}$.

As u_i is standard, none of the letters $Fd_{j+1}, \ldots, Fd_{j+r}$ can be smaller than Fu_i.

Hence, $Fd_j = Fd_{j+1} = \cdots = Fd_{j+r}$, and, accordingly,

$$\beta(u_i) = \beta(d_j d_{j+1} \cdots d_{j+r}) = \beta(d_j)\beta(d_{j+1}) \cdots \beta(d_{j+r}).$$

Remark 4.5. The inverse bijection $p \to \binom{w}{\sigma}$ is obtained

(i) by decomposing the colored m-permutation p into its colored cycles

$$\{c_1, c_2, \ldots, c_r\};$$

(ii) taking for each c_i the unique bistandard biword BIS $c_i = \binom{w_i}{\sigma_i}$ contained in c_i (Corollary 4.3);

(iii) then $\binom{w}{\sigma}$ is the juxtaposition product of the $\binom{w_i}{\sigma_i}$'s written in *decreasing order* $(\binom{w_i}{\sigma_i} > \binom{w_j}{\sigma_j}$ if and only if either $w_i > w_j$, or $w_i = w_j$ and $F\sigma_i > F\sigma_j)$.

For instance the colored m-biword (4.2) corresponds to the colored m-permutation of Fig. 1. Furthermore, the monomial $\beta(p)$ that counted the color adjacencies in p (see (3.3)) is precisely equal to $\beta_{\text{dec}}(w)$ determined in (2.4).

Acknowledgements

The author is grateful to Professor Garsia for the enthusiasm he showed during the preparation of [4], the paper that originated the present one, and especially for the content of Lemma 3.1. The author is also thankful to Professor Srivastava for his masterly editorial work.

References

[1] P. Cartier, La série génératrice exponentielle, Applications probabilistes et algébriques, Publ. I.R.M.A. Univ. Strasbourg (1971–1972) 68 pp.

[2] P. Cartier and D. Foata, Problèmes combinatoires de commutation et réarrangements, Lecture Notes in Math., 85 (Springer-Verlag, Berlin, Heidelberg, New York, 1969).

[3] K.T. Chen, R.H. Fox and R.C. Lyndon, Free differential calculus, IV. The quotient groups of the lower central series, Ann. of Math. 68 (1958) 81–95.

[4] D. Foata and A.M. Garsia, A combinatorial approach to the Mehler formulas for Hermite polynomials, in: Proceedings of Symposia in Pure Mathematics, Vol. 34 (Relations between Combinatorics and Other Parts of Mathematics) (Am. Math. Soc., Providence, RI, 1979) 163–179.

[5] D. Foata and M.P. Schützenberger, Théorie géométrique des polynômes eulériens, Lecture Notes in Math. 138 (Springer-Verlag, Berlin, Heidelberg, New York, 1970).

[6] D.M. Jackson, The combinatorial interpretation of the Jacobi identity from Lie algebras, J. Combin. Theory 23 (A) (1977) 233–256.

[7] P.A. MacMahon, Combinatory Analysis, Vol. 1 (Cambridge Univ. Press, Cambridge, 1915) (reprinted: Chelsea Publ. Co., New York, 1960).

[8] W. Miller Jr., Symmetry Groups and Their Applications (Academic Press, New York, London, 1972).

[9] M.P. Schützenberger, On a factorization of free monoids, Proc. Am. Math. Soc. 16 (1965) 21–24.

[10] G. Viennot, Algébres de Lie libres et monoïdes libres, Lecture Notes in Math., 691 (Springer-Verlag, Berlin, Heidelberg, New York, 1978).

Annals of Discrete Mathematics 6 (1980) 137–156

FROM NIM TO GO*

Aviezri S. FRAENKEL

Department of Applied Mathematics, The Weizmann Institute of Science, Rehovot, Israel

Nim, checkers, chess and Go belong to the same family of 2-player full information 0–1 games without chance moves. Yet the strategy of Nim is very simple, that of chess seemingly very hard. Why?? Some of the reasons for this are explored and new theories are demonstrated by means of sample games of increasing complexity which span a bridge from Nim to Go.

1. Nim-like games

Nim is of course a well-known game. Fig. 1 is an example with three piles: a pile with one token, a pile with two and a pile with three tokens. Two players alternately select one pile and remove from it any positive number of tokens. The player first unable to move loses, and his opponent wins. The theory of the game is very simple, though perhaps unexpected: Write down the number of tokens in each pile in binary, and sum without carry (vector addition over GF(2)): if the resulting *Nim-sum* $\sigma = 0$, then the second player can win, otherwise the first player can win. For the above example $\sigma = 0$ (Table 1). So it pays to be polite: offer your opponent to make the first move!

Nim is a member of the following family of games: Two-player, perfect information (unlike some card games where information is hidden) without chance moves (no dice) and outcome lose, win or tie only. This family is loosely known under the name *combinatorial games*, for short *games* in the sequel. We shall further assume (except at the end of the talk) that the player first unable to move loses and his opponent wins. This subset is called the subset of Last-Player-Win (LPW) games.

Incidentally, note that the class of combinatorial games is a very respectable family. No game played in Las Vegas — and more recently in Atlantic City — taints this family! Furthermore, well-known games such as checkers, chess and Go are proud members of this family. Though both Nim and chess belong to it, Nim has a very simple strategy, whereas for chess not even Bobby Fisher, Anatoly Karpov or Viktor Korchnoi *know* whether White has an opening move which guarantees a win.

What are the reasons for this discrepancy? This question motivates our investigations of combinatorial games.

* This expanded form of the talk was prepared while the author visited at the Department of Computer Science, University of Illinois, Urbana, IL, during the academic year 1978–1979.

Fig. 1. A game of Nim with three piles.

Table 1. The Nim-sum σ of the game in Fig. 1 is 0.

	1	1
	1	0
	0	1
$\sigma=$	0	0

We start with some notation for games without ties.

Let N be the set of all positions from which the Next (first) player can win (of course: irrespective of the moves of his opponent). Let P be the set of all positions such that the Previous (second) player can win (irrespective of the moves of his opponent).

With each game we associate its *game-graph* $R=(V, E)$, which is a finite acyclic digraph, where V is the set of game positions and $(u, v) \in E$ if and only if there is a move from u to v in the game. See Figs. 2 and 3.

There is a simple algorithm for determining the N, P-pattern of any game-graph R. Define the set of *followers* of $u \in V$ by $F_R(u)=F(u)=\{v \in V:(u, v) \in E\}$. The above definitions of N, P imply that $u \in P$ if and only if $F(u) \subseteq N$, whereas $u \in N$ if and only if $F(u) \cap P \neq \emptyset$. Thus for every LPW-game all *sinks* (vertices u with $F(u)=\emptyset$) are labeled P. Working backwards, one gets the N, P labels depicted in Figs. 2 and 3 in $O(|E|)$ steps. Since chess is a finite game, its game-graph can be constructed and its N, P, T-pattern can be computed, where T denotes Tie. (We shall see later that the determination of the T-positions is also easy.) Hence our first conclusion is:

chess is just as easy as Nim.

Suppose that two players are given n disjoint combinatorial games. Each player at

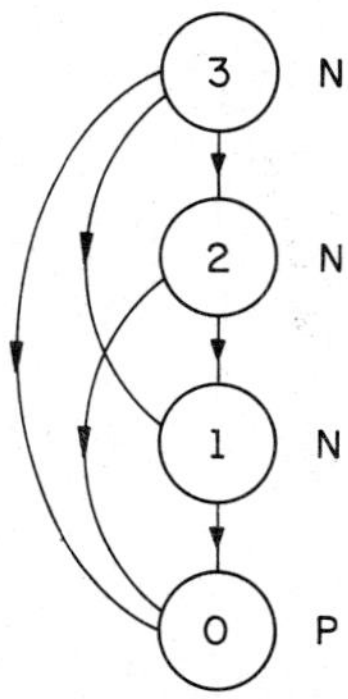

Fig. 2. Game-graph of one Nim-pile. The numbers inside the vertices denote the pile size.

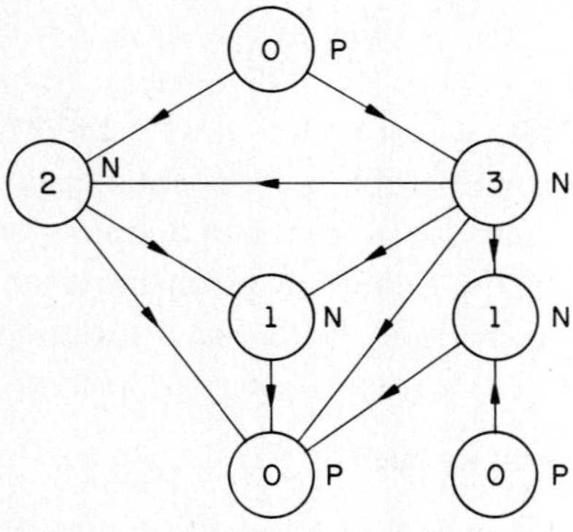

Fig. 3. Random game-graph.

his turn selects a game and makes a move in it. This game is called the *disjunctive compound* or *sum* of the n component games.

Now put a number of tokens on some vertices of Fig. 3. Each player at his turn selectes a token and moves it, along a directed edge, to a neighboring vertex (occupied or not). In particular, vertices may be multiply occupied. This is clearly a disjunctive compound. The disjunctive compound is a *Nim-like game* if the digraph on which it is played is acyclic and each vertex can be multiply occupied without interaction among tokens. Consistent with our agreement, the player first unable to move loses and the other wins.

If we place three tokens on the vertices labeled 1, 2, 3 of Fig. 2 and play according to the rules employed for Fig. 3, this disjunctive compound is nothing but the three-pile Nim game described at the beginning! Also chess can be considered as a sort of compound game, where several tokens are placed on the 8×8 board digraph.

Put a token on each of the two vertices labeled 1 of Fig. 3, both of which are N-positions. This is a P-position of the disjunctive compound, because the second player can clearly win in one move. But a token on a vertex labeled 1 and on the vertex labeled 2, both of which are also N-positions, constitutes an N-position of the disjunctive compound: the first player can obviously win by making the move $2\to 1$. Thus two N's may sometimes be a P and sometimes an N-position of the disjunctive compound, and so the N, P tool is not strong enough to settle disjunctive compounds.

Assume that $R_i=(V_i, E_i)$ are n digraphs on which a disjunctive compound is played. Even if $R_1=\cdots=R_n$, this is not a game-graph in general, but rather a digraph on which a game is played, as we just did on the digraphs of Figs 2 and 3. We may assume that there is precisely one token on each component graph. The game-graph $\bar{R}=(\bar{V}, \bar{E})$ of the disjunctive compound is the digraph with vertices $\bar{u}=(u_1,\ldots,u_n)$, where $u_i\in V_i$ is the occupied vertex of R_i $(i=1,\ldots,n)$. Let $\bar{v}=(v_1,\ldots,v_n)\in\bar{V}$. Then $(\bar{u},\bar{v})\in\bar{E}$ if and only if there exists $1\leqslant k\leqslant n$ such that $v_k\in F(u_k)$ and $v_i=u_i$ for all $i\neq k$.

For Nim where each of the n piles has at most $k=n$ tokens, the disjunctive compound game-graph has $\binom{n+k}{k}=\binom{2n}{n}\geqslant 2^n$ vertices. For chess on an $n\times n$ board with $O(\frac{1}{2}n^2)$ distinct pieces, we get $O(2^{n^2})$ vertices. Thus for both games the number of vertices of the game-graph is exponential in n, implying an exponential strategy. If the best strategy for a game is nonpolynomial, we call the game *intractable*, consistent with the definition of an intractable problem [1, 38]. The strategy of an intractable game is normally just a variation of an exhaustive search through all game positions. Because of this, our second conclusion is:

Nim is just as hard as chess.

Recall, however, that we indicated above a simple strategy for Nim. The tool which provides a strategy for Nim and Nim-like games is the *classical Sprague–Grundy function* g. Let S be a finite set of nonnegative integers, $\bar{S}$ its complement. Define mex $S=\min \bar{S}=$ least nonnegative integer not in S. Then

$$g(u)=\text{mex }g(F(u)).$$

For every finite acyclic digraph, g exists uniquely. See e.g. Berge [6]. Since $\text{mex}(\emptyset)=0$, all sinks have value $g=0$. Working backwards from the sinks, the g-values given inside the vertices of Fig. 3 are obtained.

For $\bar{u}=(u_1,\ldots,u_n)$, let

$$\sigma(\bar{u})=\sum_{i=1}^{n}{}' g(u_i),$$

where the dash on $\sum$ indicates a Nim-sum, rather than an ordinary sum. We claim that for the disjunctive compound of Nim-like LPW-games,

$$N=\{\bar{u}\in\bar{V}:\sigma(\bar{u})>0\},\qquad P=\{\bar{u}\in\bar{V}:\sigma(\bar{u})=0\}.$$

The strategy of Nim is consistent with this claim, because Fig. 2 implies that $g(k)=k$, where k is the size of the Nim-pile. Thus the rule given at the beginning of this talk is a special case of this claim.

For proving the claim, suppose that $g(u_1),\ldots,g(u_n)$ and $\sigma(\bar{u})=0$ are as given in Table 2. Any move from the position $\bar{u}=(u_1,\ldots,u_n)$ involves moving in some u_i, say from u_1 to u_1'. By the definition of g, $g(u_1')\neq g(u_1)$, hence in at least one column of Table 2 the parity of the number of bits is changed, and so $\sigma(\bar{u}')>0$, where $\bar{u}'\in F_{\bar{R}}(\bar{u})$.

Table 2. Nim-sum 0

$g(u_1)=0$	1	0	1	$\cdots$	1
$g(u_2)=1$	0	0	1	$\cdots$	0
.					
.					
.					
$g(u_n)=1$	1	0	0	$\cdots$	1
$\sigma(\bar{u})=0$	0	0	0	$\cdots$	0

Table 3. Positive Nim-sum

$g(u_1)=0$	0	0	0	$\cdots$	1
$g(u_2)=1$	0	0	1	$\cdots$	0
$\vdots$					
$g(u_n)=1$	1	0	0	$\cdots$	1
$\sigma(\bar{u})=0$	1	0	1	$\cdots$	0

If $\sigma(\bar{u})>0$, there exists a leftmost column c_0 which contains a 1-bit of $\sigma(\bar{u})$ (second column from left in Table 3). Hence in c_0 there is an odd number of 1's, so at least one. Without loss of generality, assume that this occurs in $g(u_n)$. So put the c_0th bit of $g(u_n)$ to 0, and reverse the parity of the other bits of $g(u_n)$ in those columns in which $\sigma(\bar{u})$ has a 1-bit. This results in a value t which clearly satisfies:

(i) $0 \leqslant t < g(u_n)$;

(ii) $t \oplus \sum_{i=1}^{n-1} g(u_i) = 0$, where $\oplus$ denotes Nim-sum.

Moreover, there is a move from u_n to u'_n such that $g(u'_n)=t$, because $g(u_n)$ is the mex of the g-values of its followers, and so for *every* $0 \leqslant t < g(u_n)$ there is some $u'_n \in F(u_n)$ such that $g(u'_n)=t$.

Thus the player in a position $\bar{u}_0$ with $\sigma(\bar{u}_0)>0$ can always move to some $\bar{u}_1 \in F(\bar{u}_0)$ with $\sigma(\bar{u}_1)=0$. If there is any move left, his opponent necessarily moves to some $\bar{u}_2 \in F(\bar{u}_1)$ with $\sigma(\bar{u}_2)>0$. Since $\bar{R}$ is acyclic, the player playing from $\bar{u}_0$ reaches some sink $\bar{u}_p$ of $\bar{R}$ with $\sigma(\bar{u}_p)=0$ in $O(mn)$ steps (where $\max_i |V_i| = m$), proving our claim.

Also note that the g-values of a digraph with m vertices can be computed in $O(|E|\,|V|)=O(m^3)$ steps, and thus the strategy is polynomial. (A strategy is called *polynomial* if it can be computed in $O(m^k)$ steps, where m is the size of the game, such as the number of vertices of the graph on which it is played (*not* the game-graph, which may be exponentially larger!), and k is some positive constant). Let us call a game *tractable* if it has a polynomial strategy, otherwise it is intractable. The main conclusion of the "From Nim" part of the talk is:

Nim-like games are tractable.

Remark. Computing a strategy of play for a given position u usually consists of two steps:

(i) Finding whether u is in N, P or T.

(ii) If u is in N or in T, computing the next move from u.

When we say that a strategy is tractable, we mean that both steps are polynomial. When we say that a strategy is intractable, we mean that either of the two steps is not polynomial. We like to point out already now, that the complexity results which will be mentioned in the last part of the talk — which provide *evidence* for intractability and one actually *proves* intractability — are all assertions about the complexity of step (i).

2. Extensions in various directions

The above theory was already known in the 30's (see [40, 68]). Why don't games like checkers, chess and Go submit to this theory! Below we mention some — though not all — reasons for this.

(I) *Dynamic tie positions.* The more sophisticated games, even checkers, may contain *dynamic tie positions.* These are non-sink positions from which no player can force a win, and therefore both can avoid losing. The set of such positions will be denoted by T. It is thus necessary to analyze (dynamic) ties in the family of combinatorial games. T-positions can occur only if the game-graph contains cycles. For digraphs with cycles, g behaves, unfortunately, very erratically:

(a) It may exist uniquely and determine the game's strategy, exactly as for Nim-like games (Fig. 4a).

(b) It may exist uniquely but not determine the game's strategy. In Fig. 4b the unique g-values are depicted inside the vertices. Since the digraph has no sinks, every position is a T-position, and so the g-values do not determine the strategy.

(c) It may exist non-uniquely, in which case it obviously does not determine a

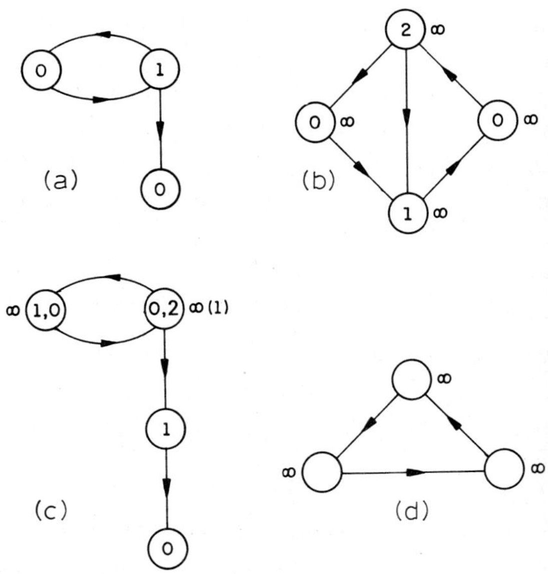

Fig. 4. a: g exists uniquely and determines the strategy. b: g exists uniquely but does not determine the strategy. c: g exists non-uniquely and does not determine the strategy. d: g does not exist.

strategy. In Fig. 4c, the two vertices in the circuit can assume two values, as indicated inside the vertices.

(d) It may not exist at all (Fig. 4d).

Bad as all this sounds, a polynomial strategy can be recovered for this case, by defining a *Generalized Sprague–Grundy* (*GSG*) *function* G which exists uniquely for every finite digraph $R=(V,E)$ (see [64, 28, 33]). Let J^0 denote the set of nonnegative integers, $V^f(R)$ the set of all vertices in V on which G is finite. A function $G: V^f \to J^0 \cup \{\infty\}$ is a GSG-function with counter function $c: V^f \to J^0$, if the following conditions are satisfied:

(**A**) If u is a vertex with finite G, then $G(u) = \text{mex } G(F(u))$.

(**B**) If v is a follower of u with larger G-value than u, then there exists a follower w of v satisfying $G(w)=G(u)$ and $c(w)<c(u)$.

(**C**) If u has value $G(u)=\infty$, then u has a follower v with value $G(v)=\infty$, such that mex $G(F(u)) \notin G(F(v))$.

Using this definition, the G-values depicted in Fig. 4 are obtained. (They are marked outside the vertices where $G \neq g$.)

If $G(u)=\infty$, we also use the notation $G(u)=\infty(K)$, where $K=\{G(v)<\infty: v \in F(u)\}$.

For handling disjunctive compounds with possible ties, a "generalized Nim-sum" is defined as an ordinary Nim-sum, but with two additional provisions:

$$a \oplus \infty(K) = \infty(K) \oplus a = \infty(K \oplus a),$$

where $K \oplus a = \{k \oplus a : k \in K\}$ (a any nonnegative integer), and $\infty(K) \oplus \infty(L) = \infty(\emptyset)$. Then for LPW-games,

$$N = \{\bar{u} \in \bar{V} : 0 < G(\bar{u}) < \infty\} \cup \{\bar{u} \in \bar{V} : G(\bar{u}) = \infty(K), 0 \in K\},$$

$$P = \{\bar{u} \in \bar{V} : G(\bar{u}) = 0\}, \qquad T = \{\bar{u} \in \bar{V} : G(\bar{u}) = \infty(K), 0 \notin K\}.$$

For example, a single token on the vertex with G-value 1 in Fig. 4a is an N-position. The same holds if we adjoin to this token one on the vertex with G-value $\infty(1)$ (Fig. 4c), since $\infty(1) \oplus 1 = \infty(0)$. (The winning move is $\infty(1) \to 1$.) A single token on a vertex with G-value 0 (Fig. 4a) is a P-position. Adjoining to this a token on $\infty(1)$ is a T-position, since $\infty(1) \oplus 0 = \infty(1)$. If another token on any vertex with G-value ∞ is adjoined, the position is still a tie.

Consider the game displayed in Fig. 5. An initial condition might be tokens on vertices A, B, C, D. Each player at his turn selects a token and moves it to an adjacent vertex (occupied or unoccupied) along a directed edge. The player first unable to move is the loser, his opponent the winner. If no player can force a last move, the outcome is declared a tie. It turns out that the given initial position is an N-position. But there is only one way to win: $D \to E$. For any other move of the first player, the second player can secure a win or at least force a tie. The analysis given in [14, Ch. 11] or the above definitions of the GSG-function and Nim-sum enable verifying this.

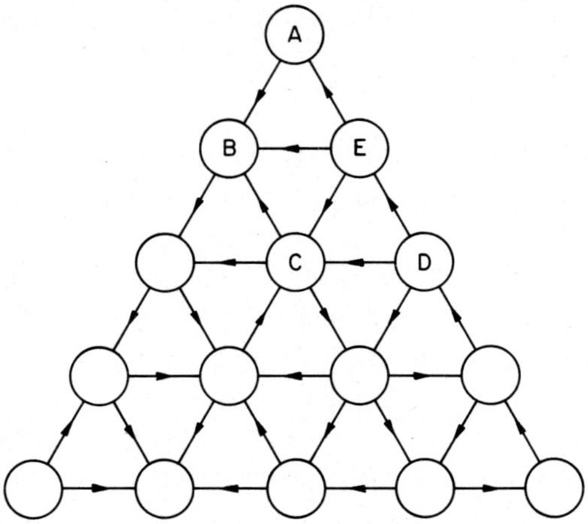

Fig. 5. Learn to beat Bobby Fisher.

Having learned to beat Bobby Fisher, we can now proceed to beat also Anatoly Karpov, the present chess champion, by determining the nature of the position A, B, C, D, E in Fig. 6. The names of these games were suggested by John Conway's question how long it would take Bobby Fisher, or any chess champion, to find the winning moves.

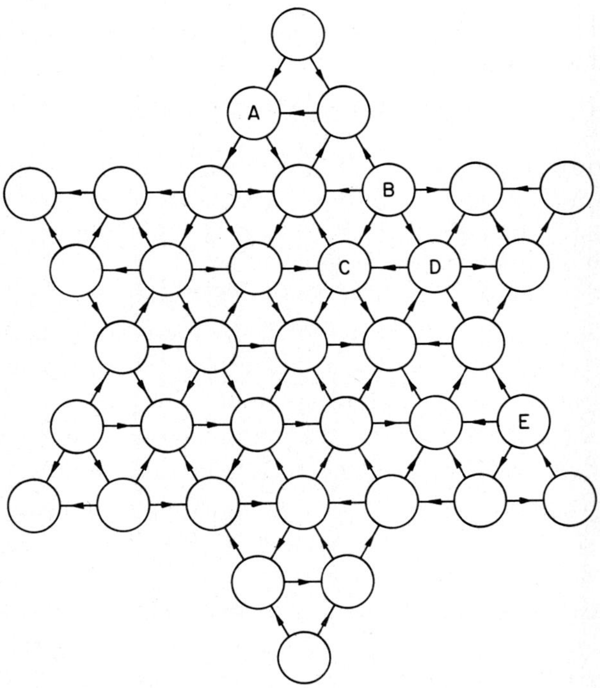

Fig. 6. Now beat Anatoly Karpov.

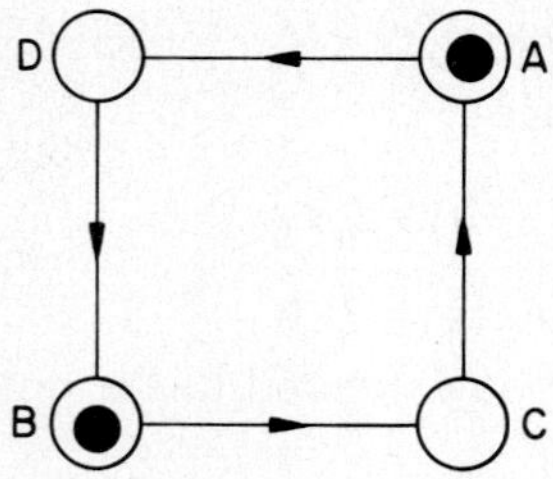

Fig. 7. A Tie-position in a non-annihilation game, which is a P-position in the annihilation-version.

(II) *Capture rules.* The second reason why chess-like games are not crackable by Nim-tools is due to interactions among tokens, such as various capture rules. A particular case is that of *annihilation games,* suggested by John Conway. These are similar to the disjunctive compounds of Nim-like games, except that all tokens are on one digraph, and:

(i) At most one token resides on any vertex.

(ii) Two tokens meeting on a vertex are annihilated.

The outcome of an annihilation game may be quite different from that of a Nim-type game. For example, consider a position consisting of two tokens on vertices A and B in Fig. 7. Since there is no sink, this is clearly a T-position in the Nim-like game in which there is no interaction between tokens. But as an LPW annihilation game it is a P-position: the first player moves, say, $B \to C$, and then the second player wins by the annihilation move $C \to A$. However, the position (A, B) in Fig. 8 is a T-position for both versions of play.

Let $V = V(R) = (z_1, \ldots, z_n)$ be any ordering of $V(R)$, where $R = (V, E)$ with $|V| = n$ is the digraph on which our annihilation game is played. Then each position in the game-graph $\bar{R} = C(R)$ of the annihilation game (called the *contrajunctive compound* of R) is defined by $\bar{u} = (u^1, \ldots, u^n)$, where $u^i = 1$ if there is a token on z_i, $u^i = 0$ otherwise. The game-graph thus contains 2^n vertices, suggesting an exponential strategy.

It turns out, however, that a "linear space" polynomial strategy can be recovered for contrajunctive compounds by noticing the following facts:

(i) $V^f(\bar{R})$ is a linear subspace of $V(\bar{R})$ under $\oplus$.

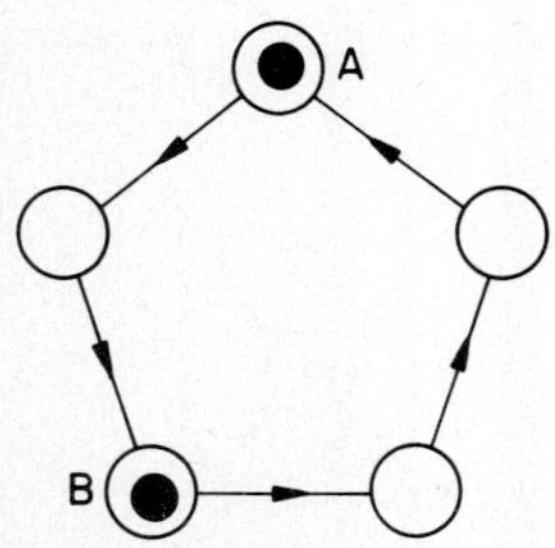

Fig. 8. A tie-position in both the annihilation and the non-annihilation game.

(ii) G is a homomorphism of $V^f(\bar{R}) \xrightarrow{\text{onto}} \mathrm{GF}(2)^t$ for suitable $t \geq 0$, with kernel $V_0(\bar{R})$ and quotient space

$$\{V_i(\bar{R}): 0 \leq i < 2^t\} = V^f(\bar{R})/V_0(\bar{R}).$$

(*Definition:* $V_i(\bar{R}) = \{\bar{u} \in V(\bar{R}): G(\bar{u}) = i\}$.)

(iii) The homomorphism can be extended to the entire space $V(\bar{R})$ to yield the ∞-values of G.

The resulting computation of the G-function on $\bar{R}$ is bounded by $O(n^6)$.

Define a strategy in the *narrow sense* to be a strategy which depends only on the *current position.* A strategy in the *wide sense* depends on possible previous positions as well. Clearly, every combinatorial game has a strategy in the narrow sense (see Smith [64]).

In Nim-like games, the computation of the N, P-pattern implies an immediate polynomial playing strategy in the narrow sense: The winner goes from his present N-position to *any* P-follower, and the loser has to go to an N-follower. After a finite number of steps, the winner reaches a sink. The situation is similar in games with possible tie-positions without annihilation. However, the winner cannot just go to *any* P-follower, but rather has to go to a P-follower with smallest counter function value c in order to avoid cycling indefinitely.

The existence of a polynomial algorithm for computing the N, P, T-pattern for a contrajunctive compound does not imply a polynomial strategy of play, however, because of possible cycling. It turns out that a polynomial strategy in the *wide sense*, bounded by $O(n^5)$ steps, can be formulated. It is not yet known whether there exists a strategy in the *narrow sense* which is polynomial in n (see [31, 34]).

As an example, consider the Innocent Marble game (Fig. 9). Put an even number of indistinguishable "marbles" on distinct vertices of each of the five components. Gill and Jean play by alternately selecting a marble and moving it to a neighboring vertex, say u, along a directed edge. If u was occupied, both

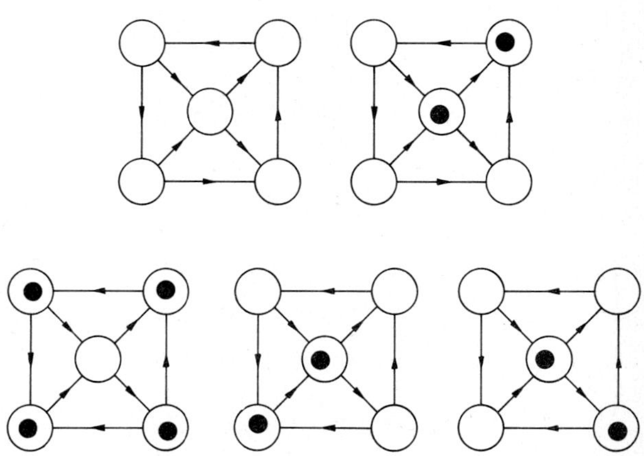

Fig. 9. The Innocent Marble game.

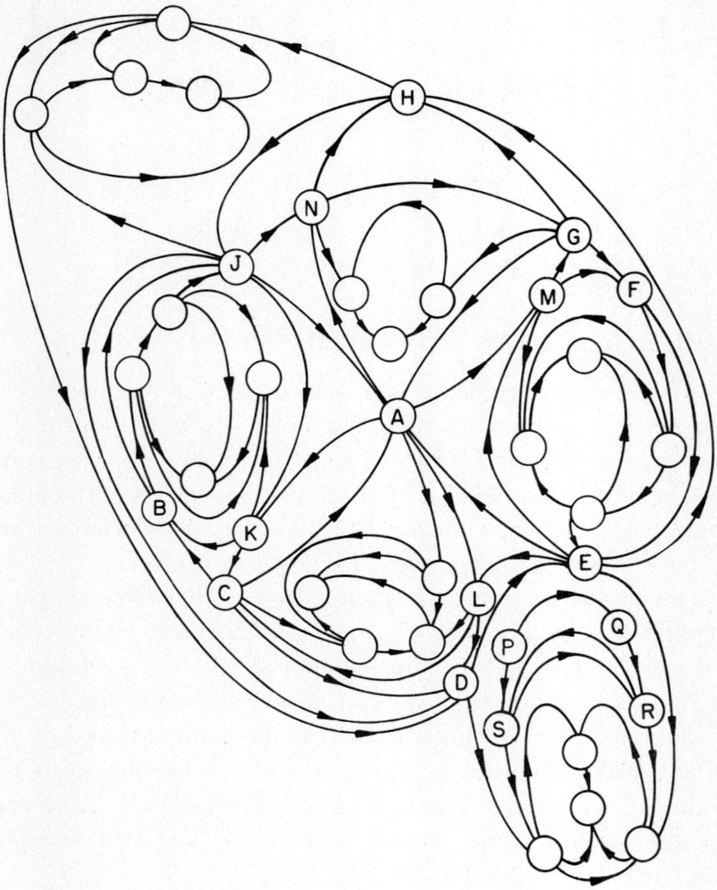

Fig. 10. Worlds in collision.

marbles go out of the game. The player first unable to move is the loser, his opponent the winner who gets all the marbles. If there is no last move, the outcome is a tie.

It turns out that the position depicted in Fig. 9 is an N-position, and there is a unique winning move (which is it?). After mastering this game, we might like to tackle Worlds in Collision (Fig. 10), in which "worlds" or "stars" are placed initially in certain distinct "space stations" (vertices). Each of two "superspacemen" can streak a star from a station to a neighboring station along a directed edge. Two colliding stars explode and vanish in a cloud of interstellar dust. In contrast to marbles, which get removed only in pairs, stars can disappear also singly in the "black holes" (sinks), where they vanish ("falling stars"), since they cannot reemerge from a black hole to streak again across the firmament!

A variation is the Particles and Antiparticles game played on the five orbits of Fig. 11. Place i electrons (−) and i positions (+) on distinct vertices of each orbit

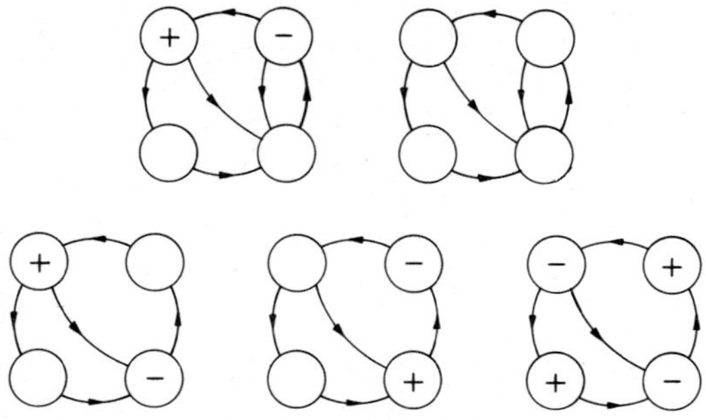

Fig. 11. The particles and antiparticles game.

($i = 0$, 1 or 2). A typical initial arrangement is given in the figure. Two physicists play the game by moving alternately. Each physicist at his turn selects a particle and moves it to an adjacent vertex u along a directed edge, provided that u is either unoccupied or occupied by a particle of opposite type. In the latter case, of course, both particles get annihilated, as physicists tell us particles and antiparticles do. The lose–win-tie outcomes are defined as usual.

There is an easy strategy for this game played on the given simple graph, or on other "simple" graphs. For example, the given position, in case you did not yet guess it, is a tie. But a careless move on your side can easily make you the loser! However, we do not know a general strategy for this game played on an arbitrary digraph with n vertices, since it is not quite covered by the theory of annihilation games. If there is a strategy which is polynomial in n, it may not be all that simple: A special case (when all particles are electrons, say) is a generalization of the game *Welter*, which has a rather complex — though polynomial — theory [7, 14 Ch. 13, 71].

(III) *Partizanship.* Another factor which prevents Nim-theory from asserting itself for the analysis of "chess-like" games, is the partizanship of the latter.

A game is *impartial* if both players have the same set of moves for all game positions. Otherwise the game is *partizan.* Nim-like games are impartial. Chess-like games are partizan, because if Jack plays the black pieces, Mike will not let him touch the white pieces.

In the sequel we shall denote a partizan game simply by the word game. The following two inductive definitions are due to Conway [14].

(i) If M^L, M^R are any two sets of games, there exists a game $M = \{M^L \mid M^R\}$. All games are constructed in this way.

(ii) If LE, RI are any two sets of numbers and no member of LE is $\geqslant$ any member of RI, then there exists a number $\{LE \mid RI\}$. All numbers are constructed in this way.

Thus the numbers constitute a subclass of the class of games.

The first games and numbers are created by putting $M^L = M^R = LE = RI = \emptyset$. Some samples are given in Fig. 12, where L (left) plays to the south-west and R (Right) to the south-east. If, as usual, the player first unable to move is the loser

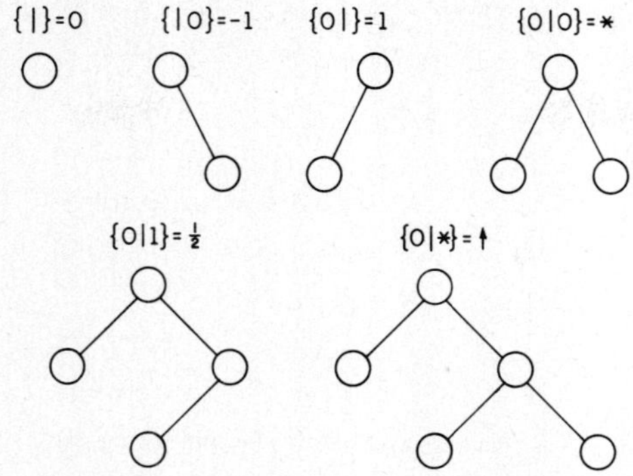

Fig. 12. A few early partizan games.

and his opponent the winner, then the examples suggest:

$M>0$ if L can win,

$M<0$ if R can win,

$M=0$ if the second player can win,

$M \parallel 0$ if the first player can win (for example in $*$),

and we shall in fact define $>$, $<$, $=$, $\parallel$ by these conditions. The relations can be combined as follows:

$M \geqslant 0$ if L can win as second player,

$M \leqslant 0$ if R can win as second player,

$M \mathrel{|\!\rhd} 0$ if L can win as first player,

$M \mathrel{\lhd\!|} 0$ if R can win as first player.

Alternative inductive definitions of $\leqslant, \parallel$ can be given. Let $x=\{x^L \mid x^R\}$. Then

$x \leqslant y$ if and only if $x^L \mathrel{\lhd\!|} y$ for all x^L and $x \mathrel{\lhd\!|} y^R$ for all y^R,

$x = y$ if and only if $x \leqslant y$ and $y \leqslant x$,

$x \parallel y$ (x is *fuzzy* with y) if and only if $x \not\leqslant y$ and $y \not\leqslant x$,

as well as a consistency proof of both definitions. This enables proving many properties of games in a simple manner. For example, define $-M=\{-M^R \mid -M^L\}$. Then $G-G=0$ (the second player can win in $G-G$ by "imitating" the moves of the first player in the other game). Also:

$$x \leqslant y \mathrel{\lhd\!|} z \quad \text{or} \quad x \mathrel{\lhd\!|} y \leqslant z \Rightarrow x \mathrel{\lhd\!|} z$$

(consider the game $z-x=(z-y)+(y-x)$, in which L can win as first player).

$$x^L \mathrel{\lhd\!|} x \mathrel{\lhd\!|} x^R$$

(clearly L can win as first player in $x-x^L$ and in x^R-x).

If x is a number, $x^L<x<x^R$.

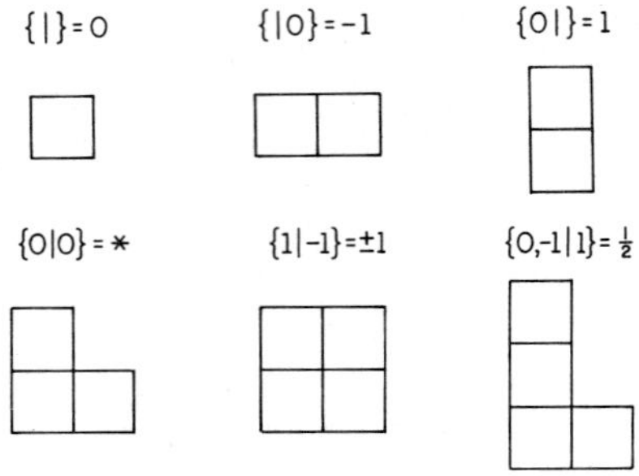

Fig. 13. A few values of the Domineering game.

Most important is that disjunctive compounds simply become *sums*, defined by:

$$M+H=\{M^L+H, M+H^L \mid M^R+H, M+H^R\}.$$

Consider for example the game of Domineering, played with dominoes each of which covers precisely two squares of an $n \times n$ chessboard. L tiles vertically, R horizontally. After playing Domineering for a while, the board may break up into several parts, whence the game becomes a disjunctive compound of these parts. The values of several small configurations are given in Fig. 13.

Since the relation of the value of a game M with 0 determines the strategy for playing it, computing the value of M is fundamental to the theory. In this direction the *Simplicity Theorem* is helpful: Let $x=\{x^L \mid x^R\}$ be any game. If there exists a *number* z such that $x^L \lhd\!\!| \; z \lhd\!\!| \; x^R$ and no option of z satisfies this relation, then $x=z$. Thus $\{-1 \mid 99\}=0$. There also exist algorithms for computing values associated with games which are not numbers. See [13–16].

3. Complexity of combinatorial games

As we gradually increase the complexities of games from Nim-like games to chess-like games, it becomes increasingly harder to find polynomial strategies. The failure to find a polynomial strategy for a family of games may be due to our stupidity. But it may also be because there is no polynomial strategy!

To show that a family of games is intractable seems to be difficult, but some beginnings in this direction were made by Chandra and Stockmeyer [11, 70]. It was recently proved, using [70], that a natural generalization of chess to an $n \times n$ board is intractable [27]. Other results, *suggesting* intractibility of various games were proved earlier, namely that finding a strategy is "just as hard" as a few hundred other problems for none of which a polynomial solution has been found, but none of which was shown to be inherently nonpolynomial. One such family, known as "NP-complete", contains problems such as the Travelling Salesman problem, the Hamiltonian Circuit problem, and the question of the satisfiability of a Boolean formula.

Let us define informally, and rather simplistically, a Deterministic Turing Machine (DM) to be a modern-day digital computer, and a Nondeterministic Turing Machine (NM) to be a DM in which the "next-move" function is multiple-valued. Thus an NM may traverse the branches of a rooted tree "in parallel", from the root to the leaves. A DM can do this only branch after branch. The set of problems which can be solved polynomially on a DM and on an NM is denoted by P and NP respectively. Clearly $P \subseteq NP$. It is an unsolved problem whether $P = NP$ or not. However, there are *complete* problems in NP: A problem Q is *NP-complete* if: (i) $Q \in NP$; (ii) existence of a polynomial algorithm for Q implies that $NP = P$. A problem is *NP-hard* if it satisfies (ii), but (i) does not necessarily hold. Any NP-hard problem is at least as hard as any NP-complete problem. The NP-complete problems are the "hardest" in NP: they lie in $NP - P$ if $NP \neq P$.

Another family of interest, called *P-space*, consists of all problems which can be solved on an NM by using a polynomial amount of memory cells. Since in a polynomial amount of time only a polynomial amount of memory can be consulted, $NP \subseteq$ *P-space*. Again it is not known whether the inclusion is proper or not. Any *Pspace-hard* problem is at least as hard as any *Pspace-complete* problem (defined analogously to NP-hard and NP-complete), and the latter is at least as hard as any NP-complete problem. For further details see e.g., Aho et al. [1, Ch. 10], and Garey and Johnson [38].

Many games have been proved to be NP-hard or even Pspace-complete, for example by Schaefer [59, 60]. Since our complexity measures are asymptotic, they are not applicable to, say, checkers as played on an 8×8 board or Go played on a 19×19 board. But it has been shown recently that both of those games, played on an $n \times n$ board are in fact Pspace-hard [25, 51]. Earlier Even and Tarjan [20] showed that the game of Hex, played on an arbitrary graph, is Pspace-complete. But whether it is also Pspace-complete on an $n \times n$ Hex-board is still open. Berlekamp has shown that Redwood Furniture, which is a special case of Left–Right Hackenbush [14, 37] is NP-hard [38]. Robertson and Munro [58] have shown that Generalized Instant Insanity is NP-complete. (Of course this game cannot belong to our respectable family of combinatorial games! It is indeed a solitaire game.)

Slight perturbations of the annihilation games introduced above are also NP-hard. For example, suppose that tokens of r different types are placed on distinct vertices of a digraph $R = (V, E)$ whose edges are distributed into r sets $E_1, \ldots, E_r$ (not necessarily disjoint), such that $\bigcup_{i=1}^{r} E_i = E$. A move consists of selecting a token of type i on some $u \in V$, and moving it to some $v \in F_i(u)$, where $F_i(u) = \{v : (u, v) \in E_i\}$ $(1 \leq i \leq r)$. If v is occupied by a token of any type, both tokens get annihilated and are removed from the game. The game is impartial, and the case $r = 1$ is the above polynomial family of annihilation games. But for $r \geq 2$ the problem of determining whether an arbitrary position is a P-position is NP-hard, even if R is acyclic and bipartite.

A partizan version of this game is obtained by letting the tokens be of one type only, but $E = E_1 \cup E_2$. Player i at his turn moves a token from some $u \in V$ to some $v \in F_i(u)$ $(i \in \{1, 2\})$. If v is occupied, both tokens get annihilated. The question whether a position is a P-position is again NP-hard, even if R is restricted as before. The special case $E_1 = E_2 = E$ is the polynomial class of annihilation games.

A superficially similar partizan game is the case where the tokens are of two types, but the edges are of one type only. Player i moves a token of type i $(i \in \{1, 2\})$ from some $u \in V$ to some $v \in F(u)$ which is not occupied by a token of the same type. If v is occupied by a token of the opposite type, the latter (only) is captured and removed from the game. The question whether any position is a P-position is also NP-hard [32]. A commercial version of this game is manufactured in Israel under the name of "Arrows", distributed in the U.S. under the name "On The Line".

There are, however, many families of games with capture rules for which neither P-membership nor NP-hardness is yet known. A case in point is the set of *cannibal games*. In this impartial family, all tokens (cannibals) and all edges are of one type. When a cannibal moves onto an occupied vertex it consumes the entire tribe residing there (greedy cannibals), or only a single victim (polite cannibals). Of course if initially there is at most one cannibal on every vertex, greedy and polite cannibals are the same. But in general they are different, and it appears that being polite is considerably more difficult than being greedy!

A particularly simple case is shown in Fig. 14. A number of cannibals are placed initially on arbitrary positions of the semi-infinite strip of squares. Each player at his turn selects a cannibal and moves it one square to the right. If the latter square is occupied, a bloody meal ensues, either greedy or polite, according to the version being played. A cannibal which reached square 0 (graveyard) succeeded in escaping violent death; it dies of natural causes. The player first unable to move loses, and his opponent wins. Even this case has not been solved polynomially. Several special cases, however, have been solved. For example, if there is only a small number of "feasting squares" close to the end of the strip, with no captures taking place on "non-feasting" squares, polynomial strategies have been found [18]. Also the game "Particles and Antiparticles" (Fig. 11), played on an arbitrary finite digraph is not known to belong to P, nor has NP-hardness been proved.

To whatever category these and other doubtful games may belong, more and more interesting games seem to join the class of NP-hard problems. Since for this

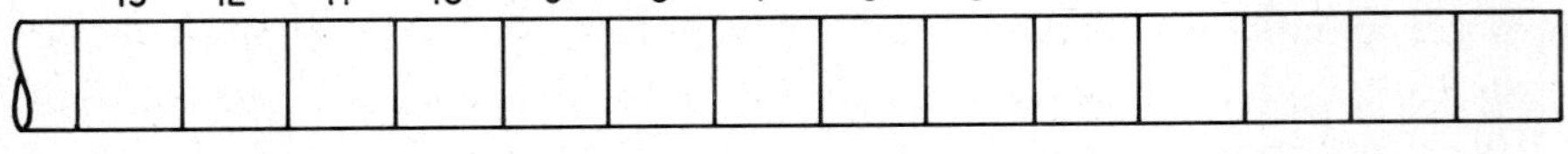

Fig. 14. Cannibals.

class it is unlikely that polynomial strategies will ever be found, the conclusion of the "To Go" part of our talk is:

chess-like games are likely to be intractable

This, admittedly, is a rather weak conclusion, and it reflects the present state of complexity theory. As weak as it is, it leads to an interesting question: Can polynomial suboptimal strategies be formulated for subclasses of NP-hard games, which give the correct strategy most of the time, or at least with a bounded number of exceptions? We do not known of any work in this direction. Of course there are heuristic programs for playing chess, checkers and other games, developed in the course of investigations in artificial intelligence. But the quality of these programs, some of which are quite monumental and ingeneous and contain self-learning elements, is largely determined a posteriori, by their actual performance in the field. No a priori error bounds are known. On the other hand, suboptimal algorithms were formulated for some specific questions in other fields, such as determining whether a given integer is prime. See Rabin [56].

One particular family of games for which such an approach might be attempted is that of Last-Player-Losing (LPL) games, i.e., games in which the player first unable to move is the winner, his opponent the loser. These games are generally much harder than LPW-games [14 Ch. 12, 41]. However, there is a subset S of LPL-games, including LPL-Nim, which can be treated by the Sprague–Grundy theory. Ferguson [21] gave a number of conditions which suffice for putting an LPL-game fulflling them into S. Some of his conditions are also necessary. Using the necessary conditions, one can easily point to a number of games in $\bar{S}$ (the complement of S).

One such game is Nimbi, invented by Piet Hein. Twelve tokens placed on the twelve vertices of a hexagonal board (Fig. 15), constitute the initial position of Nimbi. The board contains four horizontal rows, four diagonal rows in one direction and four diagonal rows in the other direction. Each player at his turn can remove any contiguous portion of tokens from a single row. Thus 159D may be removed in one move. But if 5 had been removed previously, then 19 or 1D or

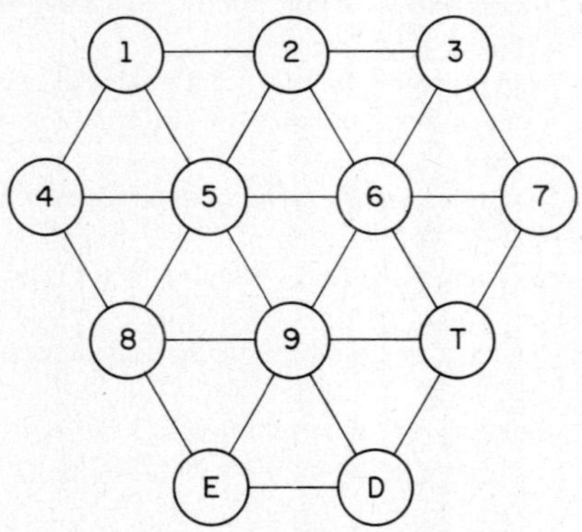

Fig. 15. The game of Nimbi.

19D cannot be removed in a single move, though 1 or 9 or D or even 9D can. The player first unable to move is the winner, his opponent the loser.

The P-positions of Nimbi and those of LPW-Nimbi have a large common intersection [26]. A similar phenomenon was observed by Fremlin, for a slightly different version of Nimbi played on a rectangular board, called by him Nim-squared, which he saw in the *Observer* in 1961 or 1962 [35]. Part of this intersection is due to positions which are essentially Nim-positions. For Nim a very large intersection between LPW-P-positions and LPL-P-positions is well-known to exist. But even if we disregard those, a sizeable intersection seems to remain, which is notable in view of Singmaster's result that for any game, "almost none" of its positions are P-positions, that is, $n(P)/n \to 0$ as $n \to \infty$, where $n(P)$ denotes the number of P-positions from among the total number n of positions [63].

If two players play an LPL-game R which belongs to $\bar{S}$ whose P-positions have a large overlap with the P-positions of R's LPW-version, the latter P-positions may be computed, if this is easy. The strategy of moving to them — except at the end of the game, which has to be analyzed separately — is likely to lead to a win most of the time.

This is but one out of a large number of exciting questions which can be asked about combinatorial games. The references below suggest an abundance of interesting research problems which wait to be tackled.

References

[1] A.V. Aho, J.E. Hopcroft and J.D. Ullman, The Design and Analysis of Computer Algorithms (Addison-Wesley, Reading, MA, 1974).

[2] R. Austin, Impartial and partisan games, Thesis, University of Calgary (1976).

[3] W.W.R. Ball, Mathematical Recreations and Essays, revised by H.S.M. Coxeter (Macmillan, New York, 1947).

[4] R.B. Banerji, Similarities in games and their use in strategy construction, in Proceedings Symposium on Computers and Automata (Polytechnic Press, Brooklyn, NY, 1971) 337–357.

[5] I. Bárány, On a class of balancing games, J. Combin. Theory 26(A) (1979) 115–126.

[6] C. Berge, The Theory of Graphs and Its Applications, translated by A. Doig (Wiley, New York, 1962).

[7] E.R. Berlekamp, Some recent results on the combinatorial game called Welter's Nim, in Proceedings 6th Annual Princeton Conference on Information Science and Systems (1972) 203–204.

[8] E.R. Berlekamp, The Hackenbush number system for compression of numerical data, Information and Control 26 (1974) 134–140.

[9] E.R. Berlekamp, J.H. Conway and R.K. Guy, Winning Ways (Harcourt, Brace and Jovanovich, New York, in preparation).

[10] C.L. Bouton, Nim, a game with a complete mathematical theory, Ann. Math., Princeton 3(2) (1902) 35–39.

[11] A.K. Chandra and L.J. Stockmeyer, Alternation, in Proceedings 17th Annual Symposium Foundations of Computer Science (Houston, TX, Oct. 1976) (IEEE Computer Society, Long Beach, CA, 1976) 98–108.

[12] V. Chvátal and P. Erdös, Biased positional games, Annals of Discrete Math. 2 (1978) 221–229.

[13] J.H. Conway, All numbers great and small, University of Calgary, Math. Res. Paper No. 149 (1972).

[14] J.H. Conway, On Numbers and Games (Academic Press, London, 1976).

[15] J.H. Conway, All games bright and beautiful, Am. Math. Monthly 84 (1977) 417–434.

[16] J.H. Conway, A gamut of game theories, Math. Mag. 51 (1978) 5–12.

[17] A.P. Domoryad, Mathematical Games and Pastimes, translated by H. Moss (Pergamon Press, Oxford, 1964).

[18] R.B. Eggleton and A.S. Fraenkel, Cannibals: A game with captures, in Abstracts of Communications, Int. Congr. Math., Vancouver, B.C. (1974) 186.

[19] P. Erdös and J.L. Selfridge, On a combinatorial game, J. Combin. Theory 14(A) (1973) 298–301.

[20] S. Even and R.E. Tarjan, A combinatorial problem which is complete in polynomial space, in: Proceedings 7th Annual ACM Symposium on Theory of Computing, (Alburquerque, NM, 1975 (Assoc. Comput. Mach., New York, 1975) 66–71.

[21] T.S. Ferguson, On sums of games with last player losing, Int. J. Game Theory 3 (1974) 159–167.

[22] A.S. Fraenkel, Combinatorial games with an annihilation rule, in Proceedings Symposium Applied Mathematics Vol. 20: The Influence of Computing on Mathematical Research and Education (Am. Math. Soc., Providence, RI, 1974) 87–91.

[23] A.S. Fraenkel, The particles and antiparticles game, Comput. Math. Appl. 3 (1977) 327–328.

[24] A.S. Fraenkel and I. Borosh, A generalization of Wythoff's game, J. Comb. Theory 15(A) (1973) 175–191.

[25] A.S. Fraenkel, M.R. Garey, D.S. Johnson, T. Schaefer and Y. Yesha, The complexity of checkers on an $N \times N$ board — Preliminary report, in: 19th Annual Symposium Foundations of Computer Science, Ann Arbor, MI, Oct. 1978 (IEEE Comput. Soc., Long Beach, CA, 1978) 55–64.

[26] A.S. Fraenkel and H. Herda, Never rush to be first in playing Nimbi, Math. Mag. 53 (1980) 21–26.

[27] A.S. Fraenkel and D. Lichenstein, Computing a perfect strategy for $n \times n$ chess requires time exponential in n, to appear.

[28] A.S. Fraenkel and Y. Perl, Constructions in combinatorial games with cycles, in: A. Hajnal, R. Rado and V.T. Sós, eds., Colloquia Mathematica Societatis János Bolyai 10: Proceedings International Colloquium on Infinite and Finite Sets, Keszthely, Hungary, 1973 Vol. 2 (North-Holland, Amsterdam, 1975) 667–699.

[29] A.S. Fraenkel and U. Tassa, Strategy for a class of games with dynamic ties, Comput. Math. Appl. 1 (1975) 237–254.

[30] A.S. Fraenkel, U. Tassa and Y. Yesha, Three annihilation games, Math. Mag. 51 (1978) 13–17.

[31] A.S. Fraenkel and Y. Yesha, Theory of annihilation games, Bull. Am. Math. Soc. 82 (1976) 775–777.

[32] A.S. Fraenkel and Y. Yesha, Complexity of problems in games, graphs and algebraic equations, Discrete Appl. Math. 1 (1979) 15–30.

[33] A.S. Fraenkel and Y. Yesha, The generalized Sprague–Grundy function and its invariance under certain mappings, to appear.

[34] A.S. Fraenkel and Y. Yesha, Theory of annihilation games — I, to appear.

[35] D. Fremlin, Well-founded games, Eureka 36 (1973) 33–37.

[36] D. Gale, A curious Nim-type game, Am. Math. Monthly 81 (1974) 876–879.

[37] M. Gardner, Mathematical games, a column in each issue of Sci. Am.

[38] M.R. Garey and D.S. Johnson, Computers and Intractability: A Guide to the Theory of NP-Completeness (W.H. Freeman, San Francisco, CA, 1979).

[39] S.W. Golomb, A mathematical investigation of games of "take-away", J. Combin. Theory 1 (1966) 443–458.

[40] P.M. Grundy, Mathematics and games, Eureka 2 (1939) 6–8.

[41] P.M. Grundy and C.A.B. Smith, Disjunctive games with the last player losing, Proc. Cambridge Philos. Soc. 52 (1956) 527–533.

[42] R.K. Guy, Partizan and impartial combinatorial games, University of Calgary, Math. Res. Paper No. 315 (1976).

[43] R.K. Guy and C.A.B. Smith, The G-values of various games, Proc. Cambridge Philos. Soc. 52 (1956) 514–526.

[44] A.W. Hales and R.I. Jewitt, Regularity and positional games, Trans. Am. Math. Soc. 106 (1963) 222–229.
[45] O. Hanner, Mean play of sums of positional games, Pacific J. Math. 9 (1959) 81–99.
[46] J.C. Holladay, Cartesian products of termination games, in: Contributions to the Theory of Games, Vol. 3, Annals of Math. Studies No. 39 (Princeton, NJ, 1957) 189–200.
[47] O. Itzinger, The South American game, Math. Mag. 50 (1977) 17–21.
[48] D.E. Knuth, Surreal numbers (Addison-Wesley, Reading, MA, 1974).
[49] J.C. Lagarias, Discrete balancing games, Bull Inst. Math. Acad. Sinica 5 (1977) 363–373.
[50] S.-Y.R. Li, Sums of Zuchswang games, J. Combin. Theory 21(A) (1976) 52–67.
[51] D. Lichenstein and M. Sipser, Go is Pspace hard, in: 19th Annual Symposium Foundations of Computer Science, Ann Arbor, MI, Oct. 1978 (IEEE Comput. Soc., Long Beach, CA, 1978) 48–54.
[52] J. Milnor, Sums of positional games, in: H.W. Kuhn and A.W. Tucker, eds., Contributions to the Theory of Games, Vol. 2, Annals of Math. Studies No. 28 (Princeton, NJ, 1953) 291–301.
[53] E.H. Moore, A generalization of the game called Nim, Annals of Math. 11 (1909) 93–94.
[54] T.H. O'Beirne, Puzzles and Paradoxes (Oxford University Press, London, 1965).
[55] M.O. Rabin, Effective computability of winning strategies, in: Contributions to the Theory of Games, Vol. 3, Annals of Math. Studies No. 39 (Princeton, NJ, 1957) 147–157.
[56] M.O. Rabin, Probabilistic Algorithms, in: J.F. Traub, ed., Proceedings Symposium on New Directions and Recent Results in Algorithms and Complexity, Carnegie-Mellon, 1976 (Academic Press, New York, 1976) 21–39.
[57] M. Richardson, Solutions of irreflexive relations, Annals of Math. 58 (1953) 573–590.
[58] E. Robertson and I. Munro, NP-completeness, puzzles and games, Utilitas Math. 13 (1978) 99–116.
[59] T.J. Schaefer, Complexity of decision problems based on finite two-person perfect information games, in: Proceedings 8th Annual ACM Symposium on Theory of Computing, Hershey, PA, 1976 (Assoc. Comput. Mach., New York, 1976) 41–49.
[60] T.J. Schaefer, On the complexity of some two-person perfect information games, J. Comput. System Sci. 16 (1978) 185–225.
[61] F. Schuh, The Master Book of Mathematical Recreations, translated by F. Göbel, edited by T.H. O'Beirne (Dover, New York, 1968).
[62] R. Silber, Wythoff's Nim and Fibonacci representations, Fibonacci Quart. 15 (1977) 85–88.
[63] D. Singmaster, Almost all games are first person games, private communication.
[64] C.A.B. Smith, Graphs and composite games, J. Combin. Theory 1 (1966) 51–81.
[65] C.A.B. Smith, Compound games with counters, J. Recreational Math. 1 (1968) 67–77.
[66] C.A.B. Smith, Simple game theory and its applications, Bull. Inst. Math. Appl. 7 (1971) 3–8.
[67] J. Spencer, Balancing games, J. Combin. Theory 23(B) (1977) 68–74.
[68] R. Sprague, Über mathematische Kampfspiele, Tôhoku Math. J. 41 (1935–36) 438–444.
[69] H. Steinhaus, Definitions for a theory of games and pursuit, Naval Res. Logist. Quart. 7 (1960) 105–108.
[70] L.J. Stockmeyer and A.K. Chandra, Provably difficult combinatorial games, SIAM J. Comput. 8 (1979) 151–174.
[71] C.P. Welter, The theory of a class of games on a sequence of squares, in terms of the advancing operation in a special group, Nederl. Akad. Wetensch. Proc. Ser. A57 = Indagationes Math. 16 (1954) 194–200.
[72] W. Wythoff, A modification of the game of Nim, Nieuw Arch. Wisk. 7 (1907) 199–202.
[73] A.M. Yaglom and I.M. Yaglom, Challenging Mathematical Problems with Elementary Solutions, Vol. II, translated by J. McCawley Jr., revised and edited by B. Gordon (Holden-Day, San Francisco, CA, 1967).

Annals of Discrete Mathematics 6 (1980) 157–174

CONFIGURATIONS IN A PLANE OF ORDER TEN*

Marshall HALL Jr.

California Institute of Technology, Pasadena, CA, U.S.A.

1. Introduction

If there is a projective plane of order ten, then there is a linear code V over GF(2) generated by the rows (taken modulo 2) of the incidence matrix where the 111 rows correspond to lines and the 111 columns correspond to points. This has been discussed in [1, 2]. Every vector v of V yields a set of points and these points together with the lines going through two or more of these form a configuration.

It is the purpose of this paper to describe some of these configurations involving only a small number of points. The weight $w(v)=n$ of a vector v over GF(2) is the number of ones in v and will correspond to a set of n points, and so in turn to a configuration M_n. If n is even every line contains an even number of the n points, while if n is odd every line contains an odd number of the n points.

A configuration M_{12} is an oval and is a set of 12 points such that no line contains as many as three of them. A configuration M_{16} has eight lines of four points each, each point being on two of these eight lines, and also 72 further lines with exactly two points. There are six non-isomorphic M_{16}'s and these are listed in Section 2. There are a number of types of M_{19}'s, also discussed in Section 2.

The main interest of this paper is in a particular kind of configuration C_{20} on 20 points. This is part of an M_{20} a configuration of 20 points in which no line contains more than four of the 20 points. Here C_{20} is a configuration of 20 points on 20 lines, each line containing four points and each point on four lines. In this respect the configurations C_{20} are a self-dual set. An M_{20} contains the C_{20} and 70 further lines each with exactly two of the 20 points. The sum of two ovals with exactly two points in common is such an M_{20}. In Section 4 the issue of the existence of such configurations is investigated. The primitive M_{20}'s certainly exist if there are no M_{16}'s or only a small number of ovals. There need not be any M_{12}'s or M_{16}'s, but the writer believes that there will always be primitive M_{20}'s in a plane, and it is hoped that a slight extension of the results of this section will prove this fact.

Given a primitive M_{20}, its extension to a full projective plane is not a trivial

*This research was supported in part by NSF Grant No. MCS75-23338 A02.

problem, but in any individual case is a reasonable one. It is hoped that this approach will be able to settle the existence of a plane of order ten with a reasonable amount of computation.

2. The binary code from a plane of order 10

Let A be the incidence matrix of a plane of order 10, PG(2, 10) so that A is a 111 by 111 matrix of 0's and 1's such that

$$AA^{\mathrm{T}} = A^{\mathrm{T}}A = 10I + J, \qquad AJ = JA = 11J. \tag{2.1}$$

Here A^{T} is the transpose of A, I is the identity and J is the matrix of all 1's.

The incidence matrix A determines a binary code V in the following way: Consider the rows of A as vectors over $GF(2)$ and let V be the vector space spanned by these vectors. In A we let the columns correspond to the points of the plane PG(2, 10) and the rows to the lines. In this respect V is the vector space spanned by the lines of PG(2, 10).

If v is a vector of V, then the *weight* of v, $w(v)$ is the number of 1's in v. If in V there are a_i vectors of weight i, then the polynomial $A(z)$

$$A(z) = \sum_{i=0}^{111} a_i z^i \tag{2.2}$$

is called the weight enumerator of V.

In MacWilliams, *et al.* [2] or in Erbach [1] the following properties of the weight enumerator are found:

(1) $a_1 = 1, \quad a_2 = \cdots = a_{10} = 0, \quad a_{11} = 111.$

(2) $a_{111-i} = a_i.$

(3) $a_{4m+1} = 0, \quad a_{4m+2} = 0.$

(4) $a_{15} = 0.$

All weights can be expressed in terms of a_{12} and a_{16}. We shall need only the following two relations:

$$\begin{aligned} a_{19} &= 24675 + 141a_{12} + 7a_{16}, \\ a_{20} &= 386010 + 354a_{12} + 5a_{16}. \end{aligned} \tag{2.3}$$

A vector v of V of weight $w(v) = n$ is a set of n points which form a configuration M_n. If n is odd every line of PG(2, 10) meets M_n in an odd number of points. If n is even, every line of PG(2, 10) meets M_n in an even number of points. By private communication I learn that Eugene Rodemich of the Jet Propulsion Laboratory has shown that $a_{12} \leqslant 64{,}033$ and $a_{16} \leqslant 6{,}434{,}864$ though this information will not be used here.

The statement above that $a_{11} = 111$ is equivalent to the assertion that every M_{11} is a line L, as every line is an M_{11} and there are exactly 111 lines.

M_{12}: Ovals. Let $P_1, \ldots, P_{12}$ be the 12 points of an M_{12}. A line L through one of the P_i must meet M_{12} in an even number of points and so meets M_{12} in at least one more point. As there are 11 lines through P_i each of these meets M_{12} in exactly one further point. Thus every line meets M_{12} in 0 or 2 points and there are $\binom{12}{2} = 66$ lines intersecting M_{12} in 2 points and 45 lines which do not intersect M_{12}. Thus an M_{12} is an *oval*. An oval in a plane of order n is a set C of $n+2$ points (n even) of $n+1$ points (n odd) such that no line contains more than 2 points of C. The numbers $n+2$ for n even and $n+1$ for n odd are maximal for sets not containing 3 points of any line.

M_{16}. For a given M_{16} let there be b_i lines which intersect M_{16} in i points. Here as 16 is even, $b_1 = b_3 = b_5 = b_7 = b_9 = b_{11} = 0$. Suppose that a line L intersects M_{16} in $t = 6$, 8, or 10 points. Then $M_{16} + L$ is a vector with $16 - t + (11 - t)$ points since in $M_{16} + L$ the t points common to M_{16} and L are removed and the remaining $11 - t$ points of L are added. But $27 - 2t = 7$ for $t = 10$ and $27 - 2t = 15$ for $t = 6$. These are both impossible as $a_7 = 0$, $a_{15} = 0$. For $t = 8$, $27 - 2t = 11$. In this case $M_{16} + L$ is a line T, and $M_{16} + L = T$, $M_{16} = L + T$. But with L and T two lines they are either identical, $L = T$, $L + T = 0$ or have exactly one point in common in which case $L + T$ is an M_{20}. Hence $b_6 = b_8 = b_{10} = 0$ for an M_{16}. Thus only b_0, b_2, b_4 remain. These satisfy equations

$$\begin{aligned} b_0 + b_2 + b_4 &= 111, \\ 2b_2 + 4b_4 &= 176, \\ b_2 + 6b_4 &= 120. \end{aligned} \tag{2.4}$$

The first of these counts the total number of lines, the second the incidence of the 16 points on lines, each point being on 11 lines. The third counts pairs of points P_iP_j there being $\binom{16}{2} = 120$ unordered pairs. These equations have the solution

$$b_0 = 31, \qquad b_2 = 72, \qquad b_4 = 8. \tag{2.5}$$

Consider the pencil of 11 lines through one of the points P_i of the 16 points. Each of these lines contains one or three further points of the 16 points and so there will be exactly two lines with three further points and 9 with one further point as $x + y = 11$, $3x + y = 15$ has $x = 2$, $y = 9$ as its unique solution.

Let us associate with M_{16} the configuration C_{16} consisting of the 8 lines each with 4 of the points each of the 16 points being on exactly two of the 8 lines. Up to isomorphism there are exactly six different configurations C_{16}. We list them here

I	II	III
1, 2, 3, 4	1, 2, 3, 4	1, 2, 3, 4
1, 5, 6, 7	1, 5, 6, 7	1, 5, 6, 7
2, 8, 9, 10	2, 5, 8, 9	2, 5, 8, 9
3, 11, 12, 13	3, 6, 8, 10	3, 6, 10, 11
4, 14, 15, 16	4, 11, 12, 13	4, 12, 13, 14
5, 8, 11, 14	7, 11, 14, 15	7, 12, 15, 16
6, 9, 12, 15	9, 12, 14, 16	8, 10, 13, 15
7, 10, 13, 16	10, 13, 15, 16	9, 11, 14, 16

(2.6)

IV	V	VI
1, 2, 3, 4	1, 2, 3, 4	1, 2, 3, 4
1, 5, 6, 7	1, 5, 6, 7	1, 5, 6, 7
2, 5, 8, 9	2, 5, 8, 9	2, 5, 8, 9
3, 6, 10, 11	3, 6, 10, 11	3, 6, 10, 11
4, 12, 13, 14,	8, 12, 13, 14	8, 12, 13, 14
8, 12, 15, 16	10, 12, 15, 16	10, 12, 15, 16
7, 10, 13, 15	4, 7, 13, 15	4, 9, 13, 15
9, 11, 14, 16	9, 11, 14, 16	7, 11, 14, 16

By extensive computer search J.L. Carter in his 1974 Thesis at Berkeley has eliminated cases II, III, V and VI and part of Case IV showing that they cannot be completed to a full plane. In Case IV it remains to consider subcases in which three of the 8 lines are concurrent. Specifically

$$\begin{matrix} 1, & 2, & 3, & 4, P \\ 7, & 10, & 13, & 15, P \\ 9, & 11, & 14, & 16, P \end{matrix} \quad \text{and} \quad \begin{matrix} 5, 12, P \\ 6, 8, P \\ \end{matrix} \tag{2.7}$$

or

$$\begin{matrix} 2, & 5, & 8, & 9, Q \\ 3, & 6, & 10, & 11, Q \\ 4, & 12, & 13, & 14, Q \end{matrix} \quad \text{and} \quad \begin{matrix} 1, 15, Q \\ 7, 16, Q \\ \end{matrix} \tag{2.8}$$

or both (2.7) and (2.8) IV has automorphisms

$$\begin{aligned} \alpha &= (1, 6)(2, 10)(3)(4, 11)(5, 7)(8, 15)(9, 13)(12, 16)(14) \quad (P, Q), \\ \beta &= (16)(2, 9)(3, 14)(4, 11)(5, 8)(6, 12)(7, 15)(10, 13) \quad (P)(Q) \end{aligned} \tag{2.9}$$

Here α interchanges (2.7) and (2.8) so that to pursue Case IV we may assume that (2.7) holds. Case I remains open.

M_{19}. Here of course $b_2 = b_4 = b_6 = b_8 = b_{10} = 0$. If a line L has 11 points of M_{19}, then $X = M_{19}+L$ has 8 points which is impossible as $a_8 = 0$. If L has 9 points of M_{19}, then $X = M_{19}+L$ has $19-9+2 = 12$ points and is an oval M_{12}. Conversely given an oval M_{12} and a line L through 2 points of M_{12}, then $M_{12}+L = M_{19}$. Let M_{12} be an oval on $P_1, P_2, \dots, P_{10}, P_{11}, P_{12}$ and suppose that L goes through P_{11}, P_{12}

$$L = \{P_{11}P_{12}A_1A_2, \dots, A_9\}. \tag{2.10}$$

Then M_{19} has the points $P_1, \dots, P_{10}, A_1, \dots, A_9$. Here every line P_iP_j, i, $j = 1, \dots, 10$ goes through one of the A's. Every line A_iP_j goes through a further P_k.

For this type we have

$$\begin{array}{lll} & b_9 = 1, & \text{Line } A_1, \dots, A_9, \\ \textit{Oval type } M_{19} & b_3 = 45, & P_iP_jA_k. \\ & b_1 = 65, & \end{array} \tag{2.11}$$

The number of Oval type M_{19}'s is $66a_{12}$ since given an oval M_{12} we may add to it any one of its 66 secants and this determines a unique line of the M_{19} with 9 points.

If a line L has 7 points of an M_{19}, then $M_{19}+L$ has $19-7+4 = 16$ points and L has 4 points. Here with $M_{19}+L = M_{16}$ then $M_{19} = M_{16}+L$ where L contains 4 points of the M_{16}. Suppose

$$L = \{1, 2, 3, 4, A_1, \dots, A_7\} \tag{2.12}$$

where 1, 2, 3, 4 are the points of L in M_{16}. If the points of M_{16} are 1, 2, 3, 4, 5, ..., 16, then the points of $M_{19} = M_{16}+L$ are $A_1, \dots, A_7$, 5, ..., 16. No line of the M_{19} except L contains more than one of the A's and no line of PG(2, 10) contains more than 4 of the points 5, ..., 16. Hence M_{19} contains exactly one line with 7 points namely L.

Our standard equations are

$$\begin{aligned} b_1 + b_3 + b_5 + b_7 &= 111 \\ b_1 + 3b_3 + 5b_5 + 7b_7 &= 209 \\ 3b_3 + 10b_5 + 21b_7 &= 171 \end{aligned} \tag{2.13}$$

and as $b_7 = 1$ we have

$$\begin{array}{ll} & b_7 = 1, \\ & b_5 = 3, \\ M_{19} \text{ of 16-type} & b_3 = 40, \\ & b_1 = 67. \end{array} \tag{2.14}$$

Given an M_{16} we may add to it any one of the 8 lines containing 4 points to obtain an M_{19}, and this will be the unique line of the M_{19} containing 7 points.

Hence the number of M_{19}'s of 16-type is $8a_{16}$.

There remain M_{19}'s for which no line contains more than 5 points. These are the primitive M_{19}'s. We have

$$\begin{aligned} b_1 + b_3 + b_5 &= 111, \\ b_1 + 3b_3 + 5b_5 &= 209, \\ 3b_3 + 10b_5 &= 171. \end{aligned} \tag{2.15}$$

From this $b_5 = 6$, $b_3 = 37$, $b_1 = 68$.

Since $a_{19} = 24675 + 141\ a_{12} + 7a_{16}$ the number of primitive M_{19}'s is P_{19} given by

$$a_{19} - 66a_{12} - 8a_{16} = 24675 + 75a_{12} - a_{16} = P_{19}. \tag{2.16}$$

Since $P_{19} \geq 0$ we have $a_{16} \leq 24675 + 75a_{12}$ and with Rodemich's bound $a_{12} \leq$ 64,033 we find $a_{16} \leq 4{,}827{,}150$ which is better than his independent bound for a_{16}.

Let us summarize our findings on M_{19}'s:

Oval type:	16-type:	Primitive type:	
$b_9 = 1$,	$b_7 = 1$,	$b_5 = 6$,	
$b_3 = 45$,	$b_5 = 3$,	$b_3 = 37$,	
$b_1 = 65$,	$b_3 = 40$,	$b_1 = 68$,	(2.17)
	$b_1 = 67$,		
Number: $66a_{12}$,	Number: $8a_{16}$,	Number: $24675 + 75a_{12} - a_{16}$.	

3. 20 point configurations

We recall from (2.3) that $a_{20} = 386{,}010 + 354a_{12} + 5a_{16}$. If there is a line L with 10 points of an M_{20}, then $M_{20} + L$ contains $20 - 10 + 1 = 11$ points and is another line K. Conversely if L and K are two different lines, then $L + K$ has $11 + 11 - 2 =$ 20 points. This gives us $\binom{111}{2} = 6105$ M_{20}'s of type $L + K$. If a line L contains 8 points of an M_{20}, then $M_{20} + L$ has $20 - 8 + 3 = 15$ points which is impossible as $a_{15} = 0$. Hence apart from the two line M_{20}'s of shape $L + K$ all others have at most 6 points on a line. Our equations (where as usual b_i is the number of lines containing i points of M_{20}) are

$$\begin{aligned} b_0 + b_2 + b_4 + b_6 &= 111, \\ 2b_2 + 4b_4 + 6b_6 &= 220, \\ b_2 + 6b_4 + 15b_6 &= 190. \end{aligned} \tag{3.1}$$

Here we may express b_0, b_2, b_4 in terms of b_6

$$\begin{aligned} b_0 &= 21 - \ \ b_6, \\ b_2 &= 70 + 3b_6, \\ b_4 &= 20 - 3b_6. \end{aligned} \tag{3.2}$$

Hence $0 \leqslant b \leqslant 6$.

We list the possible values of the parameters

	b_0	b_2	b_4	b_6
E_6:	15	88	2	6
E_5:	16	85	5	5
E_4:	17	82	8	4
E_3:	18	79	11	3
E_2:	19	76	14	2
E_1:	20	73	17	1
E_0:	21	70	20	0

(3.3)

In cases $E_1, \ldots, E_6$ there will be at least one line with 6 points and these are imprimitive types. In case E_0 we say the M_{20} is primitive. Let L be a line of the M_{20} containing 6 points. Then $M_{20}+L$ has $20-6+5=19$ points and with $M_{20}+L=M_{19}$, L has 5 points of the M_{19}. Hence from (2.17) the M_{19} is not of oval type but may be of 16 type or of primitive type. We can describe exactly those M_{20}'s arising from M_{19}'s of 16 type but can only estimate those arising from M_{19}'s of primitive type.

Suppose first that in $M_{20}+L=M_{19}$ the M_{19} is of 16-type. Then L has 5 points of the M_{19} and a further line K of the M_{19} has 7 points in the M_{19}. In M_{20}, K has at most 6 points so that from $M_{20}+L=M_{19}$, K does not contain any of the 6 points of L in the M_{20} but does contain one of the remaining points of L in the M_{19}. Hence in M_{20} both K and L contain 6 points but are disjoint in the M_{20}. Hence $M_{20}+K+L$ contains $20-6-6+5+5-2=16$ points, and K and L both contain 4 points in the M_{16} but are disjoint in the M_{16}. We may see this by

$$\begin{aligned} K &= \{1, 2, 3, 4, 5, 6, P, a_1, a_2, a_3, a_4\}, \\ L &= \{7, 8, 9, 10, 11, 12, P, b_1, b_2, b_3, b_4\} \end{aligned} \tag{3.4}$$

where P is the intersection of K and L. If M_{20} has points $\{1, \ldots, 20\}$, then $M_{20}+K+L$ has points $\{13, \ldots, 20, a_1, a_2, a_3, a_4, b_1, b_2, b_3, b_4\}$ and is an M_{16}, where K has the 4 points a_1, a_2, a_3, a_4 and L has the disjoint points b_1, b_2, b_3, b_4. Since $13, \ldots, 20$ are points of an M_{16} no line contains more than 4 of these. Thus in the M_{20} on $\{1, \ldots, 20\}$ a further line with 6 points will contain one of $\{1, \ldots, 6\}$ and one of $\{7, \ldots, 12\}$. In an M_{20} arising from an M_{16} of type I it could happen that there are two further disjoint 6 point lines.

Conversely let C_{16} be the configuration of the points of an M_{16} and the 8 lines containing 4 of these points. If $a_1a_2a_3a_4$ is one of the lines, then each of a_1, a_2, a_3, a_4 is on one further line of C_{16} and in C_{16} there remain three lines such as $b_1b_2b_3b_4$ disjoint from $a_1a_2a_3a_4$. If K is the line $a_1a_2a_3a_4$ and L the line $b_1b_2b_3b_4$, then $M_{16}+K+L$ has $16-8+12=20$ points and is an M_{20}. From a C_{16} we may choose $a_1a_2a_3a_4$ in 8 ways and then choose $b_1b_2b_3b_4$ in 3 ways. Hence as in (3.3) we may choose K and L in $8\cdot\frac{3}{2}=12$ way. Thus the number of M_{20}'s of this 16-type is at most $12a_{16}$.

An M_{19} of primitive type has 6 lines L with 5 points. Here $M_{19}+L=M_{20}$ is an M_{20} of 19 type in which L has 6 points. But if K is another line of M_{19} with 5 points which does not intersect L in M_{19}, then in $M_{19}+L=M_{20}$, K will also have 6 points and $M_{19}+L+K=M_{19}^*$ is a further primitive M_{19}. Here $M_{20}=M_{19}+L=M_{19}^*+K$ and the representation is not unique. It may also happen that in an M_{20} of 16 type there is a further 6 point line S different from the two disjoint 6 point lines K and L and $M_{20}+S$ is a primitive M_{19}. As the number of primitive 19's is $24675+75a_{12}-a_{16}=M_{19}$, there will be at most $6M_{19}=148050+450a_{12}-6a_{16}$ M_{20}'s of 19 type.

Hence except for primitive M_{20}'s we have at most

$$\begin{array}{rl} 6105 & \text{type } K+L, \\ 12a_{16} & \text{16-type,} \\ 148050+450a_{12}-\ 6a_{16} & \text{19-type.} \end{array} \tag{3.5}$$

Since $a_{20}=386010+354a_{12}+5a_{16}$ there are m_{20} primitive M_{20}'s where from (2.17)

$$m_{20}\geqslant 231855-96a_{12}-a_{16}=207180-171a_{12}+m_{19}. \tag{3.6}$$

Since $m_{19}\geqslant 0$ it follows that m_{20} will certainly be positive if $a_{12}\leqslant 1211$.

From (3.3) a primitive M_{20} has $b_0=21$, $b_2=70$, $b_4=20$ and $b_i=0$ otherwise the configuration C_{20} will consist of the 20 points and the 20 lines containing 4 points. Let P_i, $i=1,\ldots,20$ be one of the 20 points. Each of the 11 lines through P_i contains besides P_i one or three further P's. With u lines containing three further P's and v containing one further P we have $u+v=11$, $3u+v=19$ so that $u=4$, $v=7$. Hence P_i is on exactly 4 of the lines of the C_{20}. We have shown:

Theorem 3.1. *A configuration C_{20} derived from a primitive M_{20} consists of 20 lines each with 4 of the M_{20} points and each of the 20 points of the M_{20} lies on exactly 4 of these 20 lines.*

It is interesting that a C_{20} is a dual configuration of points and lines. But in the plane although every line contains an even number of these 20 points, it is not obvious that dually every point is on an even number of the 20 lines. These properties would be reversed in a C_{20} derived from the dual plane.

4. Existence of primitive 20 point configurations

In the previous section it was shown that if a plane of order 10, PG(2,10) contains at most 1211 ovals, then it will contain primitive 20 point configurations. We note that having at most 1211 ovals is sufficient for the existence of primitive M_{20}'s but by no means obviously necessary.

In this context we shall study the structure of the ovals in a PG(2,10). In any projective plane a *k-arc* is a set of k points no three on a line. If $P_1, \ldots, P_k$ are the points of a *k-arc*, a line P_iP_j is called a *secant.*

Lemma 4.1. *If P_1, P_2, P_3, P_4 are a 4-arc A, then there are* 56 *points of* PG(2,10) *not on a secant of A.*

Proof. The secants of A are

$$P_1P_2Q_1, \qquad P_1P_3Q_2, \qquad P_1P_4Q_3,$$
$$P_2P_3Q_3, \qquad P_2P_4Q_2, \qquad P_3P_4Q_1$$

where Q_1, Q_2, Q_3 are the intersections of the secants. Each of these secants contains 8 further distinct points so that the secants contain $4+3+48=55$ points. Thus there are $111-55=56$ points not on a secant.

Lemma 4.2. *If P_1, P_2, P_3, P_4, P_5 are a 5-arc A of* PG(2,10) *there are* 31 *points not on a secant of A.*

Proof. Each of the 10 secants, say P_iP_j has 3 intersection points Q_r, Q_s, Q_t with secants P_xP_y, P_xP_z, P_yP_z where i, j, x, y, z are 1, 2, 3, 4, 5 in some order. No intersection point is on three secants since we cannot have three disjoint pairs of P's. There are 15 intersection points Q, three on each secant. A secant is of the form $P_iP_jQ_rQ_sQ_t \cdots$ and has six further points. Hence the 10 secants contain $5+15+60=80$ points and 31 points remain.

For 6-arcs the result is less clear cut.

Lemma 4.3. *If $P_1, \ldots, P_6$ are a 6-arc A of* PG(2,10), *there are at most* 15 *points not on a secant of A.*

Proof. Each of $P_1, \ldots, P_6$ is on 5 secants. Any other point is on at most 3 secants, as we have at most three disjoint pairs of P's. Let B_i be the number of points on i secants. Then $B_5=6$ for the six points $P_1, \ldots, P_6$ and other points are in B_0, B_1, B_2, or B_3. Then

$$\begin{aligned} B_0+B_1+B_2+B_3+B_5&=111,\\ B_1+2B_2+3B_3+5B_5&=165,\\ B_2+3B_3+10B_5&=105.\end{aligned} \tag{4.1}$$

The first of these counts the total number of points, the second the incidences of points on the 15 secants and the third counts the intersections of the 15 secants. Noting that $B_5=6$ we have

$$\begin{aligned} B_0 &= 15-B_3, \\ B_1 &= 45+3B_3, \\ B_2 &= 45-3B_3, \\ B_5 &= 6. \end{aligned} \tag{4.2}$$

Here $B_0=15-B_3$ is the number of points not on a secant. The following is almost a corollary.

Theorem 4.1. (E. Rodemich), *Given* 6 *points* $P_1, \ldots, P_6$ *in a* PG(2, 10) *there are at most two different ovals containing them.*

Proof. Let $P_1, \ldots, P_6$ be points of PG(2, 10). If they are not a 6-arc, then there is no oval containing them. If $P_1, \ldots, P_6$ are a 6-arc A, then an oval containing $P_1, \ldots P_6$ contains six of the set of at most 15 points not on a secant of A. Two different ovals do not have as many as 7 points in common so that ovals containing the 6-arc A use disjoint sets of 6 of the points not on the secants. As there are at most 15 points available we can have at most two disjoint sets of 6 from these and so at most two ovals containing $P_1, \ldots, P_6$.

If X and Y are two point sets in PG(2, 10) it is convenient to write XY for their intersection. We also write $|X|$ for the number of points in X. In this notation $|X+Y|=|X|+|Y|-2|XY|$, where of course $X+Y$ is the sum in the code V over GF(2).

Lemma 4.4. *If* C_1 *and* C_2 *are two ovals which intersect in exactly two points, then their sum* C_1+C_2 *is a primitive* M_{20}.

Proof. C_1+C_2 has $12+12-2=20$ points, and is an M_{20}. As no line meets C_1 in more than two points or C_2 in more than two points no line meets this M_{20} in more than 4 points so that it must be a primitive M_{20}.

Hence to decide on the existence of a primitive M_{20} in the plane, we may exclude the above possibility and make the following assumption:

Assumption C. *In a plane of order* 10 *there do not exist two ovals intersecting in exactly two points.*

It is my objective to show that with Assumption C there are at most 1211 ovals in the plane and so from (3.6) that there do exist primitive M_{20}'s. In this direction we prove two lemmas, using Assumption C.

Lemma 4.5. *Given five points* 1, 2, 3, 4, 5 *there are at most two ovals containing all* 5.

Proof. Assume to the contrary that there are three ovals C_1, C_2, C_3 containing the points 1, 2, 3, 4, 5. Here necessarily $|C_1C_2|=6$, $|C_1C_3|=6$ and $|C_2C_3|=6$. From Lemma 4 C_1, C_2, C_3 do not all contain the same 6 points so that we have

$$\begin{aligned} C_1 &= 1, 2, 3, 4, 5, 6, 7, 8, 9, 10, 11, 12, \\ C_2 &= 1, 2, 3, 4, 5, 6, 13, 14, 15, 16, 17, 18, \\ C_3 &= 1, 2, 3, 4, 5, 7, 13, 19, 20, 12, 22, 23. \end{aligned} \tag{4.3}$$

Here because of Theorem 4.1. C_3 cannot contain all of $1, \ldots, 6$ and so contains 1, 2, 3, 4, 5 and 7 from C_1, 13 from C_2. But here

$$C_1+C_2=C=\{7, 8, 9, 10, 11, 12, 13, 14, 15, 16, 17, 18\}$$

is an oval and intersects C_3 in $\{7, 13\}$ contrary to assumption C.

Lemma 4.6. *Given four points* 1, 2, 3, 4 *there are at most* 12 *ovals containing all four.*

Proof. Given points 1, 2, 3, 4 there are by Lemma 4. exactly 56 points not on the secants of the 4-arc 1, 2, 3, 4. Consider the family $C_1, \ldots, C_n$ of ovals containing all of 1, 2, 3, 4. Each C_i contains 8 of the 56 points which we may designate as $5, \ldots, 60$. This immediately shows that $n \leq 14$, since with 15 ovals there would be $120 = 8 \cdot 15$ incidences of these 56 points on the ovals and as $120 > 2 \cdot 56$ there would be at least one of the 56 points on three of $C_1, \ldots, C_n$ conflicting with Lemma 4.5. But we can do better and show that $n \leq 12$.

Unless $n \leq 7$ we have at least 8 ovals $C_1, \ldots, C_8$ and 64 incidences of the 56 points on these ovals so that at least one of these is on two ovals C_1, C_2 which then intersect in 6 points

$$\begin{aligned} C_1 &= 1, 2, 3, 4, 5, 6, 7, 8, 9, 10, 11, 12, \\ C_2 &= 1, 2, 3, 4, 5, 6, 13, 14, 15, 16, 17, 18. \end{aligned} \tag{4.4}$$

Here $5, 6, \ldots, 18$ are 14 of the 56 points. By Lemma 4.5 the points 5 and 6 cannot appear on a further oval C_i. Of the $56-14=42$ further points there cannot be 11 ovals using only these 42 points as $11 \cdot 8 = 88 > 2 \cdot 42$ and one point would appear in three ovals. In this way we can get at most $2+10=12$ ovals. Now suppose that a further oval C_3 uses one of the points $7, \ldots, 18$, which we may without loss take to be 7. Then $|C_1C_3|=6$ and C_3 contains 7 and 8. But C_1+C_2 is the oval $7, \ldots, 18$ and C_3 containing 7 and 8 cannot intersect C_1+C_2 in only these two points so that C_3 also contains points from $13, \ldots, 18$ and thus $|C_2C_3|=6$ and without loss we may take

$$C_3 = 1, 2, 3, 4, 7, 8, 13, 14, 19, 20, 21, 22. \tag{4.5}$$

Here C_1, C_2, C_3 together use 18 of the 56 points 5, 6, ..., 60 and there remain $56-18=38$ points. There cannot be as many as 10 ovals using only these 38 since $10 \cdot 8 = 80 > 2 \cdot 38$ and we would have a conflict with Lemma 4.5. This would lead to at most $3+9=12$ ovals C_i. If a further oval C_4 intersects any one of C_1, C_2, C_3 in points besides 1, 2, 3, 4, then as argued for C_3 with respect to C_1 and C_2, C_4 must intersect each of C_1, C_2, C_3 in 6 points and without loss.

$$C_4 = 1, 2, 3, 4, 9, 10, 15, 16, 19, 20, 23, 24. \tag{4.6}$$

Together $C_1C_2C_3C_4$ use 20 of the 56 points. A further oval C_5 intersecting any one of C_1, C_2, C_3, C_4 in more than 4 points intersects all four in 6 points and is necessarily

$$C_5 = 1, 2, 3, 4, 11, 12, 17, 18, 21, 22, 23, 24. \tag{4.7}$$

Thus C_5 uses only the same 20 points as C_1, C_2, C_3, C_4. Any further oval cannot use any of the 20 points 5, ..., 24, and must use 8 of the remaining 36 which are 25, ..., 60.

As with $C_1, \ldots, C_5$ with the remaining 36 points one oval uses 8 of them, two six-intersecting ovals use 14, three six-intersecting use 18, and four or five six-intersecting use 20 points. With 36 points we can have at most 7 further ovals and by this have 5 six-intersecting and two others which may or may not be six-intersecting. This proves our lemma since in every case there are at most 12 ovals.

The following can probably be improved with more careful analysis.

Lemma 4.7. *A triangle is in at most* 100 *ovals.*

Proof. Let 1, 2, 3 be the points of a triangle, and let C be an oval containing 1, 2, 3.

$$C = 1, 2, 3, 4, 5, 6, 7, 8, 9, 10, 11, 12. \tag{4.8}$$

Every oval containing 1, 2, 3 intersects C in at least one further point and so contains the four points 1, 2, 3, i where i is one of 4, ..., 12. Each of these sets of 4 is in C and at most 11 further ovals from Lemma 4.6. This gives us at most $1+9 \cdot 11 = 100$ ovals containing 1, 2, 3.

The subsets of size n of a set of size v form what is called a Johnson scheme $J(n, v)$. If x and y are two of these sets which have r elements in common we define the Johnson distance $d_J(x, y)$ by

$$d_J(x, y) = n - r. \tag{4.9}$$

Under this definition of distance a Johnson scheme $J(n, v)$ is an association scheme (which we shall not define or discuss here). Delsarte [3] developed an extensive theory of association schemes in coding theory. He establishes some basic inequalities on p. 27 and discusses Johnson schemes on pp. 44–49.

Without giving further explanation let us consider a set Y of y subsets of size 12 in a set of 111 points. (Naturally we have the ovals of a plane in mind but this approach is more general.) Suppose that any two subsets intersect in 0, 2, 4, 6 or 12 points (this last for a set intersecting itself). Of the y^2 pairs $[u, v]$, $u, v \in Y$ let there be y_i which have intersection of size i. (Clearly $y_{12} = y$ for the identical pairs $[u, u]$.)

Note that

$$y_0 + y_2 + y_4 + y_6 + y_{12} = y^2 \tag{4.10}$$

counts all pairs. We now define averages for these intersections,

$$r_i = y_i/y, \quad i = 0, 2, 4, 6, 12, \qquad r_{12} = 1. \tag{4.11}$$

From (4.10) it is now immediate that

$$r_0 + r_2 + r_4 + r_6 + 1 = y. \tag{4.12}$$

Among the inequalities for this case given by Delsarte we have in decimal form

$$\begin{aligned} &48.333333r_6 + 27.777778r_4 + 7.2222222r_2 - 13.333333r_0 + 110 \geqslant 0,\\ &1017.5343r_6 + 256.59704r_4 - 55.397032r_2 + 81.551020r_0 + 5994 \geqslant 0,\\ &11669.314r_6 + 336.12319r_4 - 154.92573r_2 - 302.55838r_0 + 215709 \geqslant 0,\\ &70532.651r_6 - 6723.5554r_4 + 2243.7478r_2 + 758.36169r_0 + 5{,}767{,}189 \geqslant 0. \end{aligned} \tag{4.13}$$

These are in fact rational coefficients. Thus the first of these is exactly

$$\frac{145}{3} r_6 + \frac{250}{9} r_4 + \frac{65}{9} r_2 - \frac{40}{3} r_0 + 111 \geqslant 0. \tag{4.14}$$

Applying these inequalities to the ovals in a plane of order 10 we have $r_6 \leqslant \binom{12}{6} = 924$ since by Theorem 4.1, given an oval C a second oval C' intersecting C in 6 points is fully determined by the choice of the 6 points.

Using $r_6 \leqslant 924$ and $a_{12} = r_0 + r_2 + r_4 + r_6 + 1$, with the inequalities (4.13) Eugene Rodemich obtained the bound $a_{12} \leqslant 64{,}033$ mentioned earlier. Under Assumption C we have $r_2 = 0$ and will be able to get a much stronger bound on r_6. If we could also get an independent bound on r_4 we could use (4.14) to bound r_0 in terms of r_4 and r_6 and thus estimate a_{12}.

With two ovals intersecting in 6 points we may associate a 6 by 6 Latin square. Let $C = C_1$ and C_2 be two ovals intersecting in points 1, 2, 3, 4, 5, 6.

$$\begin{aligned} C_1 &= 1, 2, 3, 4, 5, 6, 7, 8, 10, 11, 12,\\ C_2 &= 1, 2, 3, 4, 5, 6, 13, 14, 15, 16, 17, 18. \end{aligned} \tag{4.15}$$

Here $C_1 + C_2 = C_3 = 7, \ldots, 18$ is a third oval. From C_1 the complete pencil of lines through the point 1 is 1, 2; 1, 3; 1, 4; . . . ; 1, 12. From C_2 the complete pencil consists of the lines 1, 2; 1, 3; . . . ; 1, 6; 1, 13; . . . ; 1, 18. Hence the six lines 1, 7; . . . ; 1, 12 are the same as the lines 1, 13; 1, 14; . . . , 1, 18 in some order.

Hence we will have, with appropriate numbering the six lines

$$\begin{array}{l} 1,\ 7, 13 \\ 1,\ 8, 14 \\ 1,\ 9, 15 \\ 1, 10, 16 \\ 1, 11, 17 \\ 1, 12, 18 \end{array} \tag{4.16}$$

Similarly for 2, j, $j = 7, \ldots, 12$ and 2, k, $k = 13, \ldots, 18$ these are the same 6 lines in some order. We may do the same with each of the pencils through 3, 4, 5, and 6 obtaining lines ijk, $i = 1, \ldots, 6$, $j = 7, \ldots, 12$ and $k = 13, \ldots, 18$. A possible array of this kind is the following:

$$\begin{array}{llllll} 1,\ 7, 13 & 2,\ 7, 14 & 3,\ 7, 15 & 4,\ 7, 16 & 5,\ 7, 17 & 6,\ 7,\ 8 \\ 1,\ 8, 14 & 2,\ 8, 15 & 3,\ 8, 13 & 4,\ 8, 18 & 5,\ 8, 16 & 6,\ 8, 17 \\ 1,\ 9, 15 & 2,\ 9, 13 & 3,\ 9, 14 & 4,\ 9, 17 & 5,\ 9, 18 & 6,\ 9, 16 \\ 1, 10, 16 & 2, 10, 17 & 3, 10, 18 & 4, 10, 13 & 5, 10, 14 & 6, 10, 15 \\ 1, 11, 17 & 2, 11, 18 & 3, 11, 16 & 4, 11, 15 & 5, 11, 13 & 6, 11, 14 \\ 1, 12, 18 & 2, 12, 16 & 3, 12, 17 & 4, 12, 14 & 5, 12, 15 & 6, 12, 13 \end{array} \tag{4.17}$$

We have 36 lines i, j, k, $i = 1, \ldots, 6$; $j = 7, \ldots, 12$, $h = 13, \ldots, 18$ with every possible pair i, j and i, k, and as the 36 pairs j, k are distinct we must have every possible one exactly once. If we let $1, \ldots, 6$ determine the rows, $7, \ldots, 12$ the columns of a 6 by 6 Latin square and $13, \ldots, 18$ the digits, the 36 triples i, j, k determine a Latin square on $13, \ldots, 18$. The example above yields

$$\begin{array}{l} 13, 14, 15, 16, 17, 18 \\ 14, 15, 13, 17, 18, 16 \\ 15, 13, 14, 18, 16, 17 \\ 16, 18, 17, 13, 15, 14 \\ 17, 16, 18, 14, 13, 15 \\ 18, 17, 16, 15, 14, 13 \end{array} \tag{4.18}$$

If we divide the points of (4.15) into three sets b_1, b_2, b_3

$$\begin{aligned} b_1 &= 1, 2, 3, 4, 5, 6, \\ b_2 &= 7, 8, 9, 10, 11, 12, \\ b_3 &= 13, 14, 15, 16, 17, 18 \end{aligned} \tag{4.19}$$

then $C = C_1 = b_1 + b_2$, $C_2 = b_1 + b_3$ and $C_1 + C_2 = C_3 = b_2 + b_3$. A further oval C_i will intersect b_1, b_2, b_3 respectively in e_1, e_2, e_3 points where $|C_1C_i| = e_1 + e_2$, $|C_2C_i| = e_1 + e_3$, $|C_3C_i| = e_2 + e_3$. Since $b_1 = C_1C_2$, $b_2 = C_1C_3$ and $b_3 = C_2C_3$ by

Theorem 4.1 no further oval contains all six points of b_1, b_2, or b_3. Here e_1 is the number of points common to C_1, C_2, and C_i so that by Lemma 4.5, we cannot have $e_1 = 5$. Similarly we cannot have $e_2 = 5$ or $e_3 = 5$. We note that we cannot have e_1, e_2, $e_3 = 4, 2, 0$ since then $|C_1C_3| = 2$ which is excluded by our assumption.

The possibilities for e_1, e_2, e_3 are 4, 2, 2; 2, 4, 2; 2, 2, 4; 3, 3, 3; 3, 3, 1; 3, 1, 3; 1, 3, 3; 2, 2, 2; 4, 0, 0; 0, 4, 0; 0, 0, 4; and 0, 0, 0.

If $|C_iC_1| = 6$, then for C_i we have one of the patterns 2, 4, 2; 4, 2, 2; 3, 3, 1; of 3, 3, 3 for e_1, e_2, e_3. If C_i has the pattern 2, 4, 2 then $C_1 + C_i$ has the pattern 4, 2, 2 so that it is enough to consider 2, 4, 2. Let C_iC_1 have an intersection of the following shape:

$$\begin{aligned} &i_1i_2, \quad i = 1, \ldots, 6, \\ &j_1j_2j_3j_4, \ j = 7, \ldots, 12, \\ &k_1k_2, \quad k = 13, \ldots, 18. \end{aligned} \tag{4.20}$$

In this case the lines i_1k_1, i_1k_2, i_2k_1 and i_2k_2 cannot contain any one of j_1, j_2, j_3, j_4 so that we shall have lines

$$i_1j_5k_1, \qquad i_1j_6k_2, \qquad i_2j_6k_1, \qquad i_2j_5k_2. \tag{4.21}$$

This means that in the Latin square we have the array:

$$\begin{array}{cc|cc|} & & j_5 & j_6 \ \text{columns} \\ \cline{3-4} & i_1 & k_1 & k_2 \\ \text{rows} & & & \\ & i_2 & k_2 & k_1 \\ \cline{3-4} \end{array} \tag{4.22}$$

Thus for a 2, 4, 2 intersection pattern there must be a two by two Latin subsquare of our Latin square, which is a transposition here (k_1, k_2) in the cycle pattern between two rows, here i_1 and i_2. Thus given the lines of (4.17) corresponding to the square (4.18) the intersection pattern

$$\begin{aligned} &1, 4 \\ &8, 9, 11, 12 \\ &13, 16 \end{aligned} \tag{4.23}$$

corresponds to the subsquare of (4.18) in the 1st and 4th rows

$$\begin{matrix} 13 & 16 \\ 16 & 13 \end{matrix} \tag{4.24}$$

The six by six Latin squares have been completely classified. There are up to renumbering of rows, columns, and digits a total of 22 squares. The square (4.18) has 27 transpositions, one other had 19, and the rest at most 15. In this case there are at most 27 C_i intersecting C_1, C_2, C_3 in 2, 4, 2 and 27 in 4, 2, 2 patterns.

Now suppose the intersection is 3, 3, 1 or 3, 3, 3. Then in

$$\begin{array}{lcl} i_1i_2i_3 & & i_1i_2i_3 \\ j_1j_2j_3 & \text{or} & j_1j_2j_3 \\ k_1 & & k_1k_2k_3 \end{array} \qquad (4.25)$$

we shall have lines in both cases

$$\begin{array}{l} i_1j_4k_1 \\ i_2j_5k_1 \\ i_3j_6k_1 \end{array} \qquad (4.26)$$

so that $i_1i_2i_3$ and k_1 determine $j_1j_2j_3$. As there are $\binom{6}{3}=20$ choices for i_1, i_2, i_3 and 6 choices for k_1 there are at most $20\cdot 6=120$ C_i whose intersection pattern with $C=C_1$ is given by (4.25). Apart from $C_2=7,\ldots,18$ and $C_3=C_1+C_2$ this gives at most $27+27+120=154C_i$ with $|CC_i|=6$, and so $r_6\leqslant 156$. But enough of these possible patterns conflict with Lemma 4.5 so that the number is appreciably less. In general consider cases

$$\begin{array}{ll} 1,2,3 & \\ j_1j_2j_3 & k=13,\ldots,18. \\ k & \end{array} \qquad (4.27)$$

Conceivably we may have the following sort of patterns

$$\begin{array}{cccccc} 1,2,3 & 1,\ 2,\ 3 & 1,\ 2,\ 3 & 1,\ 2,\ 3 & 1,2,\ 3 & 1,\ 2,\ 3 \\ 7,8,9 & 7,10,11 & 8,10,11 & 9,11,12 & 7,8,12 & 9,10,12 \\ 13 & 14 & 15 & 16 & 17 & 18 \end{array} \qquad (4.28)$$

If we take the first of these as C_i and the fifth as C_j, then $|CC_iC_j|=1, 2, 3, 7, 8$ in conflict with Lemma 4.5. However, the first four do not conflict with each other and it is easy to see that in (4.27) at most 4 choices of k give choices of $j_1j_2j_3$ which do not have two j's (or three) in common. Thus in general for the pattern (4.25) there are at most $20\cdot 4=80$ consistent patterns.

Similarly there will be conflicts between the 2, 4, 2; 4, 2, 2; and the 3, 3, 1, or 3, 3, 3. If we have the 2, 4, 2 pattern (4.20), then because of the lines (4.21) we will have as permissible patterns

$$\begin{array}{lcl} i_1\ \ i_2\ \ i_3, & & i_1\ \ i_2\ \ i_4, \\ j_1\ \ j_2\ \ j_3, & \text{and} & j_1\ \ j_2\ \ j_3, \\ k_1, & & k_2. \end{array} \qquad (4.29)$$

Here from (4.20) and (4.29) we have three ovals intersecting in i_1i_2, j_1, j_2, j_3 in conflict with Lemma 4.5.

A more detailed examination of (4.17) shows that we may assume $r_6\leqslant 90$ in this case and a more casual examination of the other cases shows that we can take $r_6\leqslant 100$ throughout.

Trivially from Lemma 4.6 we have $r_4\leqslant\binom{12}{4}\cdot 11=495\cdot 11=5445$ as there are

$\binom{12}{4}=495$ choices of 4 points on C and through each of these there at most 11 further ovals.

Along with (4.14) and $a_{12}=1+r_0+r_4+r_6$ this gives $a_{12}\leqslant\frac{37}{4}+\frac{37}{12}r_4+\frac{37}{8}r_6$ and with $r_4\leqslant 5445$, $r_6\leqslant 100$ we have $a_{12}\leqslant 17{,}270$. In order to have our desired bound $a_{12}\leqslant 1211$, we need to have $r_4\leqslant 239$. Under Assumption C this is probably true but would require some detailed study. We note that the sum of two ovals intersecting in 4 points is an M_{16} so that if $a_{16}=0$, then $r_4=0$ and again we conclude that a primitive M_{20} exists.

It is not known how many isomorphism types of primitive M_{20}'s there are, but Richard Anstee of the California Institute of Technology has developed a computer program to test these for completion to a full plane. Some can be eliminated almost immediately. Consider the following primitive M_{20}:

1, 2, 3, 4	4, 7, 10, 13
1, 5, 6, 7	4, 5, 15, 19
1, 8, 9, 10	4, 9, 16, 18
1, 11, 12, 13	5, 9, 17, 20
2, 5, 8, 11	6, 13, 17, 19
2, 6, 10, 14	7, 11, 15, 18
2, 12, 16, 20	8, 12, 17, 18
3, 6, 9, 12	10, 11, 16, 19
3, 7, 8, 14	14, 15, 16, 17
3, 13, 15, 20	14, 18, 19, 20

(4.30)

It can be quickly shown that this cannot be completed to a plane. Let P be the intersection of the two lines 3, 6, 9, 12 and 4, 7, 10, 13. Then every line through P contains an even number of points of the M_{20} and so a set of 0, 2, or 4 points. And of course P is joined exactly once to each of the remaining 12 points. The machine calculates the possible lines and then attempts to put them together consistently. The possible lines through P are

3, 6, 9, 12, P

4, 7, 10, 13, P

2, 5, 8, 11

14, 15, 16, 17

14, 18, 19, 20

(4.31)

1,14	2,15	5,14	8,15	11,14	1,14	1,15	1,16
1,15	2,17	5,16	8,16	11,17	5,14	2,15	5,16
1,16	2,18	5,18	8,19	11,20	11,14	8,15	8,16
1,17	2,19		8,20				
1,18				1,17	1,18	1,19	1,20
1,19				2,17	2,18	2,19	8,20
1,20				11,17	5,18	8,19	11,20

The machine then finds that there is no combination of these which includes all of the 12 remaining points each exactly once. This can also be checked by hand. For example if no one of the 4 point lines is used we see that 1, 2, 8, 11 must be paired with 15, 17, 19, 20 in some order. This then forces the use of the pairs 5, 16 and also 5, 18, clearly a conflict.

References

[1] D.W. Erbach, Code theoretic evidence concerning the plane of order 10, Dissertation, Cambridge University (1976).
[2] F.J. MacWilliams, N.J.A. Sloane and J.G. Thompson, On the existence of a projective plane of order 10, J. Combin. Theory 14 (A) (1973) 66–78.
[3] P. Delsarte, An algebraic approach to the association schemes of coding theory, Philips Res. Rep. Suppl. 10 (1973).

Annals of Discrete Mathematics 6 (1980) 175–188

CONSTRUCTION OF OPTIMAL CODES AND OPTIMAL FRACTIONAL FACTORIAL DESIGNS USING LINEAR PROGRAMMING

Noboru HAMADA

Department of Mathematics, Faculty of Science, Hiroshima University, Hiroshima, Japan

Fumikazu TAMARI

Department of Mathematics, Fukuoka University of Education, Akama, Munakata, Fukuoka, Japan

1. Introduction

Let $V(n; s)$ be an n-dimensional vector space over a Galois field GF(s) of order s, i.e.,

$$V(n; s) = \{\boldsymbol{x} : \boldsymbol{x}^{\mathrm{T}} = (x_1, x_2, \ldots, x_n), x_i \in \mathrm{GF}(s)\}$$

where n is a positive integer and s is a prime or prime power. A k-dimensional subspace C of $V(n; s)$ is said to be an $(n, k, d; s)$-*code* (or an s-ary linear code with code length n, the number of information symbols k and the minimum distance d) if the minimum distance of the code C is equal to d, that is,

$$\min\{d(\boldsymbol{\alpha}, \boldsymbol{\beta}) : \boldsymbol{\alpha}, \boldsymbol{\beta} \in C, \boldsymbol{\alpha} \neq \boldsymbol{\beta}\} = d$$

where $d(\boldsymbol{\alpha}, \boldsymbol{\beta})$ denotes the Hamming distance between two vectors $\boldsymbol{\alpha}^{\mathrm{T}} = (\alpha_1, \alpha_2, \ldots, \alpha_n)$ and $\boldsymbol{\beta}^{\mathrm{T}} = (\beta_1, \beta_2, \ldots, \beta_n)$ (i.e., the number of integers i $(1 \leq i \leq n)$ such that $\alpha_i \neq \beta_i$). Since the Hamming distance between two vectors $\boldsymbol{\alpha}$ and $\boldsymbol{\beta}$ is the number of positions in which they differ, the minimum distance d for a linear code C equals the minimum weight of its nonzero vectors, i.e.,

$$\min\{w(\boldsymbol{\gamma}) : \boldsymbol{\gamma} \in C, \boldsymbol{\gamma} \neq \boldsymbol{0}\} = d$$

where $w(\boldsymbol{\gamma})$ denotes the Hamming weight of a vector $\boldsymbol{\gamma}$ (i.e., the number of nonzero components of the vector $\boldsymbol{\gamma}$).

It is well-known [3, 20] that if the elements of an $(n, k, d; s)$-code C are used as codewords over an s-ary noisy channel, the code is capable of correcting all patterns of $[\frac{1}{2}(d-1)]$ or fewer errors where $[x]$ denotes the greatest integer not exceeding x. Hence it is desirable to obtain an $(n, k, d; s)$-code C (called an *optimal linear code*) whose minimum distance d is maximum among $(n, k, *; s)$-codes in order to obtain a linear code which is capable of correcting most errors for given integers n, k and s. This problem is known [4, 5] as "the packing problem" or "the problem of constructing maximal t-linearly independent sets" and has been investigated by many authors. It is also known that in order to obtain an optimal linear code and an optimal fractional factorial design (cf. [4] and [5]), it is sufficient to solve the following Problem A, instead of solving the above problem, for any integers k, d and s such that $k \geq 2$ and $d \geq 1$.

Problem A. Find a linear code C whose code length n is minimum among $(*, k, d; s)$-codes for given integers k, d and s.

A lower bound (cf. (2.5) in Section 2) for n of Problem A was given by Griesmer [11] for the case $s=2$ and by Solomon and Stiffler [25] for the general case. Hence in order to solve Problem A, it is sufficient to construct an $(n, k, d; s)$-code which attains the lower bound (2.5) for given integers k, d and s in the case where there exists such an $(n, k, d; s)$-code.

The purpose of this paper is to show that Problem A is equivalent to the Problem B of a linear programming which has some geometrical structure and to give a geometrical method of constructing a solution of Problem B using flats and spreads in a finite projective geometry and to obtain a necessary and sufficient condition for integers k, d and s that there exists such a geometrical solution of Problem B for given integers k, d and s. Since there is no space to give the proof of Theorem 4.2, we gave the proof of Theorem 4.2 in [17] and [18]. Using these methods, we can obtain solutions of Problems A and B for any integers k, d and s if d is sufficiently large in comparison with given integers k and s, and even if d is a small integer, we can obtain solutions of Problems A and B for many integers k, d and s.

In the following, let k and d be any given integers such that $k \geq 2$ and $d \geq 1$ and let s be any given prime or prime power and let us denote by $\theta_0 + \theta_1 s + \cdots + \theta_{k-2} s^{k-2}$ and θ_{k-1}, the remainder and the quotient of $d-1$, respectively, when it is divided by s^{k-1}, i.e.,

$$d = 1 + \theta_0 + \theta_1 s + \cdots + \theta_{k-2} s^{k-2} + \theta_{k-1} s^{k-1} \tag{1.1}$$

where θ_i's are integers such that $0 \leq \theta_i \leq s-1$ for $i = 0, 1, \ldots, k-2$ and $\theta_{k-1} \geq 0$.

2. Preliminary results

With the help of the Galois field GF(s), we can define a *finite projective geometry* PG(t, s) of t dimensions as a set of points satisfying the following conditions:

(a) A point in PG(t, s) is represented by (ν) where ν is a nonzero element of GF(s^{t+1}).

(b) Two points (ν_1) and (ν_2) represent the same point when and only when there exists a nonzero element σ of GF(s) such that $\nu_1 = \sigma \nu_2$.

(c) A μ-flat, $0 \leqslant \mu \leqslant t-1$, in PG$(t, s)$ is defined as a set of points

$$\{(a_0\nu_0 + a_1\nu_1 + \cdots + a_\mu \nu_\mu): \cdots\}$$

where a_i's run independently over the elements of GF(s) and are not all simultaneously zero and $\nu_0, \nu_1, \ldots, \nu_\mu$ are linearly independent elements of GF(s^{t+1}) over the coefficient field GF(s).

Let $C(k; s) = \{\boldsymbol{c}_1, \boldsymbol{c}_2, \ldots, \boldsymbol{c}_{v_k}\}$ be a set of $v_k = (s^k - 1)/(s-1)$ nonzero vectors $\boldsymbol{c}_j (j = 1, 2, \ldots, v_k)$ in $V(k; s)$ such that there is no pair $(\boldsymbol{u}, \boldsymbol{w})$ in $C(k; s)$ such that $\boldsymbol{u} \sim \boldsymbol{w}$ where "$\boldsymbol{u} \sim \boldsymbol{w}$" means that there exists a nonzero element σ in GF(s) such that $\boldsymbol{u} = \sigma \boldsymbol{w}$. Then v_k vectors in $C(k; s)$ can be regarded as v_k points in PG$(k-1, s)$.

Let H_i $(i = 1, 2, \ldots, v_k)$ be v_k hyperplanes in PG$(k-1, s)$, that is, let $H_i = \{\boldsymbol{c}: \boldsymbol{h}_i^{\mathrm{T}} \boldsymbol{c} = 0$ over GF(s), $\boldsymbol{c} \in C(k; s)\}$ for $i = 1, 2, \ldots, v_k$ where $\{\boldsymbol{h}_1, \boldsymbol{h}_2, \ldots, \boldsymbol{h}_{v_k}\}$ $(\equiv H(k; s))$ is a set of v_k nonzero vectors $\boldsymbol{h}_i (i = 1, 2, \ldots, v_k)$ in $V(k; s)$ such that there is no pair $(\boldsymbol{u}, \boldsymbol{w})$ in $H(k; s)$ such that $\boldsymbol{u} \sim \boldsymbol{w}$. Let $N = \|n_{ij}\|$ $(i = 1, 2, \ldots, v_k$, $j = 1, 2, \ldots, v_k)$ be the incidence matrix of v_k points $\boldsymbol{c}_j$ $(j = 1, 2, \ldots, v_k)$ in PG$(k-1, s)$ and v_k hyperplanes $H_i (i = 1, 2, \ldots, v_k)$ in PG$(k-1, s)$ where

$$n_{ij} = \begin{cases} 1, & \text{if } \boldsymbol{c}_j \in H_i, \text{ i.e., } \boldsymbol{h}_i^{\mathrm{T}} \boldsymbol{c}_j = 0 \text{ over GF}(s), \\ 0, & \text{otherwise.} \end{cases}$$

Consider the following linear programming (cf. McCluskey [19] for the case $s = 2$):

Problem B. Find a vector $\boldsymbol{x}^{\mathrm{T}} = (x_1, x_2, \ldots, x_{v_k})$ of nonnegative integers x_j $(j = 1, 2, \ldots, v_k)$ that minimizes the summation $\sum_{j=1}^{v_k} x_j$ subject to the following inequalities:

$$\sum_{j=1}^{v_k} (1 - n_{ij}) x_j \geqslant d \qquad (i = 1, 2, \ldots, v_k) \tag{2.1}$$

for given integers k, d and s.

Theorem 2.1. *Problem* A *is equivalent to Problem* B.

Proof. Let C be a solution for Problem A, that is, a linear code whose code length n is minimum among $(*, k, d; s)$-codes for given integers k, d and s, and let G be a $k \times n$ matrix whose row vectors are generator of C. Since n is minimum, any column vector of G is not equal to a zero vector $\boldsymbol{0}$. Let us denote by x_j $(1 \leqslant j \leqslant v_k)$, the number of column vectors $\boldsymbol{g}$ of G such that $\boldsymbol{g} \sim \boldsymbol{c}_j$ for a given vector $\boldsymbol{c}_j$ in $C(k; s)$. Then it follows from "$w(\boldsymbol{e}) = \sum_{i=1}^{n} w(e_i)$ for $\boldsymbol{e}^{\mathrm{T}} = (e_1, e_2, \ldots, e_n)$" that

$$w(G^{\mathrm{T}} \boldsymbol{h}_i) = \sum_{j=1}^{v_k} x_j w(\boldsymbol{c}_j^{\mathrm{T}} \boldsymbol{h}_i) = \sum_{j=1}^{v_k} x_j (1 - n_{ij}) \tag{2.2}$$

for any vector $\boldsymbol{h}_i$ in $H(k; s)$. Since $\sum x_j = n$, $G^{\mathrm{T}} \boldsymbol{h}_i \in C$ and $w(\boldsymbol{e}) \geqslant d$ for any vector

$\boldsymbol{e}$ ($\neq \boldsymbol{0}$) in C, $\boldsymbol{x}^{\mathrm{T}} = (x_1, x_2, \ldots, x_{v_k})$ is a vector of nonnegative integers x_j that minimizes the summation $\sum x_j$ subject to condition (2.1), that is, $\boldsymbol{x}$ is a solution of Problem B.

Conversely, let $\boldsymbol{x}^{\mathrm{T}} = (x_1, x_2, \ldots, x_{v_k})$ be a solution for Problem B. Let

$$G = [\overbrace{\boldsymbol{c}_1, \ldots, \boldsymbol{c}_1}^{x_1}; \overbrace{\boldsymbol{c}_2, \ldots, \boldsymbol{c}_2}^{x_2}; \cdots; \overbrace{\boldsymbol{c}_{v_k}, \ldots, \boldsymbol{c}_{v_k}}^{x_{v_k}}] \tag{2.3}$$

and let C be the subspace over GF(s) generated by column vectors of G^{T}.

In order to show that C is a solution of Problem A, that is, C is an $(n, k, d; s)$-code whose code length n is minimum among $(*, k, d; s)$-codes, it is sufficient to show that (a) $w(G^{\mathrm{T}}\boldsymbol{h}) \geq d$ for any nonzero vector $\boldsymbol{h}$ in $V(k; s)$ and (b) the rank, Rank (G), of the matrix G is equal to k since the code length of C is equal to $\sum x_j$ and $\sum x_j$ is minimum.

Since there exists a unique vector $\boldsymbol{h}_i$ in $H(k; s)$ such that $\boldsymbol{h} \sim \boldsymbol{h}_i$ for any nonzero vector $\boldsymbol{h}$ in $V(k; s)$ and $w(G^{\mathrm{T}}\boldsymbol{h}) = W(G^{\mathrm{T}}\boldsymbol{h}_i)$ if $\boldsymbol{h} \sim \boldsymbol{h}_i$, it follows from (2.2) and (2.3) that (a) holds. If Rank$(G) \neq k$, i.e., Rank$(G) \leq k-1$, there exists a vector $\boldsymbol{h}_\alpha$ in $H(k; s)$ such that $G^{\mathrm{T}}\boldsymbol{h}_\alpha = \boldsymbol{0}$ over GF(s). This implies that there exists a hyperplane H_α in PG$(k-1, s)$ which contains all column vectors of G. Hence it follows that if $x_j \neq 0$, then $n_{\alpha j} = 1$, i.e.,

$$\sum_{j=1}^{v_k} (1 - n_{\alpha j}) x_j = 0,$$

which contradicts (2.1). Hence (b) holds. This completes the proof.

It is easy to see that the lower bound for the code length n in Problem A which is due to Solomon and Stiffler [25] can be also expressed as follows:

Theorem 2.2. *For any vector $\boldsymbol{x}$ of nonnegative integers x_j $(j = 1, 2, \ldots, v_k)$ which satisfy the inequalities* (2.1),

$$\sum_{j=1}^{v_k} x_j \geq k + \theta_0 v_1 + \theta_1 v_2 + \cdots + \theta_{k-1} v_k \tag{2.4}$$

if d is expressed by (1.1) *where $v_i = (s^i - 1)/(s-1)$ for $i = 1, 2, \ldots, k$.*

Theorem 2.3. *For any $(n, k, d; s)$-code,*

$$n \geq k + \theta_0 v_1 + \theta_1 v_2 + \cdots + \theta_{k-1} v_k \tag{2.5}$$

if d is expressed by (1.1).

Theorems 2.1 and 2.2 show that in order to obtain a solution of Problem A for given integers k, d and s, it is sufficient to obtain a vector $\boldsymbol{x}$ of nonnegative integers x_j $(j = 1, 2, \ldots, v_k)$ which satisfy the inequalities (2.1) and achieve the lower bound (2.4) for given integers k, d and s in the case where there exists such a vector $\boldsymbol{x}$. In the next section, we shall give a geometrical method of constructing such a vector $\boldsymbol{x}$ using flats in a finite projective geometry.

3. Construction of a solution $\{x_j\}$ of Problem B

Let k, d and s be given integers and let $\varepsilon_i = (s-1) - \theta_i$ for $i = 0, 1, \ldots, k-2$ and let $D = \{\mu: \varepsilon_\mu \neq 0, 0 \leq \mu \leq k-2\}$ where θ_i's are integers given by (1.1). Let $\mathcal{B}$ be a set of ε_0 0-flats, ε_1 1-flats, ..., ε_{k-3} $(k-3)$-flats and ε_{k-2} $(k-2)$-flats in PG $(k-1, s)$, i.e., let

$$\mathcal{B} = \{V_i^{(\mu)}: i = 1, 2, \ldots, \varepsilon_\mu, \mu \in D\} \tag{3.1}$$

where $V_i^{(\mu)}$ $(i = 1, 2, \ldots, \varepsilon_\mu)$ denote (not necessarily distinct) ε_μ μ-flats in PG$(k-1, s)$ for each integer μ in D. In the special case $(\varepsilon_0, \varepsilon_1, \ldots, \varepsilon_{k-2}) = (0, 0, \ldots, 0)$, $\mathcal{B}$ is the empty set $\emptyset$. Let $\eta_j(\mathcal{B})$ $(j = 1, 2, \ldots, v_k)$ be the number of flats $V_i^{(\mu)}$ $(1 \leq i \leq \varepsilon_\mu, \mu \in D)$ in $\mathcal{B}$ which contain the jth point $\boldsymbol{c}_j$ in PG$(k-1, s)$.

There are many ways of choosing a set $\mathcal{B}$ of ε_0 0-flats, ε_1 1-flats, ..., ε_{k-3} $(k-3)$-flats and ε_{k-2} $(k-2)$-flats in PG$(k-1, s)$. Let us denote by $\mathcal{F}(\varepsilon_0, \varepsilon_1, \ldots, \varepsilon_{k-2}; k, s)$, a family of all sets $\mathcal{B}$ which consist of ε_0 0-flats, ε_1 1-flats, ..., ε_{k-3} $(k-3)$-flats and ε_{k-2} $(k-2)$-flats in PG$(k-1, s)$.

Theorem 3.1. *If there exists a set $\mathcal{B}$ in $\mathcal{F}(\varepsilon_0, \varepsilon_1, \ldots, \varepsilon_{k-2}; k, s)$ such that $\max\{\eta_j(\mathcal{B}) - 1: 1 \leq j \leq v_k\} \leq \theta_{k-1}$ for given integers k, d and s, the vector $\boldsymbol{x}$ whose jth component x_j $(1 \leq j \leq v_k)$ is given by*

$$x_j = \theta_{k-1} - (\eta_j(\mathcal{B}) - 1) \tag{3.2}$$

is a solution of Problem B *for given integers k, d and s where $\varepsilon_i = (s-1) - \theta_i$ for $i = 0, 1, \ldots, k-2$ and θ_i's are integers given by* (1.1).

Proof. Since any μ-flats $V_i^{(\mu)}$ $(i = 1, 2, \ldots, \varepsilon_\mu)$ in $\mathcal{B}$ consist of $v_{\mu+1}$ points in PG$(k-1, s)$, it follows that

$$\sum_{j=1}^{v_k} \eta_j(\mathcal{B}) = \sum_{\mu=0}^{k-2} \varepsilon_\mu v_{\mu+1}.$$

Hence

$$\sum_{j=1}^{v_k} x_j = (\theta_{k-1} + 1)v_k - \sum_{j=1}^{v_k} \eta_j(\mathcal{B}) = k + \sum_{i=0}^{k-1} \theta_i v_{i+1},$$

which shows that $\{x_j\}$ achieves the lower bound (2.4). If

$$\theta_{k-1} \geq \max\{\eta_j(\mathcal{B}) - 1: 1 \leq j \leq v_k\},$$

each x_j given by (3.2) is a nonnegative integer. Hence it is sufficient to show that the vector $\boldsymbol{x}$ satisfies condition (2.1).

Since $\eta_j(\mathscr{B})$ denotes the number of flats $V_l^{(\mu)}$ $(1 \leqslant l \leqslant \varepsilon_\mu, \mu \in D)$ in $\mathscr{B}$ such that $\mathbf{c}_j \in V_l^{(\mu)}$ and $n_{ij} = 0$ if and only if $\mathbf{c}_j \in H_i^c$, it follows that

$$\sum_{j=1}^{v_k} (1 - n_{ij})\eta_j(\mathscr{B}) = \sum_{\mu \in D} \sum_{l=1}^{\varepsilon_\mu} |H_i^c \cap V_l^{(\mu)}| \tag{3.3}$$

for $i = 1, 2, \ldots, v_k$ where H_i^c and $|S|$ denote the complement of H_i and the cardinality of the set S, respectively. Since $|H_i^c \cap V_l^{(\mu)}| = |V_l^{(\mu)}| - |H_i \cap V_l^{(\mu)}|$ and $|H_i \cap V_l^{(\mu)}| = v_\mu$ or $v_{\mu+1}$, we have $|H_i^c \cap V_l^{(\mu)}| \leqslant v_{\mu+1} - v_\mu = s^\mu$, i.e.,

$$\sum_{\mu \in D} \sum_{l=1}^{\varepsilon_\mu} |H_i^c \cap V_l^{(\mu)}| \leqslant \sum_{\mu=0}^{k-2} \varepsilon_\mu s^\mu.$$

Hence it follows from (3.3) that

$$\sum_{j=1}^{v_k} (1 - n_{ij}) x_j \geqslant (\theta_{k-1} + 1)s^{k-1} - \sum_{\mu=0}^{k-2} \varepsilon_\mu s^\mu = d$$

for $i = 1, 2, \ldots, v_k$. This completes the proof.

Remark 3.1. Theorem 3.1 shows that if there exists a set $\mathscr{B}$ in $\mathscr{F}(\varepsilon_0, \varepsilon_1, \ldots, \varepsilon_{k-2}; k, s)$ such that $\max\{\eta_j(\mathscr{B}) - 1 : 1 \leqslant j \leqslant v_k\} \leqslant \theta_{k-1}$ for given integers k, d and s, there exists an $(n, k, d; s)$-code which achieves the lower bound (2.5) for given integers k, d and s.

Corollary 3.1. *In the special case* $k = 2$, *there exists an* $(n, 2, d; s)$*-code which achieves the lower bound* (2.5) *for any integers* d *and* s *such that* $d \geqslant 1$.

Proof. Let us denote v_2 $(= s + 1)$ vectors in $C(2; s)$ by

$$\mathbf{c}_1^{\mathrm{T}} = (1, 0), \qquad \mathbf{c}_2^{\mathrm{T}} = (0, 1) \quad \text{and} \quad \mathbf{c}_{j+2}^{\mathrm{T}} = (1, \alpha^j) \tag{3.4}$$

for $j = 0, 1, \ldots, s-2$ where α is a primitive element of GF(s). Let $d = 1 + \theta_0 + \theta_1 s$ and $\varepsilon_0 = (s-1) - \theta_0$ and let $\mathscr{B} = \emptyset$ or $\{\mathbf{c}_j : j = 1, 2, \ldots, \varepsilon_0\}$ according as ε_0 is zero or not where $0 \leqslant \theta_0 \leqslant s-1$ and $\theta_1 \geqslant 0$. Then $\mathscr{B} \in \mathscr{F}(\varepsilon_0; 2, s)$ and $\eta_j(\mathscr{B}) = 1$ or 0 according as the jth vector $\mathbf{c}_j$ in $C(2; s)$ is contained in $\mathscr{B}$ or not. Hence it follows from Theorem 3.1 and $\max\{\eta_j(\mathscr{B}) - 1 : 1 \leqslant j \leqslant v_2\} \leqslant 0$ that the vector $\boldsymbol{x}$ whose jth component x_j $(1 \leqslant j \leqslant v_2)$ is given by

$$x_j = \begin{cases} \theta_1 & \text{if} \quad c_j \in \mathscr{B}, \\ \theta_1 + 1 & \text{otherwise} \end{cases} \tag{3.5}$$

is a solution of Problem B for any integers θ_0, θ_1 and s such that $0 \leqslant \theta_0 \leqslant s-1$ and $\theta_1 \geqslant 0$ (i.e., any integers d and s). This completes the proof.

Corollary 3.2. *In the case $k=3$, there exists an $(n, 3, d; s)$-code which achieves the lower bound (2.5) for any integers d and s such that $d \geq s^2-2s+1$.*

Proof. Let $d=1+\theta_0+\theta_1 s+\theta_2 s^2$ and $\varepsilon_i=(s-1)-\theta_i$ for $i=0, 1$ where θ_i's are integers such that $0 \leq \theta_0$, $\theta_1 \leq s-1$ and $\theta_2 \geq 0$.

(i) In the case $\theta_2 \geq 1$ and $0 \leq \theta_0$, $\theta_1 \leq s-1$ (i.e., $d \geq s^2+1$), it is sufficient to show that there exists a set $\mathfrak{A}$ of $s-1$ 0-flats and $s-1$ 1-flats such that $\eta_j(\mathfrak{A}) \leq 2$ for $j=1, 2, \ldots, v_3$ in order to show that there exists a set $\mathcal{B}$ of ε_0 0-flats and ε_1 1-flats such that $\max\{\eta_j(\mathcal{B})-1\colon 1 \leq j \leq v_3\} \leq \theta_2$ for any integers θ_0 and θ_1 such that $0 \leq \theta_0$, $\theta_1 \leq s-1$. Because

$$\max\{\eta_j(\mathcal{B})-1 : 1 \leq j \leq v_3\} \leq \max\{\eta_j(\mathfrak{A})-1 : 1 \leq j \leq v_3\} \leq 1 \leq \theta_2$$

if $\mathcal{B} \subset \mathfrak{A}$.

Let α be a primitive element of GF(s) and let $\boldsymbol{e}_i^{\mathrm{T}}=(1, \alpha^i, \alpha^{-i})$ for $i=1, 2, \ldots, s-1$. Let $V_j^{(1)}$ $(j=1, 2, \ldots, s-1)$ be 1-flats in PG(2, s) defined by $V_j^{(1)}=\{\boldsymbol{c} : \boldsymbol{e}_j^{\mathrm{T}}\boldsymbol{c}=0$ over GF(s), $\boldsymbol{c} \in C(3; s)\}$ and let $V_i^{(0)}$ $(i=1, 2, \ldots, s-1)$ be any distinct points (i.e., 0-flats) which are not contained in $\{V_j^{(1)} : j=1, 2, \ldots, s-1\}$ and let

$$\mathfrak{A}=\{V_i^{(0)} : i=1, 2, \ldots, s-1\}+\{V_j^{(1)} : j=1, 2, \ldots, s-1\}.$$

Then it is obvious that $\eta_j(\mathfrak{A}) \leq 2$ for $j=1, 2, \ldots, v_3$ in the case $s=2$ or 3. In the case $s \geq 4$, it is easy to see that any three vectors in $\{\boldsymbol{e}_1, \boldsymbol{e}_2, \ldots, \boldsymbol{e}_{s-1}\}$ are linearly independent over GF(s). This implies that

$$V_\alpha^{(1)} \cap V_\beta^{(1)} \cap V_\gamma^{(1)}=\emptyset$$

for any integers α, β and γ such that $1 \leq \alpha < \beta < \gamma \leq s-1$. Hence $\eta_j(\mathfrak{A}) \leq 2$ for $j=1, 2, \ldots, s-1$ even if $s \geq 4$.

(ii) In the case $\theta_2=0$, $\theta_1=s-1$ and $0 \leq \theta_0 \leq s-1$ (i.e., $s^2-s+1 \leq d < s^2+1$), let

$$\mathcal{B}=\{V_i^{(0)} : i=1, 2, \ldots, \varepsilon_0\}$$

where $V_i^{(0)}$ $(i=1, 2, \ldots, s-1)$ are any distinct points in PG(2, s) and $\mathcal{B}=\emptyset$ in the case $\varepsilon_0=0$. Then $\mathcal{B}$ is a set in $\mathcal{F}(\varepsilon_0, 0; 3, s)$ such that

$$\max\{\eta_j(\mathcal{B})-1 : 1 \leq j \leq v_3\} \leq 0$$

for any integer ε_0 $(0 \leq \varepsilon_0 \leq s-1)$.

(iii) In the case $\theta_2=0$, $\theta_1=s-2$ and $0 \leq \theta_0 \leq s-1$ (i.e., $s^2-2s+1 \leq d < s^2-s+1$), let

$$\mathcal{B}=\{V_1^{(1)}\}+\{V_j^{(0)} : j=1, 2, \ldots, \varepsilon_0\}$$

where $V_1^{(1)}$ is any 1-flat in PG(2, s) and $V_j^{(0)}$ $(j=1, 2, \ldots, s-1)$ are any distinct points in PG(2, s) which are not contained in $V_1^{(1)}$. Then $\mathcal{B}$ is a set in $\mathcal{F}(\varepsilon_0, 1; 3, s)$ such that

$$\max\{\eta_j(\mathcal{B})-1 : 1 \leqslant j \leqslant v_3\} = 0.$$

Hence it follows from Theorem 3.1 that Corollary 3.2 holds.

Let k, d and s be given integers such that $(\theta_0, \theta_1, \ldots, \theta_{k-2}) \neq (s-1, s-1, \ldots, s-1)$, i.e., $(\varepsilon_0, \varepsilon_1, \ldots, \varepsilon_{k-2}) \neq (0, 0, \ldots, 0)$ and let

$$D = \{\mu : \varepsilon_\mu \neq 0, 0 \leqslant \mu \leqslant k-2\} = \{l_1 - 1, l_2 - 1, \ldots, l_t - 1\}$$

where $\varepsilon_i = (s-1) - \theta_i$, $|D| = t$ and $1 \leqslant l_1 < l_2 < \cdots < l_t \leqslant k-1$. Then we have the following result which is due to Solomon and Stiffler [25].

Corollary 3.3. *If k, d and s are integers such that*

$$\theta_{k-1} \geqslant \max\{\varepsilon_\mu - 1 : 0 \leqslant \mu \leqslant k-2\} \quad \text{and} \quad \sum_{i=1}^{t} l_i \leqslant k, \tag{3.6}$$

there exists an $(n, k, d; s)$-code which achieves the lower bound (2.5).

Proof. Since $\sum_{i=1}^t l_i \leqslant k$, there exists a set of t flats W_r $(r = 1, 2, \ldots, t)$ in PG($k-1, s$) such that $W_i \cap W_j = \emptyset$ for any integers i and j $(1 \leqslant i < j \leqslant t)$ where W_r denotes an $(l_r - 1)$-flat for $r = 1, 2, \ldots, t$. Let

$$V_1^{(l_i-1)} = V_2^{(l_i-1)} = \cdots = V_{s-1}^{(l_i-1)} = W_i$$

for each i and let

$$\mathcal{B} = \{V_j^{(l_i-1)} : j = 1, 2, \ldots, \varepsilon_{l_i-1}, i = 1, 2, \ldots, t\}.$$

Then $\mathcal{B}$ is a set in $\mathcal{F}(\varepsilon_0, \varepsilon_1, \ldots, \varepsilon_{k-2}; k, s)$ such that

$$\max\{\eta_j(\mathcal{B}) - 1 : 1 \leqslant j \leqslant v_k\} = \max\{\varepsilon_\mu - 1 : 0 \leqslant \mu \leqslant k-2\} \leqslant \theta_{k-1}.$$

Hence we have the required result from Theorem 3.1.

From the actual point of view, it is desirable to obtain a solution of Problem A for comparatively small integers k, d (or θ_{k-1}) and s. On the other hand, Theorem 3.1 and Remark 3.1 show that there exists an $(n, k, d; s)$-code which achieves the lower bound (2.5) if there exists a set $\mathcal{B}$ in $\mathcal{F}(\varepsilon_0, \varepsilon_1, \ldots, \varepsilon_{k-2}; k, s)$ such that

$$\max\{\eta_j(\mathcal{B}) - 1 : 1 \leqslant j \leqslant v_k\} \leqslant \theta_{k-1}$$

for given integers k, d and s. Hence it is necessary to obtain a set $\mathcal{B}$ in $\mathcal{F}(\varepsilon_0, \varepsilon_1, \ldots, \varepsilon_{k-2}; k, s)$ such that

$$\max\{\eta_j(\mathcal{B}) - 1 : 1 \leqslant j \leqslant v_k\}$$

is minimum for given integers k, s and ε_j $(j = 0, 1, \ldots, k-2)$. In the next section, we shall obtain a necessary condition for integers k, s and ε_j $(j=0, 1, \ldots, k-2)$ that there exists a set $\mathscr{B}$ in $\mathscr{F}(\varepsilon_0, \varepsilon_1, \ldots, \varepsilon_{k-2}; k, s)$ such that

$$\max\{\eta_j(\mathscr{B})-1 : 1 \leqslant j \leqslant v_k\} \leqslant t$$

for a given nonnegative integer t. In [17] and [18], we showed that a necessary condition for integers k, d and s in Theorem 4.1 is also a sufficient condition in the case $t=0$ or 1 and gave a geometric method of constructing such a set $\mathscr{B}$ in $\mathscr{F}(\varepsilon_0, \varepsilon_1, \ldots, \varepsilon_{k-2}; k, s)$ using spreads in a finite projective geometry.

4. A necessary and sufficient condition for integers k, s and ε_i $(i=0, 1, \ldots, k-2)$ that there exists a set $\mathscr{B}$ in $\mathscr{F}(\varepsilon_0, \varepsilon_1, \ldots, \varepsilon_{k-2}; k, s)$ such that $\max\{\eta_j(\mathscr{B})-1 : 1 \leqslant j \leqslant v_k\} \leqslant t$ for a given integer t

Let k be any integer such that $k \geqslant 3$ and let us denote by $E(k, s)$, a set of ordered sets $(\varepsilon_1, \varepsilon_2, \ldots, \varepsilon_{k-2})$ of integers ε_i $(i=1, 2, \ldots, k-2)$ such that $0 \leqslant \varepsilon_i \leqslant s-1$. Let $E_t(k, s)$ $(t=0, 1, 2, \ldots)$ be a set of ordered sets $(\varepsilon_1, \varepsilon_2, \ldots, \varepsilon_{k-2})$ in $E(k, s)$ such that either

$$\text{(a)} \sum_{i=1}^{k-2} \varepsilon_i \leqslant t+1 \quad \text{or (b)} \quad \sum_{i=1}^{k-2} \varepsilon_i \geqslant t+2$$

and

$$\beta_1+\beta_2+\cdots+\beta_{t+2} \leqslant (t+1)k-(t+2)$$

for the first $t+2$ integers $\beta_1, \beta_2, \ldots, \beta_{t+2}$ in the following series:

$$\overbrace{k-2, k-2, \ldots, k-2}^{\varepsilon_{k-2}}; \overbrace{k-3, k-3, \ldots, k-3}^{\varepsilon_{k-3}}; \cdots; \overbrace{1, 1, \ldots, 1}^{\varepsilon_1}. \tag{4.1}$$

For example, consider the case $k=6$, $s \geqslant 3$ and $(\varepsilon_1, \varepsilon_2, \varepsilon_3, \varepsilon_4)=(2, 2, 0, 1)$. In this case, $\sum_{i=1}^4 \varepsilon_i=5$ and $(\beta_1, \beta_2, \beta_3, \beta_4, \beta_5)=(4, 2, 2, 1, 1)$. Hence $(2, 2, 0, 1) \notin E_0(6, s)$ but $(2, 2, 0, 1) \in E_1(6, s)$. In general,

$$E_0(k, s) \subset E_1(k, s) \subset E_2(k, s) \subset \cdots \quad \text{and} \quad E_j(k, s)=E(k, s) \tag{4.2}$$

for $j=k-2, k-1, \ldots$ since $\beta_i \leqslant k-2$ for each integer i and $\beta_1+\beta_2+\cdots+\beta_k \leqslant (k-1)k-k$ in the case $t=k-2$.

Lemma 4.1. *Let k, s, t and ε_i $(i=0, 1, \ldots, k-2)$ be any integers such that $k \geqslant 3$, $t \geqslant 0$ and $0 \leqslant \varepsilon_i \leqslant s-1$. Then there exists a set $\mathscr{B}$ in $\mathscr{F}(\varepsilon_0, \varepsilon_1, \ldots, \varepsilon_{k-2}; k, s)$ such that $\max\{\eta_j(\mathscr{B})-1 : 1 \leqslant j \leqslant v_k\} \leqslant t$ if and only if there exists a set $\mathfrak{A}$ in $\mathscr{F}(0, \varepsilon_1, \ldots, \varepsilon_{k-2}; k, s)$ such that $\max\{\eta_j(\mathfrak{A})-1 : 1 \leqslant j \leqslant v_k\} \leqslant t$.*

Proof. Suppose that there exists a set $\mathfrak{A}$ in $\mathscr{F}(0, \varepsilon_1, \ldots, \varepsilon_{k-2}; k, s)$ such that $\max\{\eta_j(\mathfrak{A})-1 : 1 \leqslant j \leqslant v_k\} \leqslant t$ for given integers k, s, t and ε_i $(i=1, 2, \ldots, k-2)$.

Since there are v_k points in $PG(k-1, s)$ and

$$v_k - \sum_{i=1}^{k-2} \varepsilon_i v_{i+1} \geqslant v_k - \sum_{i=1}^{k-2} (s-1)v_{i+1} > s-1,$$

there exist $s-1$ points $V_i^{(0)}$ $(i=1, 2, \ldots, s-1)$ in $PG(k-1, s)$ which are not contained in any flat of $\mathfrak{A}$. Let ε_0 be any integer such that $1 \leqslant \varepsilon_0 \leqslant s-1$ and let $\mathcal{B} = \mathfrak{A} + \{V_i^{(0)} : i = 1, 2, \ldots, \varepsilon_0\}$. Then $\mathcal{B}$ is a set in $\mathcal{F}(\varepsilon_0, \varepsilon_1, \ldots, \varepsilon_{k-2}; k, s)$ such that $\max\{\eta_j(\mathcal{B}) - 1 : 1 \leqslant j \leqslant v_k\} \leqslant t$. Converse is obvious.

Theorem 4.1. *A necessary condition for* ε_j $(j = 0, 1, \ldots, k-2)$ *that there exists a set* $\mathcal{B}$ *in* $\mathcal{F}(\varepsilon_0, \varepsilon_1, \ldots, \varepsilon_{k-2}; k, s)$ *such that* $\max\{\eta_j(\mathcal{B}) - 1 : 1 \leqslant j \leqslant v_k\} \leqslant t$ *for given integers k, s and t is that* $0 \leqslant \varepsilon_0 \leqslant s-1$ *and* $(\varepsilon_1, \varepsilon_2, \ldots, \varepsilon_{k-2}) \in E_t(k, s)$.

Proof. Let k, s, t and ε_i $(i = 0, 1, \ldots, k-2)$ be given integers. In the case $\sum_{i=1}^{k-2} \varepsilon_i \leqslant t+1$, it follows from the definition of $\eta_j(\mathfrak{A})$'s that

$$\max\{\eta_j(\mathfrak{A}) - 1 : 1 \leqslant j \leqslant v_k\} \leqslant \sum_{i=1}^{k-2} \varepsilon_i - 1 \leqslant t$$

for any set $\mathfrak{A}$ in $\mathcal{F}(0, \varepsilon_1, \ldots, \varepsilon_{k-2}; k, s)$. Suppose that there exists a set $\mathfrak{A}$ in $\mathcal{F}(0, \varepsilon_1, \ldots, \varepsilon_{k-2}; k, s)$ such that

$$\max\{\eta_j(\mathfrak{A}) - 1 : 1 \leqslant j \leqslant v_k\} \leqslant t$$

for given integers ε_i $(i = 1, 2, \ldots, k-2)$ such that

$$\sum_{i=1}^{k-2} \varepsilon_i \geqslant t+2.$$

Then $\bigcap_{i=1}^{t+2} W_i$ must be the empty set $\emptyset$ for any $t+2$ flats $W_1, W_2, \ldots, W_{t+2}$ in $\mathfrak{A}$. Let W_i^* be the dual space in $PG(k-1, s)$ of W_i and let W^* be the dual space of $\bigcap_{i=1}^{t+2} W_i$. Then it is easy to see that

$$\dim(W^*) + 1 \leqslant \sum_{i=1}^{t+2} \{\dim(W_i^*) + 1\},$$

$$\dim(W_i^*) + 1 = k - \{\dim(W_i) + 1\}$$

and

$$\dim(W^*) + 1 = k$$

where "$\dim(W) = \mu$" means that W is a μ-flat in $PG(k-1, s)$. Hence any $t+2$ flats $W_1, W_2, \ldots, W_{t+2}$ in $\mathfrak{A}$ must satisfy the following condition:

$$\dim(W_1) + \dim(W_2) + \cdots + \dim(W_{t+2}) \leqslant (t+1)k - (t+2). \tag{4.3}$$

Since $\mathfrak{A}$ is a set of ε_{k-2} $(k-2)$-flats, ε_{k-3} $(k-3)$-flats, $\ldots$, ε_2 2-flats and ε_1 1-flats, it follows from (4.3) that the first $t+2$ integers $\beta_1, \beta_2, \ldots, \beta_{t+2}$ in the series (4.1) must satisfy the following condition:

$$\beta_1 + \beta_2 + \cdots + \beta_{t+2} \leqslant (t+1)k - (t+2). \tag{4.4}$$

Hence we have the desired result from Lemma 4.1.

Remark 4.1. It follows from Theorems 2.1, 3.1 and 4.1 that if a necessary condition in Theorem 4.1 is also a sufficient condition for given integers k, s and t, then there exists an $(n, k, d; s)$-code which achieves the lower bound (2.5) for given integers k, d and s provided that

$$\theta_{k-1} \geq t, \quad 0 \leq \theta_0 \leq s-1 \quad \text{and} \quad (s-1-\theta_1, \ldots, s-1-\theta_{k-2}) \in E_t(k, s) \tag{4.5}$$

where θ_i's are integers given by (1.1). Since

$$d = 1 + \theta_0 + \theta_1 s + \cdots + \theta_{k-2} s^{k-2} + \theta_{k-1} s^{k-1} \tag{4.6}$$

and $\theta_{k-1} \geq 0$, it is necessary that θ_{k-1} is a small integer (i.e., t is a small integer) in order that d is a small integer.

Since there is no space to give the proof of the following theorem, we gave the proof and the constructive method of such a set $\mathscr{B}$ in [17] and [18].

Theorem 4.2. *Let k, s and ε_j $(j = 0, 1, \ldots, k-2)$ be any integers such that $k \geq 3$ and $0 \leq \varepsilon_j \leq s-1$. If $0 \leq \varepsilon_0 \leq s-1$ and $(\varepsilon_1, \varepsilon_2, \ldots, \varepsilon_{k-2}) \in E_t(k, s)$ for $t = 0$ or 1, there exists a set $\mathscr{B}$ in $\mathscr{F}(\varepsilon_0, \varepsilon_1, \ldots, \varepsilon_{k-2}; k, s)$ such that $\max\{\eta_j(\mathscr{B}) - 1 : 1 \leq j \leq v_k\} \leq t$.*

Corollary 4.1. *Let k, d and s be any integers such that $k \geq 3$ and $d \geq 1$.*

(i) *If $0 \leq \theta_0 \leq s-1$ and $(s-1-\theta_1, s-1-\theta_2, \ldots, s-1-\theta_{k-2}) \in E_0(k\ s)$, there exists an $(n, k, d; s)$-code which achieves the lower bound (2.5) for any nonnegative integer θ_{k-1} where θ_i's are integers given by (1.1).*

(ii) *If $0 \leq \theta_0 \leq s-1$ and*

$$(s-1-\theta_1, s-1-\theta_2, \ldots, s-1-\theta_{k-2}) \in E_1(k, s) - E_0(k, s),$$

there exists an $(n, k, d; s)$-code which achieves the lower bound (2.5) for any integer θ_{k-1} such that $\theta_{k-1} \geq 1$.

Example 4.1. (i) In the case $k = 4$, we have

$$E_0(4, s) = \{(\varepsilon_1, \varepsilon_2) : 0 \leq \varepsilon_1 \leq s-1, \varepsilon_2 = 0\} + \{(0, 1)\},$$
$$E_1(4, s) = \{(\varepsilon_1, \varepsilon_2) : 0 \leq \varepsilon_1 \leq s-1, \varepsilon_2 = 0, 1, 2\} \quad \text{for} \quad s \geq 3,$$
$$E_2(4, s) = E(4, s) = \{(\varepsilon_1, \varepsilon_2) : 0 \leq \varepsilon_1, \varepsilon_2 \leq s-1\} \quad \text{for} \quad s \geq 2,$$

where $E_1(4, 2) = E(4, 2)$.

(ii) In the case $k=5$, we have

$$E_0(5, s)=\{(\varepsilon_1, \varepsilon_2, \varepsilon_3): 0 \leqslant \varepsilon_1 \leqslant s-1, \varepsilon_2=0, 1, \varepsilon_3=0\}+\{(0, 0, 1)\},$$

$$E_1(5, s)=\{(\varepsilon_1, \varepsilon_2, \varepsilon_3): 0 \leqslant \varepsilon_1, \varepsilon_2 \leqslant s-1, \varepsilon_3=0, 1\}$$
$$+\{(\varepsilon_1, \varepsilon_2, \varepsilon_3): 0 \leqslant \varepsilon_1 \leqslant s-1, \varepsilon_2=0, \varepsilon_3=2\} \quad \text{for} \quad s \geqslant 3,$$

$$E_2(5, s)=\{(\varepsilon_1, \varepsilon_2, \varepsilon_3): 0 \leqslant \varepsilon_1, \varepsilon_2 \leqslant s-1, \varepsilon_3=0, 1, 2, 3\} \quad \text{for} \quad s \geqslant 4,$$

$$E_3(5, s)=E(5, s)=\{(\varepsilon_1, \varepsilon_2, \varepsilon_3): 0 \leqslant \varepsilon_1, \varepsilon_2, \varepsilon_3 \leqslant s-1\} \quad \text{for} \quad s \geqslant 2,$$

where $E_1(5, 2)=E(5, 2)$ and $E_2(5, s)=E(5, s)$ for $s=2, 3$.

Example 4.2. Consider the case $k=4$ and $s=3$. In this case, it follows from Example 4.1 that $E_1(4, 3)=E(4, 3)$ and $E_0(4, 3)=E(4, 3)-\{(1, 1), (2, 1), (0, 2), (1, 2), (2, 2)\}$. Hence it follows from Corollary 4.1 that there exists an $(n, 4, d; s)$-code which achieves the lower bound (2.5) for any integer d such that $d \geqslant 16$. Using the method in [17] and [18], we can construct these codes.

Example 4.3. Consider the case $k=5$, $s=3$ and $(\varepsilon_0, \varepsilon_1, \varepsilon_2, \varepsilon_3)=(2, 0, 0, 1)$ (i.e., $(\theta_0, \theta_1, \theta_2, \theta_3)=(0, 2, 2, 1)$). In this case, it follows from Example 4.1 that $(0, 0, 1) \in E_0(5, 3)$. Since $(\theta_0, \theta_1, \theta_2, \theta_3; \theta_4)=(0, 2, 2, 1; \mathbf{0})$, $(0, 2, 2, 1; \mathbf{1})$, $(0, 2, 2, 1; \mathbf{2})$, $(0, 2, 2, 1; \mathbf{3}), \ldots$ according as $d=52, 133, 214, 295, \ldots$, it follows from Corollary 4.1 that there exists an $(n, 5, d; 3)$-code which achieves the lower bound (2.5) for $d=52, 133, 214, 295, \ldots$.

Example 4.4. In the case $k=5$, $s=2$ and $(\varepsilon_0, \varepsilon_1, \varepsilon_2, \varepsilon_3)=(1, 1, 0, 1)$ (i.e., $(\theta_0, \theta_1, \theta_2, \theta_3)=(0, 0, 1, 0)$), it follows from Example 4.1 that $(1, 0, 1) \notin E_0(5, 2)$ but $(1, 0, 1) \in E_1(5, 2)$. Since $(\theta_0, \theta_1, \theta_2, \theta_3; \theta_4)=(0, 0, 1, 0; \mathbf{0})$, $(0, 0, 1, 0; \mathbf{1})$, $(0, 0, 1, 0; \mathbf{2})$, $(0, 0, 1, 0; \mathbf{3}), \ldots$ according as $d=5, 21, 37, 53, \ldots$, it follows from Corollary 4.1, $(1, 0, 1) \notin E_0(5, 2)$ and $(1, 0, 1) \in E_1(5, 2)$ that there exists an $(n, 5, d; 2)$-code which achieves the lower bound (2.5) for $d=21, 37, 53, \ldots$.

Example 4.5. (i) Consider the case $k=4$, $d=19$ and $s=3$. Since $(\theta_0, \theta_1, \theta_2, \theta_3)=(0, 0, 2, 0)$ and $(\varepsilon_0, \varepsilon_1, \varepsilon_2)=(2, 2, 0)$ in this case, the condition "$\theta_{k-1} \geqslant \max\{\varepsilon_\mu - 1: 0 \leqslant \mu \leqslant k-2\}$" in (3.6) does not hold. But it follows from Corollary 4.1 and $(2, 0) \in E_0(4, 3)$ that there exists an $(n, 4, 19; 3)$-code which achieves the lower bound (2.5) (i.e., $n=30$).

(ii) Consider the case $k=8$, $d=105$ and $s=2$. Since $(\theta_0, \theta_1, \ldots, \theta_7)=(0, 0, 0, 1, 0, 1, 1, 0)$, $(\varepsilon_0, \varepsilon_1, \ldots, \varepsilon_6)=(1, 1, 1, 0, 1, 0, 0)$ and $(l_1, l_2, l_3, l_4)=(1, 2, 3, 5)$ in this case, the condition "$\sum_{i=1}^{4} l_i \leqslant k$" in (3.6) does not hold. But it follows from Corollary 4.1 and $(1, 1, 0, 1, 0, 0) \in E_0(8, 2)$ that there exists an $(n, 8, 105; 2)$-code which achieves the lower bound (2.5) (i.e., $n=213$).

Example 4.6. In the case $k=2m+2$ $(m \geqslant 1)$, $s=2$ and $d=2^{2m+1}-\sum_{i=0}^{m} 2^i$ (i.e., $\varepsilon_0=\varepsilon_1=\cdots=\varepsilon_m=1$, $\varepsilon_{m+1}=\varepsilon_{m+2}=\cdots=\varepsilon_{2m}=0$ and $\theta_{k-1}=0$), it follows that $D=\{0, 1, 2, \ldots, m\}$ and $l_i=i$ for $i=1, 2, \ldots, m+1$. Hence $\sum_{i=1}^{m+1} l_i \leqslant k$ if and only if $m=1$ or 2. On the other hand, it follows from Corollary 4.1 and $(\varepsilon_1, \varepsilon_2, \ldots, \varepsilon_{k-2}) \in E_0(k, s)$ that there exists an $(n, k, d; s)$-code which achieves the lower bound (2.5) for any integers $k=2m+2$ and $d=2^{2m+1}-\sum_{i=0}^{m} 2^i$ such that $m \geq 1$. (Cf. [18, Section 4] for further details.)

Theorem 4.3. *If k, d and s are integers such that*

$$k \geq 4 \quad \text{and} \quad \theta_{k-1} \geq 1 + \sum_{i=\lambda+1}^{k-2} (i+1)\varepsilon_i/k, \tag{4.6}$$

there exists an $(n, k, d; s)$-code which achieves the lower bound (2.5) for given integers k, d and s where $\lambda = [\frac{1}{3}(2k-3)]$.

Proof. Since $(\varepsilon_1, \varepsilon_2, \ldots, \varepsilon_\lambda, 0, 0, \ldots, 0) \in E_1(k, s)$ for any integers ε_i $(i=1, 2, \ldots, \lambda)$ such that $0 \leq \varepsilon_i \leq s-1$, it follows from Theorem 4.2 that there exists a set $\mathcal{B}_1$ in $\mathcal{F}(\varepsilon_0, \varepsilon_1, \ldots, \varepsilon_\lambda, 0, 0, \ldots, 0; k, s)$ such that $\max\{\eta(\mathcal{B}_1)-1 : 1 \leq j \leq v_k\} \leq 1$. Hence it is sufficient to show that there exists a set $\mathcal{B}_2$ in $\mathcal{F}(0, 0, \ldots, 0, \varepsilon_{\lambda+1}, \ldots, \varepsilon_{k-2}; k, s)$ such that $\max\{\eta_j(\mathcal{B}_2)-1 : 1 \leq j \leq v_k\} \leq \psi$ where ψ denotes the smallest integer more than or equal to $\sum_{i=\lambda+1}^{k-2} (i+1)\varepsilon_i/k$ since $\mathcal{B}_1+\mathcal{B}_2$ is a set in $\mathcal{F}(\varepsilon_0, \varepsilon_1, \ldots, \varepsilon_{k-2}; k, s)$ such that

$$\max\{\eta_j(\mathcal{B}_1+\mathcal{B}_2)-1 : 1 \leq j \leq v_k\} \leq 1+\psi \leq \theta_{k-1}.$$

Let $\boldsymbol{b}_i$ $(i=1, 2, \ldots, k)$ be any vectors in $C(k; s)$ such that $\boldsymbol{b}_1, \boldsymbol{b}_2, \ldots, \boldsymbol{b}_k$ are linearly independent over GF(s) and let $\boldsymbol{\beta}_{(i-1)k+j} = \boldsymbol{b}_j$ for $i=1, 2, \ldots, \psi$ and $j=1, 2, \ldots, k$. For each integer μ $(\lambda+1 \leq \mu \leq k-2)$ such that $\varepsilon_\mu \neq 0$, let $V_l^{(\mu)}$ $(l=1, 2, \ldots, \varepsilon_\mu)$ be μ-flats in PG$(k-1, s)$ generated by $\mu+1$ vectors $\boldsymbol{\beta}_{\pi+i}$ $(i=1, 2, \ldots, \mu+1)$ where

$$\pi = (l-1)(k-1) \quad \text{or} \quad (l-1)(\mu+1) + \sum_{i=\mu+1}^{k-2} (i+1)\varepsilon_i$$

according as $\mu=k-2$ or not. Let $\mathcal{B}_2 = \{V_i^{(\mu)} : i=1, 2, \ldots, \varepsilon_\mu, \ \mu \in D^*\}$ where $D^* = \{\mu : \varepsilon_\mu \neq 0, \ \lambda+1 \leq \mu \leq k-2\}$. Then it is easy to see that $\mathcal{B}_2$ is a required set. This completes the proof.

Corollary 4.2. *In the case $s=2$, there exists an $(n, k, d; 2)$-code which achieves the lower bound (2.5) for any integers k and d such that*

$$k \geq 4 \quad \text{and} \quad d \geq (\zeta+1)2^{k-1}+1 \tag{4.7}$$

where $\zeta = \lceil (k+\lambda+1)(k-\lambda-2)/2k \rceil$, $\lambda = \lfloor (2k-3)/3 \rfloor$ and $\lceil x \rceil$ (or $\lfloor x \rfloor$) is the smallest integer more than or equal to x (or the greatest integer less than or equal to x).

Baumert and McEliece [2] showed that if $d \geq h2^{k-1}$ and $h = \lceil \frac{1}{2}(k-1) \rceil$, there exists an $(n, k, d; s)$-code which achieves the lower bound (2.5). Since $\zeta+1 < h$ if $k \geq 8$, our result is stronger than the Baumert–McEliece's result when $k \geq 8$.

References

[1] A. Barlotti, Some classical and modern topics in finite geometrical structures, in: J.N. Srivastava et al., eds., A Survey of Combinatorial Theory, (North-Holland, Amsterdam, 1973) 1–14.
[2] L.D. Baumert and R.J. McEliece, A note on the Griesmer bound, Trans. Information Theory 19 (1973) 134–135.
[3] I.F. Blake and R.C. Mullin, The Mathematical Theory of Coding (Academic Press, New York, 1975).
[4] R. C. Bose, Mathematical theory of the symmetrical factorial design, Sankhya 8 (1947) 107–166.
[5] R.C. Bose, On some connections between the design of experiments and information theory, Bull. Inst. Int. Statist. 38 (1961) 257–271.
[6] R.C. Bose and J.N. Srivastava, On a bound useful in the theory of factorial designs and error correcting codes, Ann. Math. Statist. 35 (1964) 408–414.
[7] P. Delsarte, An algebraic approach to the association schemes of coding theory, Philips Res. Rep. Suppl. No. 10 (1973).
[8] P. Dembowski, Finite Geometries (Springer-Verlag, Berlin, Heidelberg, New York, 1968).
[9] E.N. Gilbert, A comparison of signaling alphabets, Bell System Tech. J. 31 (1952) 504–522.
[10] J.M. Goethals, Some combinatorial aspects of coding theory, in: J.N. Srivastava et al., eds., A Survey of Combinatorial Theory (North-Holland, Amsterdam, 1973) 189–208.
[11] J.H. Griesmer, A bound for error-correcting codes, IBM J. Res. Develop. 4 (1960) 532–542.
[12] B.R. Gulati, B.M. Johnson and U.Koehn, On maximal t-linearly independent sets, J. Combin. Theory 15 (A) (1973) 45–53.
[13] B.R. Gulati and E.G. Kounias, On bounds useful in the theory of symmetrical factorial designs, J. Roy. Statist. Soc. Ser. B 32 (1970) 123–133.
[14] B.R. Gulati and E.G. Kounias, Maximal sets of points in finite projective space, no t-linearly dependent, J. Combin. Theory 15 (A) (1973) 54–65.
[15] N. Hamada and F. Tamari, Construction of maximal t-linearly independent sets, in: S. Ikeda et al., eds., Essays in "Probability and Statistics" (Keibundo Matsumoto Printing Co., Tokyo, 1976) 41–55.
[16] N. Hamada and F. Tamari, On a geometrical method of construction of maximal t-linearly independent sets, J. Combin. Theory 25 (A) (1978) 14–28.
[17] N. Hamada and F. Tamari, Construction of optimal linear codes and optimal fractional factorial designs using flats and spreads in a finite projective geometry, Tech. Rep. No. 6, Statistical Research Group, Hiroshima University, Hiroshima, Japan (1979).
[18] N. Hamada and F. Tamari, Construction of optimal linear codes using flats and spreads in a finite projective geometry, Europ. J. Combinatorics, submitted for publication.
[19] E.J. McCluskey Jr. Error-correcting codes — A linear programming approach, Bell System Tech. J. 38 (1959) 1485–1512.
[20] W.W. Peterson and E.J. Weldon Jr. Error Correcting Codes, 2nd edn (MIT Press, Cambridge, MA, 1972).
[21] M. Plotkin, Binary codes with specified minimum distance, IRE Trans. Information Theory IT-6 (1960) 445–450.
[22] B. Segre, Curve razionale e k-archi negli spazi finiti, Ann. Mat. Pura Appl. 39 (4) (1955) 357–379.
[23] B. Segre, Lectures on Modern Geometry. (Cremonese, Rome, 1961).
[24] E. Seiden, On the problem of construction and uniqueness of saturated 2_R^{K-p} designs, in: J.N. Srivastava et al., eds., A Survey of Combinatorial Theory (North-Holland, Amsterdam, 1973) 397–401.
[25] G. Solomon and J.J. Stiffler, Algebraically punctured cyclic codes, Information and Control 8 (1965) 170–179.
[26] F. Tamari, A note on the construction of optimal linear codes, J. Statist. Planning and Inference, submitted for publication.
[27] R.R. Varshamov, Estimate of the number of signals in error correcting codes, Dokl. Akad. Nauk. SSR 117 (1957) 739–741.

Annals of Discrete Mathematics 6 (1980) 189–200

COMBINATORIAL TOPOLOGY AND THE TRADE OFF METHOD IN BIB DESIGNS*

A. HEDAYAT* and Shuo-Yen Robert LI**

Department of Mathematics, University of Illinois, Chicago, IL 60680, U.S.A.

When two BIB designs based on the same set of varieties have identical parameters, one can obtain either of them from the other by trading off some blocks for an equal number of blocks. Define a *trade* to be two collections of blocks such that each of the $\binom{v}{2}$ pairs of varieties is contained in the same number of blocks from both collections. A method, based on combinatorial topology, for studying the trades with block size 3 is introduced. Using this method we give a new proof of the fact that volume of a trade can never be 5. Then we present a topological proof of the existence of a unique type of minimal trades, into which every trade can be decomposed. We also give a short proof of the fact that BIB(v, b, r, k, λ) designs, with possibly negative frequency of blocks, exist whenever the necessary conditions $rv = bk$ and $\lambda(v-1) = r(k-1)$ hold. The technique in this proof is a simple version of Graver and Jurkat [2] concerning null t-design.

0. Introduction

Let v, b, r, k, λ be positive integers related by the equations $rv = bk$ and $\lambda(v-1) = r(k-1)$. Let Σ denote the set $\{1, 2, \ldots, v\}$. A 2-element subset of Σ shall be called a *pair* and a k-element subset shall be called a *block*.

Let P denote the incidence matrix of pairs versus blocks, which is a $\binom{v}{2}$ by $\binom{v}{k}$ zero–one matrix. A $\binom{v}{k}$-dimensional integer vector D is called a BIB(v, b, r, k, λ) design, or simply a BIB(v, k, λ) design, if

$$PD = \lambda\mathbf{1},$$

where $\mathbf{1}$ is the vector with all entries equal to 1. Unless otherwise mentioned, we require the entries in the design D to be non negative. An entry in D represents the multiplicity (frequency) that the corresponding block appears in the design. An integer vector T of the same dimension is called a (v, k) *trade* if

$$PT = \mathbf{0}.$$

The sum of all positive entries in a trade is called its *volume.*

Let D be a BIB(v, k, λ) design. For every (v, k) trade T, the vector $D + T$ is

* Research supported in part by AFSOR 76-3050.
**Research supported in part by NSF Grant No. MCS 77-03533.

another BIB(v, k, λ) design provided that all of its entries are non negative. Conversely every BIB(v, k, λ) design can be written in the form of $D+T$.

As explained in [1] and [4], it is desirable to search for different BIB designs with identical parameters for the purpose of experimental designing or controlled sampling. Given a design D to start with, in order to search for designs with the same parameters it suffices to investigate the trades. The purpose of this article is to introduce a combinatorial topological method for studying trades on blocks of size 3. We shall demonstrate the convenience of this method in constructing trades and then use this method to prove two theorems characterizing the trades. It is easy to establish the nonexistence of trades of volumes 1, 2, and 3. The nonexistence of a trade of volume 5 without repeated blocks is pointed out in [5]. Hedayat and Li [4] asserts the nonexistence of a trade of volume 5 with or without repeated blocks. The proof of this nonexistence, though straightforward, has not appeared in the literature, because it contains many tedious steps in exhaustive search. In Section 2, we shall give a topological proof of the non existence of trades of volume 5. The smallest volume of a trade is 4, and there exists a unique trade with volume 4 up to isomorphism. In Section 3, we prove that every trade is a linear combination of finitely many trades of this minimal type. This fact is a special case of [2, Theorem 5.1] about t-designs.

1. The compact surface associated with a trade

For convenience the block consisting of the unordered elements $x_1, \ldots, x_k$ will be denoted by $(x_1 \cdots x_k)$. This is equivalent to a $\binom{v}{k}$-dimensional column vector with one of the entries equal to 1 and all others 0. Similarly the typical notation for a pair will be (xy).

We now direct our attention to studying $(v, 3)$ trades.

Example 1.1.

$$(125)+(146)+(234)+(356)-(124)-(156)-(235)-(346)$$

represents a trade. When this trade is added to the design

$$(124)+(137)+(156)+(235)+(267)+(346)+(457),$$

we obtain another design

$$(125)+(137)+(146)+(234)+(267)+(356)+(457).$$

In other words, from the first design the four blocks (124), (156), (235), and (346) have been traded for the blocks (125), (146), (234), and (356) to obtain the second design.

Now we introduce a geometric representation of the $(v, 3)$ trades. Given a trade T, construct a compact surface without boundary as follows. First create two collections of 2-simplexes (triangles) with their vertices labeled by elements of V. The 2-simplexis in one collection will be called the *positive triangles* and those in the other collection will be called the *negative triangles.* For every term $+(xyz)$ in T, there corresponds a positive triangle with vertices labeled by x, y, and z. If the coefficient of (xyz) in T is $m > 1$, then there are m copies of such a triangle. On the other hand, for every term $-(xyz)$ in T, there corresponds a negative triangle in the similar manner. So every pair (xy) appears on the same number of triangles in both collections. Thus, there exists a one-to-one matching between the edges of positive triangles and the edges of negative triangles so that every matched pair share the same two labels. When we identify every matched pair of edges in the natural way, we obtain a compact surface without boundary. Here we emphasize the possible *nonuniqueness of the matching. Different matchings may lead to different geometric configurations.* (See Examples 1.4 and 1.6 below.) Also the *labels on the vertices are not necessarily all distinct.*

Example 1.2. The trade in Example 1.1 is represented by the diamond-shaped topological sphere (see Fig. 1). Here in the picture the shaded regions are the negative triangles.

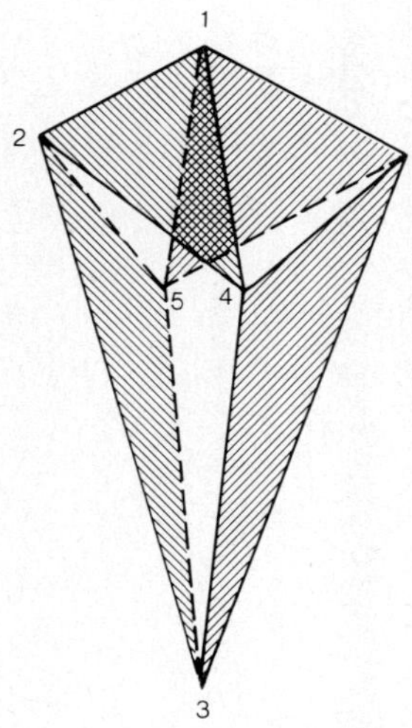

Fig. 1

In general, a trade give rise to a compact surface that is partitioned into positive triangles and negative triangles with the following two properties.

(1) Any two positive triangles can not intersect each other except possibly at their vertices. Neither can any two negative triangles.

(2) The intersection of a positive triangle with a negative triangle is vacuum, or one vertex, or two vertices, or an edge.

We shall refer to such a partition of surfaces, with or without boundary, as an *Eulerian triangulation*, although it is *not quite a triangulation in the usual sense of algebraic topology*. The edges of the triangles form an Eulerian graph[1] on the surface, i.e., a graph such that *the degree (valency) of every vertex is an even integer. Also no vertex can have degree equal to two*, because then there would be two triangles sharing two common edges.

The following example of trade is also obtained by triangulating a sphere.

Example 1.3.

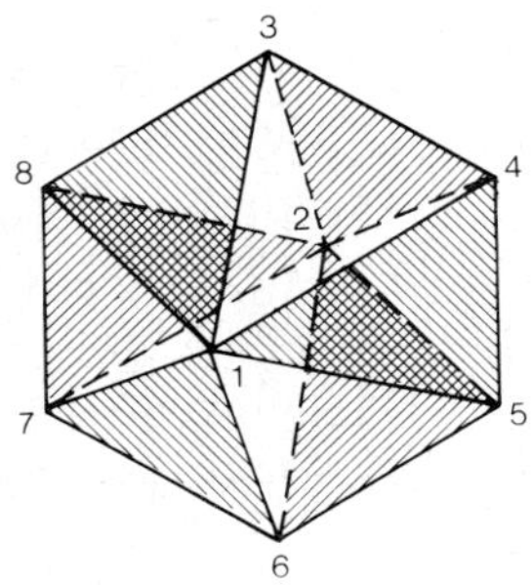

Fig. 2

Fig. 2 represents the trade

$$(134)+(156)+(178)+(238)+(245)+(267)-(138) \\ -(145)-(167)-(234)-(256)-(278).$$

It is well-known that a *compact connected surface is either a sphere, or a connected sum of tori, or a connected sum of projective planes* (see, for example, [6, Theorem 5.1]). The standard presentation of the connected sum of n tori is by identifying edges of a $4n$-gon in pairs (Fig. 3). Similarly for the connected sum of n projective planes we have Fig. 4.

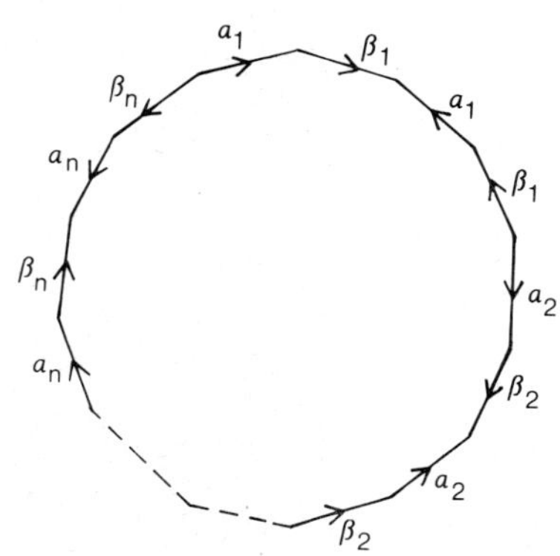

Fig. 3

[1] A more precise terminology would be Eulerian multigraph than Eulerian graph according to Harary [3].

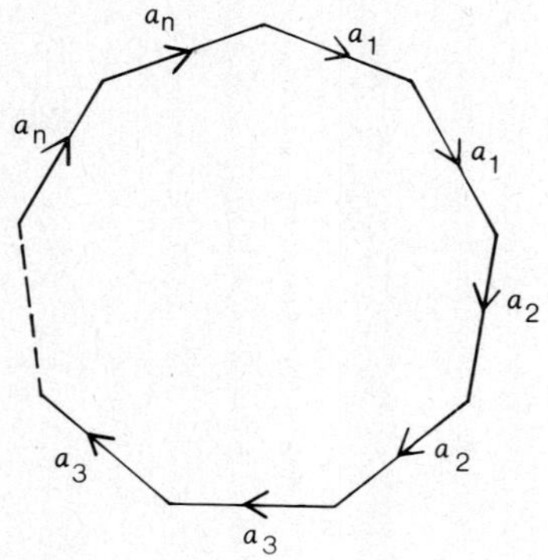

Fig. 4

Using these standard presentation of surfaces, we can easily construct more trades.

Example 1.4

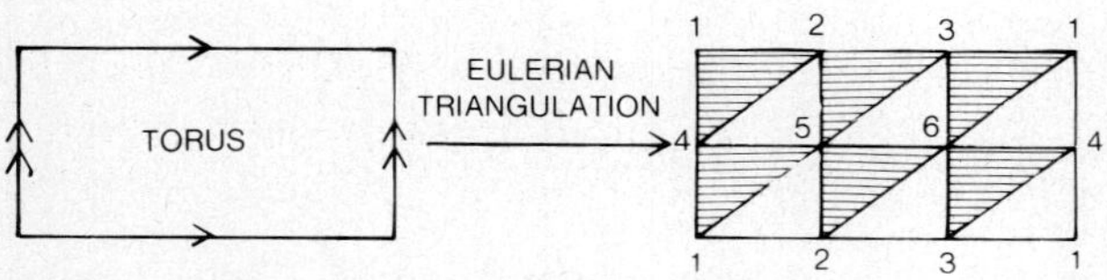

Fig. 5

Example 1.5

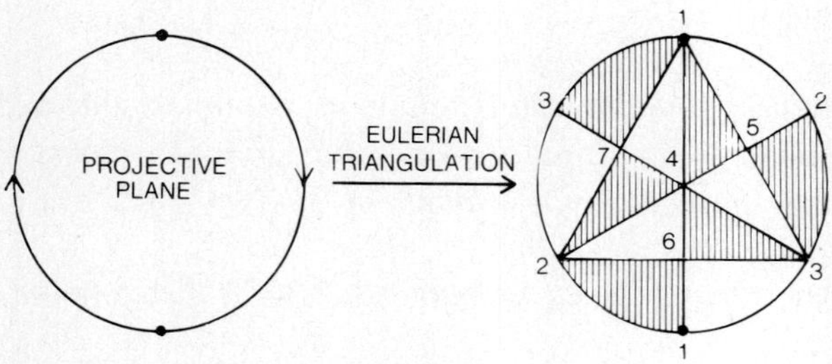

Fig. 6

Example 1.6

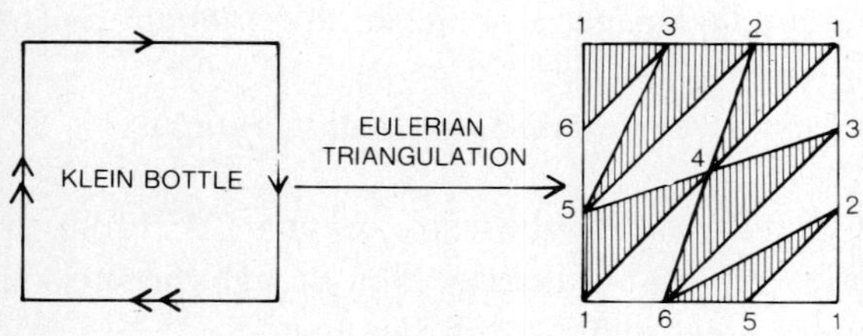

Fig. 7

Note that the figures in Examples 1.4 and 1.6 represents the same trade.

Example 1.7

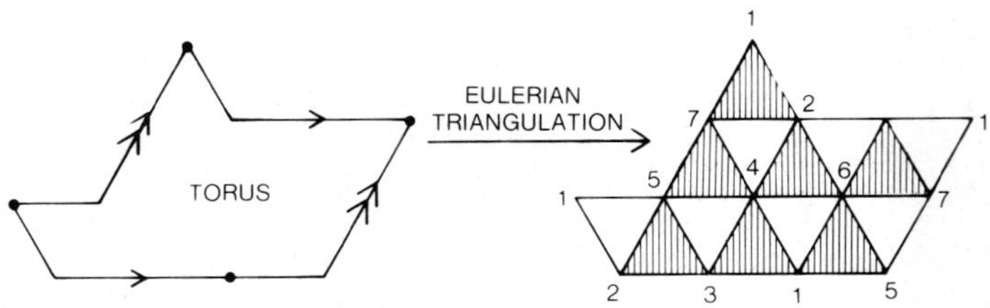

Fig. 8

Example 1.8

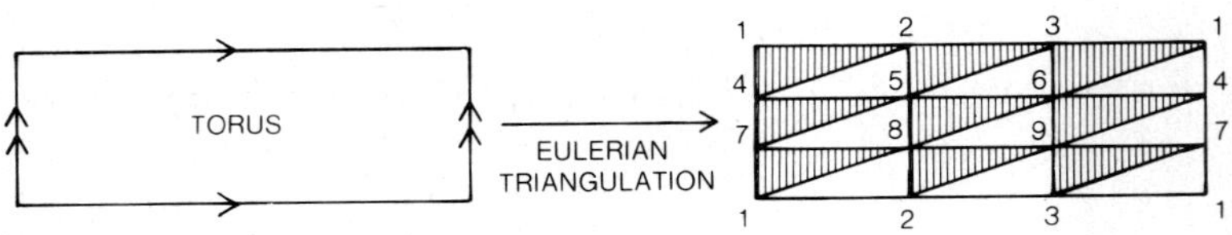

Fig. 9

2. Nonexistence of trades of volume 5

We have seen the convenience in constructing trades from the concept of Eulerian triangulation. In the proof of Theorem 2.1 below, we shall also find the same concept useful in establishing negative results. First we state a couple of selfevident lemmas.

Lemma 2.1. *For every Eulerian triangulation of a compact surface with boundary, the number of boundary edges that are on positive triangles differs from the number of those on negative triangles by a multiple of* 3.

Lemma 2.2. *There exist no trades of volume* 1, 2 *or* 3; *therefore the minimum trade volume is* 4.

Lemma 2.3. *If a disc is Eulerian triangulated with exactly* 2 *boundary edges, then*

(i) *exactly one boundary edge is on a positive triangle and the other is on a negative triangle, and*
(ii) *there are at least* 4 *positive and* 4 *negative triangles.*

Proof. Statement (i) follows directly from Lemma 2.1. From this, we know the Eulerian triangulation represents a trade, even though the surface has a boundary. The second statement now follows from Lemma 2.2.

Theorem 2.1. *There exist no trades of volume* 5.

Proof. Assuming there exists compact surface without boundary that has been Eulerian triangulated by exactly 5 positive and 5 negative triangles, we *want to derive a contradiction.* First, we know that the triangulation on every connected component of the surface represents a trade. So the surface must be *connected by Lemma* 2.2. There are 10 triangles in total, so there are 15 edges. Let n be the number of vertices. The Euler characteristic of this surface is

$$\chi = n - 15 + 10 = n - 5 \leqslant 2.$$

The inequality has been due to the connectedness. We label the vertices by $1, 2, \ldots, n$, respectively. There are three cases to examine.

Case 1. $\chi = 2$. Then $n = 7$ and the surface is a topological sphere. The edges in the triangulation form a planar graph and its degree sequence is

$$(6, 4, 4, 4, 4, 4, 4).$$

With a suitable relabeling, *the neighborhood around the vertex* of degree 6 is as in either graph below (Fig. 10). In the first graph, the six arrows are supposed to be

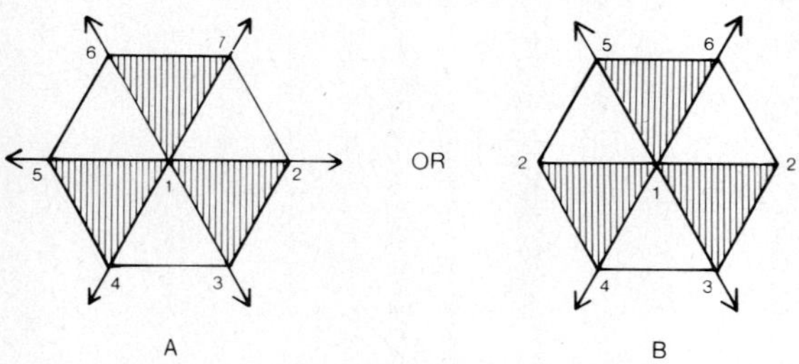

Fig. 10

linked in pairs to form a planar graph, but this is obviously impossible. After identifying the two points labeled as 2, the second graph lead to the following configuration. Again the arrows can not be linked in pairs to form a planar graph.

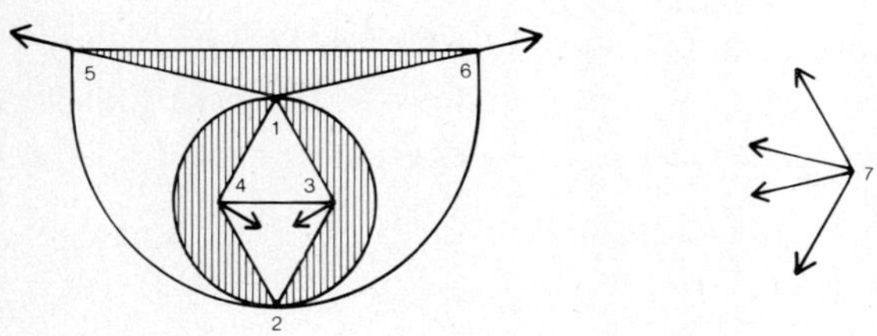

Fig. 11

Case 2. $\chi = 1$. Then $n = 6$ and the surface is a projective plane. The degree sequence has to be one of the following three:

$$(6, 6, 6, 4, 4, 4) \quad \text{or} \quad (8, 6, 4, 4, 4, 4) \quad \text{or} \quad (10, 4, 4, 4, 4, 4).$$

Since in any case some vertex has degree at least 6, we may assume that there are two edges α and β joining between vertices 1 and 2. These two edges form a cycle. Since the *fundamental group* of a projective plane is $Z/2Z$, this cycle is either trivial or is the generator of the fundamental group.

First we assume that the cycle generates the fundamental group. Then the projective plane can be drawn as a square with edges identified in pairs as in Fig. 12. So we have an Eulerian triangulation of the square disc based on Fig. 13.

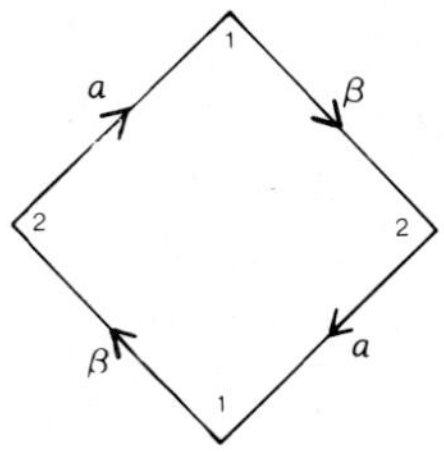

Fig. 12

Here

$$w, x, y, z \in \{3, 4, 5, 6\}$$

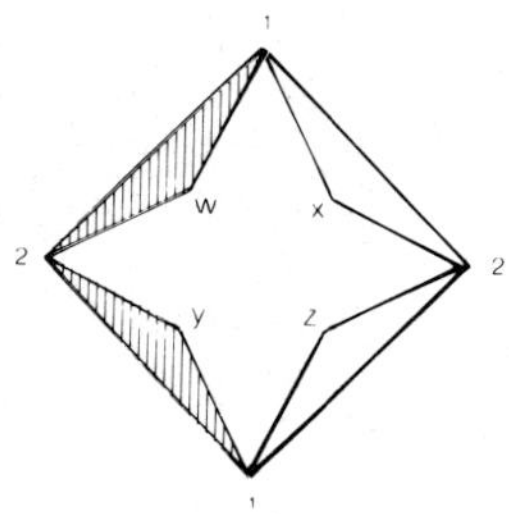

Fig. 13

and

$$w \neq x,\ w \neq z,\ x \neq y,\ y \neq z.$$

From Lemma 2.3, we also have $w \neq y$ and $x \neq z$. So w, x, y, and z are all distinct. By symmetry, let $w = 3$, $x = 4$, $y = 5$, and $z = 6$. Observe that vertex 1 must have degree more than 6 and vertex 2 has degree at least 6. Therefore the degree sequence is

$$(8, 6, 4, 4, 4, 4),$$

and the arrows in Fig. 14. should be linked in pairs to form the triangulation. But this is obviously impossible.

We now assume that α and β form a trivial cycle. The cycle then cuts the projective plane into two parts: a *disc* and a *Möbius band*. From Lemma 2.3 the Eulerian triangulation on the disc part takes at least 4 positive and 4 negative triangles. So the Möbius band is Eulerian triangulated by at most 1 positive and 1 negative triangles. This is a contradiction.

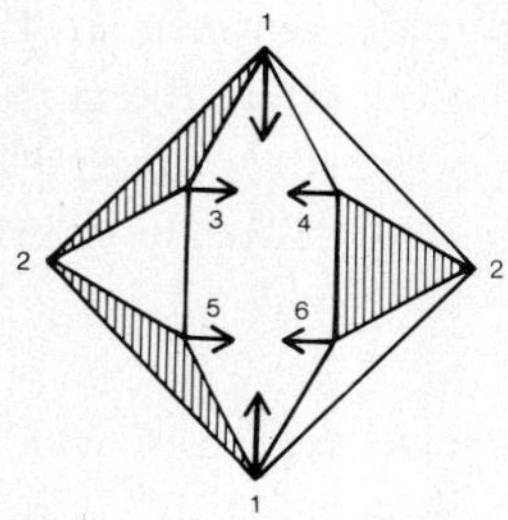

Fig. 14

Case 3. $\chi \leqslant 0$. Then $n \leqslant 5$. We need to show the nonexistence of a trade T of volume 5 on 5 or less symbols. First we may assume that T is of the form

$$(123)-(124)-(134)-(23x)+-\cdots,$$

where $x=4$ or 5. Then the coefficient of the block (145) in T must be at least 2. Thus

$$\begin{aligned} T=&(123)+2(145)-(124)-(134)-(23x) \\ &-(1y5)-(1z5)-(u45)-(v45)+-\cdots. \end{aligned}$$

But this implies that T has volume at least 7, a contradiction.

The theorem is proved.

3. Decomposition of a trade into minimal trades

The minimum volume of a trade is 4. One observes the following two easy facts.

Proposition 3.1. *If* $v \leqslant 5$, *there exists no nontrivial trade.*

Proposition 3.2. *If* $v \geqslant 6$, *there exists a unique trade of volume* 4 *up to isomorphism. This trade is represented by a diamond-shaped topological sphere as in Example* 1.2.

Let an Eulerian triangulated compact surface without boundary be given. We shall prove that by properly attaching diamond-shaped topological spheres to the surface one can obtain an Eulerian triangulation which represents the trivial trade. This is equivalent to the following.

Theorem 3.1. *Every* $(v, 3)$ *trade is a linear combination of trades of volume* 4.

The proof is by induction on v. When $v \leqslant 5$, the Eulerian triangulation to start with is representing the trivial trade because of Proposition 3.1. So we shall assume that $v \geqslant 6$ and at least one vertex on the surface is labeled by v. It suffices to show that the total degree of all vertices labeled by v can be reduced when a diamond-shaped topological sphere is properly attached to the surface. Consider two cases.

Case 1. There exists a vertex of degree 4 which is labeled by v. Say, the neighborhood around this vertex is as in Fig. 15. Choose $u < v$ such that $u \neq w, x,$

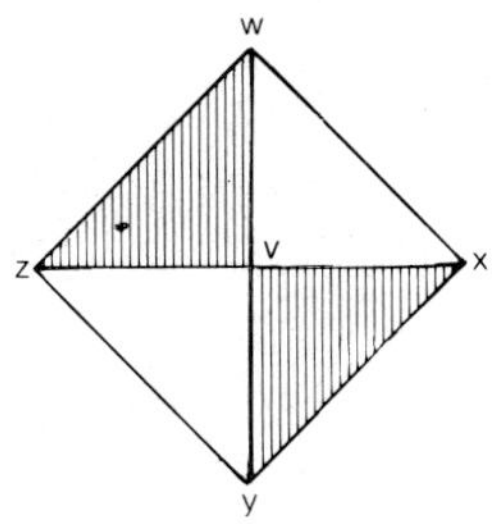

Fig. 15

y, z. Take the diamondshaped topological sphere in Example 1.2. Replace the vertices 1, 2, 3, 4, 5, 6 in it by 3, v, x, y, w, u, respectively, and then attach it to the surface by identifying the four triangles on the surface around the vertex v with the corresponding four triangles on the topological sphere. The result is a surface with the same Eulerian triangulation except that the vertex originally labeled by v receives the new label u.

Case 2. There exists a vertex of degree more than 4 which is labeled by v. We may modify the triangulation according to Fig. 16. and the resulting triangulation

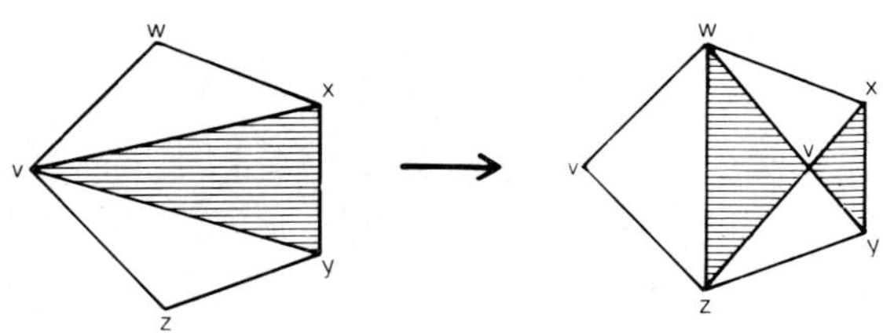

Fig. 16

represents the same trade. This creates a vertex of degree 4 which is labeled by v. The procedure in the previous case now applies and reduces the total degree of vertices which have the label v. The proof is now completed by induction.

In an algebraic setting Graver and Jurkat [2] proved that every (v, k) trade is a linear combination of trades of volume 4, where $k \geqslant 3$ is arbitrary.

4. BIB designs with possibly negative frequency of blocks

Graver and Jurkat [2] and Wilson [7] showed that t-designs, with prescribed parameters satisfying standard necessary conditions, always exist if negative frequencies of blocks in the design are allowed. For 2-designs, i.e., BIB designs, the standard necessary conditions on the parameters are the equations $rv = bk$ and $\lambda(v-1) = r(k-1)$.

Assuming these are true we shall in the following paragraphs, give a short proof the existence of a BIB(v, k, λ) design when negative frequency of blocks are allowed. From this one can construct BIB(v, k, λ) designs for a sufficiently large λ by super imposing copies of complete designs to small designs (see [7]).

Given parameters v, b, r, k, λ satisfying the two standard necessary conditions, we want to construct a BIB design with possibly negative frequencies of blocks. First we take a collection of b blocks so that every variety is repeated r times in the collection. Represent this collection by a $\binom{v}{k}$ dimensional column vector V. As before let P be the incidence matrix of pairs versus blocks. Also let Q be the incidence matrix of varieties versus pairs. Then $(1/k-1)QP$ is the incidence matrix of varieties versus blocks. Therefore

$$QPV = (k-1)r1_v = \lambda(v-1)1_v = \lambda Q1_{\binom{v}{2}}.$$

Here 1_v and $1_{\binom{v}{2}}$ are the v- and $\binom{v}{2}$-dimensional vectors with all entries equal to 1. Thus PV and $\lambda 1_{\binom{v}{2}}$ represent two collections of pairs, each of them covering every variety $\lambda(v-1)$ times. In other words, $PV = \lambda_{\binom{v}{2}}$ is a trade off between 1-designs. We may assume that $PV - \lambda_{\binom{v}{2}}$ is a linear combination of alternating sums of the form

$$(x_1\, y_1) - (y_1\, x_2) + (x_2\, y_2) - (y_2\, x_3) + - \cdots + (x_n\, y_n) - (y_n\, x_1).$$

Moreover one may assume that n is equal to 2 in every alternating sum because of further decomposition in the straightforward manner.

To avoid trivial cases, let us assume that $v \geqslant k+2$. Let W be the $\binom{v}{k}$ dimensional vector representing

$$(x_1\, y_1\, x_3 \cdots x_k) - (y_1\, x_2\, x_3 \cdots x_k) + (x_2\, y_2\, x_3 \cdots x_k) - (y_2\, x_1\, x_3 \cdots x_k),$$

where $x_1, x_2, \ldots, x_k, y_1, y_2$ are distinct varieties. Then the vector PW represents

$$(x_1\, y_1) - (y_1\, x_2) + (x_2\, y_2) - (y_2\, x_1).$$

By summing vectors of the type of W, we obtain a vector U such that

$$PU = PV - \lambda 1_{\binom{v}{2}}.$$

The vector $V - U$ is a BIB(v, b, r, k, λ) design. Some entries in $V - U$ may be negative.

References

[1] W. Foody and A. Hedayat, On theory and applications of BIB designs with repeated blocks, Ann. Statist. 5 (1977) 932–945.
[2] J.E. Graver and W.B. Jurkat, The module structure of integral designs, J. Combin. Theory 15 (A) (1973) 75–90.
[3] F. Harary, Graph Theory (Addison-Wesley, Reading, MA, 1969).
[4] A. Hedayat and S.-Y.R. Li, The trade off method in the construction of BIB designs with variable support sizes, Ann. Statist., 7 (1979) 1277–1287.
[5] C.C. Lindner and A. Rosa, Steiner triple systems having a prescribed number of triples in common, Canad. J. Math. 27 (1975) 1166–1175.
[6] W.S. Massey, Algebraic Topology: An Introduction (Brace and World, 1967).
[7] R.M. Wilson, The necessary conditions for t-designs are sufficient for something, Utilitas Math. 4 (1973) 207–215.

Annals of Discrete Mathematics 6 (1980) 201–214

BLOCK DESIGNS WITH CYCLIC BLOCK STRUCTURE

Franz HERING
University of Dortmund, Dortmund, Federal Republic of Germany

0. Introduction

For statistical applications of block designs it is sometimes advisible to exploit a combinatorial structure on the experimental units of the blocks. A well-known example for such designs are *complete latin squares* (see e.g. [1, pp.80 ff].

In this paper we start the investigation of certain very simple block designs on which the experimental units on each block carry the structure of a cyclic graph. Combinatorial problems arise by requiring also structural properties *between* these cyclic block graphs. Here we are interested in the following symmetries:

Every two-set of different treatment appears equally often as an edge in the cyclic block-graphs.

Any two different block-graphs have the same number of different edges.

A statistical application may be obtained by the following example: Suppose someone wants to estimate the effect of several treatments against tree deseases. The sick trees are the blocks of the plan and the girth of every tree is divided in sections, each being a plot for one treatment. Then we postulate a correlation between neighbouring plots, i.e. we postulate in the design the cyclical order of the plot. The combinatorial requirements above symmetrize these correlations.

In Section 1 we give the formulation of the problem and some general properties, in Section 2 we provide examples and constructions. I wish to express my gratitude to L. Danzer, who contributed several examples of this section and also found Proposition 9 independently of myself. In this paper we do not present the analysis of the corresponding linear model.

1. Definitions and general properties

Definition 1. (a) Let $\mathbf{V}, \boldsymbol{B}$ be nonempty finite sets and $I \subset (\mathbf{V}, \boldsymbol{B})$ an incidence relation between $\mathbf{V}$ and $\boldsymbol{B}$. Then $\beta = (\mathbf{V}, \boldsymbol{B}, I)$ is a *binary design* (BD), the elements of $\mathbf{V}$ are the *treatments*, the elements of $\boldsymbol{B}$ are the *blocks* of the design.

If $V \in \mathbf{V}$, $B \in \boldsymbol{B}$ we write VIB instead of $(V, B) \in I$ and

$$\mathbf{V}_B := \{V \in \mathbf{V}: VIB\},$$
$$\boldsymbol{B}_V := \{B \in \boldsymbol{B}: VIB\}.$$

(b) Now suppose that $\beta = (\mathbf{V}, \boldsymbol{B}, I)$ is a BD and that for every $B \in \boldsymbol{B}$ there is a graph $\gamma_B = (\mathbf{V}_B, \mathfrak{R}_B)$ defined, having the vertex set $\mathbf{V}_B$ and an edge set $\mathfrak{R}_B$ (where $\mathfrak{R}_B$ is a subset of the set of two-sets of elements of $\mathbf{V}_B$). Then

$$\lambda = (\beta, (\gamma_B : B \in \boldsymbol{B})) = ((\mathbf{V}, \boldsymbol{B}, I), (\gamma_B : B \in \boldsymbol{B}))$$

is a *graphic design* (GD).

(b) A GD λ is a *cyclic design* (CD), if the graph γ_B is a cycle for every $B \in \boldsymbol{B}$.

The following definition requires structural properties between the block graphs:

Definition 2. Suppose $\lambda = ((\mathbf{V}, \boldsymbol{B}, I), (\gamma_B : B \in \boldsymbol{B}))$ is a CD and $\gamma_B = (\mathbf{V}_B, \mathfrak{R}_B)$, $B \in \boldsymbol{B}$.

(a) λ is a *cyclicly balanced design* (CBD) if there exists $\rho \in \mathbf{N}$ (where $\mathbf{N}$ is the set of natural numbers) such that for every $V, W \in \mathbf{V}$, $V \neq W$ there are exactly ρ $B \in \boldsymbol{B}$ such that $\{V, W\} \in \mathfrak{R}_B$.

(b) λ is a harmonic design (HD) if λ is a CBD and there exists $\kappa \in \mathbf{N}$ such that

$$\#(\mathfrak{R}_B \cap \mathfrak{R}_C) = \kappa$$

for every $B, C \in \boldsymbol{B}$, $B \neq C$. ($\# A$ denotes the number of elements of A.)

Definition 3. The GD's

$$\lambda^{(i)} = ((\mathbf{V}^{(i)}, \boldsymbol{B}^{(i)}, I^{(i)}), \qquad (\gamma_B^{(i)} : B \in \boldsymbol{B}^{(i)})), \quad i = 1, 2$$

are *isomorphic* if there exists a bijection $\nu : \mathbf{V}^{(1)} \to \mathbf{V}^{(2)}$ and a bijection $\mu : \boldsymbol{B}^{(1)} \to \boldsymbol{B}^{(2)}$ having the following properties:

(i) $V I^{(1)} B$ is equivalent to $\nu(V) I^{(2)} \mu(B)$ for every $V \in \mathbf{V}$, $B \in \boldsymbol{B}$.

(ii) $\{V, W\} \in \mathfrak{R}_B^{(1)}$ is equivalent to

$$\{\nu(V), \nu(W)\} \in \mathfrak{R}_{\mu(B)}^{(2)} \quad \text{for every } V, W \in \mathbf{V}^{(1)}, B \in \boldsymbol{B}^{(1)}.$$

Then $\pi = (\nu, \mu)$ is an *isomorphism from* $\lambda^{(1)}$ *onto* $\lambda^{(2)}$. If $\lambda^{(1)} = \lambda^{(2)} = \lambda$, then π is an *automorphism of* λ.

Observe that for $\lambda^{(1)}, \lambda^{(2)}$ being isomorphic it is not enough to assume that (i) holds (so that the corresponding BD's $\beta^{(i)} = (V^{(i)}, B^{(i)}, I^{(i)})$ are isomorphic) and that $\gamma_B^{(1)}, \gamma_{\mu(B)}^{(2)}$ are isomorphic graphs for every $B \in \boldsymbol{B}^{(1)}$. For example, if $\lambda^{(1)}$ and $\lambda^{(2)}$ are CD's with $\beta^{(1)} = \beta^{(2)}$, then $\lambda^{(1)}$ and $\lambda_{(2)}$ are not necessarily isomorphic. An isomorphism takes care not only of the structure *within* but also *between* the graphs.

Proposition 1. *Suppose* $\lambda = ((\mathbf{V}, \boldsymbol{B}, I), (\gamma_B : B \in \boldsymbol{B}))$ *is a CD with* $\# \mathbf{V} =: v$, $\# \boldsymbol{B} =: b$ *and* $\# \mathbf{V}_B = v$ *for every* $B \in \boldsymbol{B}$ *(i.e., every treatment occurs in every block, so that* $(\mathbf{V}, \boldsymbol{B}, I)$ *is a complete binary design).*

(a) *If* λ *is a* CBD, *then*

$$\rho = \frac{2b}{v-1},$$

(b) *If* λ *is a* HD, *then*

$$\kappa = \frac{(2b-v+1)v}{(v-1)(b-1)}.$$

Proof. (a) There are $\binom{v}{2}$ edges, each in ρ of the γ_B. There are b graphs γ_B each having v edges. So

$$\rho \binom{v}{2} = v \cdot b.$$

(b) Every edge is in ρ of the γ_B, so in $\binom{\rho}{2}$ pairs of γ_B's. Every pair of graphs γ_B has κ common edges. Therefore

$$\kappa \binom{b}{2} = \binom{\rho}{2}\binom{v}{2}.$$

If $\mathbf{V}$ is a set and $\boldsymbol{U}$ a subset, then $\mathbf{C}\boldsymbol{U}$ denotes the complement of $\boldsymbol{U}$ in $\mathbf{V}$. For $V \in \mathbf{V}$ we write $\mathbf{C}V$ instead of $\mathbf{C}\{V\}$. If $r \in \mathbf{N}_0$ (the set of positive integers), then $\mathfrak{P}_r(\mathbf{V}) := \{\boldsymbol{U} \subset \mathbf{V}: \# \boldsymbol{U} = r\}$. Now, if $\mathbf{V}$ is nonempty, $\# \mathbf{V} = n$ and

$$\boldsymbol{B} := \mathfrak{P}_{n-1}(\mathbf{V}),$$
$$\varepsilon := \{\{V, B\}: V \in \mathbf{V}, B \in \boldsymbol{B}, V \in B\},$$

then

$$\beta := (\mathbf{V}, \boldsymbol{B}, \varepsilon)$$

is a rather uninteresting example of a BIBD. (If $\# V = n$, then $v = b = n$, $r = k = n-1$, $\lambda = n-1$.) However, the question, for which $n \in \mathbf{N}$, $\# \mathbf{V} = n$, every $B = \mathbf{C}V$, $V \in \mathbf{V}$ can be endowed with the structure of a cyclic graph $\gamma_{\mathbf{C}V}$ so that the so obtained CD

$$\lambda = ((\mathbf{V}, \{\mathbf{C}V: V \in \mathbf{V}\}, \varepsilon), (\gamma_{\mathbf{C}V}: V \in \mathbf{V})) \tag{1}$$

is a CBD or even a HD is not trivial.

Proposition 2. *Suppose* $\mathbf{V}$ *is a finite, nonempty set with* $\# \mathbf{V} = n$, *and* λ *is a CD given through* (1).

(a) *If* λ *is a CBD, then for every* $V, W \in \mathbf{V}$, $V \neq W$ *there exist exactly two* $B \in \boldsymbol{B}$ *such that* $\{V, W\}$ *is an edge of* γ_B *(i.e.* ρ *of Definition* 2, *a equals two).*

(b) *If* λ *is a HD, then for every* $B, C \in \mathbf{B}$, $B \neq C$ *the graphs* λ_B *and* λ_C *have exactly one edge in common (i.e.* κ *of Definition* 2, *b equals one).*

Proof. (a) Every cyclic graph $\gamma_B = (B, \mathfrak{R}_B)$ has exactly $n-1$ edges, so

$$\sum_{B \in \mathbf{B}} \# \mathfrak{R}_B = n(n-1) = 2 \# \mathfrak{P}_2(\mathbf{V})$$

i.e. every set $\{V, W\}$, $V \neq W$ is in exactly two edge sets $\mathfrak{R}_B$.

(b) Every $\{V, W\} \in \mathfrak{P}_2(\mathbf{V})$ is on exactly two edge sets $\mathfrak{R}_B$ according to (a), therefore

$$\sum_{\substack{B, C \in \mathbf{B} \\ B \neq C}} \# (\mathfrak{R}_B \cap \mathfrak{R}_C) = \# \mathfrak{P}_2(\mathbf{V}) = \binom{n}{2} = \# \mathfrak{P}_2(\mathbf{B}).$$

If π is a permutation, acting on a finite set $\mathbf{V}$ with $\# \mathbf{V} = n$, then π has a *cyclic representation*

$$\pi = (V_{1,1}, \ldots, V_{1,r_1}) \cdots (V_{s,1}, \ldots, V_{s,r_s})$$

where $\pi(V_{i,j}) = V_{i,j+1}$, $i = 1, \ldots, s$, $j = 1, \ldots, r_i - 1$ and $\pi(V_{i,r_i}) = V_{i,1}$, $r_1 < r_2 < \cdots < r_s$. Then

$$T(\pi) = (r_1, r_2, \ldots, r_s)$$

is the *type* of π (see, e.g., [2, p.53]).

Through the definition

$$\pi(\boldsymbol{U}) = \{\pi(V) \colon V \in \boldsymbol{U}\},$$

π becomes canonically a permutation on $\mathfrak{P}_r(\mathbf{V})$, too.

If $r \in \mathbf{N}$, $\boldsymbol{U}_i \in \mathfrak{P}_r(\mathbf{V})$, $i = 1, \ldots, t$ and $(\boldsymbol{U}_1, \ldots, \boldsymbol{U}_t)$ is a π-cycle on $\mathfrak{P}_r(\mathbf{V})$, then $(\mathbf{C}\boldsymbol{U}_1, \ldots, \mathbf{C}\boldsymbol{U}_t)$ is a π-cycle on $\mathfrak{P}_{n-r}(\mathbf{V})$. Therefore π is of the same type on $\mathfrak{P}_r(\mathbf{V})$ and on $\mathfrak{P}_{n-r}(\mathbf{V})$.

So we obtain immediately

Lemma 1. *Suppose* $\pi = (\nu, \mu)$ *is an automorphism of* λ, *where* λ *is a CD* (1). *Then* $\mu(\mathbf{C}V) = \mathbf{C}\nu(V)$, i.e. μ *is the canonical continuation of* ν *onto* $\mathfrak{P}_{n-1}(\mathbf{V})$. *Furthermore* ν *and* μ *are of the same type.*

The above lemma makes it reasonable to say that ν (instead of $\pi = (\nu, \nu)$) is an automorphism of λ.

Proposition 3. *Suppose* λ *is a CD defined in* (1), $n = \# \mathbf{V}$ *and* $\Gamma(\lambda)$ *is the group of automorphism of* λ. *Then*

$$\# \Gamma(\lambda) \leq 2(n-1) \cdot n.$$

Moreover, when equality holds, then every automorphism of $\gamma_{\mathbf{C}V}$, $V \in \mathbf{V}$ *is a restriction of an automorphism* $\nu \in \Gamma(\lambda)$ *onto* $\mathbf{C}V$. *Furthermore every* $\nu \in \Gamma(\lambda)$ *is of one of the following types:*

$$(q, \ldots, q), \tag{2}$$

$$(1, p, \ldots, p), \tag{3}$$

$$(1, 1, 2, \ldots, 2), \quad n \text{ even}, \tag{4}$$

$$(1, 1, 1, 2, \ldots, 2), \quad n \text{ odd}, \tag{5}$$

$$(2, 4, \ldots, 4), \quad n \text{ even}, \tag{6}$$

$$(3, 6, \ldots, 6), \quad n \text{ odd}. \tag{7}$$

Proof. Suppose $\{V_1, \ldots, V_{n-1}\} = \mathbf{C}V_n \in \boldsymbol{B}$ and $(V_1, \ldots, V_{n-1})$ is a cyclic arrangement of $\mathbf{C}V_n$ so that

$$\mathfrak{R}_{\mathbf{C}V_n} = \{\{V_i, V_{i+1}\}: i = 1, \ldots, n-1\} \quad (i \bmod n).$$

Then the automorphism group Γ_n of the cyclic graph $\gamma_{\mathbf{C}V_n}$ is generated by the automorphisms α, β where

$$\alpha(V_i) = V_{i+1}, \qquad i = 1, \ldots, n-1, \quad (i \bmod n),$$

$$\beta(V_i) = V_{n-1-i}, \qquad i = 1, \ldots, n-1.$$

A straightforward calculation shows $\beta\alpha = \alpha^{-1}\beta$ so that $\#\Gamma_n = 2(n-1)$, and every $\gamma \in \Gamma_n$ is of one of the following types

$$(p, \ldots, p),$$

$$(1, 2, \ldots, 2), \quad n \text{ even}, \tag{8}$$

$$(1, 1, 2, \ldots, 2), \quad n \text{ odd}. \tag{9}$$

Now the index i of Γ_n in $\Gamma(\lambda)$ is at most n. Therefore $\#\Gamma(\lambda) = i \cdot \Gamma_n \leq n \cdot 2(n-1)$. If $\nu \in \Gamma(\lambda)$ and ν has a fixelement V_n, say, then the restriction of ν to $\mathbf{C}V_n$ is in Γ_n, therefore ν itself is of one of the types (3), (4), (5). Now assume ν has no fixelement and $(V_1, \ldots, V_q)$ is a minimal ν-cycle. If $\nu^q = 1$, then the length l of every ν-cycle divides q and from the minimality of q it follows $l = q$, i.e. ν is of type (2). If $\nu^q \neq 1$, then again $V_1, \ldots, V_q$ are fixed under ν^q and again the restriction ν_1 of ν^q onto $\mathbf{C}V_1$ is an automorphism of the cyclic graph $\gamma_{\mathbf{C}V_1}$. Now $\nu^q \neq 1$ implies $\nu_1 \neq 1$, and $q > 1$ implies that ν_1 fixes at least one $V \in \mathbf{C}V_1$. So ν_1 is of type (8) when n is even, and of type (9) when n is odd. Consequently $q = 2$ and ν is of type (6) for odd n. Also, $q = 3$ and ν is of type (7) for even n.

Definition 4. Suppose $\lambda = ((\mathbf{V}, \boldsymbol{B}, I), (\gamma_B : B \in \boldsymbol{B}))$ is a CD, $(B_1, \ldots, B_b)$ is an arrangement of $\boldsymbol{B}$ into a sequence and $r_i := \#\mathbf{V}_{B_i}$. Then the array

$$M(\lambda) = \begin{pmatrix} V_{1,1}, \ldots, V_{1,r_1} \\ \cdot \\ \cdot \\ V_{b,1}, \ldots, V_{b,r_b} \end{pmatrix} \tag{10}$$

of $V_{i,j} \in \mathbf{V}$ is a *representation of* λ, if $\{V_{i,j}, V_{i,j+1}\} \in \Re_{B_i}$, $i = 1, \ldots, b$, $j = 1, \ldots, r_i$, j modulo r_i.

Definition 5. Suppose λ is a CD given through (1), that (10) is a representation of λ and $\pi = (\nu, \mu)$ is an automorphism of λ. Then (10) is a *representation of* π, if $M(\lambda)$ has the following properties

(i) The sequence $(B_1, \ldots, B_b)$ (defining the arrangement of the rows of $M(\lambda)$) can be subdivided such that

$\mu = (B_1, \ldots, B_{p_1})(B_{p_1+1}, \ldots, B_{p_2}) \cdots (B_{p_{s-1}+1}, \ldots, B_{p_s})$ is a cyclic representation of μ. (Here $p_s = b$.)

(ii) $\nu(V_{i,j}) = V_{i,j+1}$, $i = 1, \ldots, b$, $i \neq p_n$, $n = 1, \ldots, s$, $j = 1, \ldots, r_i$.

(iii) $\nu(V_{p_n,j}) = V_{p_{n-1}+1,j}$, $n = 1, \ldots, s$, $j = 1, \ldots, r_i$.

It is obvious from the definition that every automorphism has a representation. Also from Definition 2 follows immediately:

Lemma 2. *Suppose λ is a CD having the representation* (10).

(a) *λ is a CBD if and only if for every $i, j = 1, \ldots, b$, $i \neq j$ there exist $\rho \in \mathbf{N}$ such that there are ρ pairs (u, v), $u \in \{1, \ldots, r_i\}$, $v \in \{1, \ldots, r_j\}$ with*

$$\{V_{i,u}, V_{i,u+1}\} = \{V_{j,v}, V_{j,v+1}\} \quad (u \bmod r_i,\ v \bmod r_j).$$

(b) *λ is a HD if its representation λ has property* (a) *and moreover there exists $\kappa \in \mathbf{N}$ such that for every $V, W \in \mathbf{V}$, $V \neq W$ there are exactly κ indices $i \in \{1, \ldots, b\}$ so that for every such i there exists $j \in \{1, \ldots, v_i\}$ with $\{V, W\} = \{V_{i,j}, V_{i,j+1}\}$.*

2. Some examples and construction methods of cyclical balanced block designs and harmonic designs

Proposition 4. *When n is an odd prim, then there exists a complete HD $\lambda = ((\mathbf{V}, \boldsymbol{B}, I), (\gamma_B : B \in \boldsymbol{B}))$ with $\# \mathbf{V} = n$, $\# \boldsymbol{B} = \frac{1}{2}(n-1)$, $\rho = 1$, $\kappa = 0$ having an automorphism ν of type $T(\nu) = (n)$.*

Proof. We regard GF(n) represented by the set of residues $\mathbf{V} = \{0, 1, \ldots, n-1\}$ modulo n. Define $\boldsymbol{B} = \{1, \ldots, \frac{1}{2}(n-1)\}$. Then straightforward calculation in GF(n) shows that

$$M(\lambda) = (b \cdot v)_{b=1,\ldots,\frac{1}{2}(n-1),\ v=0,\ldots,n-1}$$

is a representation of a HD λ with the required parameters and that $\nu = (0, 1, \ldots, n-1)$ is an automorphism.

The smallest examples are

$n=3$	$n=5$	$n=7$
0 1 2	0 1 2 3 4	0 1 2 3 4 5 6
	0 2 4 1 3	0 2 4 6 1 3 5
		0 3 6 2 5 1 4

We do not know yet, whether there exists such a scheme for $n=9$, but this can be decided by a moderately small computer calculation. By m-fold repeating these schemes one obtains trivially CBD's with $\rho=m$, $\kappa=\binom{m}{2}$. We do not know any example of a complete HD with $\kappa>0$, however.

Proposition 5. *There exists a CBD of type* (1) *with* $\#\mathbf{V}=n$ *for every* $n\geq 3$.

Proof. Endow $\mathbf{V}:=\{\infty, 0, 1, \ldots, n-2\}$ with addition modulo $n-1$, where $\infty \pm i=\infty$ for every $i=0, \ldots, n-2$. Define a cyclic graph on $\mathbf{C}\infty$ by the cyclic order

$$(0, 1, \ldots, n-2)$$

and on $\mathbf{C}i$, $i=0, \ldots, n-2$ a cyclic graph by the cyclic order

$$(\infty, i, n-2+i, 1+i, n-3+i, \ldots).$$

Then it follows by an easy calculation that (1) is a CBD having the automorphism $\nu=(0, 1, \ldots, n-1)$.

E.g. for $n=6, 7$ we obtain the following CBD's

$n=6$	$n=7$
0 1 2 3 4	0 1 2 3 4 5
∞ 0 4 1 3	∞ 0 5 1 4 2
∞ 1 0 2 4	∞ 1 0 2 5 3
∞ 2 1 3 0	∞ 2 1 3 0 4
∞ 3 2 4 1	∞ 3 2 4 1 5
∞ 4 3 0 2	∞ 4 3 5 2 0
	∞ 5 4 0 3 1

Now suppose that (1) is a HD having an automorphism ν of type $T(\nu)=(1, n-1)$. Without loss of generality we assume $\mathbf{V}=\{\infty, 0, 1, \ldots, n-2\}$ and $\nu=(\infty)(0, 1, \ldots, n-2)$ so that ν has a representation of the following form:

$$\begin{array}{lllll} 0 & 1 & 2 & \cdots & n-2 \\ \infty & v_1 & v_2 & \cdots & v_{n-2} \\ \infty & v_1+1 & v_2+1 & \cdots & v_{n-2}+1 \\ \cdot & & & & \\ \cdot & & & & \\ \cdot & & & & \\ \infty & v_1+(n-2) & v_2+(n-2) & \cdots & v_{n-2}+(n-2) \end{array} \qquad (11)$$

(summation is modulo $n-1$). In addition we may assume $v_1 = 0$. Then the representation (11) of ν is determined by

$$v_2, v_3, \ldots, v_{n-2}. \tag{12}$$

In Proposition 7 we characterize sequences (12) for which (11) gives the representation of a HD having the automorphism ν. In Table 1, we give a list that contains a representation for every class of isomorphic HD's having an automorphism of type $(1, n-1)$ up to $n \leqslant 9$. For $n = 10$ the list provides only three examples. Most entries of the list, especially the detailed discussion of $n = 9$, is due to L. Danzer. He also investigated the automorphism group in every case. We do not present these details. The diagrams of Fig. 1 corresponding to Table 1 are introduced in Definition 8.

For example, from the list above for $n = 7$: 3 1 2 4, the complete representation is

$$\begin{array}{cccccc}
0 & 1 & 2 & 3 & 4 & 5 \\
\infty & 0 & 3 & 1 & 2 & 4 \\
\infty & 1 & 4 & 2 & 3 & 5 \\
\infty & 2 & 5 & 3 & 4 & 0 \\
\infty & 3 & 0 & 4 & 5 & 1 \\
\infty & 4 & 1 & 5 & 0 & 2 \\
\infty & 5 & 2 & 0 & 1 & 3
\end{array}$$

The smallest example for a HD, which does not have an automorphism of type $(1, n-1)$ is

$$\begin{array}{ccccc}
0 & 1 & 2 & 3 & 4 \\
\infty & 0 & 3 & 4 & 2 \\
\infty & 1 & 4 & 0 & 3 \\
\infty & 2 & 0 & 1 & 3 \\
\infty & 4 & 1 & 2 & 0 \\
\infty & 4 & 2 & 3 & 1
\end{array}$$

(here introducing the symbol ∞ is not meaningful). It can be shown, that in this case the automorphism group is trivial.

The announced proposition characterizing the sequences (12) that generate a CBD and a HD requires the following preparations.

For every $u \in \{0, 1, \ldots, n-2\}$ we have

$$\#(\{u, n-2-u\} \cap \{0, 1, \ldots, [\tfrac{1}{2}(n-1)]\}) = 1.$$

Hence, through

$$\{\alpha(u)\} := \{u, n-2-u\} \cap \{0, \ldots, [\tfrac{1}{2}(n-1)]\},$$

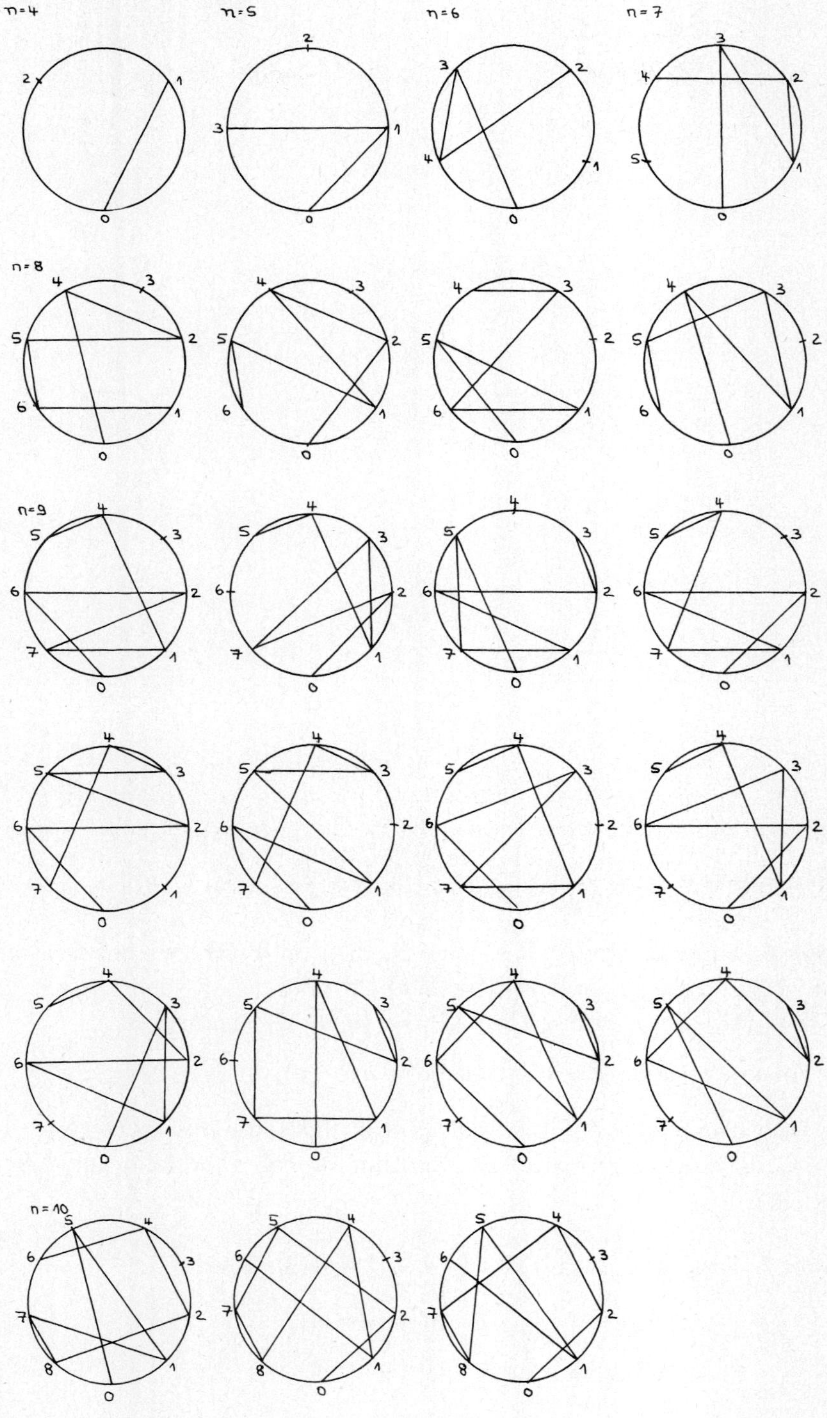

Fig. 1.

Table 1.

n	
4	1
5	1 3
6	3 4 2
7	3 1 2 4
8	4 2 5 6 1 2 4 1 5 6 5 1 6 3 4 4 1 3 5 6
9	6 2 7 1 4 5 2 7 3 1 4 5 5 7 1 6 2 3 2 6 1 7 4 5 6 2 5 3 4 7 6 1 5 3 4 7 6 3 7 1 4 5 2 6 3 1 4 5 3 1 6 2 4 5 4 1 7 5 2 3 6 4 1 5 2 3 5 1 6 4 2 3
10	5 1 7 8 2 4 6 2 5 7 8 4 1 6 2 4 7 8 5 1 6

we define a map α from $\{0, 1, \ldots, n-2\}$ into $\{0, 1, \ldots, [\frac{1}{2}(n-1)]\}$.

Definition 6. Suppose $\mathbf{W} = \{0, 1, \ldots, n-2\}$, then on $\mathfrak{P}_2(\mathbf{W})$ we define a map α through $\alpha(I) := \alpha(a-b)$ for every $I = \{a, b\} \in \mathfrak{P}_2(\mathbf{W})$.

If $I, J \in \mathfrak{P}_2(\mathbf{W})$, $I \neq J$ with $\alpha(I) = \alpha(J)$. $I = \{a, b\}$, $J = \{c, d\}$, then

$$\alpha(a-c) = \alpha(b-d) \quad \text{or} \quad \alpha(a-d) = \alpha(b-c).$$

Here both equalities hold if and only if n is odd and $\alpha(I) = \alpha(J) = \frac{1}{2}(n-1)$. In this case we have $\alpha(a-c) = \alpha(a-d)$. This remark makes the following definition meaningful:

Definition 7. Suppose $I, J \in \mathfrak{P}_2(\mathbf{W})$, $I \neq J$, $\alpha(I) = \alpha(J)$. Then

$$\beta(I, J) := \begin{cases} \alpha(a-c) & \text{if } \alpha(a-c) = \alpha(b-d), \\ \alpha(a-d) & \text{if } \alpha(a-d) = \alpha(b-c). \end{cases}$$

Proposition 6. *Suppose* (1) *is a CD with* $\mathbf{V} = \{0, 1, \ldots, n-2\}$ *having the automorphism* $\nu = (0, 1, \ldots, n-2)$ *and let* $\mathfrak{R}$ *denote the edge-set of* $\gamma_{\mathbf{C}_g}$ *for a fixed*

$g \in \mathbf{V}$. *Then* λ *is a CBD if and only if*

(i) *for every* $i \in \{1, \ldots, [\frac{1}{2}(n-1)]\}$ *there are exactly two* $I \in \mathfrak{R}$ *with* $\alpha(I) = i$,

(ii) *If* n *is even, then there is exactly one* $I \in \mathfrak{R}$ *with* $\alpha(I) = \frac{1}{2}n$.

Moreover, λ *is a HD if and only if* $\mathfrak{R}$ *has the following additional properties:*

(iii) *If* $I, J, K, L \in \mathfrak{R}$, $I \neq J$, $K \neq L$, $\alpha(I) = \alpha(J) \neq \alpha(K) = \alpha(L)$, *then*

$$\beta(I, J) \neq \beta(K, L),$$

(iv) *If* n *is even and* $I, J \in \mathfrak{R}$, $I \neq J$, *then* $\beta(I, J) \neq \frac{1}{2}n$.

Proof. If $I, J \in \mathfrak{P}_2(\mathbf{V})$, then $J \in \{I, \nu(I), \nu^2(I), \ldots\}$ if and only if $\alpha(J) = \alpha(I)$. Moreover the orbit $\{I, \nu(I), \nu^2(I), \ldots\}$ of I has the order n for every I with $\alpha(I) \neq \frac{1}{2}n$ and the order $\frac{1}{2}n$ for $\alpha(I) = \frac{1}{2}n$. Therefore the assertion is obtained directly from a representation

$$\begin{array}{l} v_{0,1}, \ldots, v_{0,n-2} \\ \cdot \\ \cdot \\ \cdot \\ v_{n-1,1}, \ldots, v_{n-1,n-2} \end{array}$$

The smallest examples for HD's with $\mathbf{V} = \{0, 1, \ldots, n-2\}$ having the automorphism $\nu = (0, 1, \ldots, n-2)$ are obtained by applying ν on the following rows:

$n = 4$: 0 1 2

$n = 5$: 0 1 3 2

$n = 7$: 0 1 2 4 6 3

We verify the conditions (i), (iii) of Proposition 6 for $n = 7$:

$$\alpha(\{0, 1\}) = \alpha(\{1, 2\}) = 1, \quad \beta(\{0, 1\}, \{1, 2\}) = 1,$$
$$\alpha(\{2, 4\}) = \alpha(\{4, 6\}) = 2, \quad \beta(\{2, 4\}, \{4, 6\}) = 2,$$
$$\alpha(\{6, 3\}) = \alpha(\{3, 0\}) = 3, \quad \beta(\{6, 3\}, \{3, 0\}) = 3.$$

It can be shown by direct calculation, that these are the unique cases for $n = 4, 5, 7$ (up to isomorphism) and that there is no such HD for $n = 6, 8$. We do not have a general conjecture. It is certainly no problem to settle more small cases up to $n \leq 15$ say, by using a computer.

Proposition 7. *Suppose* (1) *is a CD with* $\mathbf{V} = \{\infty, 0, 1, \ldots, n-2\}$ *having the automorphism* $\nu = (\infty)(0, 1, \ldots, n-2)$ *and for a fixed* $g \in \mathbf{V} \backslash \{\infty\}$ *let* $(v_1, \ldots, v_{n-2})$ *be an arrangement of* $\{0, \ldots, n-2\} \backslash \{g\}$ $(= \mathbf{C}g \backslash \{\infty\})$ *such that*

$$\mathfrak{R} := \{\{v_i, v_{i+1}\} : i = 1, \ldots, n-2\} \tag{13}$$

is the set of edges of $\gamma_{\mathbf{C}i}$ *not containing the vertex* ∞. *Then* λ *is a CBD if and only if:*

(i) *For every* $i \in \{2, \ldots, [\frac{1}{2}(n-2)]\}$ *there are exactly two* $I \in \mathfrak{R}$ *with* $\alpha(I) = i$.

(ii) *If n is odd, then there is exactly one* $I \in \mathfrak{R}$ *with* $\alpha(I) = \frac{1}{2}(n-1)$.
(iii) *There is exactly one* $I \in \mathfrak{R}$ *with* $\alpha(I) = 1$.
Moreover, λ *is a HD if and only if* $\mathfrak{R}$ *has additionally the properties* (iii), (iv) *of Proposition* 6 *(where* $\mathfrak{R}$ *is defined in* (13)) *together with*
(iv) *If* $I\, J \in \mathfrak{R}$, $I \neq J$, *then* $\beta(I, J) \neq \alpha(v_1, v_{n-2})$.

Proof. The proof follows the same line as in Proposition 6. We omit the details.

We verify the conditions (i)–(vi) of Proposition 7 on

$$n = 9: \quad \infty\ 0\ 6\ 2\ 7\ 1\ 4\ 5$$

(the first example of $n = 9$ in the list above):

$$\alpha(\{4, 5\}) = 1,$$

$$\alpha(\{6, 2\}) = 4,$$

$$\alpha(\{0, 6\}) = \alpha(\{7, 1\}) = 2, \qquad \beta(\{0, 6\}, \{7, 1\}) = 1,$$

$$\alpha(\{2, 7\}) = \alpha(\{1, 4\}) = 3, \qquad \beta(\{2, 7\}, \{1, 4\}) = 2,$$

$$\alpha(\{5, 0\}) = 3.$$

Definition 8. Let $(P_1, \ldots, P_{n-2})$ be a sequence of $n-1$ consecutive points on the unit circle, forming the vertices of a regular n-gon.

Using the assumptions and notations of Proposition 6 or Proposition 7 we join P_i, P_j by a line segment if and only if $\{i, j\} \in \mathfrak{R}$.

Then the so-obtained graph $\delta(\lambda)$ is a *diagram* of λ. A diagram is *cyclicly balanced* or *harmonic*, if the corresponding CD is.

If λ is of the type of Proposition 6, then its diagram is a *cycle*, if λ is of type of Proposition 7, then its diagram is a *linear path*. The only omitted point is P_0 in both cases. It follows from this construction immediately:

Proposition 8. *Two CD's* λ_i *both of type of Propositions* 6 *or* 7 *are isomorphic if and only if their diagrams are congruent under a rigid notation, where reflections are permitted.*

Especially in Propositions 6 and 7 the diagrams obtained from $\mathfrak{R} = \mathfrak{R}(g)$ are congruent for every g.

Definition 9. Let $\mathfrak{W} =$ be a set of $n-1$ points on the unit circle, forming the vertices of a regular $(n-1)$-gon.

If $P, Q \in \mathfrak{W}$, $P \neq Q$ and H, H^* are the two closed half-spaces defined by the line through P and Q, then

$$\alpha(P, Q) := \min\{\#(\mathfrak{W} \cap H), (\mathfrak{W} \cap H^*)\} - 1.$$

Now, if $P, Q, R, S \in \mathfrak{W}$, $P \neq Q$, $R \neq S$, $\{P, Q\} \neq \{R, S\}$, $\alpha(P, Q) = \alpha(R, S)$, then $\alpha(P, R) = \alpha(Q, S)$ or $\alpha(P, S) = \alpha(Q, R)$. Here both equalities hold if and only if n is odd and $\alpha(P, Q) = \alpha(R, S) = \frac{1}{2}(n-1)$. In this case we have $\alpha(P, R) = \alpha(P, S)$.

Therefore we may define

Definition 10. Suppose $P, Q, R, S \in \mathfrak{W}$, $P \neq Q$, $R \neq S$, $\{P, Q\} \neq \{R, S\}$, $\alpha(P, Q) = \alpha(R, S)$. Then

$$\beta(\{P, Q\}, \{R, S\}) := \begin{cases} \alpha(P, R) & \text{if } \alpha(P, R) = \alpha(Q, S), \\ \alpha(P, s) & \text{if } \alpha(P, S) = \alpha(Q, R). \end{cases}$$

The definition of α and β on the vertices of a diagram $\delta(\lambda)$ translates α and β of Definitions 6 and 7 canonically and thereby gives a geometrical characterization of cyclically balanced and harmonic diagrams.

As an example we have drawn the diagrams to the example $n = 7$, following Proposition 6 and $n = 9$, following Proposition 7.

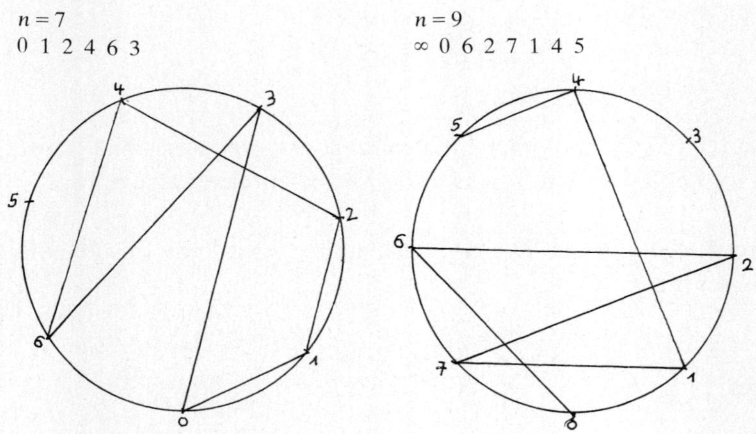

Fig. 2

Proposition 9. *Let p denote an odd prime, m a natural number,* $q = p^m$, $\mathbf{V} := GF(q)$ *and r a primitive root in* $GF(g)$. *Then there exists a HD* λ *of type* (1), λ *has the automorphism* $\nu = (0)(r, r^2, \ldots, r^{q-2})$ *and*

$$\begin{array}{lllll}
1, & r, & r^2, & \ldots, & r^{q-2} \\
0, & r-1, & r^2-1, & \ldots, & r^{q-2}-1 \\
\cdot & & & & \\
\cdot & & & & \\
\cdot & & & & \\
0, & r^t - r^{t-1}, & r^{t+1} - r^{t-1}, & \ldots, & r^{t-q-3} - r^{t-1} \\
\cdot & & & & \\
\cdot & & & & \\
\cdot & & & & \\
0, & 1 - r^{q-2}, & r - r^{q-2}, & \ldots, & r^{q-3} - r^{q-2}
\end{array}$$

is a representation of λ.

Proof. For the permutation

$$\zeta_0 = (1, r, \ldots, r^{q-2}),$$
$$\zeta_t = (0, r^t - r^{t-1}, r^{t+1} - r^{t-1}, \ldots, r^{t-q-3} - r^{t-1}), \quad 0 \neq r \in \mathrm{GF}(q)$$

we have

$$\zeta_{t+1} = \zeta_0 \zeta_t \zeta_0^{-1}.$$

This shows $\zeta_0 \in \Gamma(\lambda)$. Now

$$\zeta_0(x) = rx,$$
$$\zeta_t(x) = rx - r^{\lambda} - r^{\lambda-1}.$$

Therefore if s is defined through $r^s = 1 - r$, then

$$\begin{aligned}(\zeta_1 \zeta_0 \zeta_1^{-1})(x) &= r(r^{q-2}x + r^{q-2} - 1)) + r - 1 \\ &= rx + r - r^2 + r - 1 \\ &= \zeta_{s+1}(x).\end{aligned}$$

This shows $\zeta_1 \in \Gamma(\lambda)$. Therefore $\Gamma(\lambda)$ contains the group of affine transformations τ: $\tau(x) = ax + b$, $a \neq 0$. The rest of the proof is an easy calculation.

The proof is due to L. Danzer. The author found this result independently when q is a prime.

References

[1] J. Denes and A. D. Keedwill, Latin Squares and Their Applications (Budapest, 1974).
[2] M. Hall Jr., The Theory of Groups (New York, 1959).

Annals of Discrete Mathematics 6 (1980) 215–217

ON RESTRICTED BASES FOR FINITE FIELDS

S.A. JONI* and G.-C. ROTA**
Massachusetts Institute of Technology, Cambridge, MA 02139, USA

1. Introduction

Given an $n \times n$ matrix $B = (b_{ij})$ of zeros and ones, a permutation matrix (i.e. a matrix of zeros and ones with precisely one one in each row and column) $P = (p_{ij})$ is said to be *restricted by* B if for all i, j, $b_{ij}p_{ij} = 0$. Thus, P is restricted by B if no non-zero entry of P coincides with a non-zero entry of B. Motivated by the oft-noted analogy between finite boolean algebras and lattices of vector spaces over finite fields, we present the main results of a theory of restricted bases for finite vector spaces which parallels the classical theory of permutations with restricted position [2].

Let V_n be an n-dimensional vector space over the finite field GF(q), and let S denote a given collection of points (or vectors) in V_n. A basis B of V_n is said to be *restricted by* S if none of its elements belong to S. Two bases are considered identical if they differ only by the order of their terms.

2. Enumeration

Let $r_s(n)$ denote the number of bases of V_n restricted by S. We associate to the "forbidden set" S the simplicial complex $\Sigma(S)$ as follows: the elements of $\Sigma(S)$ are all the linearly independent subsets of S, including the empty set, ordered by inclusion. We shall call these sets the *faces* of $\Sigma(S)$. Let I_k denote the number of faces of $\Sigma(S)$ with k elements. Our enumeration theorem is obtained via Möbius inversion [3] over the simplicial complex $\Sigma(S)$.

Theorem 1.

$$r_s(n) = q^{\binom{n}{2}} \sum_{k=0}^{n} (-1)^k I_k q^{-\binom{k}{2}} (q^{n-k}-1)(q^{n-k-1}-1)\cdots(q-1)/(n-k)!.$$

---

* Research partially supported by NSF Contract No. MCS 7820264.
** Research partially supported by NSF Contract No. MCS 7701947.

3. A reduction formula

The *characteristic q-polynomial* of the simplicial complex $\Sigma(S)$ is defined by

$$p(x, S) = \sum_{k=0}^{n} (-1)^k I_k q^{-\binom{k}{2}} x^k.$$

If we set, for all non-negative integers n, k,

$$L_n x^k = \begin{cases} (q^{n-k}-1)(q^{n-k-1}-1)\cdots(q-1)/(n-k)! & \text{for } n > k, \\ 0 & \text{otherwise,} \end{cases}$$

then clearly

$$r_S(n) = q^{\binom{n}{2}} L_n(p(x, S)).$$

Therefore the characteristic q-polynomial carries all the information needed for enumeration, and is worthy of further study. Our reduction formula answers the following question. Suppose S is the disjoint union of two subsets C and D. How is the characteristic q-polynomial of S related to those of C and D? The answer depends on introducing a suitable notation. We may assume that the point 0 of V_n is not in S since this point cannot belong to any basis. The simplicial complexes $\Sigma(C)$ and $\Sigma(D)$ are the sub-simplicial complexes of $\Sigma(S)$ obtained by taking respectively the independent subsets of C and D, ordered by inclusion. For M a face of $\Sigma(C)$, we define the simplicial complex $\Sigma(D/M)$ to be the collection of all faces of $\Sigma(D)$ which are linearly independent of the subspace spanned by M. Thus, $\Sigma(D/M)$ corresponds to all possible subsets A of D such that $A \cup M$ is an independent set.

Theorem 2.

$$p(x, S) = \sum_{M \in \Sigma(C)} (-x)^m q^{-\binom{m}{2}} p(xq^{-m}, D/M)$$

where m denotes the number of elements of M.

The proof is obtained via a double Möbius inversion over the simplicial complexes $\Sigma(C)$ and $\Sigma(D)$, and the reader is refered to [1] for details.

When C consists of a non-zero one element subset of S, our reduction formula gives a "q-analog" of the classical one-square reduction formula of Kaplansky–Riordan, namely

$$p(x, S) = p(x, D) - xp(xq^{-1}, D/C).$$

4. Conclusion

We conclude with the following remarks. The simplicial complexes arising in this study are evidently of a very special kind and it would be interesting to characterize them combinatorially. A slight modification of the preceeding theory gives a similar method for computing the number of bases of *lines* (or points in projective space) of V_n restricted by the "forbidden set" of lines S. Essentially all that is needed is to replace the linear functionals L_n by M_n where

$$M_n x^k = \begin{cases} (q^{n-k}-1)\cdots(q-1)/(q-1)^{n-k}(n-k)! & \text{for } n>k, \\ 0 & \text{otherwise.} \end{cases}$$

Aesthetically this is sometimes more pleasing since we can obtain formulas in which the limit as $q \to 1$ does not collapse to zero.

A more substantial undertaking would be to develop a theory of restricted placement based on Möbius inversion over a q-simplicial complex Σ defined as follows: Σ is a family of subspaces of a finite vector space V_n, and if $W \in \Sigma$ and $U \subseteq W$, then $U \in \Sigma$. Preliminary investigations indicate that deeper techniques such as those found in the Schubert Calculus will be needed in the development of this theory.

References

[1] S.A. Joni and G.-C. Rota, A vector space analog of permutations with restricted position, J. Combin. Theory (A), 28 (1980).

[2] I. Kaplansky and J. Riordan, The problem of rooks and its applications, Duke Math. J. 13 (1946) 259–268.

[3] G.-C. Rota, On the foundations of combinatorial theory I: Theory of Möbius inversion, Z. Wahrscheinlichkeitstheorie und Vern. Gebiete 2 (4) (1964) 340–368.

Annals of Discrete Mathematics 6 (1980) 219–224

PARTITIONING OF THE MINIMUM ESSENTIAL SET CONSTRUCTION PROBLEM

Aram K. KEVORKIAN
Shell Research B.V., Amsterdam, The Netherlands

In this paper we survey two recent results on the partitioning of a strongly connected directed graph G into subgraphs such that a minimum essential set or minimum feedback vertex set of each subgraph is a subset of some minimum essential set of G.

1. Introduction

A set S of vertices is an essential set or feedback vertex set of the directed graph $G=(V, E)$ if the subgraph of G induced by the vertex set $V-S$ is an acyclic graph. If S is an essential set of G and no proper subset of S is an essential set of G, then S is called a minimal essential set. An essential set of minimum cardinality is called a minimum essential set. If a minimum essential set consists of a single vertex v, then v is called a break vertex [6]. Clearly, the notion of a break vertex is a special case of the notion of a minimum essential set.

A minimum essential set provides valuable information for analysing and shaping the structures of problems encountered in a wide variety of applications. A list of such problems given in [4] includes the solution of linear and nonlinear systems of equations, stability analysis of dynamic systems, minimization of the spike columns and maximization of the generalized upper bounding rows in linear programming. Another interesting application of the minimum essential set notion is in the use of the cyclic odd–even reduction method. Given a banded system of linear algebraic equations $Yx=b$, the cyclic odd–even reduction method requires the finding of a permutation matrix Q such that

$$QYQ^{\mathrm{T}}=\boxed{\begin{matrix} A & B \\ C & D \end{matrix}}$$

where the leading block A is a diagonal matrix of maximum order. If G is the directed graph of Y, then it can readily be verified that the vertices in G representing the rows on which the matrices C and D lie form a minimum essential set of G. This graph-theoretic interpretation of the cyclic odd–even reduction method may shed some light on some of the problems encountered in the automatic generation [1, 7] of the permutation matrix Q.

The problem of constructing a minimum essential set of an arbitrary directed graph belongs to the class of nondeterministic polynomial-time complete (NP-complete) problems [2]. Therefore, this problem is computationally intractable since all the algorithms currently known for finding a solution of an NP-complete problem require amounts of computer time that increase exponentially with the size of the problem. In view of this inherent drawback, there is need for methods which partition an NP-complete problem into similar but smaller subproblems. Our object is to survey two recent results [4] on the partitioning of the minimum essential set construction problem.

2. Notation

A graph $G=(V, E)$ consists of a nonempty set of vertices V and a set of edges E. The graphs we consider in this paper are finite, with no loops or multiple edges. If the edges in a graph G are ordered pairs (v, w) of vertices, then G is said to be a directed graph (digraph). If the edges are unordered pairs of vertices, also denoted by (v, w), then G is an undirected graph. A digraph $G=(V, E)$ is symmetric if and only if $(v, w)\in E \Rightarrow (w, v)\in E$. The digraph corresponding to an undirected graph $G=(V, E)$ is a symmetric digraph $G'=(V, E')$, where $E'=\{(v, w), (w, v) \mid (v, w)\in E\}$. If U is a set of vertices in $G=(V, E)$, then the graph $G(U)=(U, E(U))$, where $E(U)=\{(v, w)\in E \mid v, w\in U\}$, is called the subgraph of G induced by the vertex set U. The indegree $\mathrm{id}(v)$ of a vertex v is the number of edges entering v and the outdegree $\mathrm{od}(v)$ of v is the number of edges leaving v.

The adjoint graph of a matrix Y is an undirected graph $\mathscr{A}=(V, E)$ where each vertex of V corresponds to a nonzero entry of Y and $(v, w)\in E$ if and only if the entries corresponding to v and w lie in the same row or column of Y. A cycle in an adjoint graph is called a polygon [3]. If $\mathscr{P}$ is a path and P is a polygon in $\mathscr{A}$ and the subgraphs of $\mathscr{A}$ induced by the vertices of $\mathscr{P}$ and P do not contain any polygon of length 3, then $\mathscr{P}$ is called a triangulation-free path and P a triangulation-free polygon [3]. Suppose the edge set in the adjoint graph $\mathscr{A}=(V, E)$ of a matrix Y is partitioned into two sets E_c and E_r such that $(v, w)\in E_c(E_r)$ if and only if the entries corresponding to v and w lie in the same column (row). Then for a subset U of V, we define the following vertex sets.

$$\mathrm{cadj}(U)=\{v\in V-U \mid (v, w)\in E_c \text{ for some } w\in U\},$$
$$\mathrm{radj}(U)=\{v\in V-U \mid (v, w)\in E_r \text{ for some } w\in U\}.$$

A set $\mathscr{S}$ of vertices is a stable set of the undirected graph $G=(V, E)$ if no edge of E joins two vertices of $\mathscr{S}$. If $\mathscr{S}$ is not a proper subset of any stable set of G, then $\mathscr{S}$ is called a maximal stable set. A stable set of maximum cardinality is called a maximum stable set.

3. The partitioning result

In a recent paper [4] we established the following result on the partitioning of the minimum essential set construction problem.

Theorem 1. *Let $G=(V, E)$ be an undirected graph and let the symmetric digraph corresponding to G be a subgraph of the digraph $\mathcal{G}=(\mathcal{V}, \mathcal{E})$. Suppose $\mathcal{S}$ is a set of vertices satisfying the following properties.*

(1) *$\mathcal{S}$ is a maximum stable set of G.*

(2) $\mathrm{id}(v)\times\mathrm{od}(v)=0$ *for all $v\in\mathcal{S}$ in $\mathcal{G}(\mathcal{V}-(V-\mathcal{S}))$.*

Then there exists a minimum essential set S of $\mathcal{G}$ such that

$$V-\mathcal{S}\subseteq S.$$

Suppose $E=\emptyset$ in Theorem 1. Then the vertex set V is the unique maximum stable set of G. Consequently, for the class of digraphs which do not contain a cycle of length 2 Throrem 1 is superfluous since $V-\mathcal{S}=\emptyset$. However, Theorem 1 is powerful enough to contain all the currently available results that identify subsets of a minimum essential set as special cases [4].

For the class of digraphs which do not contain cycles of length 2 we have a transitive-closure type result [4] which eliminates a set of vertices $\mathbb{V}$ from a digraph $\mathcal{G}=(\mathcal{V}, \mathcal{E})$ by adding a new set of edges $\mathbb{E}$ to $\mathcal{G}(\mathcal{V}-\mathbb{V})$. In this way we obtain a new digraph $\mathbb{G}=(\mathcal{V}-\mathbb{V}, \mathcal{E}(\mathcal{V}-\mathbb{V})\cup\mathbb{E})$ which may contain cycles of length 2, thus promoting the applicability of Theorem 1. The result is as follows.

Lemma 1. *Let v be any vertex in the digraph $\mathcal{G}=(\mathcal{V}, \mathcal{E})$. Let $L(v)=\{z \mid (v, z), (z, v)\in\mathcal{E}\}$ and $\mathbb{E}=\{(u, w) \mid (u, v), (v, w)\in\mathcal{E}\}$. If*

$$\min(\mathrm{id}(v), \mathrm{od}(v))=1 \quad \textit{and} \quad L(v)=\emptyset,$$

then any minimum essential set of the digraph $\mathbb{G}=(\mathcal{V}-\{v\}, \mathcal{E}(\mathcal{V}-\{v\})\cup\mathbb{E})$ is a minimum essential set of $\mathcal{G}$.

The next natural step is to consider the case where $\min(\mathrm{id}(v), \mathrm{od}(v))>1$ and $L(v)=\emptyset$ for each vertex v in a digraph.

It is worth noting that if $L(v)\neq\emptyset$ in Lemma 1, then by Theorem 1 it follows that the set $L(v)$ of cardinality one is a subset of some minimum essential set of the digraph $\mathcal{G}=(\mathcal{V}, \mathcal{E})$.

We can illustrate Theorem 1 and Lemma 1 with the examples shown in Fig. 1. Figure 1(a) shows a strongly connected 8-vertex digraph $\mathcal{G}=(\mathcal{V}, \mathcal{E})$. Since $L(v_1)=\emptyset$ and $\mathrm{id}(v_1)=\mathrm{od}(v_1)=1$, the vertex v_1 in $\mathcal{G}$ satisfies the conditions imposed in Lemma 1. The application of Lemma 1 to $\mathcal{G}$ at the vertex v_1 yields the 7-vertex digraph $\mathcal{G}_1=(\mathcal{V}_1, \mathcal{E}_1)$ shown in Fig. 1(b), where $\mathcal{V}_1=\mathcal{V}-\{v_1\}$ and

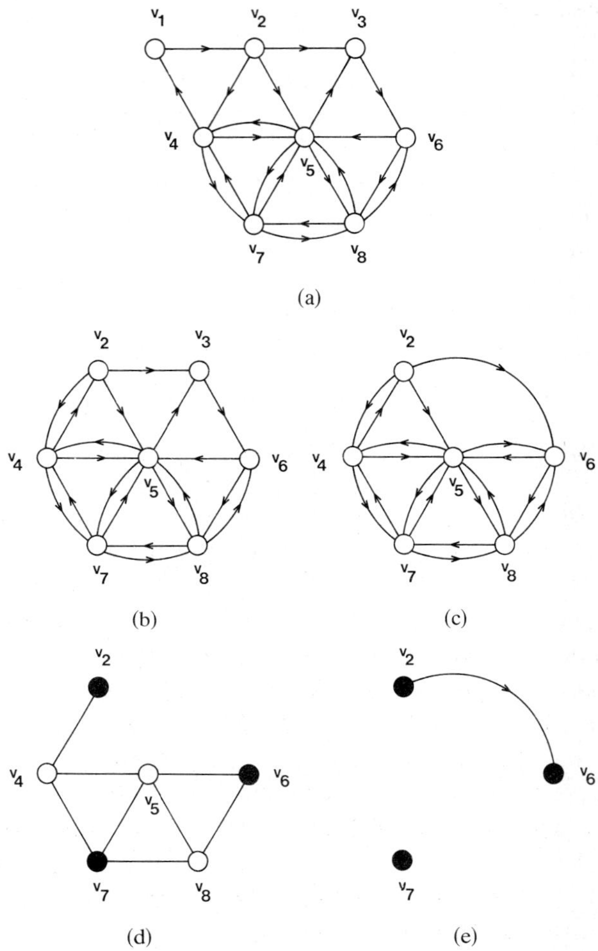

Fig. 1. (a) The digraph $\mathscr{G}=(\mathscr{V}, \mathscr{E})$. (b) The digraph $\mathscr{G}_1=(\mathscr{V}_1, \mathscr{E}_1)$. (c) The digraph $\mathscr{G}_2=(\mathscr{V}_2, \mathscr{E}_2)$. (d) The undirected graph $G=(V, E)$. (e) The digraph $\mathscr{G}_2(\mathscr{S})$.

$\mathscr{E}_1=\mathscr{E}(\mathscr{V}_1)\cup\{(v_4, v_2)\}$. Clearly, the vertex v_3 in the digraph $\mathscr{G}_1$ satisfies the conditions imposed in Lemma 1. The application of Lemma 1 to $\mathscr{G}_1$ at the vertex v_3 yields the 6-vertex digraph $\mathscr{G}_2=(\mathscr{V}_2, \mathscr{E}_2)$ shown in Fig. 1(c), where $\mathscr{V}_2=\mathscr{V}_1-\{v_3\}$ and $\mathscr{E}_2=\mathscr{E}(\mathscr{V}_2)\cup\{(v_4, v_2), (v_2, v_6), (v_5, v_6)\}$. Figure 1(d) shows the undirected graph $G=(V, E)$ corresponding to the symmetric subgraph $(\mathscr{V}_2, \mathscr{E}_2-\{(v_2, v_5), (v_2, v_6)\})$ of $\mathscr{G}_2$. It is easy to verify that the vertex set $\mathscr{S}=\{v_2, v_6, v_7\}$ shown in dark vertices is a maximum stable set of G. Thus the set $\mathscr{S}$ satisfies condition 1 in Theorem 1. Figure 1(e) shows the subgraph of $\mathscr{G}_2$ induced by the vertex set $\mathscr{V}_2-(V-\mathscr{S})=\mathscr{S}$. Since $\mathrm{id}(v)\times\mathrm{od}(v)=0$ for all $v\in\mathscr{S}$ in $\mathscr{G}_2(\mathscr{S})$, by Theorem 1 it follows that the set $S=V-\mathscr{S}=\{v_4, v_5, v_8\}$ is a minimum essential set of the digraph $\mathscr{G}_2=(\mathscr{V}_2, \mathscr{E}_2)$. Consequently, by Lemma 1 the set S is a minimum essential set of the original digraph $\mathscr{G}=(\mathscr{V}, \mathscr{E})$ shown in Fig. 1(a).

4. A result on the partitioning of the vertex set V

Suppose the vertex V in Theorem 1 is partitioned into the subsets $V_1, V_2, \ldots, V_k$ and suppose we use some exact solution method to construct maximum stable sets $\mathcal{S}_1, \mathcal{S}_2, \ldots, \mathcal{S}_k$ of the graphs $G(V_1), G(V_2), \ldots, G(V_k)$, respectively. If each of the vertices in any $\mathcal{S}_i$ is a root or a leaf in $\mathcal{G}(\mathcal{V}-(V_i-\mathcal{S}_i))$, then by Theorem 1 the problem of constructing a minimum essential set of the digraph $\mathcal{G}=(\mathcal{V}, \mathcal{E})$ is reduced to the problem of constructing a minimum essential set of $\mathcal{G}(\mathcal{V}-V_i)$. This procedure of reducing the size of the original problem is repeated until none of the remaining maximum stable sets satisfies condition 2 in Theorem 1.

As is well-known, the problem of constructing a maximum stable set of an arbitrary undirected graph is NP-complete [2]. This implies that our proposed partitioning strategy is computationally intractable in its present form since by condition 1 in Theorem 1 we require the solving of k NP-complete problems.

A way of making the proposed partitioning strategy feasible is to partition the vertex set V into subsets $V_1, V_2, \ldots, V_k$ in such a way that the k graphs induced by the V_i's have particular structural properties that make it possible to construct in a systematic way one or more maximum stable sets of each graph in polynomial-time. Evidently, a clique is a simple example of a graph which has the structural property we seek since each vertex in the clique forms a maximum stable set of the clique. For example, each of the sets $\mathcal{S}'_1=\{v_5\}$, $\mathcal{S}''_1=\{v_6\}$ and $\mathcal{S}'''_1=\{v_8\}$ is a maximum stable set of the clique $G(V_1=\{v_5, v_6, v_8\})$ in Fig. 1(d). Consequently, since v_6 is a leaf in the digraph $\mathcal{G}_2(\mathcal{V}_2-(V_1-\{v_6\}))$ by Theorem 1 the set $V_1-\{v_6\}=\{v_5, v_8\}$ is a subset of a minimum essential set of $\mathcal{G}_2$ in Fig. 1(c).

In the same paper [4] we established a result which generates in a systematic way a class of undirected graphs that contribute to the applicability of Theorem 1 as a partitioning result. Before we present the result, we need to introduce the notion of strongly adjacent polygons [5].

Definition. Let $\mathcal{F}=(P_1, P_2, \ldots, P_N)$ be an N-tuple of distinct triangulation-free polygons in an adjoint graph $\mathcal{A}$ satisfying the following property.

$$\left|\mathrm{cadj}\left(\bigcup_{i=1}^{k} P_i\right)\cap P_{k+1}\right| = \left|\mathrm{radj}\left(\bigcup_{i=1}^{k} P_i\right)\cap P_{k+1}\right| = 1 \quad \text{for } k=1,2,\ldots,N-1.$$

Then $\mathcal{F}$ is called *a family of strongly adjacent polygons.*

Let $\mathcal{A}=(V, E)$ be the adjoint graph of a matrix Y. If v is a vertex in $\mathcal{A}$ corresponding to an entry in the ith row of Y, then we shall use $R(v)$ to denote the set of vertices in $\mathcal{A}$ corresponding to the entries in the ith row of Y. We now state the result [4].

Theorem 2. *Let* $\mathcal{F}=(P_1, P_2, \ldots, P_N)$ *be any family of strongly adjacent polygons in the adjoint graph* $\mathcal{A}=(V, E)$. *Let* $\mathcal{S}=\{v_1, v_2, \ldots, v_n\}$ *be any maximum stable set of* $\mathcal{A}(\bigcup_{i=1}^{N} P_i)$. *Suppose the symmetric digraph corresponding to* $\mathcal{A}(\bigcup_{i=1}^{n} R(v_i))$ *is a subgraph of the digraph* $\mathcal{G}=(\mathcal{V}, \mathcal{E})$. *If* $\mathrm{id}(v)\times \mathrm{od}(v)=0$ *for all* $v\in\mathcal{S}$ *in* $\mathcal{G}(\mathcal{V}-((\bigcup_{i=1}^{n} R(v_i))-\mathcal{S}))$, *then there exists a minimum essential set* S *of* $\mathcal{G}$ *such that*

$$\left(\bigcup_{i=1}^{n} R(v_i)\right)-\mathcal{S}\subseteq S.$$

The adjoint graph of a matrix Y is the line graph of the bipartite graph of Y. Thus the problem of constructing a maximum stable set of an adjoint graph is an alternative formulation of the problem of finding a maximum matching in a bipartite graph. Therefore, there are many polynomial-time algorithms for constructing a maximum stable set of the graph $\mathcal{A}(\bigcup_{i=1}^{N} P_i)$ in Theorem 2.

It is interesting to note that each of the $\mathcal{A}(R(v_i))$'s in Theorem 2 consists of a clique. Therefore, since $\bigcup_{i=1}^{N} P_i\subseteq\bigcup_{i=1}^{n} R(v_i)$ it follows that the graph $\mathcal{A}(\bigcup_{i=1}^{n} R(v_i))$ in Theorem 2 represents the various ways in which $2, 3, \ldots, n$ different cliques are connected to each other through the edges of the edge set

$$E\left(\bigcup_{i=1}^{N} P_i\right)-E\left(\bigcup_{i=1}^{n} R(v_i)\right).$$

A few interesting configurations of the graph $\mathcal{A}(\bigcup_{i=1}^{n} R(v_i))$ can be found in [4].

We have used the partitioning results to provide an algorithm [4] which constructs a minimal essential set of an n-vertex symmetric digraph, a maximal stable set of an n-vertex undirected graph and an essential set of an n-vertex arbitrary directed graph in $O(n^2)$ computation steps and such that these solutions are within a known tolerance of the optimal value.

References

[1] D. Heller, A survey of parallel algorithms in numerical linear algebra, SIAM Rev. 20 (1978) 740–777.

[2] R.M. Karp, Reducibility among combinatorial problems, in: R.E. Miller and J.W. Thatcher, eds,. Complexity of Computer Computations (Plenum Press, New York, 1972) 85–104.

[3] A.K. Kevorkian, Graph-theoretic characterization of the matrix property of full irreducibility without using a transversal, J. Graph Theory 3 (1979) 151–174.

[4] A.K. Kevorkian, General topological results on the construction of a minimum essential set of a directed graph, IEEE Trans. Circuits Systems CAS-27 (1980) 293–304.

[5] A.K. Kevorkian, Simplified characterization of the matrix property of full irreducibility, to appear.

[6] S. Rao Kosaraju, On independent circuits of a digraph, J. Graph Theory 1 (1977) 379–382.

[7] G.H. Rodrigue, N.K. Madsen and J.I. Karush, Odd–even reduction for banded linear equations, J. Assoc. Comput. Mach. 26 (1979) 72–81.

Annals of Discrete Mathematics 6 (1980) 225–241

OPTIMAL DESIGN THEORY IN RELATION TO COMBINATORIAL DESIGN

J. KIEFER*
Cornell University and University of California, Berkeley, CA, U.S.A.

1. Introduction

The usefulness of combinatorial arrays for designing statistical experiments was evident to R.A. Fisher, and under his energetic leadership the construction and application of such designs enjoyed tremendous expansion. Fisher and his students and followers were responsible for much of the development in the 1930's and 1940's. Mathematicians such as R.C. Bose gave general structure to classes of these designs, and invented general methods for constructing them. It would be presumptuous of me to recite more of this history to the audience of this conference.

The spirit of much of this construction, of designs that were in some sense as symmetric as possible in their treatment of the statistical parameters of interest (for example, balanced incomplete block designs (BIBD) or latin squares (LS)), perhaps stemmed from three factors: such designs yielded "information matrices" (coefficient matrices of normal equations for least squares estimation) that, especially in that pre-computer age, made statistical calculations easy; the designs had aesthetic appeal to mathematicians and often had algebraic or geometric representations that helped one to understand and construct them; and the symmetric treatment of parameters of interest seemed a reasonable property that made such designs yield statistical estimators that looked intuitively as accurate as possible for the given number of observations.

The work of Neyman and of Wald made statisticians increasingly formulate statistical problems and criteria for "goodness" of procedures in a precise manner. A pioneering example of this in design theory is the paper by Wald [45], who considered the standard 2-way heterogeneity model with v treatments and the possibility of allowing any $v \times v$ matrix of symbols from $\{1, 2, \ldots, v\}$ as a design, and proved that a LS design was optimum in a precise sense. The result is not too surprising in view of the intuition indicated in the previous paragraph, but it turns out (Section 4) that intuition is not always so reliable; what is important, though, is that Wald gave a structure for choice of a design on well-formulated grounds that in no way rested on the motivation employed earlier for using such designs.

* Research under NSF Grant MCS75-22481 A02.

Authors such as Mood [33] extended such considerations to the setting of weighing designs; and subsequently other settings, such as that where BIBD's are often employed, were considered, as in Kiefer [23]. In a companion development, problems were also treated in which these combinatorial structures are not relevant because there is a continuum of design choices, as in curve-fitting settings in which the independent variable may be choosen to lie in an interval. This is discussed, for example, in Kiefer [24, 26] and Kiefer and Wolfowitz [30].

The present paper is devoted to indicating how, in recent years, the relationship between combinatorial and optimal design developments has begun to exhibit a motivational reversal. Instead of constructing exotic designs on intuitive grounds and then proving that they are indeed optimum, we encounter settings in which optimality considerations motivate the construction of new combinatorial structures, or the selection of a particular subset of the known ones for special emphasis and study. An interesting example (to be discussed further in Section 2) is the setting of "one-way heterogeneity", that is, of block designs. If the parameter values (block size k, number of blocks b, number of varieties v) were such that a BIBD did not exist, the early design developments sought designs that were still quite symmetric, as in the construction of PBIBD's by Bose and Nair [3]. After optimality properties of BIBD's were proved by Kiefer [23], work by Takeuchi [43, 44] showed that certain PBIBD's were also E-optimum (see (1.1)). This result (pioneering in giving the first optimality conclusion for designs lacking maximum combinatorial symmetry) was strengthened and extended by Cheng [5], who was able to delimit a particular class of PBIBD's that possess strong optimum properties. Finally, the consideration of that class led to the graph-theoretic characterization in the paper by Cheng and Gray [10] presented at this conference. Several other illustrations of such developments will be considered in this paper, but they are by no means exhaustive; for example, the work of Seiden, Hedayat, and others on various "repeated measurement" models, will not be touched upon; nor will Srivastava's search design considerations, which in part entail finding, as in Section 5, the "smallest" design that achieves an appropriate aim.

Throughout this paper we will restrict attention to a setting (slightly modified in Section 5) in which there is a class D of designs available to the statistician and in which, for each d in D, there is an $n \times m$ matrix X_d such that the expectation of the observed n-vector Y, if design d is used, is $X_d\theta$; here θ is the m-vector of unknown parameters in this "linear model". The components of Y are assumed uncorrelated, with equal variances σ^2. The "information matrix" of the design d is $X_d'X_d$ (primes denote transposes), and the normal equations (NE's) for obtaining least squares (best linear unbiased) estimators are $(X_d'X_d)t = X_d'Y$. This means that, if $c'\theta$ is a linear parametric function for which *some* linear unbiased estimator of $c'\theta$ exists when design d is used, then for any solution t of the NE's $c't$ is the best linear unbiased estimator of $c'\theta$. If X_d has rank m, then $\sigma^2(X_d'X_d)^{-1}$ is the covariance matrix of t, with a similar interpretation otherwise. Thus, we want to choose a design whose information matrix $X_d'X_d$ is "large" in some sense, and since the symmetric nonnegative definite (SND) $m \times m$ matrices are only partially ordered by the natural (and only useful definition) that $A \geq B$ if $A-B$ is SND, an "optimality functional" Φ on the SND matrices must be considered. We then choose d to minimize $\Phi(X_d'X_d)$ over D.

Often only the first v out of the m components of θ are of interest. Partitioning $X_d = (X_d^{(1)} \vdots X_d^{(2)})$ and $\theta' = (\theta^{(1)'}, \theta^{(2)'})$ accordingly, the analogue of $X_d'X_d$ is Bose's "C-matrix"

$$C_d = X_d^{(1)'}X_d^{(1)} - X_d^{(1)'}X_d^{(2)}(X_d^{(2)'}X_d^{(2)})^- X_d^{(2)'}X_d^{(1)};$$

then C_d^- is the upper $v \times v$ submatrix of $(X_d'X_d)^-$. This is the "information matrix for $\theta^{(1)}$". The optimality criterion will then typically reduce to

$$\Phi(X_d'X_d) = \Phi^{(1)}(C_d).$$

In some settings, such as that where a chemical balance is used in sufficiently many "weighings" (Section 3), C_d can be nonsingular. Typical $\Phi^{(1)}$'s in such settings (interpretable as size measures of C_d^{-1} of a "confidence ellipsoid" asserted to contain θ if Y is normal) are

$$\begin{aligned} \log \det C_d^{-1} \qquad & (D\text{-optimality}), \\ \operatorname{tr} C_d^{-1} \qquad & (A\text{-optimality}), \\ \text{max eigenvalue of } C_d^{-1} \qquad & (E\text{-optimality}). \end{aligned} \tag{1.1}$$

Sometimes it is possible to avoid specification of a particular functional. Thus, if C_{d^*} is a multiple of the identity and $\operatorname{tr} C_{d^*} = \max_{d \in D} \operatorname{tr} C_d$, it follows that d^* minimizes $\Phi^{(1)}(C_d)$ for every orthogonally invariant convex $\Phi^{(1)}$ such that $\Phi^{(1)}(bC)$ is nonincreasing in the scalar b. (The $\Phi^{(1)}$'s of (1.1) are obvious examples.) A d^* optimum for all such $\Phi^{(1)}$ is termed *universally optimum*.

In other settings C_d is singular for all d in D. For example, in the setting of one-way heterogeneity (Section 2) in which we must assign v treatments to b blocks of size k, we have $n = bk$ for all allowable designs, and $m = v + b$, an observation on treatment i in block j having expectation $\alpha_i + \beta_j$, with $\theta^{(1)} = \boldsymbol{\alpha}$ and $\theta^{(2)} = \boldsymbol{\beta}$. Then C_d is the information matrix for treatment effects. It can have rank at most $v-1$, since only "contrasts" $\sum_1^v c_i\alpha_i$ with $\sum_i^v c_i = 0$ can be estimated. In fact, C_d has zero row and column sums. In that case if P is any $(v-1) \times v$ real matrix with rows orthonormal and orthogonal to $(1, 1, \ldots, 1)$, we may replace C_d by $\bar{C}_d = PC_dP'$ in (1.1); the statistical meanings of the resulting criteria are parallel to those of (1.1), and are of course independent of P. For universal optimality, orthogonal invariance of

$$\Phi^{(1)}(C_d) = \bar{\Phi}^{(1)}(PC_dP')$$

can be replaced by invariance under (common) relabeling of components of $\theta^{(1)}$ (that is, of rows and columns of C_d), and the condition that $\bar{C}_{d^*}$ is a multiple of the identity can be rewritten conveniently as "all diagonal elements of C_{d^*} equal, all off-diagonal elements equal", which we term *complete symmetry* of C_{d^*}. When all C_d have zero row and column sums, we thus obtain

$$C_{d^*} \text{ completely symmetric, } \operatorname{tr} C_{d^*} = \max_{d \in D} \operatorname{tr} C_d \Rightarrow d^* \text{ universally optimum} \tag{1.2}$$

The use of this tool to prove optimality of BIBD's, or of LS's in their setting, is described further in Kiefer [23, 28]. We shall see that there are familiar settings where (1.2) cannot be used because the obvious candidate d^* does not satisfy one of the two conditions. The usefulness of (1.2) of course lies in the simplicity of evaluating $\operatorname{tr} C_d$ rather than $\Phi^{(1)}(C_d)$ for each d. When (1.2) cannot be applied, more difficult computations and comparisons must be made, as we shall describe.

We hereafter denote the eigenvalues of C_d by $\mu_{d1} \geqslant \mu_{d2} \geqslant \cdots \geqslant \mu_{dv}$, and $\boldsymbol{\mu}_d = (\mu_{d1}, \ldots, \mu_{d(v-1)})$, in the setting of the last paragraph where $\mu_{dv} = 0$ for all d, but where such criteria as the $\Phi^{(1)}$ corresponding to (1.1) can be expressed in terms of $\boldsymbol{\mu}_d$.

We note that, of the three optimality criteria considered explicitly in (1.1), it is generally easiest to verify that a design is E-optimum and most difficult to verify that it is D-optimum. This is not surprising, since for C_{d^*} or $\bar{C}_{d^{**}}$ a multiple of the identity ($C_{d^{**}}$ completely symmetric) it is obvious that D-optimality implies A-optimality implies E-optimality for d^* or d^{**}. The opposite implications are not valid.

Finally, we remark that many optimality proofs in the literature have diminished value because of restrictions that are put on d to simplify the proofs. For example, in the one-way heterogeneity setting with $k<v$, D is sometimes restricted to designs for which each variety appears at most once per block; of course, a BIBD d^* has this property, but in using (1.2) one need not restrict the competitors of d^*. Again, other optimality proofs in the literature make such restrictions on D as equal replication of varieties or even complete symmetry of C_d. Such restrictions assume away many of the mathematical difficulties and, in the example of (3.1) or that of the generalized Youden design setting of Section 4, also eliminate consideration of less symmetric designs that are better!

The author thanks C.S. Cheng for helpful comments.

2. Block design settings for which no BBD exists

We consider the one-way heterogeneity setting described in Section 1, with b, v, k given. We do not require $k<v$. Let n_{dij} be the number of appearances of variety i in block j, $r_{di} = \sum_j n_{dij}$, $\lambda_{dih} = \sum_j n_{dij} n_{dhj}$. A *balanced block design* has all r_{di} equal, all λ_{dih} equal, and all n_{dij} as nearly equal as possible (i.e., $|n_{dij} - k/v| < 1$); if $k<v$, this reduces to a BIBD. Construction of BBD's can easily be reduced to construction of BIBD's with block size $k - v\,\operatorname{int}(k/v)$, where $\operatorname{int}(x)$ denotes the integral part of x.

If b, v, k are such that no BBD exists, we cannot use (1.2) to find an optimum design; there are designs that maximize tr C_d, and others for which C_d is completely symmetric, but none that achieve both. Suppose $\Phi^{(1)}$ is of the form

$$\Phi_f^{(1)}(C_d) \stackrel{\text{def}}{=} \sum_1^{v-1} f(\mu_{di}) \tag{2.1}$$

where f is convex and nonincreasing. (This includes the D- and A-optimality criteria, and the E-optimality criterion can be obtained as a limit of such criteria.) Then, (1.2) can be looked at geometrically in terms of finding the value

$$\mu^* = \max_d \sum_i^{v-1} \mu_{di}$$

and then noting that, on the hyperplane $H = \{\boldsymbol{\mu}: \sum \mu_{di} = \mu^*\}$ in R^{v-1}, the "minimum" of (2.1) occurs at the point $\boldsymbol{h}$ where all μ_i are equal. But if no BBD exists, there is no d^* to which $\boldsymbol{h}$ corresponds.

One possible intuitively appealing approach in this case is to find, on H, the actual d for which $\boldsymbol{\mu}_d$ is closest to $\boldsymbol{h}$ in some sense. If the sense is Euclidean distance, this was the approach suggested by Shah [40] under the restriction (see Section 1) to a class of designs for which $\sum \mu_{di} = \text{tr } C_d$ is a constant. Without this last restriction, it is the approach taken in the more general setting of unequal block sizes by Eccleston and Hedayat [12], and which led to construction considerations. Such designs were termed (M, S)-optimum designs by them. Although the method may produce reasonably efficient designs in some case, it is clear from the fact that, in Shah's case, it amounts to minimizing tr C_d^2, that it does not guarantee optimality in any of the senses of (1.1); in fact, settings can be given in which it leads to very inefficient designs, even to a d^* with $\bar{C}_{d^*}$ singular. In geometric terms, there may be an actual d' on $\{\boldsymbol{\mu}_d: \sum \mu_{di} = \mu^* - \varepsilon\}$ for some small $\varepsilon > 0$, that is much closer to $\boldsymbol{h}$ than is any point on H; additionally, even if the resulting distance $\|\boldsymbol{\mu}_{d'} - \boldsymbol{h}\|$ is small, we would have to justify that geometric "closeness" of $\boldsymbol{\mu}_{d'}$ to $\boldsymbol{h}$ implies smallness of $\Phi^{(1)}$ at d', as is suggested by the quadratic approximation of sufficiently smooth $\Phi^{(1)}$ near $\boldsymbol{h}$. Cheng [6] has given conditions under which (M, S)-optimum designs are actually optimum in senses such as those of (1.1). The results follow from the more basic conclusions in Cheng [5], which will be discussed below.

The definition of PBIBD is extended to PBBD in an obvious way when $k \geq v$. For a group divisible (GD) PBBD with two groups, write $\lambda_{dij} = \lambda_1$ if i, j are in the same associate class, and $\lambda_{dij} = \lambda_2$ otherwise. Cheng defines a GD PBBD with 2 groups to be *most balanced* (*MB*) *of type 1* if $\lambda_2 = \lambda_1 + 1$, and *MB of type 2* if $\lambda_1 = \lambda_2 + 1 > 1$. We write MBD for MBGDPBBD.

Example. For $v = 4$, $k = 2$, let $A = \{(1, 2), (3, 4)\}$ and $B = \{(1, 3)(1, 4), (2, 3), (2, 4)\}$. Then B or $(2B) \cup A$ is type 1, and $B \cup (2A)$ is type 2.

Takeuchi's result (Section 1) is that, when $k < v$, if no BIBD exists but a MBD of type 1 exists (even with more than 2 groups), it is E-optimum. Conniffe and Stone [11] proved, when $k < v$, that a MBD of type 1 is A-optimum among designs for which all r_{di} are equal and all $n_{dij} = 0$ or 1. Despite these restrictions, their method of proof used tools that Cheng [5] extended and sharpened to obtain much stronger results, some of which we now describe. (Another result obtained under restriction to equal replications is that of D-, A-, E-optimality of linked block designs [41].)

Suppose $\Phi_f^{(1)}$ is of the form (2.1) where f is defined on $[0, \max_{d\in D} \operatorname{tr} C_d]$, is strictly decreasing and strictly convex thereon (possibly with $f(0) = +\infty$), and is continuously differentiable on the interior of that interval. If also f' is strictly concave, $\Phi_f^{(1)}$ is said to be a criterion of *type 1*; if f' is strictly convex, it is of *type 2*. *Generalized type i criteria* are pointwise limits of these. The criteria of (1.1) are all generalized type 1; although there exist type 2 criteria, they are of less statistical interest (for example, $f(x) = \varepsilon x^3 - x$ on $[0, c)$ with ε sufficiently small). A delicate calculation by Cheng replaces (1.2) by the assertion that, for $v > 2$, if C_{d^*} has two distinct nonzero eigenvalues, the larger of multiplicity 1, and if $\operatorname{tr} C_{d^*}^2 < (\operatorname{tr} C_{d^*})^2/(v-2)$ and

$$C_{d^*} \text{ maximizes} \begin{cases} \operatorname{tr} C_d \\ \operatorname{tr} C_d - \left(\dfrac{v-1}{v-2}\right)^{1/2} \left[\operatorname{tr} C_d^2 - \dfrac{(\operatorname{tr} C_d)^2}{v-1}\right]^{1/2}, \end{cases} \tag{2.2}$$

then d^* is optimum wrt all generalized type 1 criteria. An analogous result is obtained for generalized type 2 criteria if the *smaller* eigenvalue has multiplicity 1 and the last line of (2.2) is replaced by

$$\operatorname{tr} C_d - [(v-1)(v-2)(\operatorname{tr} C_d^2 - (\operatorname{tr} C_d)^2/(v-1))]^{1/2}.$$

From these results and a further computation it follows that MBD's of type i are optimum (sometimes uniquely so) wrt all generalized type i criteria. In fact, these conclusions hold for the (two corresponding types of) regular graph designs (RGD's) studied by John and Mitchell [22], and this fact (together with a study of the list of designs obtained by John and Mitchell) led to the conjecture that RGD's (or the corresponding regular graphs) with appropriate eigenvalue structure must be MBD's. This was verified by Cheng and Gray [10], as mentioned in Section 1.

A remarkable feature of this result is the separation of the two types of combinatorial structures on the basis of the optimality criterion. Moreover, in earlier work that used (1.2), only convexity of f, not the sign of f''', arose. It is interesting that when one leaves cases where a BBD exists, the set of convex f of universal optimality becomes split in this way into two (nonexhaustive) sets of criteria that depend on the third derivative. This is all in the case where, although a BBD does not exist, there is a design which is perhaps as "close" combinatorially as one can imagine to a BBD. When also a MBD does not exist, one would then expect any optimality results to be much more difficult, involving a further split among criteria relative to which *different* designs are optimum, even for the criteria of (1.1). Indeed, it is easy to give settings where this occurs, and almost no general optimality results are known in such cases.

3. Weighing designs and fractional factorial designs

In the setting of weighing with a chemical balance, $n \geqslant m = v$, we are interested in all parameters, and the (i, j)th elements x_{dij} of X_d is 1, -1, or 0 depending on whether, in the ith of n weighings, the jth object (with weight θ_j) is in the left pan, in the right pan, or absent. Sometimes x_{dij} is restricted to be ± 1. If $2^v \mid n$, one universally optimum design obviously takes $n/2^v$ observations corresponding to each v-vector of ± 1's. More interesting, if $v \mid n$ and there is a $v \times v$ Hadamard matrix, then n/v copies of it yield a universally optimum design. There is a considerable literature of optimality results obtained under the restriction that $X_d'X_d$ is completely symmetric; much of this is reviewed in Raghavarao [37]. Mood [33], besides discussing the Hadamard cases, had enumerated a few D-optimum designs for small n and v, but the first general results in the non-Hadamard case were obtained by Ehlich [13], who proved, under only the restriction $x_{dij} = \pm 1$, that if $n = v$ and $X_{d^*}'X_{d^*} = (n-1)I_n + J_n$ (where J_n consists entirely of 1's), then d^* is D-optimum. Using methods like those employed in obtaining the results of Section 2, Cheng [7] showed, more generally (and among other results), that, for general n and v, a d^* with $X_{d^*}'X_{d^*} = (n-1)I_v + J_v$ is optimum among all designs wrt all generalized type 1 criteria $\sum_1^v f(\mu_{di})$. Similarly, if one restricts x_{dij} to be ± 1, a d^* with $X_{d^*}'X_{d^*} = (n+1)I_v - J_v$ is optimum wrt all generalized type 2 criteria. Existence of designs of these last two types entails $n \equiv 1 \pmod 4$ and $n \equiv 3 \pmod 4$, respectively the former are sparse, requiring $2n-1$ to be a square if $v = n$ (see [37]; the latter exist for $n = v$ whenever there is an $(n+1)\times(n+1)$ Hadamard matrix.

Ehlich also showed that, when $n = v$ is even and all x_{dij} are restricted to be ± 1, d^* is D-optimum if $X_{d^*}'X_{d^*}$ consists of two $(\frac{1}{2}n)\times(\frac{1}{2}n)$ blocks $(n-2)I_{n/2} + 2J_{n/2}$ and all-zero off-diagonal blocks. This optimality result led to the construction problem for such designs, considered, e.g., by Yang [46]. This result also gives a good illustration, cited by Cheng, of the effect of assuming complete symmetry: If

$$X_d = \begin{bmatrix} 1 & 1 & -1 & 1 & 1 & 1 \\ 1 & 1 & 1 & 1 & 1 & -1 \\ 1 & 1 & 1 & -1 & 1 & 1 \\ 1 & 1 & 1 & 1 & -1 & 1 \\ 1 & -1 & 1 & 1 & 1 & 1 \\ 1 & -1 & -1 & -1 & -1 & -1 \end{bmatrix}, \quad X_{d'} = \begin{bmatrix} 1 & 1 & -1 & 1 & 1 & 1 \\ 1 & 1 & 1 & -1 & -1 & 1 \\ 1 & 1 & 1 & -1 & -1 & -1 \\ 1 & 1 & 1 & -1 & 1 & -1 \\ 1 & -1 & 1 & 1 & 1 & 1 \\ 1 & -1 & -1 & -1 & -1 & -1 \end{bmatrix}, \quad (3.1)$$

then $X_d'X_d = 4I_6 + 2J_6$, which is D-optimum among designs with completely symmetric information matrix; but $X_{d'}'X_{d'}$ is of the Ehlich block form, and has a 56% larger determinant.

Recently Galil and Kiefer [20, 21], using a combination of theoretical results based on work of Ehlich [13, 14, 15] and of Ehlich and Zeller [16], and machine computations using an improvement of the method of Mitchell [32], greatly extended the list of known D-optimum weighing designs. This work shows $X_d'X_d=(n+1)I_v-J_v$ is D-optimum if $2v-5\leqslant n\equiv 3 \pmod 4$, and characterizes optimum designs for all $n\geqslant v$ if $v\leqslant 12$; the previous best complete list was for $v\leqslant 5$, due to Payne [34].

This is clearly an area that invites much more combinatorial and optimality work. Another such area, closely related, is that of fractional factorial designs of the 2^s complete s-factor 2-level factorial design. Here the x_{dij} are restricted to be 0 or 1, and for the model in which interactions of more than t out of the s factors are assumed zero, there are $m=\sum_{i=0}^{t}\binom{s}{i}$ parameters. A "fractional factorial design d of resolution $2t+1$" allows all these parameters to be estimated' i.e., $X_d'X_d$ is nonsingular. The analogue of an Hadamard matrix is now an orthogonal array (OA) of size n, s constraints, 2 levels, strength $2t$; when it exists, it is universally optimum. From a statistical viewpoint, it is not even clear that the invariant criteria of universal optimality reflect the experimenter's aim; perhaps higher order interaction parameters are not as important as main effect parameters. In any event, a natural design symmetry sometimes assumed in this setting is that $X_d'X_d$ be invariant under interchange of two rows and columns representing interactions of the same order. Such designs are the balanced arrays (BA's) of Chakravarti [4], and in a number of papers Srivastava and his students (e.g., [42]) have constructed the A-optimum design among BA's of resolution V in the model with $t=2$, for a wide range of values of s and n. These designs are of considerable practical value, since prior to their construction experimenters were in the position of almost always having to restrict n to be a power of 2. Although, as in the settings treated earlier, one cannot suppose the A-optimum BA's are necessarily optimum among *all* designs, the work of Cheng [7] shows that many of them are. A BA of size n with q constraints, 2 levels, and strength L and index set $(\alpha_0, \alpha_1, \ldots, \alpha_L)$ can be considered as an $n\times q$ matrix X_d of entries 0 and 1 such that each $n\times L$ submatrix contains α_i rows equal to each L-vector with i 1's and $L-i$ 0's. (If all α_i are equal, this is an OA of strength L.) A typical result of Cheng is that a BA of strength $2t$ and index set of the form $(\alpha, \alpha, \ldots, \alpha, \alpha+1)$ is optimum wrt all generalized type 1 criteria, among all designs of resolution $2t+1$. This and the other results on optimality of BA's (e.g., for designs with $\alpha_{2t}=\alpha-1$, $\alpha\pm 2$, or $\alpha+3$ instead of $\alpha+1$ as above) again motivate construction of combinatorial arrays with particular parameter values.

4. Latin squares, Youden designs, and generalizations

In the setting of "two-way heterogeneity", for given integers $v, b_1, b_2\geqslant 2$, a design is a $b_1\times b_2$ array of entries from the set $\{1, 2, \ldots, v\}$. Here $n=b_1b_2$ and

$m = v + b_1 + b_2$. The row of X_d corresponding to cell (i, j), if that cell has entry h, consists of all zeros with the exception of a 1 in the hth, $(v+i)$th, and $(v+b_1+j)$th cell. This means variety h is planted or treated in that cell, and the expected value of the resulting observation is $\alpha_h + \beta_i + \gamma_j$, the three terms being respectively variety, row, and column effects. The vector $\theta^{(1)}$ consists of the α_h, and as in the case of one-way heterogeneity C_d can have rank at most $v-1$. The (i, j)th element of C_d is in fact

$$C_{dij} = \delta_{ij} r_{di} - \frac{\lambda^{(R)}_{dij}}{b_2} - \frac{\lambda^{(C)}_{dij}}{b_1} + \frac{r_{di} r_{dj}}{b_1 b_2} \tag{4.1}$$

where the $\lambda^{(R)}_{dij}$ are the quantities λ of Section 2 (first paragraph) computed by considering the rows of the design as blocks, and similarly for $\lambda^{(C)}$ from columns.

If $b_1 = b_2 = v$, we have the setting in which Wald [45] proved any LS to be D-optimum, and in which the tool described in Section 1 can be used to show a LS is universally optimum. It was noticed in the proof of the latter that it applied equally well (with only slightly more work to verify maximization of tr C_d in using (1.2)) to proving the optimality of a Youden square when $v \mid b_1$ and there exists a BIBD with $k = b_2$ and $b = b_1$ (and, hence, a Youden square). This suggested consideration of general $b_1 \times b_2$ arrays, and the definition of a generalized Youden design (GYD) as a $b_1 \times b_2$ array that is a BBD (not necessarily incomplete) when the rows are considered as blocks and also when the columns are considered as blocks. Thus, as described in Section 1, optimality considerations led to new combinatorial constructions. GYD's are constructed by patching together BIBD's in Kiefer [27], and by more elegant, largely geometric methods in work by Seiden and her students [38, 39]. Ash [1] constructed GYD's for essentially all practical parameter values.

When $v \mid b_1$, the setting is termed *regular* and if a GYD exists it can again be proved universally optimum by using (1.2). The situation in nonregular cases is much more difficult. In the simplest case, $v = 4$, $b_1 = b_2 = 6$ (where at least two nonisomorphic GYD's exist), it was noticed in Kiefer [23] that a GYD no longer maximizes tr C_d, so (1.2) cannot be used; later, another design was discovered that was better than any GYD, in the sense of D-optimality, in all nonregular cases $v = 4$, $b_1 = b_2$ (where necessarily $b_1 \equiv 6 \pmod{12}$).

This was perhaps surprising, an example of a most symmetric, intuitively appealing design (in the Fisherian tradition described in Section 1) which was not D-optimum. The simple universal optimality tool would not apply, and a new technique must be used to prove whatever optimum properties a GYD has. Such a device was described in Kiefer [28], and as we might expect from the last paragraph of Section 1, it is usually fairly simple to use to prove E-optimality and fairly difficult to use to prove D-optimality. The device was used in that reference to prove any GYD is always A-optimum (hence, E-optimum), and the proof announced there of the more difficult fact that a GYD is D-optimum unless $v = 4$ and the setting is nonregular, appears at the end of this section.

A natural generalization is to consider the q-way setting in which a design is a $b_1 \times b_2 \times \cdots \times b_q$ array and there are t (not necessarily distinct) variety symbols in each cell. Kishen [31] in fact constructed latin hypercubes in the case $b_1 = b_2 = \cdots = b_q$ with $t=1$ and $v = b_1^{q'}$ for some integer $q' < q$. For general q, b_i, and t, Cheng [8] extended the methods of the three previously cited papers on construction of GYD's, to construct certain Youden hyperrectangles (YHR); a YHR is an array which, for each i, is a BBD when the union of all cells with the same ith coordinate is considered to be a block. If all b_i are equal it is called a Youden hypercube (YHC). There are notions of regularity as in the GYD case, the only one of which we use in succeeding paragraphs is that $v \mid (t\prod_{j\neq i} b_j)$ for $2 \leqslant i \leqslant q$ (which specializes to the earlier definition of regularity in the GYD setting). If the above divisibility also holds for $i=1$, Cheng shows a YHR always exists; as in the GYD case, construction is easiest in regular settings. It is not yet known exactly which parameter values $v, \{b_i\}, t$ permit the construction of a YHR, even in the original GYD setting $q=2$, $t=1$.

It is then natural to ask, in the q-way setting with the parameters such that a YHR d^* exists, in what sense d^* is optimum. Cheng [9] extended the universal optimality result for regular GYD's to regular YHR's. Moreover, using the device of Kiefer [28] mentioned above, he proved that YHR's (regular or not) are always E-optimum. Among other cases considered by Cheng, perhaps most interesting is the case of Youden (not necessarily latin) hypercubes d^* (with general t), in which he used the D-optimality device of the end of this section to show that d^* is always D-optimum, if $q > 2$. Thus, the nonregular cases $q=2$, $v=4$ mentioned earlier, in which a GYD is not D-optimum, are quite special. A further comment on this is contained in the last paragraph of this section.

Another direction of generalization suggested by the above results and those of Section 2 is consideration, when v, $\{b_i\}$, t are such that no YHR exists, of arrays that are only *partially balanced* designs in each of the r directions. There has been little investigation of construction of such designs, particularly in nonregular cases. If the setting is regular and d^* is a BBD when considered as a block design in each of the last $q-1$ directions, but is a MBD of type i when considered in the first direction, Cheng used the results mentioned in Section 2 to prove optimality of d^* for all type i criteria.

We now give the proof that any GYD d^* is D-optimum when $v \neq 4$. Since the result is known in regular cases and any GYD is regular if v is prime, it suffices to consider $v \geqslant 6$. The device of Kiefer [28], replacing maximization of $\operatorname{tr} C_d$ as in (1.2), is now to show that d^* maximes $\sum_{j=1}^{v} \log c_{djj}$, from which D-optimality follows since C_d is completely symmetric. This is a more tedious computation than are those where (1.2) can be used, but again it avoids computing

$$\Phi^{(1)}(C_d) = -\sum_{1}^{v-1} \log \mu_{dj}.$$

Direct maximization of $\sum_1^v \log c_{djj}$ is also complicated because of combinatorial restrictions on the c_{dij}, but it is easily seen that it suffices to show, with

$$c(r) \stackrel{\text{def}}{=} \max_{\{d:\, r_{dj}=r\}} c_{dij},$$

that the maximum of $\sum_1^v \log c(r_j)$ subject to $\{r_j$ nonnegative integers, $\sum_1^v r_j = b_1 b_2\}$ is

$$\sum_1^v \log c_{d^*jj} = vc(\bar{r}),$$

where $\bar{r} = b_1 b_2 / v$. Noting (4.1), we have

$$g(r) \stackrel{\text{def}}{=} b_1 b_2 c(r) = b_1 b_2 r - b_1 h(r, b_1) - b_2 h(r, b_2) + r^2 \tag{4.2}$$

where

$$h(r, k) = -k[\text{int}(r/k)]^2 + (2r - k)[\text{int}(r/k)] + r. \tag{4.3}$$

We also write $\Delta(r) = g(r+1) - g(r)$. It then follows from the above development and simple properties of g (as outlined in Kiefer [28]) that it suffices to show that $\log g(r+1) - \log g(r)$ is nonincreasing ($\log g$ concave) in the "basic interval" of integers $C_0 \leqslant r < D_0$, where $[C_0, D_0]$ is the largest closed interval containing $\bar{r}$ but whose interior does not contain any integral multiples of b_1 or b_2. (The reduction to this interval is a consequence, in part, of the fairly simple fact that

$$\log[g(r+1)/g(r)] \geqslant \log[g(C_0+1)/g(C_0)] \quad \text{for} \quad r < C_0.)$$

Thus, we must show

$$g^2(r) - g(r-1)g(r+1) \geqslant 0 \quad \text{for} \quad C_0 < r < D_0. \tag{4.4}$$

Substituting $g(r-1) = g(r) - \Delta(r-1)$ and $g(r+1) = g(r) + \Delta(r)$, (4.4) thus becomes

$$C_0 < r < D_0 \Rightarrow 0 \leqslant \Delta(r)\Delta(r-1) + g(r)[\Delta(r-1) - \Delta(r)] = \Gamma_0(r) \quad \text{(say)}. \tag{4.5}$$

(Incidentally, the greater ease of proving A-optimality appears here explicitly, in that the expression $\Gamma(r)$ of (4.1) of Kiefer [28], which for A-optimality corresponds to (4.5) here, has the negative term in square brackets in (4.5) multiplied by $\frac{1}{2}$.) Define B_1, B_2, α, β by

$$\begin{aligned} C_0 &= B_1 b_1 \geqslant B_2 b_2 = b_2 \,\text{int}(C_0/b_2), \\ \alpha &= b_1^2 B_1(B_1+1) + b_2^2 B_2(B_2+1), \\ \beta &= b_1 b_2 - (2B_1+1)b_1 - (2B_2+1)b_2. \end{aligned} \tag{4.6}$$

From (4.5), we have on $[C_o+1, D_0-1]$,

$$\Gamma_0(r) = 2r^2 + 2\beta r + \beta^2 - 2\alpha - 1. \tag{4.7}$$

Considered as a quadratic function in *real* r, this has nonnegative derivative for

$r \geqslant C_0+1$ since $\beta+2r \geqslant 0$ for $r \geqslant C_0+1$. It is easy to see (as in the derivation of (4.3) of the earlier paper) that (4.5) will follow from

$$\Gamma_0(C_0+1) \geqslant 0. \tag{4.8}$$

We now proceed in a development similar to that used for $v \geqslant 6$ ("Case 1") in proving A-optimality. Simple inequalities show that (4.7) is decreased if B_2 is replaced by $B_1 b_1/b_2$. This yields

$$\Gamma_0(C_0+1) \geqslant 6C_0^2+[-4+4\sigma-6\pi]C_0+\{1+2[\pi-\sigma]+[\pi-\sigma]^2\} \tag{4.9}$$

where $\pi = b_1 b_2$ and $\sigma = b_1+b_2$ with both $b_i \geqslant 8$ (by nonregularity when $v \geqslant 6$).

The expression on the right-side of (4.9) has negative derivative with respect to C_0 for $C_0 < \frac{1}{6}\pi$, since $\sigma-\pi-1<0$. However, substitution of $\frac{1}{6}\pi$ for C_0 in that expression no longer yields a nonnegative expression (as it did in the case of the A-optimality proof, in (4.4) of the previous reference), again reflecting the greater difficulty of proving D-optimality. Instead, we must substitute the sharper upper bound for C_0,

$$C_0 \leqslant \tfrac{1}{6}(\pi-2b_1), \tag{4.10}$$

which we now justify.

Since $B_1 = C_0/b_1 < \bar{r}/b_1 = b_2/v$, (4.10) will follow from

$$\operatorname{int}(b_2/v) \leqslant \tfrac{1}{6}(b_2-2). \tag{4.11}$$

If $b_2 \equiv 0$ or 1 (mod 6), it follows from the nonregularity condition $v \nmid b_2$ that $v \geqslant 8$, and then (4.11) is satisfied for $b_2 \geqslant 8$. For other congruences (mod 6) of b_2, (4.11) is obvious. Thus, (4.10) is proved, and consequently (4.9) will follow from the nonnegativity of the expression obtained by substituting $C_0 = \frac{1}{6}(\pi-2b_1)$ into the right-side of (4.9). This expression, after mutiplication by 6 and rearrangement of terms, is

$$[b_1^2-8b_1+6]b_2^2+12(b_1-1)b_2+\{2b_1^2-4b_1+6\}, \tag{4.12}$$

which is clearly positive since both $b_i \geqslant 8$. This completes the proof.

One can see in (4.11) why this proof does not work when $v=4$. On the other hand, in Cheng's proof of the YHC D-optimality for *all* v when $q>2$, the bound analogous to (4.11) is

$$\operatorname{int}(b_1^{q-1}t/v) \leqslant (b_1^{q-1}t-q)/(2q+2),$$

which follows from $v \geqslant 2^q \geqslant 2q+2$ for $q \geqslant 3$. This is why $q=2$ yields the only cases where a GYD is not D-optimum.

5. Minimal support designs

Some design settings are of the special form of allowing repetition of rows of X_d as much as is wanted. In such problems, there is a space $\mathscr{X}$ of allowable row

m-vectors, and for a given n a design d is a choice $(\boldsymbol{x}_{d1}, \boldsymbol{x}_{d2}, \ldots, \boldsymbol{x}_{dn})$ of n not necessarily distinct elements of $\mathscr{X}$ or, equivalently, of a discrete probability *design measure* ξ on $\mathscr{X}$ defined by

$$\xi_d(\boldsymbol{x}) = n^{-1} \text{ (number of } \boldsymbol{x}_{di}\text{'s equal to } \boldsymbol{x}). \tag{5.1}$$

The settings of Sections 2 and 4 are not of this nature, but in the weighing design setting of Section 3 (unless we prohibit repetition of a weighing or factor combination, which there is no *physical* reason to do) we have the above setup with $\mathscr{X} = \{-1, 0, 1\}^m$ or $\{-1, 1\}^m$. In the fractional factorial setup, $\mathscr{X}$ consists of the range of a certain m-vector of polynomials on $\{0, 1\}^s$ of degree $\leqslant t$. In such a setting, the information matrix is

$$X'_d X_d = n \sum_{\boldsymbol{x} \in \mathscr{X}} \boldsymbol{x}'\boldsymbol{x}\xi_d(\boldsymbol{x}) \overset{\text{def}}{=} nM(\xi_d), \tag{5.2}$$

and $M(\xi_d)$ is called the *information matrix* (per observation and unit of variance) of ξ_d. For a given n, ξ_d must be a member of $\Xi_n = \{\xi\colon \xi$ is a probability measure on $\mathscr{X}$ taking only integral multiples of n^{-1} as values$\}$. If the optimality criterion Φ can be rewritten

$$\Phi(nM(\xi)) = a_n + b_n \Phi^*(M(\xi))$$

with $b_n > 0$ (as is the case for the Φ of (1.1)), we may instead consider the problem of choosing ξ in Ξ_n to minimize $\Phi^*(M(\xi))$. This, in general, is a tedious discrete programming problem, whose solution ξ^* (for a fixed $\mathscr{X}$) typically exhibits a fine structure in n, even when n is large, in which the support of ξ^* (as well as, of course, its value at each $\boldsymbol{x}$) varies. Because of this, it is useful to consider the *approximate theory* in which we choose ξ to minimize $\Phi^*(M(\xi))$ over $\Xi = \{$all discrete probability measures on $\mathscr{X}\}$. We call any solution a *Φ^*-optimum approximate design measure.* There are then useful theoretical characterizations and computer algorithms for determining an optimum ξ^*, which can be recorded for a given $\mathscr{X}$ without any dependence on n (see, e.g., [23, 26] for discussion and references); and the resulting ξ^* can often be implemented, at least for n sufficiently large compared with m, by taking an approximation to ξ^* in Ξ_n, so as to yield a design which is reasonably close to optimality for the n-observation problem.

If φ^* is strictly convex, as is the case for the D- and A-optimality criteria when $v = n$ (and often, also, when Φ^* is the E-optimality criterion), $M(\xi)$ must be the same for all Φ^*-optimum ξ in Ξ. It is obviously useful to find such a ξ that has minimal support size, since such a ξ can typically be more satisfactorily approximated in Ξ_n than one with large support, and also since (from a practical viewpoint) it may be economically advantageous to construct an experiment with as few distinct $\boldsymbol{x}_{di}$ as possible. If $\bar{\xi}$ is *some* Φ^*-optimum approximate design measure, this means finding $\tilde{\xi}$ in Ξ such that $M(\tilde{\xi}) = M(\bar{\xi})$ and $\tilde{\xi}$ has minimal support size. Assuming here that $\mathscr{X}$ is finite (as in our examples), say with N

points, list the points $\boldsymbol{x}^{(1)}, \ldots, \boldsymbol{x}^{(N)}$ (say) of $\mathscr{X}$ in any order and write ξ as an N-vector $(\xi_1, \xi_2, \ldots, \xi_N)'$ with $\xi_i = \xi(\boldsymbol{x}^{(i)})$. Write $\boldsymbol{x} = (x(1), \ldots, x(m))$ in coordinate form on $\mathscr{X}$, and list the $\frac{1}{2}m(m+1)$ $(=p$, say) functions $x(i)x(j)$ on $\mathscr{X}$, $1 \leq i \leq j \leq m$, in any order; call them $y_1, y_2, \ldots, y_p$. Let G be the $p \times N$ matrix of elements $y_i(\boldsymbol{x}^{(j)})$ $(1 \leq i \leq p, 1 \leq j \leq N)$, and let $\bar{g} = G\bar{\xi}$. The problem is then to find an N-probability vector $\tilde{\xi}$ satisfying

$$G\tilde{\xi} = \bar{g},$$
$$\operatorname{card}(\operatorname{supp}(\tilde{\xi})) = \text{minimum}. \tag{5.3}$$

Sometimes the problem can be rewritten with much-reduced p and N (see [17]).

For an example, consider the weighing problem of Section 3. Even if $\mathscr{X} = \{-1, 0, 1\}^m$, it is not hard to see that a universally optimum approximate design $\bar{\xi}$ assigns measure 2^{-m} to each point of $\{-1, 1\}^m$, and *every* design with the same information matrix $M(\bar{\xi}) = I_m$ has support a subset of $\{-1, 1\}^m$. Thus, $N = 2^m$, and G is a matrix of ± 1's, while $\bar{g}$ contains m 1's and $p - m$ 0's. If there exists an $m \times m$ Hadamard matrix, it corresponds to a $\tilde{\xi}$ satisfying (5.3) and with $\operatorname{card}(\operatorname{supp}(\tilde{\xi})) = m$, clearly a minimum. Otherwise, the solution will be supported by more than m points of $\mathscr{X}$. The existence of an $m \times m$ Hadamard matrix in this example is in fact equivalent to there being a $\tilde{\xi}$ on m points, satisfying (5.3); any such $\tilde{\xi}$ must assign measure m^{-1} to each point of support. More generally, one could ask, for each n, whether there are solutions to $G\xi = \bar{g}$ with ξ in Ξ_n. For example, when $m = 3$, any $\tilde{\xi}$ satisfying (5.3) has 4 support points, e.g., the points $(1, \pm 1, \pm 1)$ (with independent signs) with uniform measure on each point; if $n = v = m = 3$, this approximate solution of course differs from any exact Ξ_3 solution.

Another example, that has received considerable theoretical and practical effort, is that of the 3^s factorial setup with $t = 2$. In this case $m = \frac{1}{2}(s+1)(s+2)$ and $\mathscr{X}$ is the range of the m-vector F (say) of monomials of degree ≤ 2 on $\{-1, 0, 1\}^s$. (The problem that originates as "quadratic regression on the s-cube" can be reduced to this form for the Φ^*'s usually considered (see [26]).) There are no longer universally optimum designs; Φ^* matters. If $J_i^{(s)}$ consists of the points of $\{-1, 0, 1\}^s$ with i coordinates 0 and $s - i$ coordinates ± 1, it can be shown that for $s \geq 2$ there is a D-optimum approximate design supported on the union of three sets $F(J_i^{(s)})$, which in experimental terms means taking observations corresponding to a fraction of the 3^s factorial points, namely, at points in these three $J_i^{(s)}$'s. The set of three values of i for such solutions, obtained in [25] and [17], is

$$0, 1 \text{ any } i \text{ with } 2 \leq i \leq s, \text{ if } 2 \leq s \leq 5,$$
$$0, 1 \text{ or } 2, \text{ any } i \text{ with } 3 \leq i \leq s, \text{ if } 6 \leq s. \tag{5.4}$$

The corresponding design assigns equal measure to each point of $J_i^{(s)}$ for a fixed i, but that measure varies with i. It was also shown in the last reference, by rank considerations of G in (5.3), that the resulting design $\tilde{\xi}$ on 9 points for $s = 2$ and on 21 points for $s=3$ (taking $i=3$ in (5.4)), satisfies (5.3). (The case $s=1$ is trivial.) Recently Pesotchinsky [35] showed how to reduce the corresponding designs for $s=4, 5, 6$, from support on 49, 113, or 257 points, to support on 45, 61, or 89 points respectively. These reductions, which use a clever combination of OA's of strength 2, improved considerably on the reductions obtained in the earlier reference by an obvious use of OA's of strength 4. Moreover, Pesotchinshy was able to show that his designs do have minimal support, and for moderate n they are implementable by nearby designs in Ξ_n that appear to be fairly efficient compared with other actual n-observation designs. There remains the problem of extending this work to larger s; in particular, of constructing the analogues of Pesotchinsky's combinatorial arrangements.

For the E-optimality criterion, it was shown by Galil and Kiefer [19] that this problem on the 3^s-array has an arithmetically simpler solution that, for certain s, allows a solution to be obtained on only *two* $J_i^{(s)}$'s, say with $i=i_1$ and i_2. It can then be computed, from knowledge of one optimum $\bar{\xi}$ and the resulting $\bar{g}$, that i_1, i_2 must satisfy

$$5i_1i_2-3s(i_1+i_2)+s(2s+1)=0, \tag{5.5}$$

which on writing $i_1=\frac{1}{5}(3s-u)$, $i_2=\frac{1}{5}(3s+w)$ becomes $uw=s(s+5)$, from which the possible values of s and of i_1, i_2 can easily be obtained. It was also shown in Kiefer [29] that all solutions (i_1, i_2, s) lie in "infinite series" $(i, ai+b, ci+d)$ of such solutions with i in arithmetic progression, and that, with a few distinguishable exceptions, each (i_1, i_2, s) lies in such series for exactly two different pairs (a, b) and two or four (c, d)). It can occur for some s that no design with support contained in only two $J_i^{(s)}$ has minimal support, but the full picture, including an analogue of Pesotchinsky's work for D-optimality, is incomplete. In any event, this example illustrates how optimality considerations can lead to a diophantine problem.

We mention that equations of the form (5.3) have also arisen in the work of Foody and Hedayat [18]. There one is interested, for given b, k, v, in finding a BIBD with a minimum number of *distinct* blocks. $\mathscr{X}$ is replaced by the set of possible blocks, with $N=\binom{v}{k}$. The $p=\binom{v}{2}$ subsets of size 2 of $\{1, 2, \ldots, v\}$ correspond to rows of G, and the (i, j)th element of G is 1 if pair i occurs in block j and 0 otherwise. Then $\bar{g}$ is the vector $\bar{c}$ with all components equal to λ/b. Here of course one is only interested in solutions to (5.3) for which, also, $b\bar{\xi}$ has integral components. In the case $b=56$, $k=3$, $v=8$ considered in detail by Foody and Hedayat, a minimal support design has recently been determined by Pesotchinsky [36]. Aside from the combinatorial interest in nonisomorphic BIBD's produced by solutions to $G\xi=\bar{c}$ with more than minimal support, there is an "optimality" consideration: among the (universally optimum) BIBD's one may want to select the design that minimizes some secondary loss functional, which from cost considerations could well be the number of distinct blocks. The primary loss consideration here of course amounted to showing the existence of some BIBD (equivalent to showing $G\xi = \bar{c}$ has a solution in Ξ_b), or, in the absence of such a solution, to undertaking a development like that of Section 2.

If b, k, v in the block design setting are such that no BIBD, or MBD, or any similar structure exists, finding a design that minimizes $\Phi^{(1)}(C_d)$ is a very difficult computational problem. The best "search" scheme currently available is that of Mitchell [32], but Mitchell showed that, even in simple settings such as the weighing problem on $\{-1, 1\}^m$ with $m = 8$ and $n = 16$, where there is a Hadamard solution, his scheme (with a random initial design) rarely found it. There are local minima, "holes" that it is hard to climb out of. Galil and the present author have made some slight progress in trying to improve Mitchell's method, but better ideas are needed. Similarly, even when we know *some* optimum design $\bar{\xi}$, there are no satisfactory computer routines for solving (5.3). In fact, as John Hopcroft showed the author, the family of such problems is NP-complete. Both for the design optimization problem and for the minimal support problem, at least in a large class of settings, one would hope to find polynomial-time algorithms that yield solutions close to the optimum, of the type that exist in other optimization problems.

References

[1] A.S. Ash, Construction of generalized Youden designs, Thesis, Univ. of Illinois at Chicago Circle (1977).
[2] A.S. Ash, Optimality of GYD's, unpublished paper (1978).
[3] R.C. Bose and K.R. Nair, Partially balanced incomplete block designs, Sankhya 4 (1939) 337–372.
[4] I.M. Chakravarti, On some methods of construction of partially balanced arrays, Ann. Math. Statist. 32 (1961) 1181–1185.
[5] C.-S. Cheng, Optimality of certain asymmetrical experimental designs, Ann. Statist. 6 (1978) 1239–1261.
[6] C.-S. Cheng, A note on $(M.S)$-optimality, Commun. Statist., to appear.
[7] C.-S. Cheng, Optimality of some weighing and 2^n fractional factorial designs, Ann. Statist. 8 (1980) 436–446.
[8] C.-S. Cheng, Construction of Youden hypercubes, to appear.
[9] C.-S. Cheng, Optimal designs for the elimination of multi-way heterogeneity, Ann. Statist. 6 (1978) 1262–1272.
[10] C.-S. Cheng and L.J. Gray, A characterization of group-devisible designs and some related results, Ann. Discrete Math. 6 (1980).
[11] D. Conniffe and J. Stone, Some incomplete block designs of maximum efficiency, Biometrika 62 (1975) 685–686.
[12] J.A. Eccleston and A. Hedayat, On the theory of connected design characterization and optimality, Ann. Statist. 2 (1974) 1238–1255.
[13] H. Ehlich, Determinantenabschätzungen für binäre Matrizen, Math. Z. 83 (1964) 123–132.
[14] H. Ehlich, Determinantenabschätzungen für binäre Matrizen mit $m \equiv 3 \bmod 4$, Math. Z. 84 (1964) 438–447.
[15] H. Ehlich, Personal communication (1978).
[16] H. Ehlich and K. Zeller, Binäre Matrizen, Z. Angew. Math. Mech. 42 (1962) T20–T21.
[17] R.H. Farrell, J. Kiefer and A. Walbran, Optimum multivariate designs, Proc. 5th Berkeley Symp. 1 (1965) 113–138.
[18] W. Foody and A. Hedayat, On theory and applications of BIB designs with repeated blocks, Ann. Statist. 5 (1977) 932–945.

[19] Z. Galil and J. Kiefer, Comparison of designs for quadratic regression on cubes, J. Statist. Planning and Inference 1 (1977) 121–132.
[20] Z. Galil and J. Kiefer, D-optimum weighing designs, in Ann. Statist. (1980).
[21] Z. Galil and J. Kiefer, Time- and space-saving computer methods, related to Mitchell's Detmax, for finding D-optimum designs, to appear in Technometrics (1980).
[22] J.A. John and T.J. Mitchell, Optimal incomplete block designs, J. R. Statist. Soc., Ser. B 39 (1977) 39–43.
[23] J. Kiefer, On the nonrandomized optimality and randomized non-optimality of symmetrical designs, Ann. Math. Statist. 29 (1958) 675–699.
[24] J. Kiefer, Optimum experimental designs, J. R. Statist. Soc., Ser. B 21 (1959) 272–319.
[25] J. Kiefer, Optimum designs in regression problems—II, Ann. Statist. 32 (1961) 298–325.
[26] J. Kiefer, General equivalence theory for optimum designs (approximate theory), Ann. Statist. 2 (1974) 849–879.
[27] J. Kiefer, Balanced block designs and generalized Youden designs, I. Construction (patchwork), Ann. Statist. 3 (1975) 109–118.
[28] J. Kiefer, Construction and optimality of generalized Youden designs, in: J.N. Srivastava (ed.), A Survey of Statistical Designs and Linear Models (North-Holland, Amsterdam, 1975) 333–353.
[29] J. Kiefer, A diophantine problem in optimum design theory, Utilitas Math., to appear.
[30] J. Kiefer and J. Wolfowitz, Optinum designs in regression problems, Ann. Math. Statist. 30 (1959) 271–294.
[31] K. Kishen, On the construction of Latin and Hypergraeco-Latin cubes and hypercubes, J. Indian Soc. Agric. Statist. 2 (1949) 20–48.
[32] T.J. Mitchell, An algorithm for the construction of D-optimal experimental designs, Technometrics 16 (1974) 203–210.
[33] A.M. Mood, On Hotelling's weighing problem, Ann. Math. Statist. 17 (1946) 432–446.
[34] S.E. Payne, On maximizing $\det(A^TA)$, Discrete Math. 10 (1974) 145–158.
[35] L.L. Pesotchinsky, D-optimum and quasi-D-optimum second-order designs on a cube, Biometrika 62 (1975) 335–340.
[36] L.L. Pesotchinsky, Remarks on the lower bound of number of different blocks in BIB designs, to appear.
[37] D. Raghavarao, Constructions and Combinatorial Problems in Design of Experiments (John Wiley and Sons, New York, 1971).
[38] F. Ruiz, and E. Seiden, On construction of some families of generalized Youden designs, Ann. Statist. 2 (1974) 503–519.
[39] E. Seiden and C.-J. Wu, A geometric construction of generalized Youden designs for v a prime power, Ann. Statist. 6 (1978) 451–460.
[40] K.R. Shah, Optimality criteria for incomplete block designs, Ann. Math. Statist. 31 (1960) 791–794.
[41] K.R. Shah, D. Raghavarao and C.G. Khatri, Optimality of two and three factor designs, Ann. Statist. 4 (1976) 419–422.
[42] J.N. Srivastava and D.V. Chopra, Balanced optimal 2^m fractional factorial designs of resolution V, $m<6$, Technometrics 13 (1971) 257.
[43] K. Takeuchi, On the optimality of certain type of PBIB designs, Rep. Statist. Appl. Res., JUSE, 8 (1961) 140–145.
[44] K. Takeuchi, A remark added to "On the optimality of certain type of PBIB designs", Rep. Statist. Appl. Res., JUSE, 10 (1963) 47.
[45] A. Wald, On the efficient design of statistical investigations, Ann. Math. Statist. 14 (1943) 134–140.
[46] C.H. Yang, On designs of maximal $(+1,-1)$ matrices of order $n\equiv 2 \pmod 4$, Math. Comp. 22 (1968) 174–180.

Annals of Discrete Mathematics 6 (1980) 243–249

THE ASYMPTOTIC NUMBER OF LATTICES*

D.J. KLEITMAN and K.J. WINSTON
Massachusetts Institute of Technology, Cambridge, MA, USA

1. Introduction

Given any mathematical construct that can be applied to finite sets, one can raise a corresponding enumerative question: how many such constructs are there of a given cardinality?

Answers to such questions come in different forms. Most exact is an exact count for each value of n; for some problems we can count "most" of the constructs of given size and thereby have an asymptotic formula for the number in question. For still others we have only estimates of the rate of growth of the number, which amount to asymptotic formulae for the logarithm of the number in question.

Thus we know asymptotically how many partial orders [2] or finite topologies [2] or semigroups [3] there are on n elements; we have asymptotic formulae for the logarithm of the number of simplicial complices [4]. Recently Korsunov [5] has announced on asymptotic formula for this number as well.

In this paper we consider the question discussed above applied to the "lattices" as construct. Namely, we ask how many lattices are there on $n+2$ elements?

There have been some explicit numerical results for small values of n. The only previous work on the general question that we have found is by Klotz and Lucht [1]. They noticed that by taking subsets of projective planes, one could obtain

$$(2^{\sqrt{(2)}/4})^{n^{3/2}+o(n^{3/2})}$$

lattices of cardinality $n+2$. They further obtained an upper bound of $n^{n^{3/2}+o(n^{3/2})}$ for the number $L(n)$ of such lattices.

It was the purpose of the present investigation to attempt to bring these bounds closer together. Our main result is an upper bound for $L(n)$ of the form

$$\alpha^{n^{3/2}} \text{ (with } \alpha \simeq 6.11343),$$

as opposed to the $n^{n^{3/2}}$ of Klotz and Lucht [1].

It is our feeling that the lower bound obtained by them from projective planes is in fact closer to the true value.

A lattice is a partially ordered set which for each pair of elements a, b contains a unique maximal element less than both a and b, called their "meet" (written as

* Supported in part by ONR Contract No. N00014-76-C-0366.

$a \wedge b$); and a unique minimum element greater than both called their "join" (written as $a \vee b$).

This means that if a and b are both greater than c and d, then there must be at least one element e satisfying $a, b \geq e, c, d \leq e$.

A finite lattice obviously must have a greatest element, called 1, and a least element, called 0.

If the remaining elements form only two levels, so that any elements on the same level are unordered, the condition for being a lattice becomes very simple — it means that $a, b \geq c, d$ is impossible, among elements other than 0 and 1. (In such a lattice we call the lower level elements the "atoms", and the upper level elements "co-atoms".) We will find it convenient to bound the number of lattices of this kind (which include all those given by Klotz and Lucht [1]) and then use this result to bound the number of more general lattices as well.

Let $\mathcal{L}(n)$ be the set of lattices with n elements other than 0 or 1. Let $L(n) = |\mathcal{L}(n)|$. This paper is thus concerned with the asymptotic behavior of $L(n)$.

Let $S(n)$ be the number of two level lattices — those whose maximal chains have only two elements other than 0 and 1. In Section 2, we will show that

$$S(n) < \beta^{(n^{3/2} + o(n^{3/2}))} \quad (\beta \approx 1.699408\ldots). \tag{1.1}$$

In Section 3 we obtain as a corollary the upper bound

$$L(n) < \alpha^{(n^{3/2} + o(n^{3/2}))} \quad (\alpha \approx 6.111343). \tag{1.2}$$

2. Two level lattices

2.1. *Outline of proof*

We let $\mathcal{S}(n)$ be the set of two level lattices in $\mathcal{L}(n)$, and let $\mathcal{S} = \bigcup_n \mathcal{S}(n)$.

In this section we show that the inequality (1.1) holds. We proceed by induction by bounding the ratio $S(n)/S(n-1)$. This is done by bounding the number of ways to add an element (say a co-atom) to a lattice $\mathcal{S}(n-1)$. This number is in turn the number of "eligible" sets of atoms, where a set A of atoms will be *eligible* if it can be covered by a new co-atom; that is, when all x and y in A are in no other co-atom.

We say that two atoms x and y are *incompatible* when x and y are in some co-atom (so that y is not in any eligible set with x). We let $I(x)$ be the set of atoms incompatible with x. Eligibility and incompatibility are of course defined with respect to a given lattice.

We will use these concepts to construct a tree containing the eligible sets as nodes. We will then bound the size of the tree. This bound immediately puts a limitation on the ratio $S(n)/S(n-1)$, which limitation gives our main results.

2.2. *Construction of tree*

For a lattice $K \in S$ we construct a tree $T(=T(K))$ whose nodes "are" (by abuse of language) the eligible sets of K.

The root of the tree is the empty set. The empty set's offspring are the sets consisting of single atoms of K. In general, a node's offspring will have one more member than the node.

If N is a node, we choose a numbering of its offspring $N_1, N_2, \dots, N_d$ ($d = d(N) =$ out degree of N) so that if

$$\{p_i\} = N_i - N \quad \text{and} \quad P_i = \{p_i, \dots, p_d\} \quad (i = 1, \dots, d), \tag{2.2}$$

then

$$|I(p_i) \cap P_i| = \max_{y \in P_i} |I(y) \cap P_i| \quad (i = 1, \dots, d) \tag{2.2.2}$$

holds. In words, we number N's offspring so that each p_i is least compatible with the remaining pool P_i of atoms that can be added to N so as to preserve eligibility.

Under this numbering we take as N_i's ($i = 1, \dots, d-1$) offspring the eligible $N_i \cup \{p_j\}$ ($j > i$). N_d is a leaf.

It is easy to verify that T is a tree containing the eligible sets.

2.3. *The size of the tree*

We now show that after a certain point the fanout of nodes in T is limited. We need the following:

Definition. For $K \in S$, let n_1 $(=n_1(K))$ be the number of co-atoms of K. Let d_1 $(=d_1(K))$ be the smallest number of atoms less than a co-atom in K. Define n_2 and d_2 dually (i.e., reversing the roles of atom and co-atom).

Lemma 1. *Let $\mathcal{U}$ be a node of T satisfying*

$$|\mathcal{U}| > \frac{n_1}{d_2^2} \log(n_2). \tag{2.3.1}$$

Then $d(\mathcal{U}) < n_1/d_2$.

Proof. Let N be a node of T. Continue the notation of paragraph (2.2). Let $q_i = |P_i|$ ($i = 1, \dots, d$).

Draw a line from a to b if b is in a. Then there are at least $q_i d_2$ lines from the co-atoms of K to P_i. Therefore there are at least $q_i d_2(q_i d_2 - n_1)/2n_1$ pairs of elements of P_i covered by a common co-atom of K.

Thus

$$\tfrac{1}{2}\sum_{y\in P_i} |I(y)\cap P_i| \geqslant q_i d_2(q_i d_2 - n_1)/2n_1,$$

so

$$\max_{y\in P_i} |I(y)\cap P_i| \geqslant \frac{q_i d_2^2}{n_1} - d_2. \tag{2.3.2}$$

Applying (2.2.2), we obtain

$$|I(p_i)\cap P_i| \geqslant \frac{q_i d_2^2}{n_1} - d_2. \tag{2.3.3}$$

But $z\in I(p_i)\cap P_i$ implies that $N_i\cup\{z_i\}$ is not an eligible set. This implies

$$d(N_i) < q_i(1-d_2^2/n_1) + d_2. \tag{2.3.4}$$

Suppose that $d_2^2/n_1 \leqslant 1$. Then

$$d(N_i) < d(N)(1-d_2^2/n_1) + d_2 \tag{2.3.5}$$

since q_i is less than $d(N)$. Repeated application of (2.3.5) gives

$$d(\mathcal{U}) < d(\emptyset)(1-d_2^2/n_1)^{|\mathcal{U}|} + \frac{n_1}{d_2}(1-(1-d_2^2/n_1)^{|\mathcal{U}|})$$

$$< n_2 e^{-|\mathcal{U}|d_2^2/n_1} + n_1/d_2. \tag{2.3.6}$$

Applying (2.3.1) then proves the lemma.

Thus we can assume that $d_2^2/n_1 > 1$. Let $\mathcal{U}$ satisfy (2.3.1), and let N be $\mathcal{U}$'s parent. Let $\mathcal{U} = N_i$ in the numbering of paragraph (2.2). If $q_i < n_1/d_2$, then $d(\mathcal{U}) < n_1/d_2$ as desired. But if $q_i \geqslant n_1/d_2$, then (2.3.4) implies

$$d(\mathcal{U}) < n_1/d_2(1-d_2^2/n_1) + d_2 = n_1/d_2 \tag{2.3.7}$$

and the lemma is proven.

2.4. *Ways to add elements*

We now use Lemma 1 to show that the number of ways to add an element to a lattice in $\mathcal{S}(n-1)$ is bounded.

Let K be a lattice in $\mathcal{S}(n)$. Choose a co-atom of K containing d_1 atoms and remove it; call the resulting lattice K_1. Define w_1 as the maximum over all such choices of K_1 of the number of ways to add a co-atom containing $d_1(K)$ atoms to K_1. Define K_2 and w_2 dually. Then we have:

Lemma 2.

$$\min\{w_1, w_2\} < \beta^{3/2\sqrt{n}+o(\sqrt{n})} \quad (\beta \approx 1.699408\ldots).$$

Proof. (In this proof n_1, n_2, d_1, and d_2 have argument K.)

w_1 is the maximum (over choices K_1 as above) of the number of nodes of size d_1 in the tree $T(K_1)$.

Clearly

$$w_2<\binom{n_1}{d_2}<n_1^{d_2}$$

holds. Therefore we can assume that

$$d_2>\frac{\sqrt{n}\log(\beta)}{\log(n_1)} \tag{2.4.1}$$

holds.

Let $s=(n_1/d_2)\log(n_2)$; s is the right-hand side of (2.3.1). Then the inequality $\binom{n_2}{s}<n_2^s$ together with (2.4.1) implies that $\log\binom{n_2}{s}=o(\sqrt{n})$. Since $w_1<\binom{n_2}{d_1}$ holds, we can therefore assume that s is less than d_1.

Let $\mathcal{U}$ be a node of $T(K_1)$ of size s. Let $\mathcal{U}$ satisfies (2.3.1) for $T(K_1)$, so we can apply Lemma 1. Lemma 1 implies that there are at most

$$\binom{\frac{n_1-1}{d_2}}{d_1-s}$$

descendants of $\mathcal{U}$ of size d_1. Thus the following inequality holds:

$$w_1<\binom{n_2}{s}\binom{\frac{n_1-1}{d_2}}{d_1-s}<\binom{\frac{n_1}{d_2}}{d_1-s}\beta^{o(\sqrt{n})}. \tag{2.4.2}$$

If we define

$$f(x)=\frac{x^{\sqrt{x}}}{(x-1)^{(x-1)/\sqrt{x}}}, \tag{2.4.3}$$

then Stirling's formula implies

$$\binom{n_1/d_2}{d_1-s}<f(y))^{((n_1d_1)/d_2)^{1/2}} \quad \text{where } y=\frac{n_1}{d_2(d_1-s)}. \tag{2.4.4}$$

A calculation shows that

$$\max_{x\in\mathbb{R}^+} f(x)=\beta^{3/\sqrt{2}}.$$

Hence (2.4.2) gives

$$w_1<\beta^{(3/\sqrt{2})((n_1d_1)/d_2)^{1/2}+o(\sqrt{n})}. \tag{2.4.6}$$

Since $n_1+n_2=n$,

$$\min\left\{n_1\frac{d_1}{d_2}, n_2\frac{d_2}{d_1}\right\}$$

is less than $\frac{1}{2}n$.
Therefore (2.4.6) and its dual prove the lemma.

2.5. *The bound on S(n)*

Theorem.

$$S(n)<\beta^{n^{3/2}+o(n^{3/2})}.$$

Proof. By induction; this is true for small n because of the $o(n^{3/2})$ term. So assume the theorem is true for $n-1$.

Consider the following map from $\mathscr{S}(n)$ to $\mathscr{S}(n-1)$: map K to K_1 if $w_1<w_2$; otherwise map K to K_2. (K_1 and K_2 were defined in paragraph (2.4).)

Note that for each $M\in\mathscr{S}(n-1)$, the set $\{d_1(K)\mid K$ maps to $M\}$ has less than n members. Since $2n$ is negligible (i.e., $\log(2n)=o(\sqrt{n})$), Lemma 2 implies than the map is less than $\beta^{3/2\sqrt{n}+o(\sqrt{n})}$-to-one.

The binomial theorem gives

$$(n-1)^{3/2}+3/2\sqrt{n}=n^{3/2}+o(1).$$

This completes the proof of the theorem.

3. General lattices

In this section it will sometimes be convenient to think of lattices as sets of ordered pairs:

$$K\equiv\{(x,y)\mid x,\ y\in K \text{ and } x<y \text{ in } K\}.$$

Theorem.

$$L(n)<\alpha^{n^{3/2}+o(n^{3/2})}\quad(\alpha=\beta^{2+\sqrt{2}}\cong 6.11343).$$

Proof. By induction on n. Assume true for $j<n$. For $x\in K\in\mathscr{L}(n)$, let $\rho_k(x)$ be the size (minus one) of a maximal chain in K with x as greatest element (thus $\rho_k(0)=0$).

For each $k\in\mathscr{L}(n)$ there is a number j so that $(x\in K\mid\rho_k(x)<j\}\leqslant\frac{1}{2}n$ but $\{x\in K\mid\rho_K(x)\leqslant j\}>\frac{1}{2}n$. We can choose a set B consisting of all $x\in K$ with $\rho_k(x)<j$ and enough $y\in K$ with $\rho_K(y)=j$ so that $|B|=[\frac{1}{2}n]$. Let $T=\{x\in K, x\notin B\}$. Then B has the property that $x\in B$, $y\in T$ implies $x\not> y$.

Let

$$K_1 = \{(x, y) \in K \mid x \in B \text{ and } (y \in B \text{ or } y = 1)\},$$
$$K_2 = \{(x, y) \in K \mid (x = 0 \text{ or } x \in T) \text{ and } y \in T\},$$

$$K_3 = \{(x, y) \in K \mid x \in B \text{ and } y \in T \text{ and } (x = 0 \text{ or } y = 1 \text{ or } y \text{ covers } x \text{ in } K)\}.$$

The mapping $K \to (K_1, K_2, K_3)$ is 1-1 from $\mathscr{L}(n)$ into $\mathscr{L}([\frac{1}{2}n]) \times \mathscr{L}(\lceil\frac{1}{2}n\rceil)x \times \mathscr{S}(n)$. Hence

$$L(n) < \alpha^{([n/2]^{3/2} + [n/2]^{3/2} + (\log_\alpha \beta) n^{3/2} + o(n^{3/2})}$$
$$< \alpha^{(\sqrt{(2)}/2 + \log_\alpha \beta) n^{3/2} + o(n^{3/2})} < \alpha^{n^{3/2} + o(n^{3/2})}.$$

This completes the proof of the theorem.

References

[1] W. Klotz and L. Lucht, Endliche Verbände, J. Reine Angew. Math. 247 (1971) 58–68.

[2] D.J. Kleitman and B. L. Rothschild, Asymptotic enumeration of partial orders on a finite set, Trans. Am. Math. Soc. 205 (1975) 205–220.

[3] D.J. Kleitman, B.L. Rothschild and J. H. Spencer, The number of semigroups of order *n*, Proc. Am. Math. Soc. 55 (1), (1976) 227–232.

[4] D.J. Kleitman and G. Markowsky, On Dedekind's problem: The number of isotone boolean functions. ii, Trans. Am. Math. Soc. 213 (1975) 373–390.

[5] A.D. Korsunov, Solution of Dedekind's problem on the number of monotonic boolean functions, Dokl. Akad. Nauk SSR 233(4) (1977) (Transl. by Am. Math. Soc. in Soviet Math. Dokl. 18 (2) 442–445).

Annals of Discrete Mathematics 6 (1980) 251–255

A NEW STATISTICS ON WORDS

A. LASCOUX and M.P. SCHÜTZENBERGER

Départment de Mathématiques, Paris VII, Tour 55, 2 Place Jussieu, 75005 Paris (France)

The study of various statistics on words is a chapter of combinatorics and it has been recently surveyed by Foata [1]. One ot these statistics is the so-called "Major Index" of a word. It plays a role in the theory of representations of the symmetric group. We present here a generalization, the "load", which we used to prove Foulkes' conjecture, i.e. to supply a q-analogue of Kostka numbers [2].

It turns out that the load can be defined in terms of a minimization problem. We develop this approach here without entering into the algebraïc background or applications. A special feature of the problem is that a solution is given by the most simple minded algorithm, a not too frequent occurrence. Therefore, we suspect that it is a very degenerate special case of a more general question already encountered in operations research. Since we have not been able to trace it in the literature, we apologize to those authors, whose articles may have been unwillingly overlooked.

1. The minimization problem

Throughout this paper consider two finite totally ordered alphabets,

$$A=\{a<b<c<\cdots\} \quad \text{and} \quad \Gamma=\{\alpha<\beta<\cdots\}.$$

As usual the number of occurrences of a letter x in a word w is denoted by $|w|_x$, and one says that a word is *standard* iff no letter occurs more than once in it.

We also consider a fixed word v in the free monoid A^*, and we make the basic assumption that its multidegree is non-increasing, i.e. that $|v|_a \geq |v|_b \geq \cdots$.

A *lifting* of v is a standard word w in the free monoid $(A\times\Gamma)^*$ such that, on the one hand, v is its projection on A^* and, on the other hand, its minimal alphabet, B, is an initial interval of $A\times\Gamma$, i.e. $c\gamma\in B$ only if $c'\gamma'\in B$ for any $c'\leq c$ and $\gamma'\leq\gamma$. Our basic assumption on the multidegree of v insures that v has at least one lifting (and, in fact it has only one iff v is standard). Note that B is the union over all greek letters γ of sets $A_\gamma\times\{\gamma\}$ where A_γ is an initial intervalle of A.

Let w be a lifting of v. For each $c\gamma\in B$ we say that the letter $c\gamma$ is *marked* iff it is on the *right* of the letter $c_-\gamma$ where c_- denotes the (immediate) predecessor of c

in A. Thus, a being always the first letter of A, we have that no letter $a\gamma$ is marked. We deduce by two summations the *delay function* $\mu = \mu_w$ and the *integrated delay function* $\bar{\mu}$ of the lifting w. The first one has domain B and $(c\gamma)\mu$ is defined as the number of marked $c'\gamma$ with $c' \leqslant c$. The second one has domain $A \times \Gamma$ and $(c\gamma)\bar{\mu}$ is the sum of $(c'\gamma')\mu$ over all $c' \leqslant c$ and $\gamma' \leqslant \gamma$.

The problem is to choose a lifting so as to minimise the maximum of the integrated delay function, i.e. the sum of the values of the delay function itself. This minimax quantity is the *load* of the word v.

Example. Let $v = a\,b\,d\,a\,b\,c\,b\,a\,c\,d$, of length $|v| = 10$. The partition of 10 expressing the multidegree is $K = (2\,2\,3\,3)$. One of the $2!\,2!\,3!\,3!$ possible liftings is:

$$a\gamma,\ b\beta,\ d\alpha,\ a\alpha,\ \underline{b}\gamma,\ c\alpha,\ a\beta,\ b\alpha,\ \underline{c}\beta,\ \underline{d}\beta.$$

Marked letters are underlined. The delay and integrated delay functions are:

d	1	2	
c	1	1	
b	1	0	1
a	0	0	0
	α	β	γ

d	3	6	7
c	2	3	4
b	1	1	2
a	0	0	0
	α	β	γ

The natural lifting (discussed below) and corresponding functions are:

$$a\gamma,\ b\beta,\ d\alpha,\ a\beta,\ b\alpha,\ \underline{c\beta},\ a\alpha,\ \underline{b\gamma},\ \underline{c\alpha},\ \underline{\underline{d}\beta}.$$

d	1	2	
c	1	1	
b	0	0	1
a	0	0	0
	α	β	γ

d	2	5	6
c	1	2	3
b	0	0	1
a	0	0	0
	α	β	γ

We shall prove that 6 is indeed the load of v.

A case of special interest is when v is a standard word. Then the lifting and subsequent marking and functions are unique. We consider v as a permutation on the set of its letters and look at the inverse permutation v^{-1} on the places. For instance if $v = d\,a\,\underline{e}\,c\,f\,\underline{b}$, one has $v^{-1} = 2\,\underline{6}\,4\,1\,\underline{3}\,\underline{5}$. Marked letters correspond to numbers greater than their left neighbour. The load of v is by definition $0+1+1+1+2+3=8$. Up to reversal, this is the major index $5+2+1$ of v^{-1}, i.e. the sum of the lengths of the right segments of v^{-1} determined by the underlined numbers. No such connection between the major index and the load exists for non-standard words.

2. The main result

For any lifting w and $\gamma \in \Gamma$, we denote by w^{γ} the subword of w whose alphabet is $A_{\gamma} \times \{\gamma\}$. A straightforward attempt at a solution of our minimax problem consists in constructing successively these subwords for $\gamma=\alpha,\beta,\ldots$ by the following algorithm in which w is understood to be read from right to left.

Suppose that the subwords w^{β} have already been determined for $\beta<\gamma$. Then $a\gamma$ is the first occurrence of a (which has not yet been used) and, supposing that $c_{-}\gamma$ has been chosen, one takes for $c\gamma$ the first (non-used) occurrence of a c, on the left of $c_{-}\gamma$, if there is one. If none exists $c\gamma$ is the first occurrence of a (non used) c in w. This second possibility is the one which induces a marking of $c\gamma$.

The resulting lifting, marking and functions will be called *natural*. Our main result is the

Property 1. *The number of marked letters is minimal for the natural lifting; further, if $\bar{\nu}$ is the natural integrated delay function, one has $\bar{\nu}\leqslant\bar{\mu}$ for the integrated delay function $\bar{\mu}$ relative to any other lifting of v.*

Thus the load of v is given by the above algorithm.

3. A conjugacy lemma

We suppose here that $v=cv'$, where c is a letter which is not the first one of A and one of its liftings $w=c\gamma\cdot w'$. Let $v_1=v'c$. It is clear that $w_1=w'\cdot c\gamma$ is a lifting of v_1. The number of marked letters and delay function of w (resp. w_1) will be r and ν (resp. r_1 and ν_1).

Lemma 1. *The lifting w_1 is natural iff w is so. Further: $r_1=r+1$ or $=r$, depending upon whether c is or is not the last letter of A_γ; the delay functions ν and ν_1 are equal except for $(c\gamma)\nu=(c\gamma)\nu_1+1$.*

Proof. The truth of the first assertion follows immediately from the definitions. To check the other ones, note that the markings of w and w_1 are the same except, possibly for $c\gamma$ and $c_{+}\gamma$ ($c_{+}=$ the successor of c). In fact, because of $c\neq a$, the letter $c\gamma$ is marked in w_1 but not in w and if $c_{+}\in A_\gamma$, i.e. if c is not the last letter of A_γ, the contrary is true of $c_{+}\gamma$.

Corollary 2. *Assume $v=v'a^m$ where $m=|v|_a$ and that Property 1 holds for v'. Then it holds for any word $v_1=u''a^m u'$, where $u'u''=v'$.*

Proof. The definitions apply to the word v' since it has a non-increasing multidegree on the alphabet $A\backslash a$.

The natural lifting of v is deduced from that of v' by left multiplication by the word $\cdots a\beta\cdot a\alpha$. Since this does not change the marking, Property 1 also holds for v. Thus, by induction on the length of u', it suffices to check the result when u' is a single letter. But this follows from the lemma since $u'\neq a$ by hypothesis.

4. Proof of the main result

We recall that the *plactic monoid* is the quotient of the free monoid A^* by the least congruence $\equiv$ such that for any letters b, c, d, one has:

$$\text{if} \quad b<c<d, \quad \text{then} \quad cbd \equiv cdb \quad \text{and} \quad bdc \equiv dbc;$$

$$\text{if} \quad b<c, \quad \text{then} \quad cbb \equiv bcb \quad \text{and} \quad ccb \equiv cbc.$$

It has been studied by G. de B. Robinson, C. Schensted, D.E. Knuth and Curtis Greene (Cf. [3]). We shall only require the fact that all the words in a congruence class have the same multidegree and that one of them has the form $u''a^m u'$ with a not occurring in u'' or u'.

Lemma 3. *If v and v_1 are congruent there corresponds to any lifting of v a lifting of v_1 having the same marked letters.*

Proof. Consider first the case when $v = v'bdv''$ and $v_1 = v'dbv''$ where $b < d \neq b_+$. It is clear that $w' \cdot b\gamma \cdot d\gamma' \cdot w''$ is a lifting of v iff $w' \cdot d\gamma' \cdot b\gamma \cdot w''$ is a lifting of v_1 and that they have the same marked letters. This case covers the first elementary congruence defining $\equiv$. Suppose now that $v = v'cbbv''$ and $v_1 = v' = v'bcbv''$ where $c = b_+$. Let $w = w' \cdot c\gamma \cdot b\gamma' \cdot b\gamma'' \cdot w''$ be a lifting of v, where γ, γ' and γ'' are any three greek letters. If $\gamma \neq \gamma'$ it is clear that $w' \cdot b\gamma' \cdot c\gamma \cdot b\gamma'' \cdot w''$ is a lifting of v_1 having the same marked letters. If $\gamma = \gamma'$ we have $\gamma \neq \gamma''$ and we can begin by replacing w_1 by the lifting $w_1' = w' \cdot c\gamma \cdot b\gamma'' \cdot b\gamma \cdot w''$ of v without changing the marking. A similar argument applies to the other cases needed to cover the second elementary defining congruences, and the result is proved.

To conclude the proof of Property 1 we proceed by induction on the number of letters of the alphabet, the initial case being trivial.

Take any word $v_1 = u''a^m u'$ with $|u''u'|_a = 0$ in the congruence class of v. By the induction hypothesis the property holds for the word $u'u''$ which is on the smaller alphabet $A \setminus a$. Thus, by Corollary 3, it holds for v_1, i.e. one has the inequality $\bar{\nu}_1 \leq \bar{\mu}_1$ between any integrated delay function $\bar{\mu}_1$ and the natural one $\bar{\nu}_1$. Applying Lemma 3 gives the result, and a similar argument shows that the natural lifting has a minimal number of marked letters.

5. A further property of the natural delay function

We consider the natural lifting w of the word v, and two consecutive greek letters β and $\gamma = \beta_+$. We let M be the set of marked letters.

Lemma 4. *Assume that $d\beta \in M$ and $d\gamma \notin M$. There exists a letter $b < d$ such that $b\beta \notin M$ and $b\gamma \in M$ and that $c\beta, c\gamma \notin M$ for any intermediate c, i.e. $b < c < d$.*

Proof. Because of the algorithm defining the natural lifting we can assume for simplicity, that β is the first letter of the greek alphabet. The hypothesis $d\beta \in M$ implies that all the letters d are to the right of $d_{-}\beta$ and $d\beta$ is the first of them. Thus w has the subword $d_{-}\beta \cdot d\gamma \cdot d\beta$. Now if $d\gamma$ is not marked, $d_{-}\gamma$ must be to the right of $d\gamma$, and we have shown that w contains the subword $d_{-}\beta \cdot d_{-}\gamma$.

To conclude the proof we need only to verify, by induction on the number of intermediate letters, that if $c\beta \cdot c\gamma$ is a subword of w, then $c\beta$ is not marked and, either $c\gamma$ is marked or $c_{-}\beta \cdot c_{-}\gamma$ is again a subword of w. Indeed the hypothesis that $c\beta$ is to the left of $c\gamma$ prevents that it be marked, i.e. that $c_{-}\beta$ be to its left. Thus w contains the subword $c\beta \cdot c_{-}\beta \cdot c\gamma$. Suppose further that $c\gamma$ is not marked. Then $c_{-}\gamma$ must be on its right and the argument is concluded.

Corollary 5. *The natural delay function ν is non-decreasing in each of its two arguments.*

Proof. By definition, one has $(b\beta)\nu \leq (c\beta)\nu$ for $b < c$ and any β, and the inequality $(c\beta)\nu \leq (c\gamma)\nu$ follows from the above lemma by induction in the alphabet, since $(\mathrm{a}\beta)\nu = 0$ identically.

References

[1] D. Foata, Distributions eulériennes et mahonniennes, in: Proc. Advanced School on Higher Combinatorics. (M. Aigner Editor, Berlin, 1978).

[2] A. Lascoux and M.P. Schützenberger, Sur une conjecture de H.O. Foulkes, C.R. Acad. Sci. Paris 286 (1978) 323–324.

[3] G.P. Thomas, On Schensted's construction and the multiplication of Schur functions, Adv. in Math. 30 (1978) 8–32.

Annals of Discrete Mathematics 6 (1980) 257–274

SOME RESULTS ON THE EXISTENCE OF SQUARES

R.C. MULLIN*, P.J. SCHELLENBERG**, D.R. STINSON and S.A. VANSTONE***

Department of Combinatorics and Optimization, University of Waterloo, Waterloo, Ontario, Canada, N2L 3G1

In order to investigate the spectrum of skew Room squares, the authors obtain some preliminary results in preparation for a recursive attack. Some of these seem to be of interest in their own right. For example, if $N(v)$ denotes the maximum number of mutually orthogonal Latin squares of side v, it is shown, by direct construction, that $N(82) \geq 8$ and $N(100) \geq 8$ and, by using recursive constructions, that $N(v) \geq 8$ for $v \geq 9445$.

It is also shown that for $v = 2^{\alpha}t + 1$, $(2, t) = 1$, $\alpha \neq 1, 2, 6, 7$, there exists a skew Room square of side v.

1. Introduction

A *Room square* $R(v)$ of side v is a $v \times v$ array defined on the elements of a $(v+1)$-set V such that

(a) every cell of $R(v)$ is either empty or contains an unordered pair of elements of V,

(b) every element of V is contained in precisely one cell of each row and of each column of $R(v)$, and

(c) every unordered pair of elements of V is contained in precisely one cell of $R(v)$.

It is easy to show that a necessary condition for the existence of an $R(v)$ is that v is odd. In fact, it has been shown [8] that if v is odd and $v \neq 3, 5$, then there is an $R(v)$.

If an $s \times s$-subarray S of $R(v)$ has all its entries from a subset U of V with $s+1$ elements and if S is a Room square $R(s)$ defined on U, then $R(s)$ is called a *(Room) subsquare* of $R(v)$.

If the rows (columns) of a Room square are permuted in any fashion whatsoever, the result is a Room square. A Room square is said to be *standardized* if one of its objects has been distinguished and labelled ∞ and the rows and columns of the square have been permuted so as to have the symbol ∞ appearing in each cell on the main diagonal. This, of course, accounts for all the occurrences of ∞. Hereafter, whenever we refer to a Room square we mean a standardized Room square.

If whenever cell (i, j), $i \neq j$, of a (standardized) $R(v)$ contains a pair, the cell (j, i) is empty, then the Room square is said to be *skew*.

Supported in part by NRC Grants A3071*, A8509** and A9258***.

A *skew (Room) subsquare* of a skew Room square $R(v)$ is a (Room) subsquare of side s situated symmetrically about the main diagonal of $R(v)$. Clearly a skew subsquare is a skew Room square. Let

$$S=\{v\text{: there is a skew Room square of side } v\}.$$

The purpose of this paper is to study the set S.

A pairwise balanced design (PBD) of index 1 is a pair $(X, \mathfrak{A})$ where X is a set of elements and $\mathfrak{A}$ is a set of subsets (called blocks) of X such that every pair of elements, x, y of X is contained in precisely one block of $\mathfrak{A}$. When convenient, we will refer to $(X, \mathfrak{A})$ simply as X. The *order* of the PBD X is $|X|$ (the cardinality of X).

Let K be a set of natural numbers. We say that $(X, \mathfrak{A})$ is a PBD with block sizes from K if, for any $A\in\mathfrak{A}$, $|A|\in K$. Let $B[K]$ be the set of natural numbers v such that there is a PBD X of order v with block sizes from K. A set of natural numbers K is said to be *PBD-closed*, or more simply, *closed* if $K=B[K]$. Wilson [17, 18, 19] has extensively studied PBD-closure.

It is known [13] that the set S defined above is closed. Wilson's result [18, Theorem 5.1] implies that there is a skew Room square of side v for all sufficiently large odd integers v; that is, there exists a smallest integer B such that S contains all odd integers greater than B. However, "sufficiently large" is unspecified inasmuch as no upper bound for B is known. Our ultimate purpose is to establish such an upper bound and the results of this report are meant to provide a mechanism for doing so.

Our attack on this problem relies heavily on Latin squares [5], in particular, on sets of mutually orthogonal Latin squares (MOLS) [5]. For our purposes, we require sets of 7 or 8 MOLS and Sections 2, 3 and 4 are concerned with the study of such sets. In Section 2, we show that there are 8 MOLS of orders 82 and 100. In Section 3, we show that there are 8 MOLS of order v for all $v>9444$ and, in Section 4, that there are 7 MOLS of order v for $v>3620$.

In the later section, we use these results to establish that for $v=2^{\alpha}t+1$, $(2, t)=1$, $\alpha\neq 1, 2, 6, 7$, there is a skew Room square of side v.

2. Mutually orthogonal Latin squares

For the definitions of a Latin square of order n and of a set of mutually orthogonal Latin squares (MOLS) of order n the reader is referred to [5].

Let $N(n)$, $n\geqslant 2$, denote the maximum number of MOLS of order n, and let n_r denote the least integer such that for $v>n_r$, $N(v)\geqslant r$.

Table 1 contains many of the known results on n_r.

Wilson [20] has shown for sufficiently large v, $N(v)\geqslant v^{\frac{1}{17}}$.

The following four fundamental results can be found in [5].

Table 1

n_2	6	Bose et al. [3]
n_3	$\leqslant$ 14	Hanani [6], Wang and Wilson [15]
n_4	$\leqslant$ 52	Hanani [6], Wilson [20], Van Lint [12]
n_5	$\leqslant$ 62	Hanani [6]
n_6	$\leqslant$ 90	Wilson [20]
n_7[a]	$\leqslant$5036	Van Lint [12]

[a] Szajowski [11] has shown that for v odd and $v>469$, $N(v)\geqslant 7$.

Theorem 2.1. *For $n\geqslant 2$, $1\leqslant N(n)\leqslant n-1$.*

Any two Latin squares of order 1 are orthogonal. There is only one Latin square of order 0 (the null square), but it is orthogonal to itself. Thus it is not unreasonable to adopt the conventions that $N(0)=N(1)=\infty$.

Theorem 2.2. *$N(n)=n-1$ if n is a prime power.*

Theorem 2.3. *$N(nm)\geqslant \min\{N(n), N(m)\}$.*

Theorems 2.2 and 2.3 imply

Theorem 2.4. *If $n=p_1^{a_1}p_2^{a_2}\cdots p_k^{a_k}$ is the factorization of n into powers of distinct primes p_i, then*

$$N(n)\geqslant \min\{p_i^{a_i}-1:\ i=1,2,\ldots,k\}.$$

Wilson [20] has established the following three theorems.

Theorem 2.5. *If $0\leqslant u\leqslant t$, then*

$$N(mt+u)\geqslant \min\{N(m), N(m+1), N(t)-1, N(u)\}.$$

Theorem 2.6. *If $0\leqslant u$, $v\leqslant t$, then*

$$N(mt+u+v)\geqslant \min\{N(m), N(m+1), N(m+2), N(t)-2, N(u), N(v)\}.$$

Theorem 2.7. *If $t>\frac{1}{2}(l-1)(l-2)$, then*

$$N(mt+l)\geqslant \min\{N(m), N(m+1), N(m+2), N(t)-l\}.$$

Let $q=mt+1$ be a prime power and let C_0 be a multiplicative subgroup of $GF(q)\setminus\{0\}$ of order t and index m. Let the cosets of this subgroup be $C_0, C_1, C_2, \ldots, C_{m-1}$. These are called the cyclotomic classes of $GF(q)$ of index m. For $q=mt+1$ a prime power, define $A(m, t)$ to be the matrix-minus-diagonal of order $k=m+2$ with elements a_{ij}, $i \neq j$, from GF(q) satisfying the property that for $1 \leqslant j_1 < j_2 \leqslant k$, the set

$$\{a_{ij_1} - a_{ij_2} \mid 1 \leqslant i \leqslant k,\ i \neq j_1,\ i \neq j_2\}$$

is a set of distinct representatives of the cyclotomic classes $C_0, C_1, \ldots, C_{m-1}$. Wilson [21] has established

Lemma 2.8. *Let $q=mt+1$ be a prime power. If $N(t) \geqslant m$ and if there is a matrix-minus-diagonal $A(m, t)$, then $N(q+t) \geqslant m$.*

For $q=mt+1$ a prime power, define $V(m, t)$ to be a vector

$$(b_1, b_2, \ldots, b_{m+1})$$

with elements from GF(q) satisfying the property that for each $k \in \{1, 2, \ldots, m+1\}$, the set

$$\{b_i - b_j : i \in \{1, 2, \ldots, m+1\} \backslash \{m+2-k\},\ j \equiv i+k \pmod{m+2} \text{ and } i \leqslant j \leqslant m+1\}$$

is a system of distinct representatives of the cyclomatic classes $C_0, C_1, \ldots, C_{m-1}$.

Lemma 2.9. *Let $q=mt+1$ be a prime power. If $N(t) \geqslant m$ and if there is a vector $V(m, t)$, then $N(q+t) \geqslant m$.*

Proof. This lemma is based on the observation that the vector $V(m, t)$ implies the existence of a circulant matrix-minus-diagonal $A(m, t)$ whose first row is

$$\phi, b_1, b_2, \ldots, b_{m+1}$$

where ϕ represents the fact that the diagonal cell (1, 1) is empty.

More precisely, for $i \neq j$, $\{i, j\} \subset \{1, 2, \ldots, m+2\}$, define $a_{ij} = b_k$ where $k \equiv j-i \pmod{m+2}$ and $1 \leqslant k \leqslant m+1$. It can be shown that the matrix-minus-diagonal whose entries are a_{ij}, $i \neq j$, is an $A(m, t)$. Then, by Lemma 2.8, $N(q+t) \geqslant m$.

In [21], Wilson exhibits an $A(4, 7)$ and an $A(4, 9)$, both of which are circulants generated by a $V(4, 7)$ and $V(4, 9)$ respectively. However, the $A(6, 7)$ which he exhibits is not circulant. By computer search, we found the $V(6, 7)$

$$(0, 1, 3, 16, 35, 26, 36)$$

which implies the existence of a circulant $A(6, 7)$. We were also able to establish

Theorem 2.10. $N(82) \geqslant 8$ *and* $N(100) \geqslant 8$.

Proof.

$$V(8, 9) = (0, 1, 20, 70, 23, 59, 3, 8, 19).$$

$$V(8, 11) = (0, 1, 6, 56, 22, 35, 47, 23, 60).$$

3. A bound for n_8

In this section, we show that $n_8 \leqslant 9444$; in fact, $N(v) \geqslant 8$ for $v > 7768$, $v \neq 9402$, 9444. These results were obtained with the aid of a computer. Because of the large volume of data required to establish these bounds, we only include a brief description of the technique used. Full details can be found in the research report [9].

Lemma 3.1. *If* $n \geqslant 33799$, *then* $N(n) \geqslant 8$.

Proof. By Theorems 2.4 and 2.10, $N(v) \geqslant 8$ for $v = 81, 82, 83$. Thus we may construct 8 MOLS for numbers of the form $81t + u + v$ if $N(t) \geqslant 10$, $N(u) \geqslant 8$, $N(v) \geqslant 8$ and $0 \leqslant u, v \leqslant t$ by Theorem 2.6. For n a natural number, define

$$v_n = \begin{cases} \text{the least integer } i \text{ such that } \tfrac{1}{2}n \leqslant i \leqslant n,\ N(i) \geqslant 8, \\ \text{and } N(n-i) \geqslant 8 \text{ should exist;} \\ \infty, \quad \text{otherwise.} \end{cases}$$

We note that $v_n = \infty$ if $n = 3, 4, 5, 6, 7, 8$, and perhaps 15 or 21. For larger values of n, up to 2000, v_n was shown to be finite with the aid of a computer. For a given integer $n \geqslant 22$, define $D(n) + 1 = i$ to be the least integer greater than 22 such that $v_i > n$. Thus, if $t \geqslant 22$ and $N(t) \geqslant 10$, we may construct 8 MOLS of order v for $81t + 22 \leqslant v \leqslant 81t + D(t)$ by Theorem 2.6. By computer search, it is shown that $D(147) = 834 > 831$. Also, by Theorem 2.5, $N(417) \geqslant 15$ since $417 = 16 \cdot 25 + 17$, and $N(16)$, $N(17) \geqslant 15$, $N(25) = 24$. Let $n \geqslant 33799 = 81 \cdot 417 + 22$ and let $\alpha = \frac{1}{81}(n - 831)$. Now, given any integer $s \geqslant 2$, there is an integer t such that $s \leqslant t \leqslant s + 9$ and $(t, 210) = 1$. Theorem 2.4 implies $N(t) \geqslant 10$. Then there is an integer t such that $t \geqslant 417$, $\alpha \leqslant t < \alpha + 10$, and $N(t) \geqslant 10$. For such an integer t, let $w = n - 81t$. Then $22 \leqslant w \leqslant 831$, $0 \leqslant v_w \leqslant 417 \leqslant t$, and thus $N(n) \geqslant 8$ by Theorem 2.6.

Theorem 3.2. $n_8 \leqslant 9444$.

Remarks. (1) $N(n) \geqslant 8$ if $n > 7768$, $n \neq 9402$, 9444.
(2) $N(n) \geqslant 8$ if $n > 2739$ and n is odd.

This result was obtained by determining all relevant $n < 33799$ with $N(n) \geqslant 8$. The bulk of the work was done by using the sequences 16, 17; 31, 32; 81, 82, 83; and 99, 100, 101 in repeated applications of Theorems 2.5, 2.6 and 2.7. Various other higher sequences were also used.

Table 2 is a list of those orders for which 8 MOLS were not constructed.

4. A bound for n_7

In this section, we update the results on n_7 of Van Lint [12] and Szajowski [11].

Table 2

2	3	4	5	6	7	8	10	12	14
15	18	20	21	22	24	26	28	30	33
34	35	36	38	39	40	42	44	45	46
48	50	51	52	54	55	56	57	58	60
62	63	65	66	68	69	70	72	74	75
76	77	78	80	84	85	86	87	88	90
91	92	93	94	95	96	98	102	104	105
106	108	110	111	112	114	115	116	118	119
120	122	123	124	126	129	130	132	133	134
135	136	138	140	141	142	145	146	147	148
150	152	154	155	156	158	159	160	161	162
164	165	166	168	170	172	174	175	178	180
182	183	184	186	188	189	190	192	194	195
196	198	200	201	202	203	204	205	206	210
212	213	214	215	216	218	220	222	224	226
228	230	231	232	234	235	236	237	238	240
242	244	245	246	248	249	250	252	254	255
258	259	260	262	264	266	268	270	274	276
278	280	282	284	286	287	290	291	292	294
295	296	298	300	301	302	303	306	308	309
310	312	314	316	318	322	324	326	327	328
329	330	332	334	335	336	338	339	340	344
345	346	348	354	355	356	357	358	360	362
363	364	365	366	370	371	372	374	375	376
378	380	382	386	388	390	392	393	394	395
396	398	399	402	405	406	408	410	415	418
420	422	424	426	427	428	429	430	434	435
436	438	440	442	444	446	447	450	452	453
454	456	458	460	462	466	468	469	470	471
472	474	476	478	482	484	485	486	488	490
492	494	495	498	500	501	502	504	506	508
510	511	514	516	518	520	522	524	526	530
532	534	542	545	546	548	550	552	553	554
555	556	558	560	561	562	564	565	566	567
568	570	572	573	574	578	579	580	581	582
584	585	586	588	591	594	595	596	597	604
610	612	614	616	618	620	622	626	627	628
630	632	633	634	635	636	638	640	642	644
645	646	648	650	651	652	654	655	658	660
662	663	664	666	668	670	674	676	678	680
682	684	686	690	692	694	695	696	698	700
702	706	708	710	712	716	718	721	723	728
734	735	738	740	741	742	744	745	746	748
749	750	754	755	756	758	759	760	762	764
766	770	772	774	778	780	782	790	796	802
804	805	806	808	810	812	814	818	819	820
822	824	826	828	830	834	835	836	840	842
843	844	845	847	852	855	858	866	868	870
872	874	876	878	882	884	886	888	890	894
896	898	904	905	906	914	917	920	921	930
932	933	934	935	936	938	939	940	942	946
948	950	951	952	954	956	958	959	964	965
966	968	982	994	996	998	1000	1002	1004	1006
1010	1012	1014	1016	1018	1020	1022	1026	1028	1030
1032	1034	1036	1038	1042	1044	1045	1046	1048	1050
1052	1058	1059	1060	1068	1070	1074	1076	1078	1080
1082	1084	1086	1092	1094	1096	1102	1105	1106	1108
1110	1112	1114	1116	1118	1120	1122	1124	1126	1128
1130	1132	1134	1135	1138	1140	1141	1142	1144	1146
1150	1154	1162	1180	1182	1186	1188	1190	1192	1194
1196	1198	1202	1204	1206	1208	1210	1212	1214	1218
1220	1222	1224	1226	1228	1230	1234	1236	1238	1240

Table 2 (cont.)

1242	1244	1245	1246	1248	1250	1252	1253	1254	1255
1256	1257	1258	1260	1262	1263	1266	1267	1268	1270
1274	1276	1278	1285	1286	1292	1304	1324	1326	1330
1332	1336	1338	1340	1348	1354	1366	1368	1372	1385
1396	1400	1412	1414	1415	1416	1418	1419	1420	1422
1426	1428	1430	1432	1434	1436	1438	1442	1444	1445
1446	1448	1450	1452	1454	1460	1462	1463	1464	1478
1479	1490	1492	1496	1502	1508	1509	1510	1512	1514
1515	1516	1518	1522	1524	1526	1527	1533	1534	1547
1554	1570	1576	1578	1580	1592	1603	1612	1614	1615
1618	1620	1622	1624	1626	1628	1630	1636	1638	1640
1642	1646	1650	1655	1658	1676	1678	1682	1702	1704
1706	1714	1718	1720	1722	1724	1726	1732	1734	1736
1738	1740	1742	1746	1750	1752	1754	1756	1758	1762
1764	1766	1768	1770	1772	1774	1778	1780	1782	1784
1786	1788	1790	1792	1796	1798	1799	1800	1802	1804
1806	1810	1814	1816	1818	1820	1822	1828	1832	1834
1836	1850	1862	1912	1924	1925	1926	1929	1930	1935
1939	1941	1942	1946	1948	1954	1956	1958	1960	1962
1964	1966	1970	1972	1974	1976	1978	1980	1981	1982
1986	1988	1990	1992	1994	1996	1998	2002	2004	2005
2006	2008	2010	2012	2014	2020	2022	2024	2040	2046
2056	2062	2068	2070	2072	2074	2076	2084	2092	2098
2110	2116	2122	2134	2140	2142	2146	2150	2152	2154
2156	2158	2162	2164	2166	2174	2180	2181	2182	2184
2186	2234	2236	2246	2252	2258	2266	2270	2298	2308
2314	2318	2326	2328	2332	2338	2340	2342	2343	2344
2346	2348	2364	2380	2386	2396	2398	2402	2404	2406
2408	2410	2412	2418	2420	2428	2434	2436	2438	2446
2452	2454	2456	2464	2470	2471	2506	2526	2532	2562
2566	2568	2577	2580	2588	2607	2658	2660	2662	2664
2666	2668	2670	2710	2722	2724	2732	2737	2738	2740
2742	2744	2746	2748	2750	2756	2758	2762	2764	2766
2774	2780	2792	2794	2798	2810	2814	2822	2824	2828
2834	2836	2844	2846	2852	2854	2856	2858	2860	2862
2866	2870	2886	2892	2902	2908	2918	2920	2924	2926
2936	2942	2948	2950	2952	2954	2956	2958	2962	2966
2968	2970	2972	2974	2980	2984	2986	2988	2994	3012
3062	3064	3084	3106	3118	3122	3124	3126	3128	3130
3134	3146	3164	3166	3238	3244	3250	3256	3258	3262
3268	3270	3278	3286	3288	3304	3310	3320	3386	3412
3434	3436	3464	3466	3470	3474	3498	3554	3560	3566
3572	3578	3580	3590	3596	3608	3614	3620	3622	3624
3626	3628	3630	3634	3638	3640	3642	3644	3646	3652
3654	3656	3658	3660	3662	3678	3730	3734	3736	3738
3740	3742	3748	3750	3754	3756	3758	3766	3772	3784
3786	3790	3796	3802	3806	3890	3896	3908	3910	3914
3926	3930	3932	4056	4124	4126	4136	4138	4146	4152
4154	4180	4184	4190	4198	4200	4202	4204	4206	4210
4216	4218	4220	4226	4228	4230	4232	4234	4236	4238
4242	4244	4246	4250	4252	4254	4394	4396	4402	4414
4420	4424	4428	4438	4454	4466	4472	4478	4484	4488
4490	4498	4502	4508	4510	4518	4520	4524	4526	4556
4580	4582	4584	4586	4588	4590	4594	4598	4600	4602
4604	4606	4622	4626	4652	4674	4738	4776	4846	4850
4872	4918	4922	4936	5028	5030	5036	5038	5042	5044
5046	5048	5050	5052	5060	5068	5074	5086	5092	5098
5138	5144	5146	5148	5164	5324	5350	5362	5384	5398
5414	5418	5420	5542	5564	5582	5668	5704	5734	5740
5746	5766	5772	5912	6184	6202	6302	6314	6316	6328
6330	6556	6844	7156	7192	7368	7378	7384	7388	7398
7454	7460	7464	7466	7474	7528	7768	9402	9444	

In view of these papers, we need only consider MOLS of order $n \leqslant 5036$ and, for $n \geqslant 471$, we need only consider even n.

From Szajowki's list of orders for which it is not known that there are 7 MOLS, we can omit 151, since it is a prime, and $177 = 16 \cdot 11 + 1$ by Theorem 2.5. Furthermore, we can delete 56, 57, 65, 82 and 100 since

(i) Van Lint [12] establishes that $N(57) \geqslant 7$ and $N(65) \geqslant 7$;

(ii) Mills [7] has shown that $N(56) \geqslant 7$; and

(iii) $N(82) \geqslant 8$ and $N(100) \geqslant 8$ by Theorem 2.10.

Using the above facts together with Theorems 2.5, 2.6 and 2.7 we are able to show

Theorem 4.1. $n_7 \leqslant 4146$.

Again the details may be found in [9]. Table 3 is a list of orders for which we are unable to construct 7 MOLS.

5. Skew Room squares

A survey of results on skew Room squares is given in [13]. The results from that paper required here are updated and listed as lemmata below. Recall that

$$S = \{v: \text{ there is a skew Room square of side } v\}.$$

Beaman and Wallis [1] have constructed a skew Room square of side nine. Thus we have.

Lemma 5.1. *For v an odd prime power, $v \neq 3, 5$, $v \in S$.*

Lemma 5.2. *Suppose there is a skew Room square of side v_2 which contains a skew Room subsquare of side v_3.*

(i) *If $v_2 - v_3 \neq 6$ and if there is a skew Room square of side v_1, then there is a skew Room square of side $v_1(v_2 - v_3) + v_3$ which contains skew Room subsquares of sides v_1, v_2 and v_3. This result also holds for $v_3 = 0$.*

(ii) *(Beaman and Wallis [2].) If $v_3 \neq 0$ and $v_2 - v_3 \neq 12$, then there is a skew Room square of side $5(v_2 - v_3) + v_3$ which contains skew Room subsquares of sides v_2 and v_3.*

Lemmata 5.1 and 5.2 (i) imply that if $(n, 15) = 1$, then $n \in S$.

For n a positive integer, let G be an additive Abelian group of order $2n + 1$. By a *starter* in G we mean a list

$$X = (\{x_1, y_1\}, \{x_2, y_2\}, \ldots, \{x_n, y_n\})$$

of unordered pairs of elements of G such that

(i) the pairs of X are a partition of $G \setminus \{0\}$, and

(ii) the differences, $\pm(x_i - y_i)$, are a partition of $G \setminus \{0\}$.

Table 3
7 MOLS not known

2	3	4	5	6	7	10	12	14	15
18	20	21	22	24	26	28	30	33	34
35	36	38	39	40	42	44	45	46	48
50	51	52	54	55	58	60	62	63	66
68	69	70	74	75	76	77	78	84	85
86	87	90	91	92	93	94	95	98	102
106	108	110	111	114	116	118	119	120	122
123	124	126	130	132	133	134	135	138	140
142	146	148	150	154	155	156	158	159	162
164	166	170	172	174	175	178	180	182	183
186	188	189	190	194	196	198	202	204	205
206	210	212	214	215	218	220	222	226	228
230	231	234	236	237	238	242	244	246	250
252	254	258	260	262	266	268	270	274	276
278	282	284	286	290	291	292	294	295	298
300	301	302	303	306	308	310	314	316	318
322	324	326	330	332	334	335	338	340	346
348	354	356	358	362	364	366	370	372	374
378	380	382	386	388	390	394	396	398	402
406	410	418	420	422	426	427	428	430	434
436	438	442	444	446	450	452	454	458	460
462	466	468	469	470	474	476	478	482	484
486	490	492	494	498	500	502	506	508	510
514	516	518	522	524	526	530	532	534	542
546	548	550	554	556	558	560	562	564	566
570	572	574	578	580	582	586	588	594	596
604	610	612	614	618	620	622	626	628	630
634	636	638	642	644	646	650	652	654	658
660	662	666	668	670	674	676	678	682	684
686	690	692	694	698	700	702	706	708	710
716	718	734	740	742	748	750	754	756	758
762	764	766	770	772	774	778	780	796	802
804	806	810	812	814	818	820	822	826	828
830	834	836	842	844	852	866	868	876	884
914	930	946	948	950	958	964	966	982	994
996	998	1002	1006	1010	1012	1014	1020	1022	1026
1034	1076	1078	1084	1086	1094	1102	1110	1114	1118
1122	1126	1132	1180	1182	1186	1188	1190	1198	1202
1204	1206	1222	1230	1238	1242	1246	1250	1254	1258
1260	1262	1326	1330	1332	1412	1420	1422	1426	1428
1430	1438	1442	1444	1446	1454	1460	1462	1478	1492
1570	1612	1614	1618	1620	1622	1630	1706	1714	1718
1722	1724	1726	1732	1734	1740	1742	1750	1766	1774
1790	2074	2116	2122	2270	2406	2410	2732	2740	2756
2758	2762	2766	2774	2780	2834	2854	2862	3620	4146

By an *adder* for X we mean a list $A(X)$ of n distinct non-zero elements $(a_1, a_2, \ldots, a_n)$ from G such that the pairs $\{x_i + a_i, y_i + a_i\}$, $i = 1, 2, \ldots, n$, are a partition of $G \setminus \{0\}$. We say that a_i is the adder corresponding to pair $\{x_i, y_i\}$. An adder $A(X)$ is said to be skew if for each element $a \in A(X)$, $-a \notin A(X)$.

Lemma 5.3. [14]. *If an Abelian group of order $2n+1$ admits a starter X and a (skew) adder $A(X)$, then there is a (skew) Room square of side $2n+1$.*

Lemma 5.4. *If v is odd and $15 \leqslant v \leqslant 35$, then $v \in S$.*

Skew Room squares of sides divisible by 3 and 5 in this range are constructed by means of starters and skew adders. (The purported first rows for skew Room

Table 4
G is the group of integers modulo 15

Starter	3, 5	6, 11	9, 13	4, 12	8, 14	1, 2	7, 10
Adder	1	3	4	6	8	10	13

G is the group of integers modulo 21

Starter	1, 2	3, 5	9, 13	7, 15	8, 11	6, 16	10, 19	12, 17	14, 20	18, 4
Adder	1	2	3	4	7	11	12	13	15	16

G is the group of integers modulo 35

Starter	9, 26	17, 18	1, 34	3, 32	16, 19	11, 24	2, 33	15, 20	5, 30
Adder	1	2	3	4	5	6	7	8	11

Starter	14, 21	4, 31	6, 29	12, 23	10, 25	7, 28	13, 22	8, 27
Adder	12	18	19	20	21	22	25	26

squares of orders 15, 21 and 35 in [13] each contain misprints. The starters and adders in Table 4 can be used to generate skew Room squares of these orders.)

A starter X in an Abelian group of order $2n+1$ is *patterned* if $x_i + y_i = 0$ for $i = 1, 2, \ldots, n$ and X is *strong* if the sums $x_i + y_i$, $i = 1, 2, \ldots, n$, are all distinct and non-zero. Byleen [4] showed that the existence of a strong starter is equivalent to the existence of a patterned starter and corresponding adder. Although most starters used for the construction of Room squares are strong (and thus imply the existence of patterned starters X and adders $A(X)$) the weakness of this approach for the construction of skew Room squares is evident from the result, that there is no skew adder for a patterned starter in an Abelian group of order divisible by 3 whose 3-Sylow subgroup is cyclic [13]. Experience indicates that skew adders are hard to find for patterned starters in any Abelian groups whose orders are multiples of 3. However, there are (non-patterned) starters for which skew adders can be found, even in cyclic groups. For example, the starters X and skew adders $A(X)$ listed in Table 5, for cyclic groups of order 39 and 45, show that $39, 45 \in S$.

Lemma 5.5. *For v an odd integer, if $v \notin S$, then $v = 3n$, $5n$, or $75n$ where $(n, 15) = 1$.*

Proof. Let $v \in 3^\alpha 5^\beta n$ where $(n, 15) = 1$. Then $n \in S$. If $\alpha \geqslant 2$ and $\beta \geqslant 2$, then $3^\alpha \in S$,

Table 5
G is the cyclic group of integers modulo 39

Starter	1. 12	2, 11	3, 10	4, 9	5, 8	6, 7	13, 26	14, 38	15, 37
Adder	3	5	2	13	15	22	1	7	20

Starter	16, 36	17, 35	18, 34	19, 33	20, 32	21, 31	22, 30	23, 29
Adder	16	14	6	31	10	27	4	18

Starter	24, 28	25, 27
Adder	9	11

G is the cyclic group of integers modulo 45

Starter	1, 14	2, 13	3, 12	4, 11	5, 10	6, 9	7, 8	15, 30	16, 44	17, 43
Adder	3	5	2	11	14	17	20	1	30	36

Starter	18, 42	19, 41	20, 40	21, 39	22, 38	23, 37	24, 36	25, 35	26, 34
Adder	23	39	16	27	32	10	19	7	4

Starter	27, 33	28, 32	29, 31
Adder	24	12	8

$5^\beta \in S$ and thus $v \in S$. If $\alpha > 2$ and $\beta = 1$, then $3^\alpha \cdot 5^\beta = 3^{\alpha-1} \cdot 15$. Since $3^{\alpha-1} \in S$ and $15 \in S$, $v \in S$. If $\alpha = 2$ and $\beta = 1$, then $3^\alpha \cdot 5^\beta = 45 \in S$ which implies $v \in S$. If $\alpha \geqslant 2$ and $\beta = 0$, then $3^\alpha \cdot 5^\beta = 3^\alpha \in S$ which implies $v \in S$. If $\alpha = 1$ and $\beta > 2$, $3^\alpha \cdot 5^\beta = 15 \cdot 5^{\beta-1}$. Since $15 \in S$ and $5^{\beta-1} \in S$, $v \in S$. If $\alpha = \beta = 1$, $3^\alpha \cdot 5^\beta = 15 \in S$ which implies $v \in S$. If $\alpha = 0$ and $\beta \geqslant 2$, then $3^\alpha 5^\beta = 5^\beta \in S$ which implies $v \in S$.

Corollary 5.6. *For v an odd integer, if $v \notin S$, then $v = 3n$, $5n$, or $75n$ where $(n, 3 \cdot 5 \cdot 7) = 1$.*

Proof. Let $n = 7m$, where $(m, 15) = 1$. Then $m \in S$. Since $3n = 21m$ and $21 \in S$, $3n \in S$. Since $5n = 35m$ and $35 \in S$, $5n \in S$. Since $75n = 21 \cdot 25 \cdot m$ and $21, 25 \in S$, $75n \in S$.

Historically, constructing Room squares of sides v of the form $2^a + 1$ has presented special problems. The utility of such squares is illustrated in the following.

Theorem 5.7. *If* 2^a+1, $3\cdot 2^a+1\in S$, *then* $2^a\cdot n+1\in S$ *for all odd integers n.*

Proof. Let $2^a+1\in S$ and $3\cdot 2^a+1\in S$. If $n\in S$, then by Lemma 5.2(i) (with $v_1=n$, $v_2=2^a+1$, $v_3=1$), $2^a\cdot n+1\in S$.

By Lemma 5.5, if n is an odd integer and $n\notin S$, then $n=3k$, $5k$ or $75k$ where $k\in S$. Now, by Lemma 5.2(i) (with $v_1=k$, $v_2=3\cdot 2^a+1$, $v_3=1$), $3k\cdot 2^a+1\in S$; by Lemma 5.2(ii) (with $v_2=k\cdot 2^a+1$, $v_3=1$), $5k\cdot 2^a+1\in S$; and by Lemma 5.2(i) (with $v_1=25k$, $v_2=3\cdot 2^a+1$, $v_3=1$), $75k\cdot 2^a+1\in S$.

As mentioned, it is our ultimate aim to produce an upper bound B such that for all odd integers $v>B$, $v\in S$.

We wish first to show that if $v=2^at+1$, where $(t,2)=1$, and if $v\notin S$, then $a\in\{1,2,6,7\}$. The next step is to find v_8 such that if $v>v_8$ and $v\equiv 1 \bmod 8$, then $v\in S$. We will also lay the foundations for this investigation.

6. Heads and generalized replication numbers

By a *9-head* of order v we mean a PBD, say D, of order v with block size from S containing an ideal element ∞ which occurs only in blocks of size 9 in D. Clearly the order v of a 9-head satisfies the congruence $v\equiv 1 \pmod 8$. By the *generalized replication number* of a 9-head of order v we mean the integer $\frac{1}{8}(v-1)$. The following is an extension of a result of Hanani.

Theorem 6.1. *The set* $\mathrm{GR}=\{\frac{1}{8}(v-1)$: *v is the order of a 9-head*$\}$ *is PBD-closed.*

Proof. Let D be a PBD of order v on a set V with block sizes from GR. Let the blocks of D be $B_1, B_2, \ldots, B_t$, let $N=\{1,2,\ldots,8\}$, and let ∞ be an ideal point disjoint from $V\times N$. For each block B_i of D construct a 9-head $N(B_i)$ on the set of elements $(B_i\times N)\cup\{\infty\}$ in such a fashion that ∞ occurs precisely in the set of blocks

$$M(B_i)=\{\{\infty\}\cup\{(x,i):\ i\in N\}:\ x\in B_i\}.$$

Let

$$F=\{N(B_i)-M(B_i):\ i=1,2,\ldots,t\}.$$

Then $F\cup\{\{\infty\}\cup\{(x,i):\ i\in N\}:\ x\in V\}$ is a 9-head of order $8v+1$ on the set $(V\times N)\cup\{\infty\}$. Thus $v\in\mathrm{GR}$ which means that GR is closed.

Since any BIBD with $k=9$ and $\lambda=1$ is a 9-head, and since there is a PG(8, 2) and an EG(9, 2), we have $\{9,10\}\subset\mathrm{GR}$. If there exist k MOLS of size n, $k=7$ or 8, then the associated group divisible design GD(k, n) (see below) implies the existence of a PBD with block sizes n and $k+2$. Thus, if $n\in\mathrm{GR}$, $(k+2)n\in\mathrm{GR}$.

For the definition of an orthogonal array $OA(n, s)$ of order n and depth s the reader is referred to [5]. The existence of $s-2$ MOLS of order n is equivalent to the existence of an $OA(n, s)$. Define

$$OA(s)=\{n: \text{there exists an } OA(n, s)\}.$$

A *group divisible design* $GD(s, n)$ is a triple $(X, \mathfrak{G}, \mathfrak{A})$ where X is a set of cardinality sn,

$$\mathfrak{G}=\{G_1, G_2, \ldots, G_s\}$$

is a partition of X into subsets G_i (called groups) each containing n elements and $\mathfrak{A}$ is a class of subsets (called blocks) of X such that each block $A \in \mathfrak{A}$ contains precisely one element from each group and each pair x, y of elements not contained in the same group occur together in precisely one block of $\mathfrak{A}$. The existence of an $OA(n, s)$ is equivalent to the existence of a $GD(s, n)$.

Lemma. 6.2. $\{7, 9, 10, 17, 19, 137, 144, 337\} \subset GR$.

Proof. As noted above $\{9, 10\} \subseteq GR$. Since $8 \in OA(7)$, there exists a $GD(7, 8)$. By adjoining a new element ∞ to the groups of this $GD(7, 8)$ one obtains a 9-head on 57 varieties; hence, $7 \in GR$.

Seiden [10] has shown that there exists a resolvable BIBD (120, 255, 17, 8, 1), say D. Such a design has 17 resolution classes $R_1, R_2, \ldots, R_{17}$. Let $\{\infty_1, \infty_2, \ldots, \infty_{17}\} = T$ be a set of 17 elements disjoint from the variety set of D. By adjoining ∞_i to each block of R_i, and adjoining T as a block, one obtains a PBD with blocks of size 9 and 17. Moreover, any of the original varieties occurs only in blocks of size 9. Thus, we have a 9-head on 137 varieties, and hence $17 \in GR$. By PBD-closure, $137 \in GR$. H. Hanani (private communication) has shown that there exists a BIBD (153, 323, 19, 9, 1). Hence $19 \in GR$. Wilson [16] has established the existence of a BIBD with $\lambda = 1$, $k = 9$ and $r = 144$. Hence $144 \in GR$. Finally, Wilson [16] has shown that there exists a BIBD (337, 2696, 56, 7, 1). Since $7 \in GR$, we have, by PBD closure, that $337 \in GR$.

Theorems 6.3 and 6.5 follow easily from the definition of PBD-closure.

Theorem 6.3. *Let K be a PBD-closed set. Suppose that m, $m+1$, $m+2, \ldots, m+l \in K$ and $a_1, a_2, \ldots, a_l \in K$. If $t \in OA(m+l) \cap K$ and $0 \leq a_i \leq t$, for $i = 1, 2, \ldots, l$, then $mt + \sum_{i=1}^{l} a_i \in K$.*

Corollary 6.4. *Let $t \in OA(10) \cap GR$ and let $a \in GR$, $0 \leq a \leq t$. Then, $9t + a \in GR$.*

Proof. By Theorem 6.1, GR is PBD-closed. Also, $\{9, 10\} \subset GR$. Taking $l = 1$ and $m = 9$ in the above theorem yields the result.

Theorem 6.5. *Let K be a PBD-closed set. If $s, t \in K$ and $s \in OA(t)$, then $st \in K$.*

Theorem 6.6. *Let K be a PBD-closed set. If* $s, t \in K$, *and* $s-1 \in \mathrm{OA}(t)$, *then* $(s-1)t+1 \in K$.

Proof. Adjoin a new element to the groups of a $\mathrm{GD}(s-1, t)$.

Theorem 6.7. *Let X be a PBD with block sizes from* $K = L \cup M$ *where* $L \cap M = \emptyset$. *Suppose* $M \subseteq S$ *and for every* $l \in L$, *there is a 9-head of order l. Furthermore, suppose there is an element, say* ∞, *which is only contained in blocks of sizes from L. Then, there is a 9-head of order* $|X|$ *and hence* $\frac{1}{8}(|X|-1) \in \mathrm{GR}$.

Proof. Let B be a block of X such that $|B| \in L$. If $\infty \in B$, then replace the block B by a 9-head whose ideal element is ∞. If $\infty \notin B$, replace B by any PBD with block sizes from S. The result is a 9-head of order $|X|$.

Corollary. 6.8. *Let* $t \in \mathrm{GR}$ *and* $s \in S$. *If* $8t \in \mathrm{OA}(s)$, *then* $st \in \mathrm{GR}$.

Proof. X is the PBD obtained by adjoining an ideal element ∞ to each group of a $\mathrm{GD}(s, 8t)$. Then $L = \{8t+1\}$ and $M = \{s\}$ and Theorem 6.7 implies the result.

7. Skew squares of side $2^a t+1$, $a \neq 1, 2, 6, 7$

The following special case of Lemma 6.3 is of special interest for this section.

Lemma 7.1. *Suppose v and* 2^a *belong to* GR, *and* $v \leq 2^a$. *Then,* $v2^a \in \mathrm{GR}$.

Proof. In Theorem 6.3 put $l = 0$, $m = 2^a$ and $t = v$.

Let $A = \{a\colon 2^a \in \mathrm{GR}\}$ and $B = \{a\colon 3 \cdot 2^a \in \mathrm{GR}\}$.

Lemma 7.2. $\{6, 14, 16, 17, 19, 21\} \subset A$.

Proof. (i) By taking $t = 7$, $s = 10$ in theorem 6.6, $64 \in \mathrm{GR}$ and $6 \in A$.

(ii) Since

$$2^{14} = 9 \cdot 1783 + 337,$$
$$1783 = 9 \cdot 181 + 154,$$
$$181 = 9 \cdot 19 + 10,$$
$$154 = 9 \cdot 17 + 1$$

and since 1, 10, 17, 19, $337 \in \mathrm{GR}$, then by Corollary 6.4, $2^{14} \in \mathrm{GR}$ and $14 \in A$.

(iii) Since

$$2^{16} = 9 \cdot 7281 + 7,$$
$$7281 = 9 \cdot 793 + 144,$$
$$793 = 9 \cdot 81 + 64$$

and since 7, 64, $144 \in GR$, then Corollary 6.4 implies that $2^{16} \in GR$ and hence $16 \in A$.

(iv) Since

$$\begin{aligned} 2^{17} &= 9 \cdot 14551 + 113, \\ 113 &= 16 \cdot 7 + 1, \\ 14551 &= 9 \cdot 1609 + 70, \\ 1609 &= 9 \cdot 171 + 70, \\ 70 &= 7 \cdot 10, \\ 171 &= 9 \cdot 19 \end{aligned}$$

by Corollaries 6.4 and 6.8 and Theorem 6.6, $2^{17} \in GR$ and $17 \in A$.

(v) Since

$$\begin{aligned} 2^{19} &= 9 \cdot 58239 + 137, \\ 58239 &= 9 \cdot 6461 + 90, \quad 90 = 9 \cdot 10, \\ 6461 &= 9 \cdot 667 + 458, \quad 458 = 9 \cdot 49 + 17, \\ 667 &= 9 \cdot 73 + 10, \\ 73 &= 9 \cdot 8 + 1, \end{aligned}$$

Corollary 6.4 and Theorem 6.6 imply $2^{19} \in GR$ and, thus, $19 \in A$.

(vi) Since

$$\begin{aligned} 2^{21} &= 9 \cdot 233015 + 17, \\ 233015 &= 9 \cdot 25798 + 833, \quad 833 = 17 \cdot 49, \\ 25798 &= 9 \cdot 2779 + 787, \\ 787 &= 9 \cdot 82 + 57, \quad 82 = 9 \cdot 9 + 1, \\ 2779 &= 9 \cdot 289 + 178, \quad 178 = 9 \cdot 19 + 7, \\ 289 &= 17 \cdot 17, \end{aligned}$$

then Corollary 6.4 and Theorem 6.6 again imply that $2^{21} \in GR$ and $21 \in A$.

Lemma 7.3. *If a is a positive integer and $a \notin \{1, 2, 3, 4, 5, 7, 8, 9, 10, 11, 13, 15\}$, then $a \in A$.*

Proof. By Lemma 7.1, A is closed under addition. This, coupled with Lemma 7.2 yields the result.

Lemma 7.4. *If a is a positive integer and $a \notin \{1, 2, 3, 4, 5, 6, 7, 8, 10, 11, 12, 13, 14, 16, 18\}$, then $2^a + 1 \in S$.*

Proof. If $a \in A$, then $2^a \in GR$, and there exists a 9-head of order $2^{a+3} + 1$ and hence a skew square of that order.

Lemma 7.5. *If a is a positive integer and* $a \notin \{1, 2, 6, 7, 11, 13, 14, 18\}$, *then* $2^a + 1 \in S$.

Proof. By the preceding lemma, we need only consider $a \in \{3, 4, 5, 8, 10, 12, 16\}$. By the results of Section 6, we need not consider $a = 3$ or 5. Moreover if $a \equiv 0 \bmod 4$, then $(2^a + 1, 15) = 1$, we need not consider 4, 8, 12 or 16 further. Also, $2^{10} + 1 = 25 \cdot 41$.

Lemma 7.6. *If there exists a skew Room square of order* $t \cdot 2^{a+d} + 1$ *which contains a skew subsquare of side* $t2^a + 1$, *and if there exists a skew square of side* $2^d + 1$, *then there exists a skew square of side* $t2^{a+nd} + 1$ *for* $n = 0, 1, 2, \ldots$.

Proof. Consider the identity

$$t2^{a+2d} + 1 = (2^d + 1)[t2^{a+d} + 1 - (t2^a + 1)] + t2^a + 1.$$

Lemma 7.7. *If a is a positive integer and* $a \notin \{1, 2, 6, 7\}$, *then* $2^a + 1 \in S$.

Proof. We need only consider $a \in \{11, 13, 14, 18\}$. Note the equations

$$257 = 5(57 - 7) + 7, \qquad 57 = 7(9 - 1) + 1.$$

Since $33 \in S$, if $a \equiv 3 \bmod 5$, $2^a + 1 \in S$. This leaves only $a = 11$ and $a = 14$. But $2^{11} + 1 = 15(145 - 9) + 9$, $145 = 9(17 - 1) + 1$. Hence $2^{11} + 1 \in S$. Moreover, $2^{14} + 1 = 113 \cdot 145$, $145 = 9(17 - 1) + 1$.

Lemma 7.8. $3 \cdot 2^{12} \in \mathrm{GR}$ *and* $12 \in B$.

Proof. Again the proof is based on Corollary 6.4. Since $3 \cdot 2^{12} = 9 \cdot 1359 + 57$

$$1359 = 9 \cdot 144 + 63, \qquad 63 = 7 \cdot 9$$

and $144 \in \mathrm{GR}$, then $3 \cdot 2^{12} \in \mathrm{GR}$.

Lemma 7.9. *If b is a positive integer and* $b \notin \{1, 2, 3, 4, 5, 6, 7, 8, 9, 10, 11, 13, 14, 15, 16, 17, 18, 19, 20, 21, 22, 23, 24, 25, 27\}$, *then* $b \in B$.

Proof. This is a consequence of Lemmata 7.1, 7.3 and 7.8.

Lemma 7.10. *If a is a positive integer and* $a > 30$, *or if* $a \in \{15, 29\}$, $3 \cdot 2^a + 1 \in S$.

Proof. If $a \in B$, then $3 \cdot 2^a \in \mathrm{GR}$ and there is a 9-head of order $3 \cdot 2^{a+3} + 1$ and hence a skew Room square of that order.

Lemma 7.11. *If a is a positive integer and if* $a \notin \{11, 27\}$, *then* $3 \cdot 2^a + 1 \in S$.

Proof. Clearly $(3 \cdot 2^a + 1, 3) = 1$, and if $a \not\equiv 3 \bmod 4$, then $(3 \cdot 2^a + 1, 5) = 1$. Hence we need only consider the case $a \equiv 3 \bmod 4$. Moreover, there is a skew square of side 49 with a skew subsquare of side 7, and since $9 \in S$, this implies that $3 \cdot 2^a + 1 \in S$ for $a \equiv 1 \bmod 3$ in view of Lemma 7.6. Hence we need only consider $v \equiv 3, 11 \bmod 12$, and in view of Lemma 7.9 we need only consider $a \leqslant 30$. Moreover, if $a \equiv 3 \bmod 20$, $3 \cdot 2^a + 1 \equiv 0 \bmod 25$, and clearly this implies that $3 \cdot 2^a + 1 \in S$ in this case. This exhausts all values except for the values of $a \in \{11, 27\}$ in doubt.

Lemma 7.12. *If a is a positive integer, then* $3 \cdot 2^a + 1 \in S$.

Proof. By the previous lemma we need only consider $a = 11$, and 27.

To show the existence of a skew Room square of side $3 \cdot 2^{11} + 1$, we need only observe the following equations

$$3 \cdot 2^{11} + 1 = 9(689 - 7) + 7,$$
$$689 = 7(113 - 17) + 17,$$
$$113 = 7(17 - 1) + 1.$$

Then by Lemma 5.2, $3 \cdot 2^{11} + 1 \in S$. By Lemma 7.7, $2^{13} + 1 \in S$. To show that $3 \cdot 2^{27} + 1 \in S$, it is sufficient to show that there exists a skew square of side $3 \cdot 2^{14} + 1$ containing a skew subsquare of side $3 \cdot 2 + 1$ by Lemma 7.6. But

$$3 \cdot 2^{14} + 1 = 13 \cdot 3781,$$
$$3781 = 7(541 - 1) + 1,$$
$$(541, 15) = 1.$$

The above leads immediately to the following.

Theorem 7.13. *If* $v = 2^a t + 1$ *is a positive integer such that* $a > 0$ *and* $(t, 2) = 1$, *and if* $a \notin \{1, 2, 6, 7\}$, *then* $v \in S$.

Proof. This is a direct consequence of Lemmata 7.7, 7.12 and Theorem 5.7.

8. Conclusion

Clearly by finding v^* such that for $v > v^*$, $v \in \mathrm{GR}$ one can establish a bound for v_8 where v_8 is the least integer such that for $v > v_8$ and $v \equiv 1 \bmod 8$, $v \in S$. Since $3 \cdot 2^6 + 1$, $3 \cdot 2^7 + 1 \in S$, if it could be shown that $2^6 + 1 \in S$ and $2^7 + 1 \in S$, then it would follow that $v_8 = 1$.

This also establishes a point of departure for finding a bound for v_0, where v_0 is the least integer such that $v > v_0$, v odd, implies that $v \in S$. This will be pursued in a later paper.

References

[1] I.R. Beaman and W.D. Wallis, A skew Room square of side nine, Utilitas Math. 8 (1975) 382.
[2] I.R. Beaman and W.D. Wallis, On skew Room squares, preprint.
[3] R.C. Bose, S.S. Shrikhande and E.T. Parker, Further results on the construction of mutually orthogonal Latin squares and the falsity of Euler's conjecture, Can. J. Math. 12 (1960) 189–203.
[4] K. Byleen, On Stanton and Mullin's construction of Room squares, Ann. Math. Statist. 41 (1970) 1122–1125.
[5] M. Hall Jr., Combinatorial Theory (Blaisdell, Waltham, MA, 1967).
[6] H. Hanani, On the number of orthogonal Latin squares, J. Combin. Theory 8 (1970) 247–271.
[7] W.H. Mills, Some mutually orthogonal Latin squares, in: Proc. 8th Southeastern Conf. on Combinatorics, Graph Theory and Computing, Baton Rouge, LA (1977) 473–487.
[8] R.C. Mullin and W.D. Wallis, The existence of Room squares, Aequationes Math. 13 (1975) 1–7.
[9] R.C. Mullin, P.J. Schellenberg, D.R. Stinson and S.A. Vanstone, On the existence of 7 and 8 mutually orthogonal Latin squares, Combinatorics and Optimization Research Report, CORR 78–14 (1978).
[10] E. Seiden, A method of construction of resolvable BIBD, Sankhyā 25A (1963) 393–394.
[11] K. Szajowski, The number of orthogonal Latin squares, Zastosowania Matematyki Applicationes Mathematicae XV (1976) 85–102.
[12] J.H. van Lint, Combinatorial Theory Seminar Eindhoven, Lecture Notes in Mathematics, 382 (Springer-Verlag, Berlin, 1974).
[13] W.D. Wallis and R.C. Mullin, Recent advances in complementary and skew Room squares, in: Proc. 4th Southeastern Conf. on Combinatorics, Graph Theory and Computing, Boca Raton, FL (1973) 521–532.
[14] W.D. Wallis, A.P. Street and J.S. Wallis, Combinatorics: Room Squares, Sum-free Sets, Hadamard Matrices, Lecture Notes in Mathematics, 292 (Springer-Verlag, Berlin, 1972).
[15] S.M.P. Wang and R.M. Wilson, A few more squares II, in: Proc. 9th Southeastern Conf. on Combinatorics, Graph Theory and Computing, Boca Raton, FL (1978), to appear.
[16] R.M. Wilson, Cyclotomy and difference families in elementary Abelian groups, J. Number Theory 4 (1972) 17–47.
[17] R.M. Wilson, An existence theory for pairwise designs, I: Composition theorems and morphisms, J. Combin. Theory 13 (1972) 220–245.
[18] R.M. Wilson, An existence theory for pairwise balanced designs, II: The structure of PBD-closed sets and the existence conjectures, J. Combin. Theory 13 (1972) 246–273.
[19] R.M. Wilson, An existence theory for pairwise balanced designs, III: Proof of the existence conjectures, J. Combin. Theory 18 (A) (1975) 71–79.
[20] R.M. Wilson, Concerning the number of mutually orthogonal Latin squares, Discrete Math. 9 (1974) 181–198.
[21] R.M. Wilson, A few more squares, in: Proc. 5th Southeastern Conf. on Combinatorics, Graph Theory and Computing, Boca Raton, FL (1974) 675–680.

Annals of Discrete Mathematics 6 (1980) 275–279

INTERSECTION GRAPHS OF k-UNIFORM HYPERGRAPHS

R. NAIK and S.S. SHRIKHANDE

Department of Mathematics, University of Bombay, Bombay 400 098, India

S.B. RAO

Statistical Mathematical Division, Indian Statistical Institute, Calcutta 700 035, India

N.M. SINGHI

School of Mathematics, Tata Institute of Fundamental Research, Homi Bhabha Road, Bombay 400 005, India and Department of Mathematics, University of Wyoming, Laramie, WY 82071, U.S.A.

Some recent result of authors and others on characterizations of intersection graphs of k-uniform hypergraphs are discussed.

1. Statement of the problem

A k-uniform hypergraph H is a pair (X, E) such that X is a finite set and E is a subset of $\mathbb{P}_k(X)$ the set of all k-subsets of x, where k is an integer $k \geq 2$. Elements of X are called *vertices* while those of E are called the *edges* of H. For $x, y \in X$, by $m(x, y)$ we will denote the number of edges containing both x and y. If $x = y$, $m(x, x)$ is the *degree* of vertex x in the hypergraph. The maximum of $m(x, y)$, $x \neq y$, $x, y \in X$ is called the *multiplicity* of the hypergraph. If $m(x, y) \leq 1$ for all x, y, the hypergraph is said to be *linear*. A 2-uniform linear hypergraph is called a *graph*. For a graph G, the set of vertices is denoted by $V(G)$ and that of edges is denoted by $E(G)$. The *intersection graph* of a hypergraph $H = (X, E)$ denoted by $G(H)$ is the graph whose vertex set $V(G(H)) = E$ and two distinct vertices e, e' are joined in $G(H)$ if and only if they are intersecting edges of H. The intersection graphs of graphs are also called *line graphs*. For various other definitions related to graphs etc. see [3].

We will denote by $I_L(k)$, the set of all graphs which are intersection graphs of some k-uniform hypergraph with multiplicity at most L. Thus $I_1(k)$ is the set of graphs which are intersection graphs of some k-uniform linear hypergraphs. We will write $I(k)$ for the set $\bigcup_{L \geq 0} I_L(k)$.

A family of graphs M is said to be *hereditary* if $G \in M$ implies all vertex induced subgraphs of G are also in M. The families $I_L(k)$'s or $I(k)$'s are clearly hereditary families of graphs. Now let M be a hereditary family of graphs. If

$G \notin M$, then clearly G is not an induced subgraph of any graph in M. A graph $G \notin M$ is said to be a minimal forbidden graph for M if all vertex induced subgraphs of G are in M. Let $F(M)$ denote the family of all minimal forbidden graphs for M. Clearly then M can be characterized by saying that $G \in M$ *if and only if any graph of* $F(M)$ *is not an induced subgraph of* G. The basic problem, with which we are concerned in this paper is the following:

Problem 1. Describe the families $F(I_L(k))$ for all L and k.

A well-known result of Beineke [1] shows that $F(I_1(2))$ consists of nine graphs.

Theorem 1 (Beineke). *A graph G is a line graph if and only if G has no induced subgraph isomorphic to one of the following nine graphs (see Fig. 1).*

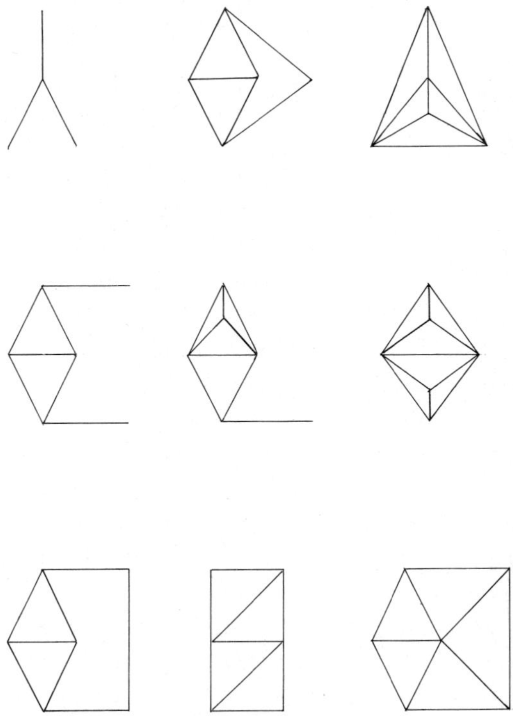

Fig. 1. $F(I_1(2))$.

However, the description for families $F(I_L(k))$, for $k \geqslant 3$ does not seem to be simple. We will give examples in the next section to show that these families are infinite. Then we will state various results recently proved. We will also describe an analogous problem related to eigen values and state a beautiful theorem of Hoffman, which in a way interconnects these two problems.

2. Statement of theorems

Let $G_1(t)$ be a graph obtained by arranging $t+2$ copies of $K_4 - e$, the complete graph on four vertices less an edge e, in the form of a chain and attaching two pendent edges at each of the two degree two vertices as shown in Fig. 2.

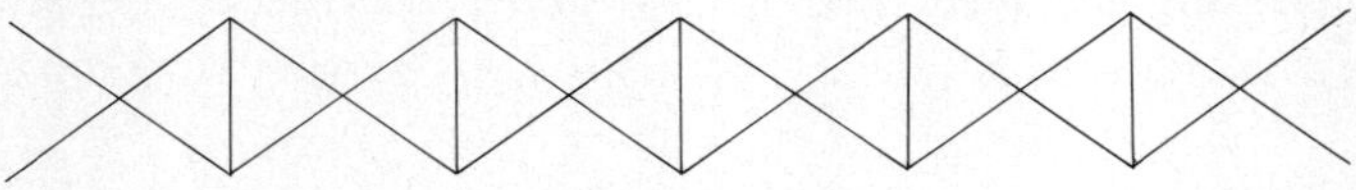

Fig. 2. The graph $G_1(t)$.

Now let $\mathcal{G}_3 = \{G_1(t) \mid t$ is a positive integer$\}$. It is not hard to show that $\mathcal{G}_3 \in F(I_1(3))$. Thus $I_1(3)$ can not be characterized by finitely many forbidden graphs. However, graphs of large minimum degrees in $I_1(3)$ can indeed be characterized by finitely many forbidden subgraphs.

Theorem 2. *There is a finite family F of graphs such that any graph G with minimum degree at least 75 belongs to $I_1(3)$ if and only if G has no induced subgraph isomorphic to a member of F.*

Remark. The graphs in the family F described in the statement of the theorem have at most 64 vertices.

The following corollary is immediate in view of the above remark.

Corollary 1. *If $G \in F(I_1(3))$, then minimum degree of G is at most 75.*

Define inductively a family $\mathcal{G}_k$, $k \geqslant 3$, as follows. $\mathcal{G}_3$ is already defined. For $k > 3$

$$\mathcal{G}_k = \{G \mid G \text{ is obtained by adding a pendent edge at every vertex of degree } k \text{ in } G_1 \text{ where } G_1 \in G_k\}.$$

It can be easily shown that G_k is contained in $F(I_1(k))$. For certain other infinite families of forbidden graphs for $I_L(k)$ or $I(k)$ see [2, 4, 7]. However, all these graphs have small minimum degrees. This raises an interesting question, whether Theorem 2 can be generalized to arbitrary k. We conjecture that this can not be done.

Conjecture 1. *$F(I_1(k))$, $k \geqslant 4$ contains graphs with arbitrarily large minimum degree.*

A *triangle* in a graph is an induced subgraph isomorphic to K_3. The *edge-degree* $\delta(e)$ in G of an edge e of a graph G is the number of triangles of G containing e.
The minimum edge-degree is the minimum of $\delta(e)$, $e \in E(G)$.
The following result has been proved recently.

Theorem 3. *There is a polynomial f of degree at most 3, with property that given any k, there exists a finite family* $\mathbb{F}(k)$ *of graphs such that any graph G with minimum edge degree at least* $f(k)$ *is a member of* $I_1(k)$ *if and only if G has no induced subgraph isomorphic to a member of* $\mathbb{F}(k)$.

The above result has been proved by taking $f(k)=k^3-k$. The family $\mathbb{F}(k)$ contains graphs of order at most $2(k^2-k+1)$. The following corollary is immediate.

Corollary 2. *There exists a polynomial f of degree at most 3 such that for any* $G\in F(I_1(k))$, $k\geq 3$, *minimum edge-degree of G is at most* $f(k)$.

Proofs of Theorems 2 and 3 are quite lengthy, the details of proofs will appear in a paper of authors [7] submitted for publication. We will not try to prove these theorems here. We also omit description of families F or $\mathbb{F}(k)$ (detailed description is given in [7]). However, we will give two simple criteria to determine whether a given graph $G\in I_1(k)$.

A clique in a graph G is set S of vertices such that the induced subgraph $G(S)$ of G with S as vertex set is a complete graph.

Theorem 4. *If* $G=(X, E)$ *is a graph, then* $G\in I_1(k)$ *if and only if in G there exists a set* $S=\{K_1, K_2, \ldots, K_r\}$ *of cliques with* $|K_i|\geq 2$, $1\leq i\leq r$, *such that*

(i) *every vertex of G is in at most k cliques of S*
(ii) *every edge of G is in a unique clique* K_i *of S.*

Theorem 5. *If G is a graph, then* $G\in I_1(k)$ *if and only if G has a set T of triangles satisfying the following two conditions:*

(i) *If abc and abd are in T with* $c\neq d$, *then cd* $\in E(G)$ *and abc, bcd are also in T.*

(ii) *Given any* $k+1$ *distinct edges of G all having a vertex in common, at least two of them are in a triangle which is an element of T.*

The above two results have been used to prove Theorems 2 and 3. Proofs of Theorems 4 and 5 are not difficult. In fact known proofs for $k=2$ can be easily extended to all k. Finally we discuss a related problem of eigen values.

Eigen values of a graph G are the eigen values of its (0, 1) adjacency matrix (see [3]). We will denote by $\alpha(G)$ the minimum eigen value of G. For any real number α define $E_\alpha\{G \mid G \text{ is a graph with } \alpha(G)>\alpha\}$. It can be easily seen that E_α is a hereditary family and that $I_1(k)\subseteq E_\alpha$ for all $\alpha<-k\leq -2$. We can now state the problem.

Problem 2. Describe the sets $F(E_\alpha)$ for all real nos. α.

Since E_α for $\alpha < -k$ is a larger family than $I_1(k)$, the set $F(E_\alpha)$ may have simpler structure than $F(I_1(k))$. This is also suggested by Hoffman's theorem [5, 6]. Hoffman's theorem in fact describes the families E_α in terms of $I_L(k)$'s. We will need a few more definitions before we can state Hoffman's theorem.

Suppose G_1 and G_2 are two graphs on the same set X of vertices. For any $x \in X$, define e_x to be the larger of two numbers {number of edges of G_1 on x which are not in G_2, number of edges of G_2 on x which are not in G_1}. Then $d(G_1, G_2)$ is defined to be $\max_{x \in X} e_x$.

For any family M of graphs and any positive real number l define $M[l]$ by

$$M[l] = \{G \mid G \text{ is a graph such that there exists a graph } G_1 \in M \text{ with } d(G, G_1) < l\}.$$

Finally define H_n to be the graph on $2n+1$ vertices in which one vertex is not joined to n other vertices, otherwise all other pairs of vertices are joined.

We can now state Hoffman's Theorem.

Theorem. *For any infinite family M of graphs the following are equivalent.*

(i) *There exists a real number l such that* $M \subseteq E_l$.

(ii) *There exists an integer* n_0 *such that for all* $G \in M$, K_{1,n_0} *is not an induced subgraph of* G, H_{n_0} *is not an induced subgraph of* G.

(iii) *There exists an integer L such that* $M \subseteq (I_L(L))[L]$.

Hoffman states the theorem a bit differently [5], however, it is not hard to see that his statement (5.3) in [5] is equivalent to statement (iii) given here (this follows by arguments similar to those used to prove Theorem 4 in [7]). In fact, by using Hoffman's ideas in [6], perhaps one may get a complete description of $F(E_l)$'s.

References

[1] L.W. Beineke, Derived graphs and digraphs, in: (H. Sachs et. al., Eds.,) Beitrage zur Graphentheorie (Tuebner, Leipzig, 1968) 17–33.

[2] J.C. Bermond, M.C. Heydman and D. Sottean, Line graphs of hypergraphs I, Discrete Math. 18 (1977) 235–241.

[3] F. Harary, Graph Theory (Addison-Wesley, Reading, MA, 1972).

[4] M.C. Heydemann and D. Sotteau, Line graphs of hypergraphs II, in: Proc. Coil Kesztlely (North-Holland, Amsterdam, 1977).

[5] A.J. Hoffman, Eigen values of graphs, in: D.R. Fulkerson , Ed., Studies in Graph Theory Part II (M.A.A. Publications, 1975) 225–245.

[6] A.J. Hoffman, On Spectrally Bounded Graphs, A Survey of Combinatorial Theory (North-Holland, 1973) Amsterdam, 277–284.

[7] R. Naik, S.B. Rao, S.S. Shrikhande and N.M. Singhi, Intersection graphs of k-uniform linear hypergraphs, submitted for publication.

Annals of Discrete Mathematics 6 (1980) 281–288

ON SOME ENUMERATION PROBLEMS IN GRAPH THEORY

Pushpa N. RATHIE

Institute of Mathematics, Statistics and Computer Science, State University of Campinas, Campinas, S.P., Brazil

In this paper the following enumeration problems in Graph Theory are solved:
(a) enumeration of simple border maps;
(b) enumeration of inequivalent 3-connected Hamiltonian rooted maps;
(c) enumeration of certain rooted triangular maps;
(d) enumeration of rooted 3-connected bicubic maps.

1. Introduction

There have been always difficulties in getting explicit expressions for enumerating various types of maps. In the past it was possible to give closed form expressions for some enumeration problems in Graph Theory using Langrange's expansion method or other methods involving series manipulations. A few such problems tackled by the author can be seen from the articles of Rathie [5, 6] and Rathie and Mathai [7]. Once again we employ the same methods to solve the following problems raised by Tutte [9, 10] and Mullin [2, 3]:

Problem 1 (Mullin [2]). To enumerate simple border maps of type $[n, m]$.

Problem 2 (Tutte [9]). To enumerate inequivalent 3-connected Hamiltonian rooted maps of $2n$ vertices.

Problem 3 (Mullin [3]). To enumerate the triangular maps with n non-root faces and a root face which is a (non-singular) digon.

Problem 4 (Tutte [10]). To enumerate rooted 3-connected bicubic maps of $2n$ vertices.

Results, in each case, can be tabulated by using the recurrence relations obtained. Some closed form expressions, in each case, are also given in particular cases.

2. Enumeration of simple border maps

In this section we enumerate simple border maps. A triangular map or map is called a simple border map if no pair of external vertices is joined by more than one edge. Let $s_{n,m}$ be the number of simple border maps of type $[n, m]$ and let the generating function $S(x, y)$ as the formal power series be

$$S(x, y) = \sum_{n=0}^{\infty} \sum_{m=0}^{\infty} s_{n,m} x^n y^m. \tag{2.1}$$

Since every map of type $[n, m]$ can be obtained uniquely from a simple border map by replacing some of the external edges by lunes (maps of type $[n, -1]$), we have (see [2]):

$$\sum_{n=0}^{\infty} \sum_{m=0}^{\infty} t_{n,m} x^n y^m = \sum_{n=0}^{\infty} \sum_{m=0}^{\infty} s_{n,m} x^n y^m (1+L(x))^{m+3} \tag{2.2}$$

where

$$t_{n,m} = \text{number of rooted triangular maps of type } [n, m]$$
$$= \frac{2^{n+1}(2m+3)!\,(3n+2m+2)!}{(m+1)!^2\, n!\,(2n+2m+4)!}, \quad n \geqslant 0,\ m \geqslant -1,\ (n, m) \neq (0, -1),$$
$$t_{0,-1} = 0, \tag{2.3}$$

and $L(x)$ which enumerates lunes is given by

$$L(x) = \sum_{n=1}^{\infty} t_{n,-1} x^n. \tag{2.4}$$

(2.1) and (2.2) give

$$S(x, y) = (1+L(x))^{-3} \sum_{n=0}^{\infty} \sum_{m=0}^{\infty} t_{n,m} x^n (y/(1+L(x))^m,$$

which on using [2, p. 376 (2.5)] yields

$$S(x, y) = \sum_{m=0}^{\infty} \frac{2^{m+3}(2m+2)!\, y^m}{(m+1)!\,(m+3)!} \{(m+3) - 2(2m+3)xw^3\} w^m (3-w)^{-m-3}, \tag{2.5}$$

where

$$w = 1 + 2xw^3. \tag{2.6}$$

Applying Lagrange's expansion theorem to (2.6), we obtain

$$w^m (3-w)^{-m-3} = \sum_{n=0}^{\infty} A_n x^n, \tag{2.7}$$

where

$$A_0 = 2^{-m-3},$$
$$A_n = \frac{3(3n+m-1)!}{n!\,(m+3)!} \sum_{r=0}^{n-1} \binom{n-1}{r} \times \frac{(m+r+3)!\,(m^2+2mn+mr+2m+3n)}{2^{m+r+4-n}(2n+m+r+1)!}, \quad n \geqslant 1, \tag{2.8}$$

and

$$xw^{m+3}(3-w)^{-m-3}=\sum_{n=0}^{\infty} B_n x^n, \tag{2.9}$$

where

$$\begin{aligned} &B_0=0,\\ &B_1=2^{-m-3},\\ &\cdot\\ &\cdot\\ &\cdot\\ &B_n=\frac{3(3n+m-1)!}{(n-1)!\,(m+2)!}\sum_{r=0}^{n-2}\binom{n-2}{r}\frac{2^{n-m-r-5}(m+r+3)!}{(2n+m+r+1)!}, \quad n\geqslant 2. \end{aligned} \tag{2.10}$$

Hence (2.5), (2.7) and (2.9) yield

$$S(x, y)=\sum_{n=0}^{\infty}\sum_{m=0}^{\infty}\frac{2^{m+3}(2m+2)!}{(m+1)!\,(m+3)!}C_{n,m}x^n y^m, \tag{2.11}$$

where

$$C_{n,m}=(m+3)A_n-2(2m+3)B_n. \tag{2.12}$$

Thus

$$s_{n,m}=\frac{2^{m+3}(2m+2)!}{(m+1)!\,(m+3)!}C_{n,m}, \tag{2.13}$$

where

$$\begin{aligned} &C_{0,m}=2^{-m-3}(m+3), \quad m\geqslant 0,\\ &C_{1,m}=2^{-m-3}(3m^2+8m+3), \quad m\geqslant 0,\\ &\cdot\\ &\cdot\\ &\cdot\\ &C_{n,m}=\frac{3(3n+m-1)!}{n(n-1)(m+2)!}\sum_{r=0}^{n-1}\frac{(m+r+3)!\,\{r(3mn-m+3n)+(n-1)m(m+2)\}}{2^{r+m+4-n}r!\,(n-1-r)!\,(2n+m+r+1)!}\\ &\qquad \text{for} \quad n\geqslant 2, \ m\geqslant 0. \end{aligned} \tag{2.14}$$

Hence (2.13) explicitly gives the number of simple border maps of type $[n, m]$ for all values of n and m. Thus

$$\begin{aligned} S(x, y)=1+2y+5y^2+\cdots+x(1+7y+31y^2+\cdots)\\ +x^2(6+48y+255y^2+\cdots)+x^3(46+398y+2345y^2+\cdots)+\cdots. \end{aligned} \tag{2.15}$$

Some of the closed form expressions for $s_{n,m}$ for $n=0$, 1 2, 3 and for all m are given below:

$$s_{0,m}=(2m+2)!/\{(m+1)!\,(m+2)!\},\tag{2.16}$$

$$s_{1,m}=(2m+2)!\,(3m^2+8m+3)/\{(m+1)!\,(m+3)!\},\tag{2.17}$$

$$s_{2,m}=3(2m+2)!\,(3m^2+9m+4)\{2!\,(m+1)!\,(m+2)!\},\tag{2.18}$$

$$s_{3,m}=(2m+2)!\,(27m^3+270m^2+621m+276)/\{3!\,(m+1)!\,(m+2)!\}.\tag{2.19}$$

From (2.13) and (2.14) it is easy to see that

$$F_n(m)=\{(m+1)!\,(m+3)!/(2m+2)!\}s_{n,m}\tag{2.20}$$

is a polynomial in m of degree $(n+1)$.

3. Enumeration of inequivalent 3-connected Hamiltonian rooted maps

In this section, the problem of finding the number of inequivalent 3-connected Hamiltonian rooted maps of $2n$ vertices is solved. Let

$$v(x)=\sum_{n=1}^{\infty} v_n x^n\tag{3.1}$$

where v_n stands for the number of inequivalent 3-connected Hamiltonian rooted maps of $2n$ vertices. Then (see [9])

$$u(x)=v(x\{1+2u(x)\}^2),\tag{3.2}$$

where

$$u(x)=\sum_{n=1}^{\infty} u_n x^n,\tag{3.3}$$

with

$$u_n=(2n)!\,(2n+2)!/\{2(n!)(n+1)!^2\,(n+2)!\}.\tag{3.4}$$

Writing

$$U(x)=\sum_{n=0}^{\infty} U_n x^n,\quad U_n=2u_n,\ n\geqslant 0,\tag{3.5}$$

$$V(x)=\sum_{m=0}^{\infty} V_m x^m,\quad V_m=\begin{cases}2v_m, & m\geqslant 1,\\ 1, & m=0,\end{cases}\tag{3.6}$$

(3.2) takes the following form

$$U(x)=V\{xU^2(x)\}.\tag{3.7}$$

Applying [8, p. 189] to $U^{2m}(x)$, we have

$$U^{2m}(x)=\sum_{n=0}^{\infty} x^n A_{2m,n}(U_1,U_2,\ldots),\tag{3.8}$$

where

$$A_{2m,n}(U_1, U_2, \ldots) = \sum_{k=1}^{2m} \binom{2m}{k} B_{n,k}(U_1, U_2, \ldots) \tag{3.9}$$

with

$$B_{n,k}(U_1, U_2, \ldots) = \sum_{\pi(n)} \frac{k!}{k_1! \cdots k_n!} U_1^{k_1} \cdots U_n^{k_n}, \tag{3.10}$$

where $\pi(n)$ denotes a partition of n such that

$$\begin{aligned} k &= k_1 + \cdots + k_n, \\ n &= k_1 + 2k_2 + \cdots + nk_n. \end{aligned} \tag{3.11}$$

In particular, (3.9) gives

$$A_{2m,0} = 1, \tag{3.12}$$

$$A_{2m,1} = 4m, \tag{3.13}$$

$$A_{2m,2} = 8m(m+2), \tag{3.14}$$

$$A_{2m,3} = (\tfrac{4}{3}m)(8m^2 + 48m + 79), \tag{3.15}$$

$$A_{2m,4} = (\tfrac{2}{3}m)(16m^3 + 192m^2 + 824m + 1302). \tag{3.16}$$

Hence (3.7) and (3.8) on using [4, p.57]

$$\sum_{u=0}^{\infty} \sum_{v=0}^{\infty} f(u, v) x^{u+v} = \sum_{s=0}^{\infty} \sum_{t=0}^{s} f(t, s-t) x^s \tag{3.17}$$

yield

$$U(x) = \sum_{s=0}^{\infty} \left\{ \sum_{t=0}^{s} V_t A_{2t,s-t}(U_1, U_2, \ldots) \right\} x^s. \tag{3.18}$$

Comparing the coefficients of x^s in (3.18) and using (3.12), we get

$$V_s = U_s - \sum_{t=0}^{s-1} V_t A_{2t,s-t}(U_1, U_2, \ldots). \tag{3.19}$$

Thus

$$\begin{aligned} v(x) = {} & x + x^2 + 3x^3 + 14x^4 + 80x^5 + 518x^6 + 3647x^7 + 27274x^8 \\ & + 213480x^9 + \cdots. \end{aligned} \tag{3.20}$$

4. Enumeration of triangular maps

In this section we will determine M_n, the number of triangular maps with n non-root faces and a root face which is a (non-singular) digon. Let

$$M(x) = \sum_{n=1}^{\infty} M_n x^n \tag{4.1}$$

then from [3] we find that

$$M(x)=(1-r)(1+8r)^{\frac{1}{2}} \tag{4.2}$$

with

$$r=x^2(1+8r)^{\frac{3}{2}}. \tag{4.3}$$

Applying Lagrange's expansion theorem (see [8, p. 146]) to (4.3), we have

$$\begin{aligned} M(x) &= \sum_{n=1}^{\infty} (x^n/n!) \frac{\mathrm{d}^{n-1}}{\mathrm{d}r^{n-1}} \left\{ \frac{\mathrm{d}M(x)}{\mathrm{d}r} (1+8r)^{3n/2} \right\} \bigg|_{r=0} \\ &= \sum_{n=1}^{\infty} \frac{3\Gamma\{\frac{1}{2}(3n+1)\}}{4\Gamma\{\frac{1}{2}(n+5)\} n!} (8x)^n. \end{aligned} \tag{4.3}$$

Thus

$$M_n = 3 \cdot 2^{3n-2} \Gamma\{\tfrac{1}{2}(3n+1)\}/[\Gamma\{\tfrac{1}{2}(n+5)\} n!]. \tag{4.4}$$

Hence

$$M(x)=3x+24x^2+256x^3+3168x^4+\cdots. \tag{4.5}$$

5. Enumeration of rooted 3-connected bicubic maps

In this section we deal with the problem of enumerating rooted 3-connected bicubic maps of $2n$ vertices posed by Tutte [10]. The problem was tackled by Mullin [1] in a different manner. Let

$$G(x)=\sum_{m=1}^{\infty} g_m x^m \tag{5.1}$$

be the generating function for the enumeration of g_m, the rooted 3-connected bicubic maps of $2m$ vertices. Then (see [10, p. 270])

$$F(x)=G[x\{1+F(x)\}^3] \tag{5.2}$$

where

$$F(x)=\sum_{n=1}^{\infty} f_n x^n \tag{5.3}$$

with

$$f_n=\frac{3(2n-1)!\, 2^n}{(n-1)!\,(n+2)!}, \quad n \geqslant 1. \tag{5.4}$$

Applying [8, p.189] to $(1+F(x))^{3m}$, we get

$$(1+F(x))^{3m}=\sum_{n=0}^{\infty} x^n A_{3m,n}(f_1, f_2, \ldots) \tag{5.5}$$

where

$$A_{3m,n}(f_1, f_2, \ldots) = \sum_{k=1}^{3m} \binom{3m}{k} B_{n,k}(f_1, f_2, \ldots) \tag{5.6}$$

with

$$B_{n,k}(f_1, f_2, \ldots) = \sum_{\pi(n)} \frac{k!}{k_1! \cdots k_n!} f_1^{k_1} \cdots f_n^{k_n} \tag{5.7}$$

where $\pi(n)$ denotes a partition of n such that

$$\begin{aligned} k &= k_1 + \cdots + k_n, \\ n &= k_1 + 2k_2 + \cdots + nk_n. \end{aligned} \tag{5.8}$$

In particular, we have

$$A_{3m,0} = 1, \tag{5.9}$$

$$A_{3m,1} = 3m, \tag{5.10}$$

$$A_{3m,2} = 3m(3m+5)/2!, \tag{5.11}$$

$$A_{3m,3} = 3m(9m^2+45m+56)/3!, \tag{5.12}$$

$$A_{3m,4} = 3m(27m^3+270m^2+897m+1014)/4!. \tag{5.13}$$

Hence (5.2) and (5.5) on using (3.17) yield

$$\begin{aligned} F(x) &= \sum_{m=0}^{\infty} \sum_{n=0}^{\infty} g_m A_{3m,n}(f_1, f_2, \ldots) x^{m+n} \\ &= \sum_{u=0}^{\infty} \sum_{v=0}^{u} g_v A_{3v,u-v}(f_1, f_2, \ldots) x^u, \end{aligned} \tag{5.14}$$

where $g_0 = 0$.

Hence equating the coeffients of x^u in (5.14) and using (5.9), we have

$$g_u = f_u - \sum_{v=1}^{u-1} g_v A_{3v,u-v}, \quad u \geq 1. \tag{5.15}$$

The expression (5.15) is a recurrence relation giving g_u for various values of u. Thus

$$G(x) = x + x^4 + 3x^6 + 7x^7 + 15x^8 + 63x^9 + \cdots. \tag{5.16}$$

References

[1] R.C. Mullin, Thesis, University of Waterloo, Waterloo, Ont. (1963).
[2] R.C. Mullin, On counting rooted triangulations, Can. J. Math. 17 (1965) 373–382.
[3] R.C. Mullin, On the average number of trees in certain maps, Can. J. Math. 18 (1966) 33–41.
[4] E.D. Rainville, Special Functions (Chelsea Publishing Co., New York, (1960).
[5] P.N. Rathie, The enumeration of Hamiltonian polygons in rooted planar triangulations, Discrete Math. 6 (1973) 163–168.

[6] P.N. Rathie, A census of simple planar triangulations, J. Combin. Theory 16 (1974) 134–138.
[7] P.N. Rathie and A.M. Mathai, Enumeration of almost cubic maps, J. Combin. Theory 13 (1972) 83–90.
[8] J. Riordan, Combinatorial Identities (Wiley, New York, 1968).
[9] W.T. Tutte, A census of Hamiltonian polygons, Can. J. Math. 14 (1962) 402–417.
[10] W.T. Tutte, A census of planar maps, Can. J. Math. 15 (1963) 249–271.

Annals of Discrete Mathematics 6 (1980) 289–291

SOME REMARKS ON AMICABLE ORTHOGONAL DESIGNS

Jennifer SEBERRY

Applied Mathematics Department, University of Sydney, Sydney, N.S.W. 2006 (Australia)

We present some new results on amicable orthogonal designs. We obtain amicable Hadamard matrices of order $2^4 \cdot 211$ and a skew Hadamard matrix of order $2^4 \cdot 295$ which were previously not known.

1. Introduction

For all definitions we refer the reader to the book of Geramita and Seberry [1].

Our aim in writing this note has been to try to fill gaps in the tables of [2]. A greater knowledge of amicable orthogonal designs in small orders would have been useful. In fact, little is known for orders >8.

2. A note on amicable orthogonal designs

Theorem 1. *There are amicable orthogonal designs of types*

(a) $((1, 1, 2, 4, \ldots, 2^{t-1}); (2^t))$,

(b) $((1, 2, 4, \ldots, 2^{t-1}); (1, 2, 4, \ldots, 2^{t-1}))$,

(c) $((1, 1, 2, 4, \ldots, 2^{t-2}, 2^{t-1}-1); (2^{t-1}-1, 2^{t-1}))$,

(d) $((1, 1, 2, 4, \ldots, 2^{t-2}); (1, 2, 4, \ldots, 2^{t-2}, 2^{t-1}))$

in order 2^t.

Proof. In Geramita and Seberry [1, Corollary 5.128], it is established that for order 2^t, there exist matrices P, Q, H such that P and Q are skew-symmetric orthogonal designs of type $(1, 2, 4, \ldots, 2^{t-1})$, and H is a symmetric orthogonal design of type (2^t), and such that $PQ^T = QP^T$, $PH^T = HP^T$, and $QH^T = HQ^T$.

Then the required matrices are

(a) P and H,

(b) P and Q,

(c) $\begin{bmatrix} P+xI & Q \\ Q & -P+xI \end{bmatrix}, \begin{bmatrix} H & Q \\ -Q & H \end{bmatrix}$ (using the matrices from order 2^{t-1}),

(d) $\begin{bmatrix} P+xI & 0 \\ 0 & -P+xI \end{bmatrix}, \begin{bmatrix} H & Q \\ -Q & H \end{bmatrix}.$

Corollary 2. *All amicable orthogonal designs of types* $((1, a); (b))$, *where* $1 \leqslant a \leqslant 2^{t-1}$ *and* $1 \leqslant b \leqslant 2^t$, *exist in order* 2^t. *All amicable orthogonal designs of types* $((1, e); (b))$, *where e is even, and* $((1, c); (d))$ *where* $d \in \{2^{t-2}, 2^{t-1}-1, 2^{t-1}, 2^{t-1}+2^{t-2}, 2^t-1, 2^t\}$ *and* $c \in \{1, 2, \ldots, 2^t-2\}$, *exist in order* 2^t. *In addition, design of type* $((1, 2^t-1); (f))$, *where* $f \in \{2^{t-2}, 2^{t-1}, 2^{t-1}+2^{t-2}, 2^t\}$ *exist in order* 2^t.

3. Two constructions for amicable Hadamard matrices

Theorem 3. *Suppose there exist amicable Hadamard matrices of order b of the form*

$$X = \begin{bmatrix} x & 1 \cdots 1 \\ - & \\ \cdot & \\ \cdot & xI+P \\ \cdot & \\ - & \end{bmatrix}, \qquad Y = \begin{bmatrix} 1 & 1 \cdots 1 \\ 1 & \\ \cdot & \\ \cdot & D \\ \cdot & \\ 1 & \end{bmatrix}.$$

Further, suppose there are amicable orthogonal designs of types $((1, a, ab-a-1); (a, ab-a))$ *in order ab. Then there exist amicable Hadamard matrices of order* $ab(b-1)$.

Proof. Since $XY^T = YX^T$ we have $PJ=0$, $DJ=-J$, $D^T=D$, $P^T=-P$ and $PD^T = DP^T$. We replace the variables of the amicable orthogonal designs of types $((1, a, ab-a-1); (a, ab-a))$ in order ab by $xI+P$, J, D, J and D respectively, then since

$$DD^T = bI-J \quad \text{and} \quad PP^T = (b-1)I-J$$

and the matrices pairwise satisfy $AB^T = BA^T$ we have amicable orthogonal designs of types $((1, ab(b-1)-1); (ab(b-1)))$ in order $ab(b-1)$. We use [1, Theorem 5.43] now to obtain the required structure of the amicable Hadamard matrices.

We use the fact that amicable orthogonal designs of types $((1, p); (1, p))$ exist in order $p+1$, $p \equiv 3 \pmod 4$ a prime power to obtain the next result.

Corollary 4. *Let* $p \equiv 3 \pmod 4$ *be a prime power. If there exist amicable orthogonal designs of types* $((1, a, ap-1); (a, ap))$ *in order* $a(p+1)$ *there exist amicable Hadamard matrices of order* $ap(p+1)$.

Example. If there exist amicable orthogonal designs of types $((1, a, 3a-1); (a, 3a))$ in order $4a$ there exist amicable Hadamard matrices of order $12a$.

Comment. At this stage more knowledge of the existence of amicable orthogonal designs would be invaluable.

We obtain one more new construction by observing that if there are amicable orthogonal designs of types $((a, b); (a+b))$ in order $a+b$, then there exists amicable orthogonal designs of types $((a, b, a+b); (a+b, a+b))$ in order $2(a+b)$. This is now used with [2, Theorem 5.85] to see

Theorem 5. *Let* $p \equiv 3 \pmod 4$ *be a prime power. Let* $q \equiv 5 \pmod 8$ *be a prime power with proper representation* $q = s^2 + 4(2p+1)^2$. *Then there exist amicable Hadamard matrices of order* $2(p+1)(q+1)$.

Proof. Since $p \equiv 3 \pmod 4$ is a prime power, there are amicable orthogonal designs of types $((1, p); (1, p))$ in order $p+1$ [1, Theorem 5.52] and types $((1, p, p+1); (p+1, p+1))$ in order $2(p+1)$. Now using $|t| = 2p+1$ in [1, Theorem 5.85] we have the result.

Example. With $q = 421 = 15^2 + 4 \cdot 49^2$ we see there are amicable Hadamard matrices of order $2^4 \cdot 211$ which is new. This construction also gives amicable Hadamard matrices of orders $2^4 \cdot 279$ and $2^4 \cdot 411$.

4. A result on skew Hadamard matrices

Using the fact that amicable Hadamard matrices exist for orders which are powers of two and [3, Theorem 4.20, p. 337] we have

Lemma 6. *Let n be the order of amicable Hadamard matrices. Suppose there exists a skew Hadamard matrix of order* $(2^t - 1)n$. *Then there exists a skew Hadamard matrix of order* $2^t n(2^t n - 1)(2^t - 1)$. *In particular, if there are amicable Hadamard matrices of order n there is a skew Hadamard matrix of order* $2n(2n-1)$.

Example. $m = 4$ and $n = 20$ in the corollary gives a skew Hadamard matrix of order $2^4 \cdot 295$. $n = 36$ and $n = 44$ in the last part of the corollary give skew Hadamard matrices of orders $2^3 \cdot 639$ and $2^3 \cdot 957$. This improves a result of [2].

References

[1] A.V. Geramita and J. Seberry, Orthogonal Designs: Quadratic Forms and Hadamard matrices (Marcel Dekker, New York, 1979).
[2] J. Seberry, On skew Hadmard matrices, Ars Combinatoria 6 (1978) 255–275.
[3] J. Seberry Wallis, Hadamard matrices, in: Combinatorics: Room Squares, Sum Free Sets, Hadamard Matrices, W. D. Wallis, A. Penfold Street and J. Seberry Wallis, Lecture Notes in Mathematics, 292 (Springer-Verlag, Berlin, 1972).

Annals of Discrete Mathematics 6 (1980) 293–300

ON THE NUMBER OF NONISOMORPHIC DESIGNS IN A REPEATED MEASUREMENT EXPERIMENT

Esther SEIDEN*

Hebrew University, Givat-Ram, Jerusalem, Israel

Suppose that an experimenter is interested in applying successively n distinct treatments to each experimental unit in a case when the effects are permanent. To examine all $2^n - 1$ treatment combinations a minimum of $\binom{n}{n/2}$ experimental units is both required and sufficient. It is shown here that for $n = 5$ there are 304 nonisomorphic solutions achieving this goal.

1. Summary

The problem considered here is a special case of the following. An experimenter is interested in applying successively n distinct treatments to each experimental unit. At the tth stage of the experiment ($t \leq n$), he wishes to compare the effects of all $\binom{n}{t}$ treatment combinations. It is assumed that the resulting effect of t treatments applied at state t does not depend on the order in which they were applied. Clearly such an experiment requires at least $\max_k \binom{n}{k}$ experimental units for any n. Hence the minimum number of experimental units will equal $\binom{n}{(n-1)/2} = \binom{n}{(n+1)/2}$ for n odd and $\binom{n}{n/2}$ for n even. It can be shown using well-known theorems on systems of distinct representatives that such a design can in fact be constructed for any n. The problem arises as to how many distinct ways such a construction can be carried out. For n odd, the crucial point of construction is how to proceed from the stage $\frac{1}{2}(n-1)$ to the stage $\frac{1}{2}(n+1)$. This problem is very complex. It is shown here that for $n = 5$ there are 304 nonisomorphic designs. Two designs are considered isomorphic when they can be obtained from each other by relabeling some of the treatments.

2. Introduction

This research was stimulated by a medical problem which was considered in the Statistical Laboratories of Michigan State University. An orthopedic surgeon was concerned with statistical evaluation of resultant laxity which is caused by each of the possible combinations of ligamenteneous injuries of a knee. Since the experimental units were human cadaver joints it was decided that the number of them be as small as possible.

*Research supported by NSF Grant MCS77-26366.

Similar situations may arise in other fields of application such as education agriculture and others. Suppose for example that students satisfied prerequisites for n future courses which are unrelated in the sense that none of them is a prerequisite for another. Nevertheless each course is beneficial for consecutive studies and makes a specific contribution to the success of the student in his further studies. At each stage of the experiment it is desirable to compare the combined effect of all possible courses taken by the students up to and including this stage and to investigate which sequence should be preferred, if any.

In many fields of applications the major problem to be solved by the experimenter could be how to formulate meaningful commensurate measures for the effects of each of the treatments under consideration. It seems that the prospects of applying such types of designs to operations research are from this standpoint more promising. Suppose that some machine has n components and the failure of each of them will effect the performance of the machine in some specific way which does not depend on the performance of the remaining components. Also the effect of failure of any subset of the components depends solely on the structure of the subset. The experimenter is interested in finding a scheme for the order of checking the components to insure the maximum reliability of the performance of the machine. He will make the decision after investigating the performance of $\max_k \binom{n}{k}$ machines in which all n components will be checked in accordance with a type of the design discussed here. In this area of research it seems natural to introduce also a cost function and a probability of failure for each component. Such ramifications may influence the choice among the designs described here.

The problem discussed here is in how many distinct ways can one design an experiment addressing itself to the situations discussed here when the number of treatments is equal to 5. Since $\binom{5}{2} = \binom{5}{3} = 10$ is the maximum value of $\binom{5}{t}$, $t \leqslant 5$ it is clear that at least 10 experimental units are required.

The first stage of the experiment presents no difficulty. One should clearly apply at random each treatment to two experimental units. At the second stage of the experiment the aim is to add one more treatment to each of the experimental units in such a way that each pair of treatments appears once. At the third stage the problem will be to apply another treatment to each unit in such a way that all possible triples are obtained. The remaining pair of treatments to be applied to each unit at the fourth and fifth stage of the experiment is determined by the previously applied three treatments since ultimately all five treatments have to be applied to each of the units. However, the order of each pair has to be determined in such a way that in the fourth as well as in the fifth stage each individual treatment is applied twice.

3. Main result

For $n = 5$, the number of nonisomorphic designs is equal to 304.

To prove this result we start with enumerating the number of distinct ways one

(Table III).

Table III

Element of the third column	Representatives	
1	23	24
2	34	35
3	14	45
4	15	25
5	12	13

The three patterns have distinct groups of automorphisms whose generators are listed below.

Pattern	Generators of automorphism
I	(12453) (2534)
II	(2534)
III	(12345)

We shall obtain next the number of equivalence classes of the designs at the third stage for each of the patterns and compute their automorphisms.

Pattern I gives two equivalence classes. The representatives are obtainable by choosing two proper sets out of the twelve distinct sets of pairs and attaching the third element in accordance with this pattern.

We represent two representatives. They are

α: 125 134 412 513 231 243 524 345 352 451
β: 125 314 142 513 321 243 254 435 532 451

The automorphism group of design α is the identity. The automorphisms of the design β have the generator (12453). Hence α represents 20 designs and β four designs and all designs are accounted for.

Pattern II gives six equivalence classes. Proceeding as before we list again a representative for each class. They are

α	β	γ	δ	γ'	δ'
125	125	125	125	215	215
134	314	314	314	134	134
412	142	142	142	412	412
513	513	513	513	153	153
231	231	321	231	231	321
245	245	245	425	425	245
523	523	253	253	523	523
342	432	432	432	342	342
354	354	534	354	354	534
451	451	451	541	541	451

As seen γ' and δ' are obtainable from γ, δ, respectively, by interchanging the first and second column. The automorphism of each of these representatives is the identity. Hence each represents four designs.

Pattern III gives eight equivalence classes whose representatives are presented.

α	β	α'	β'	γ	δ	γ'	δ'
125	125	215	215	125	125	215	215
135	315	315	135	135	135	315	315
413	143	143	413	413	413	143	143
514	514	154	154	514	514	154	154
231	231	321	321	321	321	231	231
241	421	421	241	241	241	421	421
524	254	254	524	254	254	524	524
342	342	432	432	342	432	432	342
352	532	532	352	532	352	352	532
453	453	543	543	453	543	543	453

Here again α', β', γ', δ' are obtained from α, β, γ, δ by switching the first and second columns. Designs α β α' and β' have the automorphism (12345) while γ δ, γ' and δ' the identity. Hence α β, α' and β' represent just themselves while γ δ, γ' and δ' represent each an equivalence class of five designs.

To complete each design we have to add the fourth and fifth column. Because of the symmetry of the design each element of the third column can be considered also represented by two pairs of the fourth and fifth column. It is easy to see that the three representative patterns discussed before determine the three representative patterns corresponding to the fourth and fifth column. Moreover pattern I will be the same as before. Pattern II determines the corresponding pattern for the last two columns to be:

II′		
1	23	45
2	35	15
3	24	14
4	25	12
5	34	13

(The isomorphism of II and II′ is established through the permutation (45)).

Corresponding to pattern III we shall obtain now

III′		
1	45	35
2	15	14
3	25	12
4	23	13
5	34	24

(The isomorphism of III and III′ is obtained through the permutation (25)(43)).

Main result

The total number of designs corresponding to patterns I, II and III respectively is equal to 32, 144 and 128. Hence the total number of nonisomorphic designs is equal 304.

Within each pattern there are 576 cases to be considered. The pattern fixes the third column. The 576 from a direct product of 24×24 distinct sets of 10 pairs of the first two and last two columns.

The main tool for counting is a well-known theorem in group theory. To facilitate the reading of the paper we shall state this theorem as formulated in [2, p. 45].

Theorem. *Let X be a finite set and let G be a group of permutations on X. If N is the number of orbits of G, then*

$$N = \frac{1}{|G|} \sum_{t \in G} F(t)$$

where F(t) is the number of x in X that are fixed by t.

Thus to count the number of orbits within each pattern we have to determine the numbers $F(t)$ for each t belonging to the automorphism group of the corresponding pattern. For each pattern we have to compute the number of designs within and between each of its equivalence classes.

We start now counting the number of orbits for pattern I. Here α represents 20 subdesigns and β 4. Hence the 576 complete designs split into four disjoint sets of types (α, α), (α, β), (β, α), (β, β). Their cardinalities are respectively 400, 80, 80, and 16. The first three sets have the automorphism the identity while the last set has the automorphism the subgroup generated by (12453). Hence

$$N_{\mathrm{I}} = \frac{1}{20}(400 + 80 + 80 + 5 \times 16) = \frac{640}{20} = 32$$

where N_{I} denotes the number of orbits within pattern I. Analogous calculations yield

$$N_{\mathrm{II}} = \frac{576}{4} = 144,$$

$$N_{\mathrm{III}} = \frac{1}{5}(400 + 80 + 80 + 5 \times 16) = 128.$$

As seen from the calculations there are 32 designs which have an automorphism group of order five. The remaining designs have the automorphism the identity.

Research is in progress to construct suitable linear models for these designs which make use of their specific structure.

It may be worthwhile noticing that all the designs considered here are regular GYD and hence, universally optimal if considered as designs for two way classification with no interaction.

Acknowledgements

We wish to acknowledge fruitful conversation on the problem of enumerating designs investigated here with Dr. K.B. Gross. Gross noticed that there seem to be just three ways of representing triples by pairs. He lost interest in the problem and did not want to continue research in this subject.

We also want to thank Dr. J. Stapleton of Michigan State University for pointing out to the author the problem which stimulated this research.

References

[1] P.J. Laycock and E. Seiden, On a problem of repeated measurement design with treatment additivity, RM 379 MSU (1978).

[2] J.J. Rotman, The Theory of Groups (Allyn and Bacon, Inc., Boston, MA, 1973).

Annals of Discrete Mathematics 6 (1980) 301–313

INTERSECTION THEOREMS ON STRUCTURES

Miklós SIMONOVITS and Vera T. SÓS
Eötvös Loránd University, Budapest, Hungary

1. Introduction

The basic topics of this survey paper are intersection theorems on graphs and on integers. Before turning to these topics let us mention a few things in connection with combinatorial intersection theorems.

The most elementary intersection theorems are the ones on sets. In these cases we fix a set S and a family $\mathcal{A}=\{A_1,\ldots,A_N\}$ of subsets of S, and assume that the sets $A_1,\ldots,A_N$ have some intersection property P. Then we ask for the maximum N in terms of $|S|$ or other parameters, specified in P. More generally instead of an intersection property one can consider any Boole-algebraic property (involving intersection, union, disjointness, complement, containment, rank or size) and ask for maximal or minimal sized families of subsets satisfying the given conditions.

Perhaps the first intersection-type theorem is the Fisher-inequality [11], [14] (which originated from a statistical-block-design-problem) and asserts — in a dual form — that if $A_1,\ldots,A_N$ are k-element subsets of a given n-element set and $|A_i\cap A_j|=\lambda$ for $1\leq i<j\leq N$, then $N\leq n$.

Later de Bruijn and Erdős [2] for $\lambda=1$ and Ryser for arbitrary λ proved the following

Theorem. *Let $\mathcal{A}=\{A_1,\ldots,A_N\}$ be a family of subsets of an n-element set S. If*

$$|A_i\cap A_j|=\lambda \quad \textit{for} \quad 1\leq i<j\leq N,$$

then $N\leq n$.

(The difference between this theorem and the Fisher inequality is that the condition $|A_i|=k$ is dropped.)

The extremal systems are known only for $\lambda=1$ and are isomorphic to one of the following three:

(a) $\{1\}, \{1, 2\},\ldots,\{1, n\}$,

(b) $\{2,\ldots,n\}, \{1, 2\},\ldots,\{1, n\}$,

(c) the lines of a finite geometry on n elements when such a geometry exists at all.

Another old, well-known intersection theorem (published only in 1961) is the Erdős–Ko–Rado theorem [8]:

Theorem. *Let* $1 \leq l \leq \frac{1}{2}n$. *If* $\mathscr{A} = \{A_1, \ldots, A_N\}$ *is a family of subsets of an* n-*element set* S *and* $|A_i| = l$ $(1 \leq i \leq N)$, $A_i \cap A_j \neq \emptyset$ $(1 \leq i < i \leq N)$, *then*

$$N \leq \binom{n-1}{l-1}. \tag{1}$$

Again, one can easily see that this theorem is sharp: Fixing an $x_0 \in S$ and taking all the l-element subsets $A_1, \ldots, A_N$ containing x_0 $(N = \binom{n-1}{l-1})$ we get a system such that $x_0 \in A_i \cap A_j$ $(i \neq j)$ and hence (1) is sharp. A more general question is the following one (see [23]). Let $\mathscr{L}_1$ and $\mathscr{L}_2$ be two given sets of integers and $A_1, \ldots, A_N$ be subsets of $\{1, \ldots, n\}$ satisfying

(a) $|A_i \cap A_j| \in \mathscr{L}_1$ for $1 \leq i < j \leq N$,

(b) $|A_i| \in \mathscr{L}_2$ for $1 \leq i \leq N$.

How large can N be under this condition (for fixed $\mathscr{L}_1$, $\mathscr{L}_2$ and n)? This question has been widely investigated recently and many interesting results have been proved in this area e.g. by Ray-Chaudhuri and Wilson [17], Deza and Frankl [6], Deza and Singhi [5].

The purpose of this paper is not to give a survey on intersection theorems on sets. The two theorems above are just two important examples. The reader interested in the details is referred e.g. to the survey papers of Erdős and Kleitman [9], Katona [15] or the "open problem" paper [7] of Erdős. We mention only one more well-known theorem thich is also a characteristic and initiating theorem of this field; Sperner's Theorem [24].

Theorem. *Let* $|S| = n$ *and* $A_i \subset S$ *for* $1 \leq i \leq N$ *be such that* $A_i \nsubseteq A_j$ *for* $1 \leq i < j \leq N$. *Then*

$$N \leq \binom{n}{[\frac{1}{2}n]}.$$

This theorem is sharp: the family of subsets of $[\frac{1}{2}n]$ elements of S satisfy the condition $A_i \nsubseteq A_j$ and their number is

$$\binom{n}{[\frac{1}{2}n]}.$$

The above examples concerned two elements of the set-system. This is not necessarily so. A typical result concerning more than two A_i's is the following (Erdős and Kleitman [10]):

Let $A_1, \ldots, A_N$ be a family of subsets of an n-element set S and assume that

$A_i \cup A_j = A_k$, $A_i \cap A_j = A_l$ is excluded (except if $i=j=k=l$). Then

$$N = O\left(\frac{2^n}{n^{1/4}}\right)$$

and this estimate is sharp.

Here we give a brief survey on intersection theorems of structural type, where the structures will be either graphs or subsets of integers and the intersection properties will be given in graph theoretical or arithmetical terms.

Before turning to these questions in details we remark that another type of intersection theorems on structures was considered by Deza and Frankl [6], namely, "intersection theorems" on permutations.

The basic problem considered by them is the following one:

Problem. Let $P_1, \ldots, P_N$ be permutations on $\{1, 2, \ldots, n\}$. The distance of P_i and P_j: $d(P_i, P_j)$ is the number of non fixed elements of $P_i^{-1}P_j$, that is, the number of elements on which P_i and P_j act differently. How large can N be if

(a) $\quad d(P_i, P_j) \leq \lambda \quad (1 \leq i < j \leq N)$,

or if

(b) $\quad d(P_i, P_j) \geq \lambda \quad (1 \leq i < j \leq N)$.

Some of their results were published in [6], some others can be found in this volume.

2. Graph intersection problems, introductory examples

In this chapter we shall consider graphs without loops or multiple edges.

Definition. If G and H are graphs on the same vertex set V, their intersection $G \cap H$ is the graph whose edge-set $E(G \cap H) = E(G) \cap E(H)$ and whose vertices are the elements of V incident to some edges in $E(G \cap H)$.

Given a family $\mathcal{L}$ of graphs, $f(n; \mathcal{L})$ is the maximum number of graphs $G_1, \ldots, G_N$ defined on the same n-element vertex-set V for which

$$G_i \cap G_j \in \mathcal{L} \quad (1 \leq i < j \leq N).$$

Remark 2.1. Observe that $G_i \cap G_j$ has no isolated vertices.

Below we shall consider some special cases of the general problem. In Example 2.3 $\mathcal{L}$ is the family of stars and $f(n, \mathcal{L}) = 2^{n-1}$, in Theorem 4.1 $\mathcal{L}$ is the family of paths and $f(n, \mathcal{L}) = o(n^5)$, that is, much smaller. One could ask: both the path and the star are trees, why is $f(n, \mathcal{L})$ "very large" in the first case and "very small" in the second one. This is, how we arrived at the following non-trivial (and perhaps too general?) problem.

Problem 2.1. Under which conditions on $\mathcal{L}$ is $f(n, \mathcal{L})$ polynomially bounded (i.e., when do we know that there is an r such that $f(n, \mathcal{L}) = o(n^r)$?

Theorem 2.1. *Let $\mathcal{L}$ be the family of graphs isomorphic to some of k given graphs $L_1, \ldots, L_k$. Then*

$$f(n, \mathcal{L}) = o(n^{v+2})$$

for $v = \max_{1 \leq i \leq k} v(L_i)$ (where $v(G)$ denotes the number of vertices of G).

Theorem 2.1 is very easy to prove. Its main content is that for finite families $\mathcal{L}$ $f(n, \mathcal{L})$ is always polynomially bounded. From now on we shall restrict our investigations to infinite $\mathcal{L}$'s.

Theorem 2.2 [19]. *Let $\mathcal{L}$ be a family of graphs with minimum valency ≥ 2 and maximum valency $\leq K$, for which the number of components and the number of vertices of valency $\neq 2$ are also $\leq K$. There exists an $r = r_K$ such that*

$$f(n, \mathcal{L}) = o(n^r).$$

Remark. Another way of formulating the condition of this theorem is to assume the existence of a finite family $\{L_1, \ldots, L_k\}$ of graphs with minimum valence ≥ 2 such that the family of graphs topologically equivalent with some of $L_1, \ldots, L_k$ forms $\mathcal{L}$. By the way, Theorem 2.2 is sharp in more than one way.

(a) The condition on the boundedness of the number of components in $\mathcal{L}$ is necessary: let $\mathcal{L}$ be the family of graphs consisting of vertex disjoint K_3's. All the other conditions of Theorem 2.2 are satisfied, still $f(n, \mathcal{L}) \geq 2^{[n/3]}$. Indeed, let us fix $[\frac{1}{3}n]$ vertex-disjoint triangles on an n-element set V and let $G_1, \ldots, G_N$ ($N = 2^{[n/3]}$) be the family of all graphs on V containing all the edges of some of the considered triangles and none edges of the other ones. Clearly, $G_i \cap G_j \in \mathcal{L}$.

(b) The condition on the maximum valency is also necessary: let $K_2(p;\, q)$ denote the complete bipartite graph of p and $q \geq p$ vertice in its classes. Let $\mathcal{L} = \{K_2(2, q);\, q = 2, 3, 4, \ldots\}$. We assert that

$$f(n; \mathcal{L}) \geq 2^{n-4}.$$

Indeed, let $V = \{1, \ldots, n\}$ and $G_1, \ldots, G_N$ ($N = 2^{n-4}$) be the family of $K_2(2, q)$'s the first class of which is $\{1, 2\}$, the second one is in $\{3, \ldots, n\}$ and it contains $\{3, 4\}$. Clearly, we have 2^{n-4} such graphs and $G_i \cap G_j \in \mathcal{L}$. Thus the condition on the maximum valency is really necessary.

(c) The condition on the number of vertices of valence $\neq 2$ is also necessary: let $\mathcal{L}$ be the family for all the graphs which have only valences 2 and 3.

Let G be a cubic graph on n vertices with a one-factor P. For each $P' \subseteq P$ we form $G_{P'} =: G - P'$. Any two of these $2^{[n/2]}$ graphs intersect in a graph $\in \mathscr{L}$. Thus the bound on the number of vertices of valence $\neq 2$ is also needed.

The examples above show not only that Theorem 2.2 is sharp. They show that $f(n, \mathscr{L})$ is very often exponentially large. We give a few more illustrations of that.

Example 2.1. Let $\mathscr{L}$ be the family of all complete graphs, $K_0 = \emptyset$ *included.* It is easy to see that

$$f(n, \mathscr{L}) = 2^n \quad (n > n_0).$$

Example 2.2. Let $\mathscr{L}$ be the family of all complete graphs, $K_0 = \emptyset$ *excluded.* Obviously,

$$f(n, \mathscr{L}) \geq 2^{n-2}$$

(which can be achieved e.g. by fixing 2 vertices x, y and taking all the complete graphs containing (x, y)).

Example 2.3. Let $\mathscr{L}$ be the family of stars:

$$\mathscr{L} = \{K_2(1, q): q = 0, 1, 2, \ldots\}.$$

Clearly,

$$f(n, \mathscr{L}) = 2^{n-1} \quad (n \geq n_0)$$

and the only extremal system can be described as follows: We fix a $G = K_2(1, n-1)$ and each G_i $(i = 1, \ldots, N)$ has the same n vertices as G and some edges of G. This example is very similar to that of (b) in the above remark, still it is important since it shows that if we assume that the intersection of any two graphs is a *tree*, the extremal system will be exponentially large.

Example 2.4. Let $\mathscr{L}$ be the family of all the connected graphs. Then

$$f(n, \mathscr{L}) \geq 2^{\binom{n-1}{2}}.$$

To show this we fix a tree T on $V = \{1, \ldots, n\}$ and consider all the graphs $G_i \supseteq T$. This immediately proves (2.1). However, (2.1) is not the best. To improve it we consider the following family $G_1, \ldots, G_N$. $V = \{1, \ldots, n\}$ is fixed and we choose an arbitrary graph G^{n-1} with $V(G^{n-1}) = \{1, \ldots, n-1\}$. Then for each edge (i, j) of G^{n-1} we join $\max(i, j)$ to n by an edge. Finally we add an arbitrary number of edges (i, n) to this graph. Let $G_1, \ldots, G_N$ be the family of all the graphs obtained in this way. Obviously $G_i \cap G_j$ is always connected. It is not difficult to compute that

$$N \geq (4.6) 2^{\binom{n-1}{2}}.$$

Open problem. Let $\mathscr{L}$ be the family of connected graphs. Is

$$f(n, \mathscr{L}) = O(2^{\binom{n-1}{2}})?$$

If not, what is the proper order of magnitude of $f(n, \mathscr{L})$?

3. Intersection theorems on graphs where the intersections are cycles

In this and the next chapters we try to give a description of the cases, when $\mathscr{L}$ consists of cycles or paths. Let $\mathscr{A}_1$ resp. $\mathscr{A}_2$ be the family of all cycles, $\emptyset$ included in $\mathscr{A}_1$ but excluded from $\mathscr{A}_2$.

Definition. Given an $\mathscr{L}$, the family $G_1, \ldots, G_N$ will be called an $\mathscr{L}$-intersection family if $G_i \cap G_j \in \mathscr{L}$ whenever $1 \leq i < j \leq N$.

Definition. A family $G_1, \ldots, G_N$ is a strong Δ-system if

$$G_i \cap G_j = \bigcap_{l \leq N} G_l \quad 1 \leq i < j \leq N$$

i.e. $G_i \cap G_j$, is independent of i and j. $\bigcap_{i \leq N} G_i = K$ is called the kernel.

In [19] we proved

Theorem 3.1. *If* $n \geq 4$, *then* $f(n, \mathscr{A}_2) = \binom{n}{2} - 2$ *and the only extremal system, that is, the only* $\mathscr{A}_2$*–intersection system* $G_1, \ldots, G_N$ *for* $N = f(n, \mathscr{A}_2)$ *is the following one*: $E(G_1)$ *form a triangle and* $E(G_i)$ *contains* $E(G_1)$ *and exactly one additional edge for* $i = 2, \ldots, \binom{n}{2} - 2$.

It is worth noticing, that the extremal system of Theorem 3.1 is a strong Δ–system. We also prove the following theorem:

Theorem 3.2. *If* $G_1, \ldots, G_N$ *form an* $\mathscr{A}_2$*-intersection system on n vertices but not a strong* Δ*-system, then*

$$N \leq \frac{1}{\sqrt{6}} n^2 + n.$$

The basic idea behind proving Theorem 3.2 is the following one: this theorem shows, that if a system is almost extremal, say, it has $\geq \binom{n}{2} - o(n^2)$ graphs, then it is already a strong Δ-system. This means that in some sense the property of being strong Δ-system is an important (or at least a stable) feature of the extremal system.

Actually Theorem 3.1 is a simple consequence of Theorem 3.2. We do not think that Theorem 3.2 is sharp. On the contrary:

Conjecture. *Let $G_1, \ldots, G_N$ be graphs on n vertices forming an $\mathscr{A}_2$-intersection family but not a strong Δ-system. Then*

$$N \leq \tfrac{1}{6}n^2 + o(n^2).$$

The above conjecture is sharp if true: on n points we can find $s \approx \frac{1}{6}n^2$ triangles $T_1, \ldots, T_s$ such that no two of them have a common edge. Let $E(G_i) = E(T_s) \cup E(T_i)$ for $i < s$ and $E(G_m) = \bigcup_{j<s} E(T_j)$. One can easily see that $G_i \cap G_j$ is always a triangle but it does depend on (i, j). Another more symmetric construction of this type can be found in [19].

Our next result concerns the case then $G_i \cap G_j = \emptyset$ is also allowed. In [19] we have proved that:

Theorem 3.3. *Let $n \geq 10$, $\binom{x}{2} =: \frac{1}{2}x(x-1)$ for any real x and $s =: \frac{1}{3}\binom{n}{2}$. Then*

$$\binom{s-n}{2} < f(n, \mathscr{A}_1) \leq \binom{s}{2} + s + 1.$$

The upper bound is the best possible if $n = 6k+1$ or $n = 6k+3$.

Remark. Theorem 3.3 implies that

$$f(n, \mathscr{A}_1) \approx \tfrac{1}{72}n^4.$$

Remark. If $n = 6k+1$ or $n = 6k+3$, we can easily describe the only extremal system. We choose a Steiner triple system $T_1, \ldots, T_s$ on $\{1, \ldots, n\}$ and $\{G_{(i,j)};\ i, j \leq s\}$ are the $\binom{s}{2} + s$ graphs defined by $V(G_{(i,j)}) = \{1, \ldots, n\}$ and $E(G_{(i,j)}) = E(T_i) \cup E(T_j)$ where $1 \leq i \leq j \leq s$. (Here T_i is considered as a K_3!) Finally we add a graph G^n with no edges.

Remark. Independently from us V. Rödl also proved some of the results of [19], which include

$$f(n, \mathscr{A}_1) = o(n^4)$$

and the estimate on $f(n, \mathscr{A}_4)$ from the next chapter (unpublished).

4. Intersection problems, where the intersections are paths

The intersection problems of the paths are more difficult than the intersection problems of the cycles and the reason for that is probably, that the union of two cycles is never a cycle, while the union of two paths can easily be a path.

Notation. Let $\mathscr{A}_3$ be the family of paths, $\emptyset$ included, while $\mathscr{A}_4$ be the family of nonempty paths.

In [20] we have proved that

Theorem 4.1.

$$f(n, \mathscr{A}_3) = o(n^5),$$
$$f(n, \mathscr{A}_4) = o(n^4).$$

The following constructions show that (apart from the multiplicative constant) Theorem 4.1 is sharp:

Construction 4.1. Let us divide n vertices into five classes $C_1, \ldots, C_5$ of $\approx\frac{1}{5}n$ vertices each. Let us take all the pentagons $(x_1, \ldots, x_5)$ where $x_i \in C_i$ for $1 \leq i \leq 5$. Obviously, any two pentagons intersect in a path and their number is $\geq (\frac{1}{5}n)^5$. To get $G_1, \ldots, G_N$ we add $n-5$ isolated vertices to each of them. Thus we see that

$$f(n, \mathscr{A}_3) \geq \frac{1}{5^5} n^5.$$

The multiplicative constant $1/5^5$ is not sharp.

Open questions. What is $\overline{\lim} f(n, \mathscr{A}_3)/n^5$? Does the limit exist? Is it true that if $G_1, \ldots, G_N$ is an extremal system for $\mathscr{A}_3$, then all but $o(n^4)$ graphs are pentagons? (If yes, then the above limit does exist.)

Construction 4.2. Let the disjoint classes C_1, C_2, C_3, C_4 have $\approx\frac{1}{4}(n-k)$ vertices each and take all the paths P^{k+4} of form $(x_1, x_2, u_1, \ldots, u_k, x_3, x_4)$ where $x_i \in C_i$ and $u_1, \ldots, u_k$ are fixed vertices outside $\bigcup C_i$. Obviously, the intersection of any two such paths is a path of $\geq k$ vertices. Thus

$$f(n, \mathscr{A}_4) \geq \tfrac{1}{64} n^4 - O(n^3).$$

This construction shows that the second assertion of Theorem 4.1 is also sharp. It also answers the question posed by A. Frankel after our lecture, namely, it shows that if we allow only long paths as intersections, that will not diminish the original $O(n^4)$. The first two questions asked above in connection with $\mathscr{A}_3$ are also interesting in connection with $\mathscr{A}_4$: what is $\overline{\lim}\, f(n, \mathscr{A}_4)/n^4$ and does the limit exist?

5. Intersection properties of subset of integers

Here we consider problems of the following type:

Let $\mathscr{L}$ be a given family of sets of integers and let $\{A_1, \ldots, A_N\}$ be a family of subsets of $\{1, \ldots, n\}$. How large can N be if

$$A_i \cap A_j \in \mathscr{L} \quad (1 \leqslant i < j \leqslant N).$$

The maximum will again be denoted by $f(n, \mathscr{L})$. The simplest case is the case of intervals. The set $\{a, a+1, \ldots, a+k-1\}$ will be called an interval of length k and I_k will denote the family of intervals of length $\geqslant k$. One can easily see that

$$f(n, I_0) = \binom{n}{2} + n + 1.$$

An extremal system is the family of subsets of $\leqslant 2$ elements, $\emptyset$ included. (Other extremal systems can be obtained by replacing *some* pairs $\{a_i, b_i\}$ by the corresponding intervals.) Further,

$$f(n, I_1) = [\tfrac{1}{4}(n+1)^2]$$

and an extremal system can be obtained by considering all the intervals containing $m = [\tfrac{1}{2}(n+1)]$.

These results can be found in the paper of Graham, Simonovits and Sós [13] and the method used there also shows that

$$f(n, I_k) = [\tfrac{1}{4}(n-k+2)^2]$$

if $k \geqslant 2$, $n \geqslant k$.

The arithmetic progression-problem is already more difficult. Let P_k denote the family of all arithmetic progressions of $\geqslant k$ terms. Then

$$f(n, P_0) = \binom{n}{3} + \binom{n}{2} + n + 1.$$

The extremal system is uniquely determined: it consists of all the subsets of $\{1, \ldots, n\}$ of $\leqslant 3$ elements, [13]. The basic idea to prove this is that if $A_1, \ldots, A_N$ is a family such that $A_i \cap A_j \in P_0$, then the first, second and last elements of an A_i will uniquely determine it.

In some sense the elements of P_0 are not what we wanted: in the case of the extremal system $|A_i \cap A_j| \leqslant 2$ and this is, why it is a (trivial) arithmetic progression.

It is natural to ask, how large $f(n, P_k)$ is for $k \geqslant 3$, when the fact that $A_i \cap A_j$ is an arithmetic progression, really makes sense. We proved that [21]:

Theorem 5.1.

$$f(n, P_k) = (\tfrac{1}{24}\pi^2 + o(1))n^2 \quad (k \geqslant 2).$$

To show that

$$f(n, P_k) \geqslant (\tfrac{1}{24}\pi^2 + o(1))n^2$$

we consider all the arithmetic progressions $A_1, \ldots, A_N$ of form

$$A_i = \{[\tfrac{1}{2}n] + jd: \quad j = -a, -a+1, \ldots, -1, 0, 1, \ldots, b\}$$

where $d \leqslant n^{1/3}$, $\sqrt{n} \leqslant b \leqslant n/2d$ and $a \leqslant (n-1)/2d$. One can easily see that $A_i \cap A_j \in P_m$ for some $m \geqslant \frac{1}{3}n^{1/6}$ and

$$N = \tfrac{1}{4}n^2\left(\sum_{d=1}^{\infty} \frac{1}{d^2} + o(1)\right) = (\tfrac{1}{24}\pi^2 + o(1))n^2.$$

The upper part is much deeper. Actually we deduced Theorem 5.1 from Theorem 5.2:

Theorem 5.2. *Let $k \geqslant 2$ be fixed and $A_1, \ldots, A_N$ be subsets of $\{1, \ldots, n\}$ such that $A_i \cap A_j \in P_k$ for $1 \leqslant i < j \leqslant N$. Let us assume that none of the A_i's is an arithmetic progression. Then*

$$N = o(n^{5/3} \log^3 n). \tag{5.1}$$

Remarks. (i) Obviously Theorem 5.2 implies Theorem 5.1, since the number of arithmetic progressions is only $(\frac{1}{24}\pi^2 + o(1))n^2$. It also shows that all the almost extremal systems of Theorem 5.1 consist primarily of arithmetic progressions.

(ii) We can improve the exponent of $\log n$ in (5.1) but we do not know, whether $n^{5/3}$ can be replaced by some smaller power of n.

Though a one-term arithmetic progression is not a real arithmetic progression, we wanted to determine $f(n, P_1)$ as well.

Conjecture. $f(n, P_1) = \binom{n}{2} + 1$ *and one extremal system $\{A_1, \ldots, A_N\}$ is the system of sets of form $\{c, x, y\}$ for some fixed c (where $x = y$ or $x = c$ or $x = y = c$ are also allowed).*

There are also other, slightly different extremal systems if the conjecture is true. We could not prove it, but we can prove e.g. that

$$f(n, P_1) \leqslant \binom{n-1}{2} + \tfrac{1}{24}\pi^2 n^2 + o(n \log n).$$

Open problem. Let $A_1, \ldots, A_N$ be a extremal system in Theorem 5.1, i.e. for P_k $(k \geqslant 2)$. Is it true that all the sets A_i are arithmetic progressions?

6. Connections between Ramsey theory and graph intersection problems

As we mentioned and tried to "prove by some examples", the theory of intersection theorems of sets or other structures is a very wide and rapidly growing field of investigation. It has also many interesting, sometimes fairly deep connections to some other fields of mathematics. Obviously, the theory of block designs is such a field, [23, 14]. Some extremal graph problems, or combinatorial problems of finite geometries are also very strongly connected to intersection problems on sets [23]. An extensive part of hypergraph theory is nothing but intersection theorems in another language. (Thus e.g. the survey paper of Katona on intersection theorems has the title "Extremal problems for hypergraphs" [15].) Many questions in combinatorial number theory finally boil down to intersection theorems.

In this paragraph we have a closer look at the connection between the intersection theorems and another rapidly growing field of combinatorics, namely the (finite) Ramsey theory. We restrict ourselves to the case of ordinary graphs and 2 colours. It is well-known, that there exist two constants $c_1>0$ and $c_2>0$ such that if we colour the edges of a K_n by two colours, say red and blue, then K_n contains either a red or a blue (i.e. monochromatic) K_m for $m=[c_1 \log n]$. On the other hand, it is easy to see that if we colour the edges of K_n in a random way, i.e. colouring each edge of K_n by red with probability $\frac{1}{2}$, independently from the colouring of the other edges, then the largest monochromatic $K_m \subseteq K_n$ will have size $m \leqslant [c_2 \log n]$ with probability tending to 1 (as $n \to \infty$). This means that we can show by probabilistic methods the sharpness of the lower bound $m=[c_1 \log n]$ in this Ramsey problem. It is a longstanding and famous unsolved problem, how can one give a constructive upper bound:

Let us construct a two-colouring of the edges of K_n such that the largest monochromatic K_m has $m \leqslant [c_3 \log n]$ vertices. One can construct easily a 2-colouring with $m=[\sqrt{n}]+1$: the n vertices are divided into $\{\sqrt{n}\}$ classes as uniformly as possible and (x, y) is blue or red depending on whether they belong to the same class or not.

The sharper $m \sim n^{1/3}$ was done by Zs. Nagy and the basic idea was to use two simple intersection theorems. The vertices of G^n are the triples of a $k \sim (6n)^{1/3}$ set and $\{a, b, c\}$ is joined by a red edge to $\{a', b', c'\}$ iff $|\{a, b, c\} \cap \{a', b', c'\}|$ is even. It is almost trivial that the largest monochromatic K has $m=O(k)=O(n^{1/3})$ vertices in this colouring.

The next step was due to Abbott's 1, who proved that if one has a fixed graph G^n with the maximum $m=n^{\varepsilon}$, then one can very easily construct infinitely many such graphs. Abbott's construction combined with Erdõs's probabilistic method yields $m=n^{\varepsilon}$ in a semiconstructive way, also "showing" that the notion of constructiveness is not quite well-defined in combinatorics.

The latest development for this question is the following.

Frankl [12] constructed for $h=2k^2$ two sets of integers

$$\mathscr{L}_1=\{2ak+b \mid 0 \leqslant a, b \leqslant k-1\}$$

and

$$\mathscr{L}_2=\{0, \ldots, h-1\}-\mathscr{L}_1$$

so that if we take all the $2k^2$-element set of an n-element set and join two of them, say A_i and A_j, by a red edge if $|A_i \cap A_j| \in \mathscr{L}_1$ and by a blue edge otherwise (i.e. if $|A_i \cap A_j| \in \mathscr{L}_2$), then, using two appropriate intersection theorems, he could prove that the largest monochromatic K_m has $m \leqslant cn^{1/k}$ vertices in this colouring.

Obviously, the basic idea in the above construction is that $f(n, \mathscr{L}_1)$, $f(n, \mathscr{L}_2)$ are much smaller than $f(n, \mathscr{L}_1 \cup \mathscr{L}_2) = 2^n$. We arrived originally to the same problem for graphs from the other direction:

Open problem. Given two families of graphs: $\mathscr{L}_1$ and $\mathscr{L}_2$, is it true that if $f(n, \mathscr{L}_1)$ and $f(n, \mathscr{L}_2)$ are polynomially bounded, then $f(n, \mathscr{L}_1 \cup \mathscr{L}_2)$ is also polynomially bounded?

(If the answer is "yes", that is interesting in itself, but if the answer is "no", then one could hope for a sharpening of the Frankl construction. Further, if in the Ramsey theorem we had only say n^{ε} instead of $c \log n$ (for some fixed $\varepsilon > 0$) which is definitely not the case, that would imply the solution of the above problem.)

7. Some further open problems

The basic problem, we are interested in in connection with graph intersection problem is, as we stated, on which properties of $\mathscr{L}$ does the fact depend, whether $f(n, \mathscr{L})$ is polynomially bounded or not. The first two problems below may seem very special, we picked them because we feel that if one can solve them, one gets a step nearer to the solution of the main problem.

Problem 7.1. Let L_k denote the graph on k^2 vertices whose edges form k independent paths of k vertices (each). Let $\mathscr{L} = \{L_3, \ldots\}$. Is $f(n, \mathscr{L})$ polynomially bounded?

Problem 7.2. What type of "graph products" produces "polynomial families" from "polynomial families". E.g. let $L_k = P^k \times P^k$ be the graph whose vertices are (i, j) $1 \leq i \leq k$; $1 \leq j \leq k$ and (i, j) is joined to (i', j') if $|i - i'| = 1$ and $j = j'$ or $|j - j'| = 1$ and $i = i'$. (L_k is a "square subdivided into smaller squares".) Is $f(n, \{L_2, \ldots\})$ polynomially bounded?

The next problem concerns trees, where it is obvious that $f(n, \mathscr{L})$ is exponentially large (see Section 3.).

Problem 7.3. Let $\mathscr{L}$ be now the family of all the trees. How large $f(n, \mathscr{L})$ can be?

(Here we would like to have an asymptotically sharp estimate of $\log f(n, \mathscr{L})$.) Finally we formulate a general problem.

Problem 7.4. Is it true that if $f(n, \mathscr{L})$ is polynomially bounded for a family of graphs, then there exists an *integer* r such that $\lim_{n\to\infty} f(n, \mathscr{L})/n^r$ exists and is *positive*?

Note added in proof

(1) The Open Question in Section 4 was solved in the affirmative by Z. Füredi.

(2) V. Rödl has published his results in Comment Math. Univ. Carolinae 19 (1) (1978) 135–140.

References

[1] H.L. Abbott, Lower bounds for some Ramsey numbers, Discrete Math. 2 (1972) 289–294.
[2] N.G. de Bruijn and P. Erdős, On a combinatorial problem, Nederl. Akad. Wetensch. Proc. 51 (1948) 1277–1279.
[3] M. Deza, On maximal sets of permutations with given maximal distance, to appear.
[4] M. Deza, P. Erdős and P. Frankl, Intersection properties of systems of finite sets, Proc. London Math. Soc. 36 (1978) 369–384.
[5] M. Deza, P. Erdős and N.M. Singhi, Combinatorial problems on subsets and their intersections, Advances in Math.
[6] M. Deza and P. Frankl, On the maximum number of permutations with given maximal or minimal distance, J. Combin. Theory 22(A) (1977) 352–360.
[7] P. Erdős, Problems and results in combinatorial analysis, in: Proc. Coll. Int. Sulle Teorie Combinatorie, Roma, 1973, Accad. Nazionale dei Lincei (1973).
[8] P. Erdős, Chao Ko and R. Rado, Intersection theorems for systems of finite sets, Quart. J. Math. Oxford Ser. 12 (2) (1961) 313–318.
[9] P. Erdős and D.J. Kleitman, Extremal problems among subsets of a set, in: Combinatorial Mathematics and Applications, Proc. Chapel Hill Conference (1970)146–170.
[10] P. Erdős and D.J. Kleitman, On collections of subsets containing no 4-member Boolean algebra, Proc. Am. Math. Soc. 28 (1971) 507–510.
[11] R.A. Fisher, An examination of the different possible solutions of a problem in incomplete blocks, Ann. Eugenics 10 (1940) 52–75.
[12] P. Frankl, A constructive lower bound for some Ramsey numbers, Ars Combinatoria 3 (1977) 297–302.
[13] R.L. Graham, M. Simonovits and V.T. Sós, A note on the intersection properties of subsets of integers, to appear.
[14] M. Hall, Combinatorial Theory (Blaisdell Publ. Company, 1967).
[15] G. Katona, Extremal problems for hypergraphs, in: M. Hall and Von Lindt (eds), Combinatorics, Vol. 2, Math. Centre Tracts, 56 (Math. Centre, Amsterdam, 1974) 13–42.
[16] D. Lubell, A short proof of Sperner's lemma, J. Combin. Theory 1 (1966) 299.
[17] D.K. Ray-Chaudhuri and R.M. Wilson, On t-designs, Osaka J. Math. 12 (1975) 737–744.
[18] H.J. Ryser, On extension of a theorem of De Bruijn and Erdős on combinatorial designs, J. Algebra 10 246–261.
[19] M. Simonovits and V.T. Sós, Intersection theorems for graphs II, Coll. Math. Soc. J. Bolyai 18 Combinatorics, Keszthely, Hungary (1976) 1017–1029.
[20] M. Simonovits and V.T. Sós, Intersection theorems for graphs, Coll. Internationaux C.N,R.S. No. 260, Problemes Combinatoires et Theorie des Graphes, 389–391.
[21] M. Simonovits and V.T. Sós, Intersection properties of subsets of integer, to appear.
[22] V.T. Sós, Some remarks on the connection of graph theory, finite geometry and block designs, in: Proc. Coll. Internazionale sulle Teorie Combinatorie, Roma, 1973, Accademia Nazionale dei Lincei (1973) 223–233.
[23] E. Sperner, Ein Satz über Untermenge einer endlichen Menge, Math. Z. 27 (1928) 544–548.

Annals of Discrete Mathematics 6 (1980) 315–332

ENUMERATION AND REPRESENTATION OF NONISOMORPHIC BIPARTITE GRAPHS*

J.N. SRIVASTAVA and S. GHOSH

Colorado State University, Fort Collins, CO 80523, USA

We consider bipartite graphs whose point sets consist of two subsets V and U containing 4 and m ($\geqslant 4$) points respectively. The degree of each point in V is at least 3 and no two points in V are joined to the same set of points in U. Let S_4 and S_m be permutation groups on V and U, respectively, where S_n is the symmetric group of degree n. We obtain the number of orbits of bipartite graphs determined by $S_4 \times S_m$, by using the generalized Pólya's Theorem and Burnside's Lemma. A general method of listing the representatives of orbits is given. For $m = 5, 6$, the number of orbits is calculated and a list of representatives is presented. The importance of the above problem in the theory of statistical search linear models [3] is also discussed.

1. Introduction

Consider bipartite graphs whose point sets are partitioned into two subsets V and U such that

(a) The subset V has q points, $v_1, v_2, \ldots, v_q$,
(b) The subset U has m points, $u_1, u_2, \ldots, u_m$, (1.1)
(c) $\deg v_i \geqslant 3$, $i = 1, 2, \ldots, q$,
(d) no two points in V are joined to the same set of points in U.

Note that a point in U may or may not be joined to a point in V. There are 2^m possible subsets of the set of points in U, and the number of distinct subsets containing exactly l points of U is $\binom{m}{l}$. From (c) in (1.1), it follows that the points in any one of the $(2^m - 1 - \binom{m}{1} - \binom{m}{2})$ subsets, containing 3 or more points of U, can be joined to a point in V. Thus, the conditions (a) and (d), in (1.1), imply $[2^m - 1 - \binom{m}{1} - \binom{m}{2}] \geqslant q$.

The cartesian product of S_q and S_m, $S_q \times S_m$, is a permutation group on $V \times U$, and is given by $S_q \times S_m = \{(\omega, \gamma) \mid \omega \in S_q, \gamma \in S_m\}$, where (ω, γ) is acting on each element (v, u) of $V \times U$ as follows:

$$(\omega, \gamma)(v, u) = (\omega v, \gamma u) \tag{1.2}$$

Two bipartite graphs are isomorphic if one is obtainable from the other by a permutation in $S_q \times S_m$. We want to find the number of orbits of bipartite graphs determined by $S_q \times S_m$ and furthermore, to list the representatives of orbits. In this paper, we consider $q = 4$ and $m \geqslant 4$.

* This research was supported by the NSF Grant Number MCS-7722985, and also by previous NSF Grants given to the first author.

Let $D=\{1, 2, \ldots, m\}$, $A=(A_1, A_2, A_3, A_4)$, and A be unordered. Under the conditions

$$\begin{aligned}&\text{(a) } A_i \subseteq D, \text{ and } A_i\text{'s are distinct,}\\ &\text{(b) } |A_i| \geq 3;\ i=1,2,3,4,\end{aligned} \tag{1.3}$$

there is a natural (1 − 1) correspondence between the A's and the bipartite graphs. For example, for $m=5$, $A=(1345, 345, 12345, 2345)$ represents the bipartite graph in Fig. 1. Note that v_1 is joined to u_1, u_3, u_4, u_5; v_2 is joined to u_3, u_4, u_5; etc.

The total number of bipartite graphs is the same as the total number of A's. Two bipartite graphs are isomorphic under $S_4 \times S_m$ if and only if (iff) the corresponding A's are isomorphic under S_m and S_4. Thus the number of orbits of bipartite graphs is the same as the number of orbits of A's.

The problem considered in this paper does arise in the area of statistical search linear models [3]. Consider a 2^m factorial experiment with m factors each at 2 levels. The factorial effects are denoted as follows: μ stands for the general mean, F_i $(i=1, 2, \ldots, m)$ for the m main effects, F_iF_j $(i<j,\ i, j=1, 2, \ldots, m)$ for the $\binom{m}{2}$ 2-factor interactions, etc. For simplicity, we denote the l-factor interaction $F_{i_1}F_{i_2}\ldots F_{i_l}$ by $F(i_1, i_2, \ldots, i_l)$. In the theory of balanced 2^m factorial designs of resolution V which permit search and estimation of k extra unknown effects (BARE 5.k plans), certain matrices $W(2k\times 2k)$ play an important role [4]. The rows and columns of W correspond to $2k$ distinct factorial effects chosen out of the 3-factor and higher order interactions, whose number is $2^m-1-\binom{m}{1}-\binom{m}{2}$ $(=v_2$, say). Consider the case $k=2$. The number of $W(4\times 4)$ matrices is $\binom{v_2}{4}$. The problem is to check whether the W matrices are of full rank. Clearly, each unordered 4-plet (A_1, A_2, A_3, A_4), defined earlier, represents 4 distinct factorial effects involving 3-factor and higher order interactions. Thus, corresponding to each 4-plet, we have a $W(4\times 4)$ matrix. If two 4-plets $A_1^*=(A_{11}, A_{21}, A_{31}, A_{41})$ and $A_2^*=(A_{12}, A_{22}, A_{32}, A_{42})$ are isomorphic under $S_m\times S_4$, the corresponding W matrices are equal except for a permutation of rows and columns. We take advantage of this fact. Two W matrices corresponding to 4-plets $A_i^*=(A_{1i}, A_{2i}, A_{3i}, A_{4i})$, $i=1,2$, are called permutation-equivalent, iff A_1^* and A_2^* are isomorphic under $S_m\times S_4$. Therefore, the problem is to find a representative of the orbits of A's determined by $S_m\times S_4$. Once this is done, we need to calculate W matrices and check their rank, only for one representative A in each orbit.

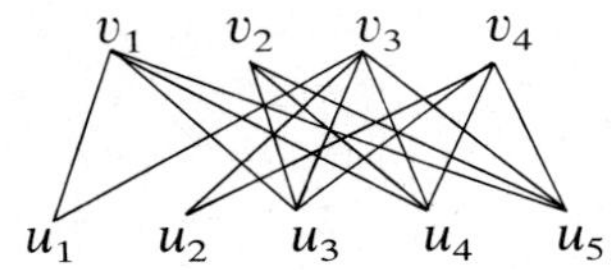

Fig. 1.

Table 1. Number of orbits

m	$\binom{\nu_2}{4}$	Number of orbits
5	1820	42
6	111930	402

This reduces the work enormously. For example, Table 1 shows for $m = 5, 6$, the values of $\binom{\nu_2}{4}$, and the number of orbits.

2. The generalized Pólya's theorem

Let D and R be finite object sets, and let A^* be a mapping of D into R. Denote the set of all maps A^* by R^D. For each $d \in D$, there are $|R|$ (the cardinality of R) possible functions, and these choices are distinct. Thus $|R^D| = |R|^{|D|}$. Consider two permutation groups: Γ on D, and Ω on R. Two functions A_1^* and A_2^* are equivalent ($A_1^* \sim A_2^*$) if, for a $\gamma \in \Gamma$ and $\omega \in \Omega$, $A_1^*\gamma = \omega A_2^*$. The relation $\sim$ is an equivalence relation, and it splits R^D into "equivalence classes", also called "orbits" or "patterns". For each $A^* \in R^D$, we assume a weight $W(A^*)$, which is an element of a commutative ring. We also assume $A_1^* \sim A_2^*$ implies $W(A_1^*) = W(A_2^*)$. Denote the weight $W(P)$ of a pattern P by the common weight of all maps A^* in P. The pattern inventory is defined as the sum of the weights of all patterns.

Theorem 2.1 [1]. *The pattern inventory is given by*

$$\sum_P W(P) = |\Gamma|^{-1}|\Omega|^{-1} \sum_{\gamma \in \Gamma} \sum_{\omega \in \Omega} \sum_{A^*}^{(\gamma,\omega)} W(A^*), \tag{2.1}$$

where $\sum_{A^*}^{(\gamma,\omega)} W(A^*)$ *equals the sum of* $W(A^*)$ *extended over all* A^* *that satisfy* $A^*\gamma = \omega A^*$.

In particular, if all weights are unity, then from (2.1), we find that the number of patterns $= |\Gamma|^{-1}\,|\Omega|^{-1}$ {number of A^*'s which satisfy $A^*\gamma = \omega A^*$}. Since Theorem 2.1 enumerates the orbits of functions, we now show a correspondence between the bipartite graphs, defined in Section 1, and the functions of the above type.

We had seen earlier that there is a $(1-1)$ correspondence between bipartite graphs and unordered sets $A^* = (A_1, A_2, A_3, A_4)$. Also, let $R = \{0;\ 1, 2, 3, 4;\ 12, 13, 14, 23, 24, 34;\ 123, 124, 134, 234;\ 1234\}$. Then a set $A = (A_1, A_2, A_3, A_4)$, with $A_i \subseteq D$, may be considered as a mapping from D to R as follows. If d belongs to $\bar{A}_1 \cap \bar{A}_2 \cap \bar{A}_3 \cap \bar{A}_4$, then $A(d)$ is the element 0 of R. If $(i_1, \ldots, i_q)$ is a subset of integers $1, 2, 3, 4$, where $1 \leq q \leq 4$, such that $d \in A_j$ iff

$j \in (i_1, \ldots, i_q)$, then d is said to be mapped to the element $(i_1, i_2, \ldots, i_q)$ of R.

Example 1. Let $m=5$, $D=\{1,2,3,4,5\}$, and let $A=(125, 134, 124, 135)$. The function A maps the elements of D to R as follows: $A(1)=1234$, $A(2)=13$, $A(3)=24$, $A(4)=23$, $A(5)=14$.

Clearly, every possible set corresponds to a mapping of D to R, and all possible $|R|^{|D|}$ functions are permissible. Also, the symmetric group S_m is acting on D. On R, there is acting a permutation group (say, S_4^*) which is induced by S_4 acting on $\{1,2,3,4\}$. To see this, consider a permutation say $(2,4,3,1)$ of $(1,2,3,4)$. This induces a permutation of (A_1, A_2, A_3, A_4) to (A_2, A_4, A_3, A_1). This, in turn, induces the permutation of R to $R^*=\{0, 2, 4, 3, 1;\ 24, 23, 12, 34, 14, 13;\ 234, 124, 123, 134;\ 1234\}$. In what follows, permutations in S_4^* will be denoted by the same symbols as in S_4.

The equivalence of mappings (A_1, A_2, A_3, A_4) from D to R is defined both with respect to S_m and S_4^*. Thus, henceforth, we shall assume that the groups Γ and Ω considered at the beginning of this section are identical with S_m and S_4^* respectively. With this understanding, the following example illustrates the equivalence of two functions in R^D.

Example 2. In Example 1, let $A^0=(124, 125, 235, 234)$. Consider $\gamma=(12)(3)(45)\in S_5$ and $\omega=(1)(23)(4)\in S_4$. We have

$$
\begin{aligned}
A\gamma(1) &= A(2) = 13 = \omega(12) = \omega A^0(1),\\
A\gamma(2) &= A(1) = 1234 = \omega(1234) = \omega A^0(2),\\
A\gamma(3) &= A(3) = 24 = \omega(34) = \omega A^0(3),\\
A\gamma(4) &= A(5) = 14 = \omega(14) = \omega A^0(4),\\
A\gamma(5) &= A(4) = 23 = \omega(23) = \omega A^0(5).
\end{aligned}
$$

therefore, $A\gamma(d)=\omega A^0(d)$, for all $d\in D$ and hence $A \sim A^0$.

For any $A=(A_1, A_2, A_3, A_4)$, we define the weight by

$$
\begin{aligned}
W(A) &= 0 \quad \text{if at least one of (a) and (b), in (1.3), do not hold,}\\
&= 1 \quad \text{otherwise.}
\end{aligned}
$$

We now calculate the number of orbits for $m=5$. Take, for illustration, $\gamma=(1)(2,4,5,3)$ and $\omega=(1)(2)(3,4)$. It can be checked that only the following A's satisfy $A\gamma=\omega A$: $(12345, 2345, 125, 134)$, $(2345, 12345, 125, 134)$, $(12345, 2345, 134, 125)$, $(2345, 12345, 134, 125)$. The permutation γ has cycle structure 14 and ω has $1^2 2$, where $1^{b_1}2^{b_2}3^{b_3}4^{b_4}$ indicates (for $i=1,2,3,4$) b_i cycles of length i, where $b_1, b_2, b_3, b_4 \geqslant 0$; and $b_1+2b_2+3b_3+4b_4=4$. Table 2 shows the number of A's satisfying $A\gamma=\omega A$, for γ and ω with different cycle structure.

Table 2. Number of A which satisfies $A\gamma = \omega A$

Cycle structure	No. of $\omega \in S_4$ \ No. of $\gamma \in S_5$	5	14	23	$1^2 3$	12^2	$1^3 2$	5
		24	30	20	20	15	10	1
4	6	0	12	0	0	0	0	0
13	8	0	0	12	48	0	0	0
2^2	3	0	0	0	0	120	48	0
$1^2 2$	6	0	4	4	0	144	448	0
1^4	1	0	0	0	24	24	1680	43680

Thus, the total number of A's which satisfy $(A\gamma = \omega A) = (6\cdot 12\cdot 30 + 12\cdot 8\cdot 20 + 48\cdot 8\cdot 20 + 120\cdot 3\cdot 15 + 48\cdot 3\cdot 10 + 4\cdot 30\cdot 6 + 4\cdot 20\cdot 6 + 144\cdot 15\cdot 6 + 448\cdot 10\cdot 6 + 24\cdot 20 + 24\cdot 15 + 16800 + 43680 = 120960)$. Therefore, the number of orbits $= (4!\,5!)^{-1}\cdot 120960 = 42$.

It is an enormous amount of work to find the entries in Table 2. The work is reduced considerably by using Burnside's Lemma. However, none of DeBruijn's theorem and Burnside's Lemma gives the list of representatives of orbits.

3. Application of Burnside's lemma

We first find the number of orbits of Γ^* as it acts on D^*.

Theorem 3.1 (Burnside's Lemma). *The number $N(\Gamma^*)$ of orbits of Γ^* is given by*

$$N(\Gamma^*) = |\Gamma^*|^{-1} \sum_{\gamma \in \Gamma} \chi_{D^*}(\gamma^*), \tag{3.1}$$

where $\chi_{D^*}(\gamma^*)$ denotes the number of elements of D^* that are invariant under γ, i.e.,

$$\chi_{D^*}(\gamma^*) = |\{d^* \in D^* \mid \gamma^*(d^*) = d^*\}|. \tag{3.2}$$

We now apply the Burnside's Lemma to find the number of orbits of A's, (and hence, of bipartite graphs) defined in Section 1.

Let $D = \{1, 2, \ldots, m\}$, $B = \{A_0 \subset D \mid |A_0| \geq 3\}$, and consider S_m acting on D. For a $\gamma \in S_m$, define $\chi_B(\gamma) = |\{A_0 \in B \mid \gamma(A_0) = A_0\}|$. Note that by $\gamma(A_0)$, we mean the set obtained by applying γ to the different elements of A_0. Suppose $\mathscr{A} = \{(A_1, A_2, A_3, A_4) \mid A_i \in B,\ i = 1, 2, 3, 4\}$, where the A_i are all distinct, and (A_1, A_2, A_3, A_4) are unordered classes of sets. Clearly, $|\mathscr{A}| = \binom{|B|}{4}$. For each $\gamma \in S_m$, define $x(\gamma) = |\{A^* \in \mathscr{A} | \gamma(A^*) = \omega(A^*)$, for a $\omega \in S_4\}|$. Here, again, $\gamma(A^*)$ means

$(\gamma(A_1^*), \ldots, \gamma(A_4^*))$, where $A^* = (A_1^*, \ldots, A_4^*)$. Let $\omega \in S_4$. We denote the set, obtained by applying ω to the subscripts of the members of A^*, by $\omega(A^*)$. For example, if $A^* = (A_1^*, A_2^*, A_3^* A_4^*)$, $\omega = (2, 3, 4, 1)$, then $\omega(A^*) = (A_2^*, A_3^*, A_4^*, A_1^*)$. Now, any $\omega \in S_4$ has one of the following cycle structures: 1^4, 2^2, $1^2 2$, 13, 4. If $\gamma \in S_m$, $\omega \in S_4$ and $A^* \in \mathcal{A}$, and if furthermore, $\gamma(A^*) = \omega(A^*)$, then γ is said to be acting on A^* with cycle structure which is identical with that of ω. For example, if ω has cycle structure $1^2 2$, then γ is said to be "acting on A^* as $1^2 2$."

Example 3. Let $m = 5$, and $\gamma = (1)(2)(3)(4, 5)$. Here $B = \{123, 124, 125, 134, 135, 145, 234, 235, 245, 345; 1234, 1235, 1245, 1345, 2345; 12345\}$. Take $A^* = (123, 145, 124, 125) \in A$. Now, $\gamma(A^*) = (123, 145, 125, 124)$. Suppose $\omega = (1)(2)(3\,4)$. Then, $\gamma(A^*) = \omega(A^*)$. Thus γ is acting as $1^2 2$.

We shall call a cycle of length i an i-cycle. It is clear that $\chi_B(\gamma)$ is the number of 1-cycles in γ, where γ is acting on B. The number of 2-cycles, 3-cycles and 4-cycles in γ acting on B are $\frac{1}{2}(\chi_B(\gamma^2) - \chi_B(\gamma))$, $\frac{1}{3}(\chi_B(\gamma^3) - \chi_B(\gamma))$ and $\frac{1}{4}(\chi_B(\gamma^4) - \chi_B(\gamma^2))$, respectively. The proof goes as follows. Each 1-cycle in γ acting on B corresponds to a 1-cycle in γ^2 acting on B. Each 2-cycle in γ gives two 1-cycles in γ^2, and no other cycles in γ give any 1-cycle in γ^2. Therefore, $\frac{1}{2}(\chi_B(\gamma^2) - \chi_B(\gamma))$ is the number of 2-cycles in γ acting on B. Similarly for 3-cycles. Each 4-cycle in γ gives four 1-cycles in γ^4, each 2-cycle in γ gives two 1-cycles in γ^4, each 1-cycle in γ remains a 1-cycle in γ^4, and no other cycles in γ give any 1-cycle in γ^4. In other words, only each 4-cycle in γ^2 give 1-cycles in γ^4. Hence $\frac{1}{4}(\chi_B(\gamma^4) - \chi_B(\gamma^2))$ is the number of 4-cycles in γ acting on B.

Example 4. In Example 3, γ is acting on B as a permutation

$$(123)(145)(245)(345)(1245)(1345)(2345)(12345)$$
$$(124, 125)(134, 135)(234, 235)(1234, 1235).$$

Thus $\chi_B(\gamma) = 8$. Now, γ^2 is acting as

$$(123)(145)(245)(345)(1245)(1345)(2345)(12345)$$
$$(124)(125)(134)(135)(234)(235)(1234)(1235).$$

Therefore, $\chi_B(\gamma^2)$ is 16 and the number of 2-cycles in γ acting on $B = \frac{1}{2}(\chi_B(\gamma^2) - \chi_B(\gamma)) = 4$. Clearly, in this case, $\chi_B(\gamma^3) = \chi_B(\gamma)$, and $\chi_B(\gamma^4) = \chi_B(\gamma^2)$. Thus, the number of 3 and 4-cycles in γ acting on B are zero.

Clearly, the number of members of $\mathcal{A}$ on which γ is acting as a permutation having cycle structure 1^4, 2^2, $1^2 2$, 13 or 4 are

$$\binom{\chi_B(\gamma)}{4}, \quad \binom{\frac{1}{2}(\chi_B(\gamma^2) - \chi_B(\gamma))}{2}, \quad \binom{\frac{1}{2}(\chi_B(\gamma^2) - \chi_B(\gamma))}{1}\binom{\chi_B(\gamma)}{2}$$
$$\binom{\frac{1}{3}(\chi_B(\gamma^3) - \chi_B(\gamma))}{1}\binom{\chi_B(\gamma)}{1}, \quad \text{and} \quad \binom{\frac{1}{4}(\chi_B(\gamma^4) - \chi_B(\gamma^2))}{1}$$

respectively. Thus,

$$\chi_{\mathscr{A}}(\gamma)=\sum_{\sigma\in\{1^4,2^2,1^22,13,4\}}(\text{the number of members of } \mathscr{A} \text{ on which } \gamma \text{ has cycle structure } \sigma)$$

$$=\binom{\chi_B(\gamma)}{4}+\binom{\frac{1}{2}(\chi_B(\gamma^2)-\chi_B(\gamma))}{2} \tag{3.3}$$

$$+\left[\frac{\chi_B(\gamma^2)-\chi_B(\gamma)}{2}\right]\binom{\chi_B(\gamma)}{2}+\chi_B(\gamma)\left[\frac{\chi_B(\gamma^3)-\chi_B(\gamma)}{3}\right]$$

$$+\left[\frac{\chi_B(\gamma^4)-\chi_B(\gamma^2)}{4}\right]$$

Hence, we obtain

Theorem 3.2. *Consider the bipartite graphs, defined in Section* 1. *Then the number of orbits of bipartite graphs determined by* $S_4\times S_m$ *is given by*

$$|S_m|^{-1}\sum_{\gamma\in S_m}\chi_{\mathscr{A}}(\gamma), \tag{3.4}$$

where $\chi_A(\gamma)$ *is given in* (3.3).

We now calculate the number of orbits for $m=5,6$.

Case 1. $m=5$. The values of $\chi_A(\gamma)$, $\gamma\in S_5$, are given in Table 3. Therefore, the number of orbits

$$=(5!)^{-1}[1820+(188)\cdot 10+(17)\cdot 20+(30)\cdot 4+(15)\cdot 52+(20)\cdot 5]=42.$$

Case 2. $m=6$. The values of $\chi_A(\gamma)$, $\gamma\in S_6$, are in Table 4. Therefore, the number of orbits

$$=(6!)^{-1}[120+(18)\cdot 90+(90)\cdot 30+(120)\cdot 51+(40)\cdot 39+(40)\cdot 225$$
$$+(15)\cdot 6990+(45)\cdot 1050+(15)\cdot 286+111930]=402.$$

Table 3. The value of $\chi_{\mathscr{A}}(\cdot)$ for $m=5$

Cycle structure	1^5	1^32	1^23	14	12^2	23	5
Number of $\gamma\in S_5$	1	10	20	30	15	20	24
$\chi_B(\gamma)$ on							
5 sets	1	1	1	1	1	1	1
4 sets	5	3	2	1	1	0	0
3 sets	10	4	1	0	2	1	0
$\chi_B(\gamma)$	16	8	4	2	4	2	1
$(\chi_{B_4}(\gamma))$	1820	70	1	0	1	0	0
$\frac{1}{2}(\chi_B(\gamma^2)-\chi_B(\gamma))$	0	4	0	1	6	1	0
$(\frac{1}{2}(\chi_b(\gamma^2)-\chi_B(\gamma)))$	0	6	0	0	15	0	0
$\frac{1}{2}(\chi_B(\gamma^2)-\chi_B(\gamma))\binom{\chi_B(\gamma)}{2}$	0	112	0	1	36	1	0
$\frac{1}{3}(\chi_B(\gamma))[\chi_B(\gamma^3)-\chi_B(\gamma)]$	0	0	16	0	0	4	0
$\frac{1}{4}(\chi_B(\gamma^4)-\chi_B(\gamma^2))$	0	0	0	3	0	0	0
$\chi_{\mathscr{A}}(\gamma)$	1820	188	17	4	52	5	0

Table 4. The value of $\chi_{\mathscr{A}}(\cdot)$ for $m=6$

Cycle structure	6	15	24	1^24	123	3^2	1^33	1^42	1^22^2	2^3	1^6
Number of $\gamma \in S_6$	120	144	90	90	120	40	40	15	45	15	1
$\chi_B(\gamma)$ on											
6 sets	1	1	1	1	1	1	1	1	1	1	1
5 sets	0	1	0	2	1	0	3	4	2	0	6
4 sets	0	0	1	1	1	0	3	7	3	3	15
3 sets	0	0	0	0	2	2	2	8	4	0	20
$\chi_B(\gamma)$	1	2	2	4	5	3	9	20	10	4	42
$\binom{\chi_B(\gamma)}{4}$	0	0	0	1	5	0	126	4845	210	1	111930
$\frac{1}{2}(\chi_B(\gamma^2)-\chi_B(\gamma))$	1	0	4	3	2	0	0	11	16	19	0
$\binom{\frac{1}{2}(\chi_B(\gamma^2)-\chi_B(\gamma))}{2}$	0	0	6	3	1	0	0	55	120	171	0
$\frac{1}{2}(\chi_B(\gamma^2)-\chi_B(\gamma)\binom{\chi_B(\gamma)}{2}$	0	0	4	18	20	0	0	2090	720	114	0
$\frac{1}{3}(\chi_B(\gamma)\chi_B(\gamma^3)-\chi_B(\gamma)$	1	0	0	0	25	39	99	0	0	0	0
$\frac{1}{4}(\chi_B(\gamma^4)-\chi_B(\gamma))$	0	0	8	8	0	0	0	0	0	0	0
$\chi_{\mathscr{A}}(\gamma)$	1	0	18	30	51	39	225	6990	1050	286	111930

4. List of representatives of orbits

We present here a method of obtaining a representative of each orbit of bipartite graphs. (Incidentally, throughout this paper, for brevity, we shall write the word sub-subset as sub^{2}set, sub-sub-subset as sub^{3}set. etc.)

From Section 1, recall that the isomorphism of two bipartite graphs is equivalent to the isomorphism of the corresponding A's. We now define, for an unordered set $A=(A_1, A_2, A_3, A_4)$, the vector $\boldsymbol{h}$ given by

$$\boldsymbol{h}' = (h(0), h(1), h(2), h(3), h(4), h(123), h(124), h(134), h(234), h(12), h(13), h(23), h(14), h(24), h(34), h(1234)) \tag{4.1}$$

where $h(\cdot)$'s are defined analogously to $R(\cdot)$ as follows. Thus, $h(0)=|\bar{A}_1 \cap \bar{A}_2 \cap \bar{A}_3 \cap \bar{A}_4|$. Also, for $1 \leqslant q \leqslant 4$, let $(i_1, \ldots, i_q)$ be a subset of distinct integers chosen from $\{1, 2, 3, 4\}$. Then we define $h(i_1 \cdots i_q)$ as the number of elements in D which are contained in $A_{i_1}, \ldots, A_{i_q}$ and simultaneously not contained in A_j for every $j \notin (i_1, i_2, \ldots, i_q)$. Clearly every element $h(\cdot)$ lies between 0 and m (both inclusive), and $\sum_{\text{all } u} h(u)=m$. Notice that for a given A, there is a unique $\boldsymbol{h}$. Moreover, for two A's isomorphic under S_m the corresponding $\boldsymbol{h}$'s are identical. So we consider all possible distinct $\boldsymbol{h}$'s and then define two vectors $\boldsymbol{h}_1$ and $\boldsymbol{h}_2$ to be isomorphic if $\boldsymbol{h}_1 = w(\boldsymbol{h}_2)$, for a $w \in S_4^*$. Here, the meaning of w and S_4^* is the same as that in connection with R. Now, S_4^* is induced by S_4. Thus, two $\boldsymbol{h}$'s can be isomorphic only under the permutation group S_4^* induced by S_4. For simplicity, if two $\boldsymbol{h}$'s are isomorphic under a permutation in S_4, then we indicate this only by referring to the corresponding permutation in S_4.

We now introduce some further notation which will be useful in the sequel. By

$S_2(1, 2)$, we mean a permutation group acting on the elements 1 and 2 taken from the set of integers $\{1, 2, 3, 4\}$ using which $\boldsymbol{h}$ in (4.1) is defined. Thus, $S_2(1, 2)$ acting on $\boldsymbol{h}$ in (4.1) would transform this $\boldsymbol{h}$ to $\boldsymbol{h}_1' = (h(0), h(2), h(1), h(3), h(4), h(123), h(124), h(234), h(134), h(12), h(23), h(13), h(24), h(14), h(34), h(1234))$. Similarly, we define $S_2\ (i, j)$, $i \neq j$, $i, j \in \{1, 2, 3, 4\}$. Next, in the same manner we define permutation group $S_3(i_1, i_2, i_3)$, where i_1, i_2, i_3 are 3 distinct integers taken out of $\{1, 2, 3, 4\}$. Thus, $S_3(2, 3, 4)$ acts on the elements 2, 3 and 4, and if $w = (4, 3, 2) \in S_3(2, 3, 4)$, then $w(\boldsymbol{h}) = (h(0), h(1), h(4), h(2), h(3), h(124), h(134), h(123), h(234), h(14), h(12), h(24), h(13), h(34), h(23), h(1234))$. Finally, the group $S_4(1, 2, 3, 4)$, which we shall briefly denote by S_4, is similarly defined. In addition to the above groups, we will occasionally use the group $S_2(1, 2) \times S_3(3, 4)$, which is the product of the two groups $S_2(1, 2)$ and $S_2(3, 4)$. We now find all non-isomorphic $\boldsymbol{h}$'s. Each of these $\boldsymbol{h}$'s corresponds to many A's which are isomorphic under S_m. Corresponding to a fixed $\boldsymbol{h}$, we take one such A and the corresponding bipartite graph. We, thus, get the representative bipartite graph of each orbit.

The conditions ((1.1)b, c) on bipartite graphs naturally impose conditions on the vectors $\boldsymbol{h}$. Consider the set of all possible distinct vectors $\boldsymbol{h}$ satisfying those conditions. Our goal is to find a set H of vectors $\boldsymbol{h}$ such that no two $\boldsymbol{h}$ in H are isomorphic and any $\boldsymbol{h}$, not in H, is isomorphic to an $\boldsymbol{h}$ in H. The set H will be called a "nonisomorphic set". We first find a set of vectors $\boldsymbol{h}$ such that any $\boldsymbol{h}$ not in the set is isomorphic to an $\boldsymbol{h}$ in the set. We next divide the set into subsets (branches) such that no two $\boldsymbol{h}$ of different subsets are isomorphic. If it so happens that no two $\boldsymbol{h}$ in a particular subset are isomorphic, then we leave that subset aside and consider the other subsets. Each of these (other) subsets is then further divided into sub^{2}sets, such that the vector $\boldsymbol{h}$ corresponding to different sub^{2}sets are nonisomorphic, and each vector in a subset is isomorphic to a vector in one of its sub^{2}sets. Again we divide each sub^{2}set into sub^{3}sets (sub^{2}branches) such that no two $\boldsymbol{h}$ of different sub^{3}sets are isomorphic. We continue the process until we end up with sub^{i}sets ($i = 1, 2, \ldots$), such that no $\boldsymbol{h}$ belonging to one of these is isomorphic to an $\boldsymbol{h}$ belonging to any other, and such that every possible $\boldsymbol{h}$ is isomorphic to an $\boldsymbol{h}$ in some sub^{i}set, for some i. The totality of all vectors $\boldsymbol{h}$ in all such sub^{i}sets ($i = 1, 2, \ldots$) gives the required "non-isomorphic set" H. The above method of sorting out the non-isomorphic vectors $\boldsymbol{h}$ will be called "non-isomorphic branching".

From now on, for simplicity, an element $h(u)$ of $\boldsymbol{h}$ will sometimes be denoted by u. (Thus, the elements of $\boldsymbol{h}$ are denoted by the same symbols as the corresponding elements of R.) We also denote the elements 123, 124, 134, and 234 of $\boldsymbol{h}$ by $\bar{4}, \bar{3}, \bar{2}$, and $\bar{1}$, respectively. Let $\boldsymbol{h}^{(1)} = (1, 2, 3, 4)$, $\boldsymbol{h}^{(2)} = (12, 13, 23, 14, 24, 34)$, and $\boldsymbol{h}^{(3)} = (\bar{4}, \bar{3}, \bar{2}, \bar{1})$. The permutations in S_4^* will now be denoted by the same symbols as those in S_4 acting on $\{1, 2, 3, 4\}$. Therefore, in a permutation in S_4^*, which is induced by the permutation (123)(4) in S_4, only the symbol 4 is fixed and the others are not fixed (i.e., are free).

Lemma 4.1. *For a* $w \in S_4^*$, *two vectors* $\boldsymbol{h}_1$ *and* $\boldsymbol{h}_2$ *satisfy* $\boldsymbol{h}_1 = w\boldsymbol{h}_2$, [*i.e., they are isomorphic*] *iff* $\boldsymbol{h}_1^{(i)} = w(\boldsymbol{h}_2^{(i)})$, $i = 1, 2, 3$, *and* $h_1(j) = wh_2(j) = h_2(j)$, $j = 0, 1234$.

Proof. The above follows from the nature of S_4^*.

We perform the "nonisomorphic branching" by putting conditions *sequentially* on the elements of $\boldsymbol{h}^{(1)}$, $\boldsymbol{h}^{(2)}$, and $\boldsymbol{h}^{(3)}$, respectively. We shall start out with the first state by imposing restrictions on the 4 elements of $\boldsymbol{h}^{(1)}$. We assume

$$1 \leq 2 \leq 3 \leq 4, \quad \text{i.e.,} \quad h(1) \leq h(2) \leq h(3) \leq h(4) \tag{4.2}$$

Clearly, any $\boldsymbol{h}$ not satisfying (4.2) is isomorphic to an $\boldsymbol{h}$ satisfying it. Suppose E_1 = the number of equalities in (4.2). Note that (4.2) gives rise to 8 cases, namely 3 cases corresponding to each of $E_1 = 1$ and $E_1 = 2$, and one for each of $E_1 = 0$ and $E_1 = 3$. Thus, we divide the set of all vectors $\boldsymbol{h}$ satisfying (4.2) into 8 disjoint subsets satisfying the conditions in the 8 cases.

Theorem 4.1. *Consider the* 8 *disjoint subsets of the set of vectors* $\mathbf{h}$ *satisfying* (4.2). *Then*

(a) *No two vectors* $\mathbf{h}$ *of different subsets are isomorphic.*

(b) *No two vectors* $\mathbf{h}$ *of the subset corresponding to the case* $E_1 = 0$ *are isomorphic.*

(c) *Consider any one of the* 3 *subsets corresponding to* $E_1 = 1$. *Two vectors* $\mathbf{h}$ *belonging to such a subset can be isomorphic only under those permutations in* S_4^* (*or equivalently, in* S_4) *for which* 2 *symbols* (*out of the* 4 *symbols*) *are fixed.*

(d) *Two vectors* $\mathbf{h}$ *belonging to any one of the* 2 *subsets corresponding to* $E_1 = 2$, *namely* $1 = 2 = 3 < 4$ *and* $1 < 2 = 3 = 4$, *can be isomorphic only under permutations in* S_4^* (*i.e.,* S_4) *for which at least one symbol is fixed.*

(e) *Two vectors* $\mathbf{h}$ *of the subset corresponding to the case* $E_1 = 2$, *namely* $1 = 2 < 3 = 4$, *can be isomorphic only under permutations in* S_4^* *which interchanges the symbols* (1, 2) *and/or* (3, 4).

(f) *Two vectors* $\mathbf{h}$ *of the subset corresponding to* $E_1 = 3$, *can possibly be isomorphic under any permutation in* S_4^*.

Proof. Part (a) is true because none of the conditions in 8 cases can be obtained from the others by a permutation in S_4^*.

The condition $1 < 2 < 3 < 4$ fixes the 4 symbols in the sense that under any permutation in S_4^*, except the identity permutation, the condition would be violated. Hence, the part (b) follows.

Consider any case corresponding to $E_1 = 1$. The condition in that case fixes 2 symbols in the sense that under any permutation in S_4^* for which those 2 symbols are not fixed the condition would be violated. For example, the condition $1 = 2 < 3 < 4$ fixes the 2 symbols (3, 4). This completes the proof of part (c).

Part (d) is similar to (c). Note that the condition $1 = 2 = 3 < 4$ fixes the symbol 4 and $1 < 2 = 3 = 4$ fixes the symbol 1.

Part (e) is clear because any permutation in S_4^* except for those interchanging the symbols (1, 2) and/or (3, 4), would violate the condition $1=2<3=4$.

Notice that the condition $1=2=3=4$ will not be disturbed by any permutation in S_4^*. Thus, in the subset of vectors $\boldsymbol{h}$ satisfying $1=2=3=4$, isomorphism is possible under any permutation in S_4^*. This completes the proof of the theorem.

It follows from the above theorem that all vectors $\boldsymbol{h}$ in the subset corresponding to the case $E_1=0$, are non-isomorphic. We leave this subset aside and consider the 7 other subsets. In view of the above theorem, in these subsets, at least one of the pairs of the symbols (1, 2), (2, 3), or (3, 4) are not fixed (i.e., are free). For example, the condition $1=2<3=4$, leaves the pairs of symbols (1, 2) and (3, 4) free, i.e. we can interchange either the symbols 1 and 2, or 3 and 4, or both. The condition $1=2=3<4$ leaves (1, 2), (1, 3), and (2, 3) free, i.e. we can do all possible interchanges of the symbols 1, 2, and 3. Similarly, the condition $1=2=3=4$ leaves (1, 2), (2, 3), and (3, 4) free, i.e., all possible interchanges of the symbols 1, 2, 3, and 4 are possible. We now go to the second stage of our sequential method by putting conditions on the elements of $\boldsymbol{h}^{(3)}$.

Theorem 4.2. *Consider the subsets of the first stage. We assume the conditions* $\bar{1}\geqslant\bar{2}, \bar{2}\geqslant\bar{3}$, *or* $\bar{3}\geqslant\bar{4}$ *according as the pairs of symbols* (1, 2), (2, 3), *or* (3, 4) *are free in the subsets. Then, in a subset, a vector* $\boldsymbol{h}$ *not satisfying the condition(s) must be isomorphic to a vector* $\boldsymbol{h}$ *satisfying the condition(s).*

Proof. Suppose in a subset the symbols (1, 2) are free. Then, any $\boldsymbol{h}$ vector not satisfying the condition $\bar{1}\geqslant\bar{2}$ can be obtained from an $\boldsymbol{h}$ vector satisfying the condition by interchanging 1 and 2. Similarly, if in a subset the symbols (1, 2), (2, 3) and (3, 4) are free, then any $\boldsymbol{h}$ vector not satisfying the conditions $\bar{1}\geqslant\bar{2}, \bar{2}\geqslant\bar{3}$, and $\bar{3}\geqslant\bar{4}$, i.e., $\bar{1}\geqslant\bar{2}\geqslant\bar{3}\geqslant\bar{4}$, must be isomorphic to an $\boldsymbol{h}$ vector satisfying the conditions by interchanging the symbols 1, 2, 3, and 4. The same argument holds for the other subsets. This completes the proof.

In each subset of the first stage (except the one corresponding to $E_1=0$), we take the set of all vectors $\boldsymbol{h}$ satisfying the conditions stated in Theorem 4.2. We now break up each condition into two parts, namely $\bar{1}=\bar{2}, \bar{1}>\bar{2}$; $\bar{2}=\bar{3}, \bar{2}>\bar{3}$; $\bar{3}=\bar{4}, \bar{3}>\bar{4}$, giving us a total of 2^3 $(=8)$. This will divide the vectors $\boldsymbol{h}$ within each subset into (one or more) disjoint sub^{2}sets. For example, the vectors $\boldsymbol{h}$ in the subset satisfying the condition $1=2=3=4$ are divided into 2^3 $(=8)$ sub^{2}sets, corresponding to the above. Clearly, no two vectors $\boldsymbol{h}$ of different sub^{2}sets (out of these 8 cases) are isomorphic. There is one sub^{2}set of each subset whose members are all nonisomorphic. For example, for the subset of vectors $\boldsymbol{h}$ satisfying $1=2=3=4$ and the sub^{2}set of vectors $\boldsymbol{h}$ satisfying $\bar{1}>\bar{2}>\bar{3}>\bar{4}$, this statement holds. We single these sub^{2}sets out, and consider nonisomorphic branching in the remaining sub^{2}sets in each subset. In these remaining sub^{2}sets, at least one of the pairs of the symbols (1, 2), (2, 3), and (3, 4) are free. For example, consider the subset of vectors $\boldsymbol{h}$ satisfying $1=2=3=4$, and the sub^{2}set of vectors $\boldsymbol{h}$ which further satisfies $\bar{1}=\bar{2}>\bar{3}=\bar{4}$. In this sub^{2}set, the pairs of symbols (1, 2) and (3, 4) are free, i.e., interchanges of (1 and 2) and/or (3 and 4) are possible. We, therefore, proceed to the next higher state of our sequential method of nonisomorphic branching by restricting the members of $\boldsymbol{h}^{(2)}$.

Theorem 4.3. *Consider the sub^{2}sets in which either none, or one or two pairs of symbols* (1, 2), (2, 3), *and* (3, 4) *are free. Then the conditions in Table* 5 *give all sub^{3}sets, sub^{4}sets, and sub^{5}sets.*

Table 5. Tabulation of nonisomorphic $\boldsymbol{h}$'s except for $E_1=E_2=3$

E_1	E_2	Set-type	Conditions
0	—	(0.1)	$1<2<3<4$
1	—	(1.1)	$1=2<3<4$
		(1.11)	$\bar{1}<\bar{2}$
		(1.12)	$\bar{1}=\bar{2}$
		(1.121)	$13<23$
		(1.122)	$13=23,\ 14\leq 24.$
1	—	(1.2)	$1<2=3<4$
	—	(1.3)	$1<2<3=4$
			(Further nonisomorphic branching for each of the sets (1.2) and (1.3) is similar to that for the set (1.1).)
2	—	(2.1)	$1=2=3>4$
		(2.11)	$\bar{1}<\bar{2}<\bar{3}$
		(2.12)	$\bar{1}<\bar{2}=\bar{3}$
		(2.121)	$12<13$
		(2.122)	$12=13,\ 24\leq 34$
		(2.13)	$\bar{1}=\bar{2}<\bar{3}$
		(2.131)	$13<23$
		(2.132)	$13=23,\ 14\leq 24.$
		(2.14)	$\bar{1}=\bar{2}=\bar{3}$
		(2.141)	$14<24<34$
		(2.142)	$14=24<34,\ 13\leq 23$
		(2.143)	$14<24=34,\ 12\leq 13$
		(2.144)	$14=24=34,\ 12\leq 13\leq 23.$
2	—	(2.2)	$1<2=3=4$
			(Further branching is similar to that for set type (2.1).)
2	—	(2.3)	$1=2<3=4$
		(2.31)	$\bar{1}<\bar{2},\ \bar{3}<\bar{4}$
		(2.32)	$\bar{1}=\bar{2},\ \bar{3}<\bar{4}$
		(2.321)	$13<23$
		(2.322)	$13=23,\ 14\leq 24$
		(2.33)	$\bar{1}<\bar{2},\ \bar{3}=\bar{4}$ (Similar to (2.32))
		(2.34)	$\bar{1}=\bar{2},\ \bar{3}=\bar{4}$
		(2.341)	$13<14,\ 13<23$

Table 5. (Cont.)

E_1	E_2	Set-type	Conditions
		(2.3411)	Either $24\geq 23$, or $24\geq 14$, or both.
		(2.3412)	$24<23, 24<14$
		(2.34121)	$23<14$
		(2.34122)	$23=14, 13\leq 24$.
		(2.342)	$13=14, 13<23, 23\leq 24$
		(2.343)	$13<14, 13=23, 14\leq 24$
		(2.344)	$13=14, 13=23, 23\leq 24, 14\leq 24$
3	—	(3.1)	$1=2=3=4$
3	0	(3.0.11)	$\bar{1}<\bar{2}<\bar{3}<\bar{4}$
3	1	(3.1.12)	$\bar{1}<\bar{2}<\bar{3}=\bar{4}$
		(3.1.121)	$13<14$
		(3.1.122)	$13=14, 23\leq 24$
		(3.1.13)	$\bar{1}<\bar{2}=\bar{3}<\bar{4}$ (Similar to (3.12))
		(3.1.14)	$\bar{1}=\bar{2}<\bar{3}<\bar{4}$ (Similar to (3.12))
3	2	(3.2.15)	$\bar{1}<\bar{2}=\bar{3}=\bar{4}$
		(3.2.151)	$12<13<14$
		(3.2.152)	$12=13<14, 24\leq 34$
		(3.2.153)	$12<13=14, 23\leq 24$
		(3.2.154)	$12=13=14, 23\leq 24\leq\leq 34$
3	2	(3.2.16)	$\bar{1}=\bar{2}=\bar{3}<\bar{4}$ (Similar to 3.2.15)
3	2	(3.2.17)	$\bar{1}=\bar{2}<\bar{3}=\bar{4}$
		(3.2.171)	$13<14, 13<23$.
		(3.2.1711)	Either $24\geq 23$, or $24\geq 14$, or both.
		(3.2.1712)	$23>24, 14>24$
		(3.2.17121)	$13<24$
		(3.2.17122)	$13=24, 23<14$.
		(3.2.172)	$13=14, 13<23, 23\leq 24$
		(3.2.173)	$13<14, 13=23, 14\leq 24$
		(3.2.174)	$13=14, 13=23, 14\leq 24, 23\leq 24$.

We first consider the notation in Table 5. The symbols E_1 (as before) and E_2 respectively denote the number of equalities in the two statements (i) $1\leq 2\leq 3\leq 4$, and (ii) $\bar{1}\leq\bar{2}\leq\bar{3}\leq\bar{4}$. Thus the case excluded in this theorem is the one where $E_1=E_2=3$, i.e. where $1=2=3=4$ and $\bar{1}=\bar{2}=\bar{3}=\bar{4}$.

The "set-type" is named as follows. We consider the values of E_1 and E_2. If $E_1\leq 2$, we use the value of E_1 only. In this case, the set-type starts with the value of E_1 followed by a period. If $E_1=3$, the set-type starts with 3 followed by a period followed by the value of E_2 and again a period. In the symbol for any set-type, consider the numbers just following the last period. If there are i ($i=1,\ldots,5$) digits here, then the symbol denotes a sub^{i}set. Thus (3.2.152) denotes a sub^{3}set, since 152 has 3 digits. If s_1 and s_2 are two set-types, such that the symbol for s_1 is contained in that of s_2 (i.e. s_2 contains all of s_1 plus some extra digits at the right end), then the set of values of $\boldsymbol{h}$ corresponding to s_1 contains the set of $\boldsymbol{h}$ corresponding to s_2. Thus, for example, the set of $\boldsymbol{h}$ for (3.2.17122) is contained in that for (3.7.171).

Next, we come to the conditions defining the various set-types. Thus, (0.1) is the set of all $\boldsymbol{h}$ such that $1<2<3<4$ is satisfied. Similarly, (1.2) is the set for which we have $1<2=3<4$. Also, if a set-type symbol s_1 is contained in a set-type symbol s_2, then the conditions defining s_1 also apply to s_2. For example, the total set of conditions defining (2.132) is $1=2=3<4$, $\bar{1}=\bar{2}<\bar{3}$, $13=23$ and $14\leqslant 24$. Finally, if s is any set-type which is a sub^{i}set ($i=0,\ldots,4$), then the sub^{i+1} sets contained in s are respectively denoted simply by putting $1, 2, 3, \ldots,$ after the symbol for s. Thus, the next generation sets in (2.3) are (2.31), (2.32), (2.33), and (2.34).

With the above explanations, an inspection of Table 5 should reveal that it provides the nonisomorphic branching under consideration. However, we illustrate the proof by considering two cases in detail.

(i) consider a sub^{2}set (say (1,12)) in which the symbols (1, 2) are free and the vectors $\boldsymbol{h}$ in which satisfy the condition $13\leqslant 23$. Clearly, any vector $\boldsymbol{h}$ not satisfying the condition is isomorphic to a vector $\boldsymbol{h}$ satisfying it. Now consider the two sub^{3}sets of vectors $\boldsymbol{h}$ satisfying $13<23$ and $13=23$. Obviously, all vectors $\boldsymbol{h}$ in the sub^{3}set satisfying $13<23$ are nonisomorphic. In the sub^{3}set of vectors $\boldsymbol{h}$ satisfying $13=23$, the symbols (1, 2) are still free. In this sub^{3}set, we find a sub^{4}set of vectors $\boldsymbol{h}$ satisfying the condition $14\leqslant 24$. Any vector $\boldsymbol{h}$ in the sub^{3}set not in the sub^{4}set is isomorphic to a vector $\boldsymbol{h}$ in the sub^{4}set. Moreover, all vectors $\boldsymbol{h}$ in the sub^{4}set are nonisomorphic. This is clear if $14<24$ because the condition fixes the symbols (1, 2). If $14=24$, then clearly a vector $\boldsymbol{h}$ in the sub^{4}set satisfying it remains unaffected under the permutation of the symbols 1 and 2.

(ii) Consider a sub^{2}sset (for example, (3.2.17)) in which the pairs of symbols (1, 2) and (3, 4) are both free. We find a set of vectors $\boldsymbol{h}$ satisfying $13\leqslant 23$ and $13\leqslant 14$. Obviously any vector $\boldsymbol{h}$ not satisfying the condition is isomorphic to a vector $\boldsymbol{h}$ satisfying the condition. (To see this, notice that isomorphism can occur under any of the four permutations in $S_2(1,2)\times S_2(3,4)$.) We now have 4 sub^{3}sets satisfying the conditions (a)–(d) in Table 5. In case of (b), the symbols (3, 4) are initially free and the sub^{4}set of vectors $\boldsymbol{h}$ is obtained by satisfying $23\leqslant 24$. Similarly for the cases (c) and (d). For (a), there are two cases (i.e., two sub^{3}sets), namely (I) $24<14$, $24<23$ and (II) otherwise. In case of (II), all vectors $\boldsymbol{h}$ in the sub^{4}set are nonisomorphic. But for (I), the interchange of the symbols in the two pairs (1, 2) and (3, 4) is simultaneously possible because under interchange, the conditions remain the same. We, therefore, consider the set of vectors $\boldsymbol{h}$ within the sub^{3}set satisfying the condition $13\leqslant 24$. Any vector $\boldsymbol{h}$ not satisfying the condition is isomorphic to a vector $\boldsymbol{h}$ satisfying the condition. We consider the 2 sub^{4}sets of vectors $\boldsymbol{h}$ satisfying the conditions $13<24$ and $13=24$. The sub^{4}set satisfying the condition $13<24$ is nonisomorphic. But for the sub^{4}set with $13=24$, the simultaneous interchange in the pairs (1, 2) and (3, 4) is again possible. To meet this situation, we restrict ourselves to the sub^{5}set of vectors $\boldsymbol{h}$ satisfying the condition $23\leqslant 14$. Clearly, any vector $\boldsymbol{h}$ of the sub^{4}set under consideration not belonging to this sub^{5}set, is isomorphic to a $\boldsymbol{h}$ vector in this sub^{5}set and, furthermore, this sub^{5}set is nonisomorphic.

The other cases are treated similarly. This completes the proof of the Theorem.

In the sub^{2}set of vectors $\boldsymbol{h}$ satisfying $E_1 = E_2 = 3$, the 4 symbols 1, 2, 3, and 4 are free, i.e., two vectors $\boldsymbol{h}$ can be isomorphic under permutations in S_4^*. At the third stage of our sequential method, we put conditions on the elements of $\boldsymbol{h}^{(2)}$. However, we cannot find conditions as earlier, because the ordering of the elements of $\boldsymbol{h}^{(2)}$ is not possible as we did for $\boldsymbol{h}^{(1)}$ and $\boldsymbol{h}^{(3)}$.

We, therefore, apply a new idea. Let α_1, α_6, and α_i, $2 \leqslant i \leqslant 5$, be the minimum, maximum and ith minimum of 12, 13, 23, 14, 24, 34. Thus

$$\alpha_1 \leqslant \alpha_2 \leqslant \alpha_3 \leqslant \alpha_4 \leqslant \alpha_5 \leqslant \alpha_6 \tag{4.3}$$

The α_i can be any of 12, 13, 23, 14, 24, 34. Define $\boldsymbol{\alpha}' = (\alpha_1, \alpha_2, \ldots, \alpha_6)$. For any vector $\boldsymbol{h}$, let E_3 denote the number of equalities in (4.3).

Lemma 2. *Two* $\boldsymbol{h}$ *vectors satisfying* $E_1 = E_2 = 3$ *are isomorphic under a permutation in* S_4^* *only if the corresponding* $\boldsymbol{\alpha}$ *vectors are isomorphic.*

Proof. This follows immediately from the nature of the sub^{2}set concerned, and the definition of $\boldsymbol{\alpha}$.

We choose a set of $\boldsymbol{h}$ vectors from the sub^2 set satisfying the condition $\alpha_1 = 12$. Any vector $\boldsymbol{h}$ not satisfying this condition is isomorphic to one satisfying it under some permutation in S_4^*. We next divide the set into $2^5 (= 32)$ sub^{3}sets corresponding to the different values of E_3. Notice that there are $\binom{5}{i}$ sub^{3}sets corresponding to the $\binom{5}{i}$ cases for $E_3 = i$, $i = 0, 1, \ldots, 5$. It is clear that two $\boldsymbol{h}$ vectors of different sub^{3}sets are nonisomorphic. Thus the problem is now to find nonisomorphic sub^{4}sets in the above sub^{3}sets. Observe that 12 and 34 have no symbols in common, but each of 13, 23, 14, 24 has one symbol in common with each of 12 and 34. In sub^{3}sets, α_i $(2 \leqslant i \leqslant 6)$ can be any of 13, 23, 14, 24 and 34. To find the nonisomorphic sub^{4}sets in a sub^{3}set, we prescribe conditions on different α_i's sequentially, until all the symbols (1, 2, 3 and 4) are fixed. In Table 6 we list the conditions to find the sub^{4}sets for each sub^{3}set and the number of subcases under each.

If we follow the spirit of the notation in Table 5, then the symbols for the set-type in Table 6 should all precede with (3.3.). Thus [1.21] should be written (3.31.21). However, for simplicity, we drop (3.3.) and indicate this by changing the bracket ( ) to [ ]. The entries in Table 6 are arranged with respect to values of E_3 so as to correspond to increasing complexity. Also, the last column gives the number of nonisomorphic $\boldsymbol{h}$'s for each set-type; this has been added to provide clarity in understanding.

Theorem 4.4. *Consider the vectors* $\boldsymbol{h}$ *satisfying* $E_1 = E_2 = 3$. *Then, Table 6 gives all the nonisomorphic cases.*

Table 6. Tabulation of nonisomorphic $\boldsymbol{h}$'s, for $E_1 = E_2 = 3$

E_3	Set-type	Conditions on α_i $(i = 2, \ldots, 6)$, given $\alpha_1 = 12$	Number of cases
5	[5.1]	None	1
0	[0.1]	$\alpha_2 = 13, \alpha_6 = 34$	6
	[0.2]	$\alpha_6 = 13$	24
1	[1.1]	$\alpha_1 = \alpha_2$	15
	[1.11]	$\alpha_2 = 34, \alpha_6 = 13$	3
	[1.12]	$\alpha_2 = 13, \alpha_6 = 23, 24 < 34$	3
	[1.13]	$\alpha_2 = 13, \alpha_6 = 14, 24 < 34$	3
	[1.14]	$\alpha_2 = 13, \alpha_6 = 34$	6
	[1.2]	$\alpha_2 = \alpha_3$	18
	[1.21]	$\alpha_2 = 34, \alpha_3 = 13$	6
	[1.22]	$\alpha_2 = 13, \alpha_3 = 24$	6
	[1.23]	$\alpha_2 = 13, \alpha_3 = 14, 23 < 24$	3
	[1.24]	$\alpha_2 = 13, \alpha_3 = 23, 14 < 24$	3
	[1.3, 4, 5]	$\alpha_i = \alpha_{i+1}$ $(i = 3, 4, 5)$ (Similar to [1.2])	18 each
4	[4.1]	$\alpha_1 < \alpha_2$	1
	[4.2]	$\alpha_2 < \alpha_3$	2
	[4.21]	$\alpha_2 = 34$	1
	[4.22]	$\alpha_2 = 13$	1
	[4.3]	$\alpha_3 < \alpha_4$	3
	[4.31]	$\alpha_2 = 34, \alpha_3 = 13$	1
	[4.32]	$\alpha_2 = 13, \alpha_3 = 14$	1
	[4.33]	$\alpha_2 = 13, \alpha_3 = 23$	1
	[4.4]	$\alpha_4 < \alpha_5$	2
	[4.41]	$\alpha_5 = 24, \alpha_6 = 13$	1
	[4.42]	$\alpha_5 = 34, \alpha_6 = 13$	1
	[4.5]	$\alpha_5 < \alpha_6$	1
	[4.51]	$\alpha_6 = 34$	1
2	[2.1]	$\alpha_1 = \alpha_2 = \alpha_3$	5
	[2.11]	$\alpha_2 = 34, \alpha_3 = 13$	3
	[2.12]	$\alpha_2 = 13, \alpha_3 = 23, \alpha_4 = 14, \alpha_5 = 24, \alpha_6 = 34$	1
	[2.13]	$\alpha_2 = 13, \alpha_3 = 14, \alpha_4 = 23, \alpha_5 = 24, \alpha_6 = 34$	1
	[2.2]	$\alpha_2 = \alpha_3 = \alpha_4$	6
	[2.21]	$\alpha_2 = 34, \alpha_3 = 13, \alpha_4 = 23, \alpha_5 = 14, \alpha_6 = 24$	1
	[2.22]	$\alpha_2 = 34, \alpha_3 = 13, \alpha_4 = 24, \alpha_5 = 23, \alpha_6 = 14$	1
	[2.23]	$\alpha_2 = 34, \alpha_3 = 13, \alpha_4 = 14$	2
	[2.24]	$\alpha_2 = 13, \alpha_3 = 23, \alpha_4 = 14$	2
	[2.3.4]	$\alpha_i = \alpha_{i+1}$ $(i = 3, 4)$ (Similar to [2.2])	6 each
	[2.5]	$\alpha_1 = \alpha_2, \alpha_3 = \alpha_4$	9
	[2.51]	$\alpha_2 = 34, \alpha_3 = 13, \alpha_4 = 14, \alpha_5 = 23, \alpha_6 = 24$	1
	[2.52]	$\alpha_2 = 34, \alpha_3 = 13, \alpha_4 = 24, \alpha_5 = 23, \alpha_6 = 14$	1
	[2.53]	$\alpha_2 = 13, \alpha_3 = 14, \alpha_4 = 34$	2
	[2.54]	$\alpha_2 = 13, \alpha_3 = 23, \alpha_4 = 14, \alpha_5 = 24, \alpha_6 = 34$	1
	[2.55]	$\alpha_2 = 13, \alpha_3 = 23, \alpha_4 = 24$	2
	[2.56]	$\alpha_2 = 13, \alpha_3 = 24, \alpha_4 = 34$	2
	[2.6, 7]	$\alpha_1 = \alpha_2, \alpha_i = \alpha_{i+1}$ $(i = 4, 5)$ (Similar to [2.5]).	9 each
	[2.8]	$\alpha_2 = \alpha_3, \alpha_4 = \alpha_5$	9
	[2.81]	$\alpha_2 = 34, \alpha_3 = 13$	3

Table 6 (Cont.)

E_3	Set-type	Conditions on α_i $(i=2,\ldots,6)$, given $\alpha_1=12$	Number of cases
	[2.82]	$\alpha_2=13, \alpha_6=34$	3
	[2.83]	$\alpha_4=34, \alpha_5=13$	3
	[2.9]	$\alpha_2=\alpha_3, \alpha_5=\alpha_6$	9
	[2.10, 11]	$\alpha_3=\alpha_4, \alpha_5=\alpha_6$ (Similar to [2.8])	9
3	[3.11]	$\alpha_1<\alpha_2<\alpha_3$	2
	[3.11]	$\alpha_2=13$	1
	[3.12]	$\alpha_2=34$	1
	[3.2]	$\alpha_1<\alpha_2=\alpha_3<\alpha_4$	4
	[3.21]	$\alpha_2=34, \alpha_3=13$	1
	[3.22]	$\alpha_2=13, \alpha_3=23, \alpha_4=34$	1
	[3.23]	$\alpha_2=13, \alpha_3=14, \alpha_4=34$	1
	[3.24]	$\alpha_2=13, \alpha_3=24, \alpha_4=34$	1
	[3.3]	$\alpha_1<\alpha_2, \alpha_4<\alpha_5$	4
	[3.31]	$\alpha_2=34, \alpha_5=13, \alpha_6=23$	1
	[3.32]	$\alpha_2=34, \alpha_5=13, \alpha_6=14$	1
	[3.33]	$\alpha_2=34, \alpha_5=13, \alpha_6=24$	1
	[3.34]	$\alpha_5=34, \alpha_6=13$	1
	[3.4]	$\alpha_1<\alpha_2, \alpha_5<\alpha_6$	2
	[3.41]	$\alpha_2=34$	1
	[3.42]	$\alpha_6=34$	1
	[3.5]	$\alpha_2<\alpha_3<\alpha_4$	4
	[3.51]	$\alpha_2=34, \alpha_3=13$	1
	[3.52]	$\alpha_2=13, \alpha_3=34$	1
	[3.53]	$\alpha_2=13, \alpha_3=14$	1
	[3.54]	$\alpha_2=13, \alpha_3=23$	1
	[3.6]	$\alpha_2<\alpha_3, \alpha_4<\alpha_5$	6
	[3.61]	$\alpha_2=34, \alpha_3=13, \alpha_4=14$	1
	[3.62]	$\alpha_2=34, \alpha_3=13, \alpha_6=24$	1
	[3.63]	$\alpha_2=13, \alpha_3=34, \alpha_6=23$	1
	[3.64]	$\alpha_2=13, \alpha_3=34, \alpha_4=14$	1
	[3.65]	$\alpha_2=13, \alpha_3=34, \alpha_4=24$	1
	[3.66]	$\alpha_2=13, \alpha_3=14, \alpha_4=23$	1
	[3.7]	$\alpha_1=\alpha_2, \alpha_3=\alpha_4=\alpha_5$	4
	[3.71]	$\alpha_2=34, \alpha_6=13$	1
	[3.72]	$\alpha_2=13, \alpha_6=14$	1
	[3.73]	$\alpha_2=13, \alpha_6=23$	1
	[3.74]	$\alpha_2=13, \alpha_6=34$	1
	[3.8]	$\alpha_3<\alpha_4<\alpha_5$	4
	[3.81]	$\alpha_2=34, \alpha_3=13, \alpha_4=23$	1
	[3.82]	$\alpha_2=34, \alpha_3=13, \alpha_4=24$	1
	[3.83]	$\alpha_2=13, \alpha_3=23, \alpha_4=34$	1
	[3.84]	$\alpha_2=13, \alpha_3=14, \alpha_4=34$	1
	[3.9]	$\alpha_3<\alpha_4, \alpha_5<\alpha_6$	4
	[3.91]	$\alpha_2=34, \alpha_3=13, \alpha_6=23$	1
	[3.92]	$\alpha_2=34, \alpha_3=13, \alpha_6=24$	1
	[3.93]	$\alpha_2=13, \alpha_3=23, \alpha_6=34$	1
	[3.94]	$\alpha_2=13, \alpha_3=14, \alpha_6=34$	1
	[3.10]	$\alpha_4<\alpha_5<\alpha_6$	2
	[3.101]	$\alpha_5=34, \alpha_6=13$	1
	[3.102]	$\alpha_5=24, \alpha_6=13$	1

Proof. Consider the case $E_3 = 5$. Here, all the α's are equal. Hence, there can be only one vector satisfying this requirement. This explains the entry for the set-type [5.1].

Next, take $E_3 = 0$. Clearly, only two nonisomorphic values of α_6 are possible, namely, 13 and 34. When $\alpha_6 = 13$, it is clear that the remaining α's can take any of the values 14, 23, 24 and 34, so that the number of possible cases is 4! (= 24). When $\alpha_6 = 34$, when w.l.o.g., we can have $\alpha_2 = 13$; beyond this the other α's are free, and hence take 3! (= 6) value.

Now, consider [1.1]. Notice that α_2 can be either 34 or 13, because the other cases are isomorphic to one of these. If $\alpha_2 = 34$, then w.l.o.g., we can take $\alpha_6 = 13$. Now α_i ($i = 3, 4, 5$) can be any of 23, 14, and 24, giving 6 possible values. Among these, only (23, 14, 24), (23, 24, 14), and (24, 23, 14) are nonisomorphic, because the other 3 cases can be obtained by interchanging the symbols (1, 3) and (2, 4). This give the entry for [1.11]. When $\alpha_2 = 13$, we get 3 entries corresponding to the 3 nonisomorphic values of $\alpha_6 = 23$, 14, and 34. If $\alpha_2 = 13$, $\alpha_6 = 23$, then the interchange of the symbols 2 and 3 is possible without violating the given conditions; hence we need to add the further condition ($24 < 34$). It is now easy to see that 3 cases arise for the values of the remaining three α's. This explains [1.12]. The other two subcases of [1.1] are similar.

Next, we consider [2.8] to obtain an example for $E_3 = 2$. If $\alpha_2 = 34$, then w.l.o.g. $\alpha_3 = 13$; these conditions fix the symbols 1, 2, 3 and 4. However, because $\alpha_4 = \alpha_5$, the vector $(\alpha_4, \alpha_5, \alpha_6)$ takes only 3 distinct values. This explains [2.81]. Other cases are similar.

Above, we have verified a few entries of the Table 6. The proof can be completed by similarly verifying the other cases.

We conclude this section and the paper with the remark that the way we have presented the entries in Tables 5 and 6 is very easily programable.

References

[1] N.G. De Bruijn, Polya's theory of counting, in: E.F. Bechenback, ed., Applied Combinatorial Mathematics (Wiley, New York, 1964) 144–184.

[2] F. Harary and E.M. Palmer, Graphical Enumeration (Academic Press, New York, 1973).

[3] J.N. Srivastava, Designs for searching nonnegligible effects, in: J.N. Srivastava, ed., A Survey of Statistical Design and Linear Models (North-Holland, Amsterdam, 1975) 507–19.

[4] J.N. Srivastava and S. Ghosh, Balanced 2^m factorial designs of resolution V which allow search and estimation of one extra unknown effect, $4 \leq m \leq 8$, Commun. in Statistics.

Annals of Discrete Mathematics 6 (1980) 333–342

DECOMPOSITIONS OF RATIONAL CONVEX POLYTOPES

Richard P. STANLEY*

Department of Mathematics, Massachusetts Institute of Technology, Cambridge, MA 02139, U.S.A.

1. Decompositions

Let $\mathcal{P}$ be a d-dimensional convex polytope (or d-polytope) in $\mathbb{R}^m$ whose vertices have rational coordinates. We call $\mathcal{P}$ a *rational* convex polytope. Let $\partial\mathcal{P}$ denote the boundary of $\mathcal{P}$. If n is a positive integer, define $i(\mathcal{P}, n)$ (resp. $j(\mathcal{P}, n)$) to be the number of points $\alpha \in \mathcal{P}$ (resp. $\alpha \in \mathcal{P} - \partial\mathcal{P}$) such that $n\alpha \in \mathbb{Z}^m$. These functions have been studied by Ehrhart[1] [3], Macdonald [8, 9], McMullen [10, 11], and others. Here we will develop some new properties of $j(\mathcal{P}, n)$ and $i(\mathcal{P}, n)$. Proofs for the most part will be omitted.

A word on notation: $\mathbb{N}$ denotes the non-negative integers, $\mathbb{P}$ the positive integers, $[n]$ the set $\{1, 2, \ldots, n\}$ where $n \in \mathbb{N}$. $\mathbb{Z}$, $\mathbb{Q}$, $\mathbb{R}$ have their usual meanings. If $\alpha \in \mathbb{Q}^m$, then $l(\alpha)$ denotes the least positive integer t for which $t\alpha \in \mathbb{Z}^m$. In particular, if $\alpha \in \mathbb{Q}$, then $l(\alpha)$ is the denominator of α when written in lowest terms.

It will be more convenient to work not with $j(\mathcal{P}, n)$ and $i(\mathcal{P}, n)$ themselves, but with the formal power series

$$J(\mathcal{P}, \lambda) = 1 + \sum_{n=1}^{\infty} i(\mathcal{P}, n)\lambda^n,$$

$$\bar{J}(\mathcal{P}, \lambda) = \sum_{n=1}^{\infty} j(\mathcal{P}, n)\lambda^n.$$

Let $V = V(\mathcal{P})$ denote the set of vertices of $\mathcal{P}$, and let $\omega = (\alpha_1, \ldots, \alpha_\nu)$ be a linear ordering of the elements of V. For each non-void face F of $\mathcal{P}$, define $\delta(F) = \delta_\omega(F)$ to be that vertex α_i of F for which i is minimal. For instance, $\delta(\mathcal{P}) = \alpha_1$. Now let $\Phi = (F_0, F_1, \ldots, F_d)$ be a *flag* of faces of $\mathcal{P}$, i.e., F_i is an i-dimensional face (or i-face) of $\mathcal{P}$ and $F_0 \subset F_1 \subset \cdots \subset F_d = \mathcal{P}$. Call Φ a *full flag* if $\delta(F_i)$ is not a vertex of F_{i-1} for $1 \leq i \leq d$. If Φ is a full flag, then define $\Delta(\Phi)$ to be the d-simplex with vertices $F_0, \delta(F_1), \delta(F_2), \ldots, \delta(F_d)$.

* Partially supported by NSF Grant MCS 7701947.

[1] Our only reference to the work of Ehrhart will be [3], which is an exposition of his work over a period of many years. Further references may be found in [3].

Lemma 1.1. *The set of simplices $\Delta(\Phi)$, as Φ ranges over all full flags of $\mathcal{P}$, form the maximal faces of a triangulation Γ_ω of $\mathcal{P}$.*

An elementary Euler characteristic argument, basically that given in [9, Section 1], expresses $J(\mathcal{P}, \lambda)$ and $\bar{J}(\mathcal{P}, \lambda)$ in terms of simpler generating functions, as follows.

Proposition 1.2. *Let ω be an ordering of $V(\mathcal{P})$ with corresponding triangulation $\Gamma = \Gamma_\omega$. Let $\partial\Gamma$ denote the set of those faces $\mathcal{S} \in \Gamma$ which lie on $\partial\mathcal{P}$. Then*

$$J(\mathcal{P}, \lambda) = \sum_{\mathcal{S} \in \Gamma - \partial\Gamma} (-1)^{d - \dim \mathcal{S}} J(\mathcal{S}, \lambda) = \sum_{\mathcal{S} \in \Gamma} \bar{J}(\mathcal{S}, \lambda),$$

$$\bar{J}(\mathcal{P}, \lambda) = \sum_{\mathcal{S} \in \Gamma} (-1)^{d - \dim \mathcal{S}} \bar{J}(\mathcal{S}, \lambda) = \sum_{\mathcal{S} \in \Gamma - \partial\Gamma} J(\mathcal{S}, \lambda).$$

In order to obtain explicit expressions for $J(\mathcal{P}, \lambda)$ and $\bar{J}(\mathcal{P}, \lambda)$, it remains to evaluate $J(\mathcal{S}, \lambda)$ when $\mathcal{S}$ is a simplex. This is a result of Ehrhart, briefly discussed in [3, p. 54] (see also [8, Section 1]). Suppose that $\mathcal{S}$ is a k-simplex in $\mathbb{R}^m$ with rational vertices $\beta_0, \ldots, \beta_k$. Let $G \subset \mathbb{Z}^{m+1}$ be the abelian group generated by the set

$$S = \{(\beta, r) : \beta \in \mathbb{Z}^m,\ r \in \mathbb{P},\ \text{and } \beta/r \in \mathcal{S}\}.$$

Let H be the subgroup of G generated by the $k+1$ vectors $\gamma_i = (l(\beta_i)\beta_i, l(\beta_i))$. Then H is a subgroup of G of finite index t; in fact, t is equal to the greatest common divisor of all $(k+1) \times (k+1)$ minors of the matrix whose rows are the vectors γ_i. Now define the "half-open" parallelopipeds

$$\Delta_H = \{a_1\gamma_1 + \cdots + a_{k+1}\gamma_{k+1} : 0 \leq a_i < 1\},$$

$$\bar{\Delta}_H = \{a_1\gamma_1 + \cdots + a_{k+1}\gamma_{k+1} : 0 < a_i \leq 1\}.$$

The sets $\Delta_H \cap \mathbb{Z}^{m+1}$ and $\bar{\Delta}_H \cap \mathbb{Z}^{m+1}$ are each a set of coset representatives for H in G. Let

$$\Delta_H \cap \mathbb{Z}^{m+1} = \{(\varepsilon_i, r_i) : 0 \leq i < t\}, \qquad \bar{\Delta}_H \cap \mathbb{Z}^{m+1} = \{(\zeta_i, s_i) : 0 \leq i < t\}.$$

For instance, if we choose (ε_0, r_0) and (ζ_0, r_0) to be in H, then we have $(\varepsilon_0, r_0) = 0$ and $(\zeta_0, s_0) = \gamma_1 + \cdots + \gamma_{k+1}$.

Theorem 1.3. *With the above notation, we have*

$$J(\mathcal{S}, \lambda) = \left(\sum_{i=0}^{t-1} \lambda^{r_i}\right) \prod_{j=0}^{k} (1 - \lambda^{l(\beta_j)})^{-1},$$

$$\bar{J}(\mathcal{S}, \lambda) = \left(\sum_{i=0}^{t-1} \lambda^{s_i}\right) \prod_{j=0}^{k} (1 - \lambda^{l(\beta_j)})^{-1}.$$

Moreover, the s_i*'s are just the numbers* $\sum l(\beta_j) - r_i$ *in some order, so that*

$$\bar{J}(\mathcal{S}, \lambda) = (-1)^{k+1} J(\mathcal{S}, 1/\lambda).$$

Example 1.4. Let $\mathcal{S}$ have vertices $(0, 0)$, $(1, \frac{1}{2})$, $(1, \frac{3}{2})$. Then $G = \mathbb{Z}^3$, $\gamma_1 = (0, 0, 1)$, $\gamma_2 = (2, 1, 2)$, $\gamma_3 = (2, 3, 2)$, $t = 4$, $(\varepsilon_0, r_0) = (0, 0, 0)$, $(\varepsilon_1, r_1) = (1, 1, 1)$, $(\varepsilon_2, r_2) = (2, 2, 2)$, $(\varepsilon_3, r_3) = (3, 3, 3)$, $(\zeta_0, s_0) = (4, 4, 5)$, $(\zeta_1, s_1) = (1, 1, 2)$, $(\zeta_2, s_2) = (2, 2, 3)$, $(\zeta_3, s_3) = (3, 3, 4)$, $J(\mathcal{S}, \lambda) = (1 + \lambda + \lambda^2 + \lambda^3)/(1 - \lambda)(1 - \lambda^2)^2$, $\bar{J}(\mathcal{S}, \lambda) = (\lambda^2 + \lambda^3 + \lambda^4 + \lambda^5)/(1 - \lambda)(1 - \lambda^2)^2$.

Remark. Following Ehrhart [3, p. 47], we say that the rational convex d-polytope $\mathcal{P}$ in $\mathbb{R}^m$ is *reticular* if the lattice $\mathbb{Z}^m \cap A$ of integer points of the affine space A spanned by $\mathcal{P}$ has rank d, i.e., $\mathbb{Z}^m \cap A \cong \mathbb{Z}^d$. This is equivalent to the statement that $\mathbb{Z}^m \cap A \neq \emptyset$ (see [3, Proposition 25]). If $\mathcal{P}$ is reticular, then let $\phi : A \to \mathbb{R}^d$ be an affine transformation which is a bijection between $A \cap \mathbb{Z}^m$ and $\mathbb{Z}^d$. (Such a ϕ clearly exists.) The image $\phi(\mathcal{P})$ of $\mathcal{P}$ is a d-polytope in $\mathbb{R}^d$, and hence has a positive Euclidean volume $\psi(\mathcal{P})$, called the *relative volume* of $\mathcal{P}$. It is easy to see that $\psi(\mathcal{P})$ depends only on $\mathcal{P}$, not on ϕ. If $d = m$, then $\psi(\mathcal{P})$ is just the volume of $\mathcal{P}$. As an example, let $\mathcal{P}$ be the line segement connecting $(0, 0)$ to $(1, 1)$. Define $\phi(a, a) = a$. Then $\phi(\mathcal{P}) = [0, 1]$, so $\psi(\mathcal{P}) = 1$. Now let $\mathcal{S}$ be a k-simplex as in Theorem 1.3, and assume that $\mathcal{S}$ is reticular. It is then easily seen that the integer $t = [G : H]$ appearing in Theorem 1.3 is given by

$$t = k!\, l(\beta_0) l(\beta_1) \cdots l(\beta_k) \psi(\mathcal{S}).$$

It then follows from Proposition 1.2 that the Laurent expansion of $J(\mathcal{P}, \lambda)$ about $\lambda = 1$ begins

$$J(\mathcal{P}, \lambda) = d!\, \psi(\mathcal{P})(1 - \lambda)^{-d-1} + \cdots. \tag{1}$$

It is also easy to give a direct proof of (1).

Combining Proposition 1.2 and Theorem 1.3, we obtain an explicit expression for $J(\mathcal{P}, \lambda)$ and $\bar{J}(\mathcal{P}, \lambda)$, showing that they are rational functions of λ with denominator

$$\prod_{\beta \in V(\mathcal{P})} (1 - \lambda^{l(\beta)}),$$

a result due originally to Ehrhart [3, p. 53]. An immediate corollary of Proposition 1.2 and Theorem 1.3 is the following "reciprocity theorem", due essentially to Ehrhart [3, p. 30] and Macdonald [9, Theorem 4.6] (see also [14]).

Theorem 1.5. *As rational functions we have*

$$\bar{J}(\mathcal{P}, \lambda) = (-1)^d J(\mathcal{P}, 1/\lambda).$$

Because of Theorem 1.5, we will now confine our attention to $J(\mathcal{P}, \lambda)$. While we now have an explicit expression for this generating function, we would like a more informative expression. Toward this end, suppose that the maximal faces of the triangulation Γ_ω of $\mathcal{P}$ can be ordered $G_1, G_2, \ldots, G_s$ so that for each $2 \leq i \leq s$, the set $G_i \cap (G_1 \cup G_2 \cup \cdots \cup G_{i-1})$ is a union of facets $(=(d-1)$-faces) of G_i. Then $(G_1, G_2, \ldots, G_s)$ is called a *shelling* of Γ_ω, and Γ_ω is called *shellable.*

Theorem 1.6. *Let ω be an ordering of $V(\mathcal{P})$, and suppose that Γ_ω has a shelling $(G_1, G_2, \ldots, G_s)$. Then*

$$J(\mathcal{P}, \lambda) = \sum_{i=1}^{s} \left(\sum_{j=0}^{t_i-1} \lambda^{r_{ij}+q_{ij}} \right) \prod_{k=0}^{d} (1-\lambda^{l(\beta_{ik})})^{-1}, \tag{2}$$

where $\beta_{i0}, \beta_{i1}, \ldots, \beta_{id}$ are the vertices of G_i, where

$$J(G_i, \lambda) = \left(\sum_{j=0}^{t_i-1} \lambda^{r_{ij}} \right) \prod_{k=0}^{d} (1-\lambda^{l(\beta_{ik})})^{-1}$$

(as in Theorem 1.3), and where q_{ij} is the sum of certain of the numbers $l(\beta_{ik})$, $0 \leq k \leq d$. The precise definition of q_{ij} is not important here, but we note that

$$q_{i0} = \sum_{\beta \in V(F)} l(\beta)$$

where F is the unique minimal face of G_i whose interior is disjoint from $G_i \cap (G_1 \cup G_2 \cup \cdots \cup G_{i-1})$. In particular, $q_{10} = 0$.

In order for Theorem 1.6 to be of any value, we must show that the triangulation Γ_ω of $\mathcal{P}$ is shellable. It is easy to see that the triangulation Γ_ω has the following alternative description. Let $\omega = (\alpha_1, \ldots, \alpha_\nu)$. Pull the vertices of $\mathcal{P}$ (as defined in [6, p. 80] or [12, pp. 116–117]) in the order $\alpha_1, \ldots, \alpha_\nu$. This yields a triangulation Λ of $\partial\mathcal{P}$. Then the maximal faces of Γ_ω are those of the form $\{\alpha_1\} \cup F$, where F is a maximal face of Λ not containing α_1. In particular, $\partial\Gamma_\omega = \Lambda$.

Using [1] and the above description of Γ_ω, it is easy to deduce the following result.

Theorem 1.7. *If ω is an ordering of $V(\mathcal{P})$, then Γ_ω is shellable. Hence $J(\mathcal{P}, \lambda)$ can always be written in the form* (2).

2. Consequences

We now use the results of Section 1 to examine the generating function $J(\mathcal{P}, \lambda)$ in more detail. To save space we will not state our results in the fullest possible generality.

Theorem 2.1. *Suppose every vertex of $\mathcal{P}$ has integer coordinates. Then*

$$J(\mathcal{P}, \lambda) = \frac{W(\mathcal{P}, \lambda)}{(1-\lambda)^{d+1}}, \tag{3}$$

where $W(\mathcal{P}, \lambda)$ is a polynomial of degree at most d with non-negative integer coefficients.

Proof. Each $l(\beta_{ik}) = 1$ in (2), from which the proof follows.

Theorem 2.1 implies that $i(\mathcal{P}, n)$ is a polynomial in n of degree d, as was shown by Ehrhart [3, p. 50] (repeated in [8]) and McMullen [10, 11]. What these persons had not shown was that the coefficients of $W(\mathcal{P}, \lambda)$ are non-negative. This fact was first established in [16, Proposition 4.5] by algebraic techniques (Cohen–Macaulay rings), but now we have a more geometric proof. We remark that it follows from (1) that the leading coefficient of $i(\mathcal{P}, n)$ is $\psi(\mathcal{P})$. For further ramifications of this fact, see [3, 8, 14, pp. 209–211].

It is natural to ask whether the polynomial $W(\mathcal{P}, \lambda)$ has a more combinatorial interpretation than afforded by (2). In certain cases we can indeed say more.

Definition 2.2. Let $\mathcal{P}$ be a d-polytope in $\mathbb{R}^m$ with integer vertices, and let ω be an ordering of $V(\mathcal{P})$. We say that ω is *compressed* if the following property holds: If $\mathcal{S}$ is any face (equivalently, any maximal face) of the triangulation Γ_ω, then the relative volume $\psi(\mathcal{S})$ of $\mathcal{S}$ is equal to $1/(\dim \mathcal{S})!$. We say that $\mathcal{P}$ itself is *compressed* if every ordering ω is compressed.

Theorem 2.3. *Let $\mathcal{P}$ be a d-polytope in $\mathbb{R}^m$ with integer vertices. Let ω be an ordering of $V(\mathcal{P})$, and let $\delta = \delta_\omega$ be defined as in Section 1. Suppose that for any rational point α in the relative interior of some face F of $\mathcal{P}$, if c is the unique positive number (necessarily rational) such that $(\alpha - c\delta(F))/(1-c)$ lies on ∂F, then $l(\alpha)c \in \mathbb{Z}$. Then ω is compressed.*

Example 2.4. (a) Suppose that one of the vertices of $\mathcal{P}$ is the origin O, and that the matrix whose rows are the vertices of $\mathcal{P}$ is totally unimodular. Let $\omega = (\alpha_1, \ldots, \alpha_\nu)$ be any ordering of $V(\mathcal{P})$ for which $\alpha_1 = O$. Then ω is compressed. (The proof is omitted.) However, other orderings ω need not be compressed. For instance, if

$$V(\mathcal{P}) = \{(0,0,0), (0,0,1), (0,1,0), (1,0,0), (0,1,1), (1,1,0), (1,1,1)\},$$

then any ordering ω with $\alpha_1 = (0,1,0)$ is not compressed.

(b) Let Ω_n be the convex polytope of all $n \times n$ doubly stochastic matrices. Then Ω_n is compressed. For let M be a rational matrix in Ω_n. Then $l(M)M$ is a sum of permutation matrices. Let k be the greatest integer for which $l(M)M - k\delta(F)$ has non-negative entries. Then $l(M)c = k \in \mathbb{Z}$.

(c) Let $\mathcal{P}$ be the 3-simplex with vertices $(0,0,0)$, $(1,1,0)$, $(1,0,1)$, $(0,1,1)$. Then *no* ordering ω is compressed.

Corollary 2.5. *Let $\mathcal{P}$ be as in Definition 2.2, and let ω be a compressed ordering of*

$V(\mathcal{P})$. Let f_i be the number of i-faces of Γ_ω. Then

$$i(\mathcal{P}, n) = \sum_{i=0}^{d} f_i \binom{n-1}{i}. \tag{4}$$

It follows that if $W(\mathcal{P}, \lambda) = h_0 + h_1\lambda + \cdots + h_d\lambda^d$ (as defined in (3)), then $(h_0, h_1, \ldots, h_d)$ is the h-vector of Γ_ω, as defined in [15, p. 136].

Note that it follows from the definition of Γ_ω that the numbers f_i in the above corollary depend only on ω and the lattice $L(\mathcal{P})$ of faces of $\mathcal{P}$ (even if ω is not compressed). Hence, if ω is compressed we can compute $i(\mathcal{P}, n)$ from ω and $L(\mathcal{P})$ alone. In particular, if $\mathcal{P}$ is compressed then $L(\mathcal{P})$ determines $i(\mathcal{P}, n)$.

Corollary 2.6. *Preserve the notation of Corollary 2.5. Let $(G_1, G_2, \ldots, G_s)$ be a shelling of Γ_ω. Then h_i is equal to the number of integers $j \in [r]$ for which $G_j \cap (G_1 \cup \cdots \cup G_{j-1})$ is a union of i facets of G_j.*

Corollary 2.6 provides a kind of combinatorial interpretation to the coefficients of $W(\mathcal{P}, \lambda)$ when $\mathcal{P}$ possesses a compressed ordering ω, since by Theorem 1.6 the triangulation Γ_ω is always shellable.

Corollary 2.7. *If $\mathcal{P}$ is compressed and ω is an ordering of $V(\mathcal{P})$, then the f-vector $(f_0, f_1, \ldots, f_d)$ of Γ_ω (with f_i as in Corollary 2.5) depends only on $\mathcal{P}$, not on ω.*

Proof. Immediate from (4), since $i(\mathcal{P}, n)$ has a unique representation in the form (4).

Corollary 2.7 is a purely combinatorial statement (i.e., involving only the structure of the lattice of faces) about $\mathcal{P}$. Thus for instance from Example 2.4(b) we get restrictions on the facial structure of the polytopes Ω_n. These polytopes have been studied by various authors (see, e.g., [5] and the references given there), and it may be interesting to see how Corollary 2.7 ties in with previously known results.

Remark. Triangulations Γ of rational polytopes $\mathcal{P}$ whose simplices have small volume have been considered elsewhere, such as [4] and [7]. (In the latter reference, see especially Theorem 12* on p. 95 and Chapter III. It would be interesting to interpret the theory developed in [7] in a purely combinatorial manner.) These writers, however, do not consider the restriction that the vertices of Γ must coincide with the vertices of $\mathcal{P}$. Such restrictions have apparently been considered in some unpublished work of Lovasz, but this work is not available to this author.

Let us now briefly consider polytopes $\mathcal{P}$ whose vertices don't necessarily have integer coordinates, but only rational coordinates. Let M be the least common multiple of the numbers $l(\alpha)$, $\alpha \in V(\mathcal{P})$. Let $\zeta = e^{2\pi i/M}$. It follows from Proposition 1.2 and Theorem 1.3 (or from the stronger Theorems 1.6 and 1.7) that there are polynomials $P_0, P_1, \ldots, P_{M-1}$ such that

$$i(\mathcal{P}, n) = \sum_{k=0}^{M-1} \zeta^{kn} P_k(n).$$

Define

$$\gamma(\mathcal{P}) = \max_{1 \leq i \leq M-1} (1 + \deg P_i),$$

where we set $\deg 0 = -1$. The number $\gamma(\mathcal{P})$ measures how close $i(\mathcal{P}, n)$ is to being a polynomial. We have $\gamma(\mathcal{P}) = 0$ if and only if $i(\mathcal{P}, n)$ is a polynomial. (The number $\gamma(\mathcal{P})$ is called by Ehrhart [3, p. 12] the *grade* of $i(\mathcal{P}, n)$.) The next result establishes a conjecture of Ehrhart [3, p. 53].

Theorem 2.8. *If every j-face of $\mathcal{P}$ is reticular, then $\gamma(\mathcal{P}) \leq j$.*

Sketch of proof. Fix j. If $\mathcal{R}$ is a rational convex polytope such that every j-face of $\mathcal{R}$ is reticular, then define $\eta(\mathcal{R})$ to be the number of j-faces of R which have no integer vertices. The proof is by induction on $\eta(\mathcal{P})$. First suppose $\eta(\mathcal{P}) = 0$, i.e., every j-face has an integer vertex. Choose an ordering ω of $V(\mathcal{P})$ such that all the integer vertices come first. It follows easily from Theorem 1.5 that $\gamma(\mathcal{P}) \leq j$.

Now let $\eta(\mathcal{P}) > 0$ be arbitrary, and assume the theorem for all $\mathcal{R}$ with $\eta(\mathcal{R}) < \eta(\mathcal{P})$. Let F be a j-face of $\mathcal{P}$ with no integer vertex. Let β be an integer point on the affine span A of F but not on F. (Such a point exists since $\mathbb{Z}^m \cap A \cong \mathbb{Z}^j$. Let $\mathcal{L}$ be the convex hull of $\mathcal{P}$ and β. Every j-face of $\mathcal{L}$ is either a j-face of $\mathcal{P}$ or has β as a vertex. Moreover, F itself is a *proper* subset of a j-face of $\mathcal{L}$. Hence $\eta(\mathcal{L}) < \eta(\mathcal{P})$. By the induction hypothesis, $\gamma(\mathcal{L}) \leq j$.

Now $\mathcal{L}$ has an obvious cellular decomposition whose maximal cells consist of $\mathcal{P}$ together with the convex hull of β with each facet of $\mathcal{P}$ "visible" from β. Denoting these maximal cells different from $\mathcal{P}$ by $\mathcal{L}_1, \ldots, \mathcal{L}_t$, it is easily seen as above that $\eta(G) < \eta(\mathcal{P})$ for any face G of any $\mathcal{L}_i$ (including $G = \mathcal{L}_i$). Now an Euler characteristic argument (as in Proposition 1.2) expresses $J(\mathcal{P}, \lambda)$ as a linear combination of $J(\mathcal{L}, \lambda)$ and the $J(G, \lambda)$'s where G is a face of some $\mathcal{L}_i$. Since $\gamma(\mathcal{L}) \leq j$ and each $\gamma(G) \leq j$, there follows $\gamma(\mathcal{P}) \leq j$, as was to be proved.

Remark. Theorem 2.8 does not determine $\gamma(\mathcal{P})$, in the sense that $\mathcal{P}$ may have a j-face which is not reticular, yet $\gamma(\mathcal{P}) = j$. For instance, let $\mathcal{P}$ be the convex 3-polytope in $\mathbb{R}^3$ with vertices $(0, 0, 0)$, $(1, 0, 0)$, $(0, 1, 0)$, $(1, 1, 0)$, and $(\frac{1}{2}, 0, \frac{1}{2})$. Thus $\mathcal{P}$ has a 0-face (vertex) which is not reticular, yet $i(\mathcal{P}, n) = \binom{n+3}{3}$ so $\gamma(\mathcal{P}) = 0$.

Remark. Using Theorem 2.8, one can give non-algebraic proofs of some further results in [16], viz., Theorems 3.3 and 5.5.

3. Examples

For what classes of polytopes $\mathcal{P}$ can $i(\mathcal{P}, n)$ or $J(\mathcal{P}, \lambda)$ be explicitly computed? This interesting question has not been systematically investigated. Here we give two examples to show that further work may be worthwhile. We know of several other classes besides the two below, but for brevity's sake they will not be discussed.

Example 3.1. Let $\gamma_1, \ldots, \gamma_t \in \mathbb{Z}^m$. Let

$$\mathcal{P} = \mathcal{P}(\gamma_1, \ldots, \gamma_t) = \{a_1\gamma_1 + \cdots + a_t\gamma_t : 0 \leq a_i \leq 1\}.$$

Thus by definition $\mathcal{P}$ is a vector sum of line segments and therefore a *zonotope*. Using the techniques of [13, Section 5], it can be shown that

$$i(\mathcal{P}, n) = \sum_X f(X) n^{|X|}, \tag{5}$$

where X ranges over all linearly independent subsets of $\{\gamma_1, \ldots, \gamma_t\}$ and where $f(X)$ denotes the greatest common divisor of all minors of size $|X|$ of the matrix whose rows are the elements of X. For instance, if the matrix whose rows are $\gamma_1, \ldots, \gamma_t$ is *totally unimodular* (every minor equals 0 or ± 1), then the coefficient of n^i in $i(\mathcal{P}, n)$ is equal to the number of i-element linearly independent subsets of $\{\gamma_1, \ldots, \gamma_t\}$. In other words, these coefficients are the "independent set numbers" of the geometry or matroid (in the sense of [2] or [17]) determined by $\gamma_1, \ldots, \gamma_t$. An especially interesting case occurs when $\mathcal{P}$ is the convex hull of all $m!$ vectors $(b_1, \ldots, b_m)$ whose entries are the distinct integers $1, 2, \ldots, m$ in some order. (Although $\mathcal{P}$ is not of the form $\{a_1\gamma_1 + \cdots + a_t\gamma_t : 0 \leq a_i \leq 1\}$, there is an integral unimodular affine transformation which transforms $\mathcal{P}$ into such a polytope.) $\mathcal{P}$ is known as a *permutohedron*, and it can be deduced from (5) that

$$i(\mathcal{P}, n) = \sum_{i=0}^{m-1} g_i n^i$$

where g_i is the number of forests on m vertices with i edges. In particular, $g_{m-1} = m^{m-2}$, so the relative volume of $\mathcal{P}$ is m^{m-2}. On the other hand, Zaslavsky has shown (private communication) that $i(\mathcal{P}, 1)$ is equal to the number of distinct *ordered* score vectors $(s_1, s_2, \ldots, s_m)$ of an m-vertex tournament, i.e., if T is a tournament on the vertex set $\{v_1, \ldots, v_m\}$, then s_i is the outdegree of v_i. Hence the number of ordered score vectors of length m is equal to the number of forests on m vertices. More generally, it can be shown that if G is any (undirected) graph, then the number of distinct outdegree sequences of orientations of G is equal to the number of spanning forests of G. It would be interesting to prove this result by finding a simple one-to-one correspondence. (We can give a messy, inductive one-to-one correspondence.)

Example 3.2. Let $\mathcal{P}_d$ be the convex d-polytope in $\mathbb{R}^d$ of all points $(x_1, \ldots, x_d)$ satisfying

$$x_i \geq 0, \qquad 1 \leq i \leq d,$$

$$x_i + x_{i+1} \leq 1, \qquad 1 \leq i \leq d-1.$$

The following facts can be proved. The generating functions $J(\mathcal{P}_d, \lambda)$ (denoted $J_d(\lambda)$ for short) are determined by

$$J_0(\lambda) = 1/(1-\lambda), \qquad J_1(\lambda) = 1/(1-\lambda-\lambda^2),$$

$$J_d(\lambda) = 1/(2-\lambda^2 - J_{d-2}(\lambda)), \quad d \geq 2.$$

We have the explicit formulas

$$J_{2d}(\lambda) = \frac{(1-\lambda-\bar{\phi})\phi^{d-1} + (-1+\lambda+\phi)\bar{\phi}^{d-1}}{(1-\lambda-\bar{\phi})\phi^d + (-1+\lambda+\phi)\phi^{-d}},$$

$$J_{2d-1}(\lambda) = \frac{(1-\lambda-\lambda^2-\bar{\phi})\phi^{d-1} + (-1+\lambda+\lambda^2+\phi)\bar{\phi}^{d-1}}{(1-\lambda-\lambda^2-\bar{\phi})\phi^d + (-1+\lambda+\lambda^2+\phi)\bar{\phi}^d}$$

where

$$\phi = \tfrac{1}{2}(2-\lambda^2+\lambda(\lambda^2-4)^{\frac{1}{2}}), \qquad \bar{\phi} = \tfrac{1}{2}(2-\lambda^2-\lambda(\lambda^2-4)^{\frac{1}{2}}).$$

Moreover, the coefficient of λ^j in $W(\mathcal{P}_d, \lambda)$ is equal to the number of permutations $(b_1, b_2, \ldots, b_d)$ of $[d]$ satisfying:

(i) if $d = 2e$, then i and $i+1$ must precede $i+e$ for $1 \leq i \leq e-1$, while e must precede $2e$;

(ii) if $d = 2e+1$, then i and $i+1$ must precede $i+e+1$ for $1 \leq i \leq e$;

(iii) the number of integers $i \in [d-1]$ for which $b_i > b_{i+1}$ is equal to j.

From this one can show that the relative volume (in this case, the actual volume) $\psi_d = \psi(\mathcal{P}_d)$ is given by

$$\sum_{d=0}^{\infty} \psi_d x^d = \sec x + \tan x.$$

Note added in proof

(1) An independent proof of Theorem 2.8 appears in P. McMullen, Lattice invariant valuations on rational polytopes, Arch. Math. (Basel) 31 (1978/79) 509–516.

(2) The one-to-one correspondence asked for at the end of Example 3.1 has been found by D. Kleitman and K. Winston, Forest and score vectors, to appear.

References

[1] H. Bruggesser and P. Mani, Shellable decompositions of cells and spheres, Math. Scand. 29 (1971) 197–205.
[2] H.H. Crapo and G.-C. Rota, On the Foundations of Combinatorial Theory: Combinatorial Geometries (preliminary edition). (M.I.T. Press, Cambridge, MA, 1970).
[3] E. Ehrhart, Polynômes arithmétiques et méthode des polyèdres en combinatoire, International Series of Numerical Mathematics, Vol. 35 (Birkhäuser, Basel, 1977).
[4] H. Freudental, Simplizialzerlegungen von beschränkter Flachkeit, Annals of Math. 43 (3) (1942) 580–582.
[5] P.M. Gibson, Facets of faces of transportation polytopes, in: Proc. 7th Conf. Combinatorics, Graph Theory, and Computing, Louisiana State Univ. (1976) 323–333.
[6] B. Grünbaum, Convex Polytopes (Wiley-Interscience, New York, 1967).
[7] G. Kempf, F. Knudsen, D. Mumford and B. Saint-Donat, Toroidal Embeddings I, Lecture Notes in Math., 339 (Springer, Berlin, 1973).
[8] I.G. Macdonald, The volume of a lattice polyhedron, Proc. Camb. Philos. Soc. 59 (1963) 719–726.
[9] I.G. Macdonald, Polynomials associated with finite cell-complexes, J. London Math. Soc. 4 (2) (1971) 181–192.
[10] P. McMullen, Metrical and combinatorial properties of convex polytopes, Proc. Int. Congr. of Mathematicians (Vancouver, B.C., 1974) (Canadian Math. Congress, 1975) 491–495.
[11] P. McMullen, Valuations and Euler-type relations for certain classes of convex polytopes, Proc. London Math. Soc. 35 (1977) 113–135.
[12] P. McMullen and G.C. Shephard, Convex Polytopes and the Upper Bound Conjecture, London Math. Soc. Lecture Note Series, Vol. 3 (Cambridge Univ. Press, Cambridge 1971).
[13] G.C. Shephard, Combinatorial properties of associated zonotopes, Can. J. Math. 26 (1974) 302–321.
[14] R. Stanley, Combinatorial reciprocity theorems, Advances in Math. 14 (1974) 194–253.
[15] R. Stanley, The upper-bound conjecture and Cohen–Macaulay rings, Studies in Appl. Math. 54 (1975) 135–142.
[16] R. Stanley, Magic labelings of graphs, symmetric magic squares, systems of parameters, and Cohen–Macaulay rings, Duke Math. J. 43 (1976) 511–531.
[17] D.J.A. Welsh, Matroid Theory (Academic Press, London, 1976).

Annals of Discrete Mathematics 6 (1980) 343–347

ROTORS IN GRAPH THEORY

W.T. TUTTE
University of Waterloo, Waterloo, Ontario, Canada

By a "rotor" in a graph G is meant a subgraph of G having rotational symmetry, and whose vertices of attachment are equivalent with respect to this symmetry. We discuss which properties of a general graph G are invariant under the operation of replacing a rotor by its mirror image.

1. Definitions

Let n be an integer not less than 3. A *rotor* of *order* n is a triple $R=(K, \theta, J)$, where K is a connected finite graph, θ, is an automorphism of K of period n, and J is an orbit of θ consisting of n distinct vertices. The powers of θ are the *rotations* and J is the *principal orbit* of the rotor R.

We enumerate the vertices of J as $u_0, u_1, \ldots, u_{n-1}$, regarding the suffixes as residues mod n. We arrange also that

$$\theta^j(u_0)=u_j \tag{1}$$

for each integer j.

For each residue j we define the *reflection* f_j as a 1–1 mapping of J onto itself such that

$$f_j(u_{j+i})=u_{j-i} \tag{2}$$

for each residue i. If we restrict our attention to J any reflection can be regarded as equivalent to f_0, either preceded or followed by a suitable rotation.

We assume in what follows that K has no *reflective symmetry*, that is no automorphism that reduces to a reflection on J.

Let $R=(K, \theta, J)$ be a rotor of order n, and let L be a graph having no edges or vertices in common with K. Let g be a mapping of J into the vertex-set $V(L)$ of L. Let G be the graph formed from the union of K and L by identifying u_j with gu_j, for each residue j. We say that G is a *rotor-stator combination* with *rotor* R and *stator* L, and that it is *cemented* by g.

From the same rotor and stator we can get a second combination G' by using gf_0 as the cementing mapping. Because K has no reflective symmetry we may expect G and G' to be non-isomorphic in non-trivial cases. If we replaced gf_0 by gf_j, for any residue j, we would get the same combination G', to within an isomorphism. The change rrom G to G' is often said to be effected by "flipping" the rotor R, or by "replacing R by its mirror-image". A second flipping changes G' back into G.

2. Rotor-invariance

The author first encountered rotor-stator combinations in the researches leading to [1]. Graphs were regarded as networks of unit resistances. Current was supposed to enter such a network at one vertex, called the *positive pole*, and to leave at another, called the *negative pole*. Currents in the various wires of the network were to be determined by Kirchhoff's laws.

Consider a network G which is a rotor-stator combination with rotor $R = (K, \theta, J)$ and stator L. Let the positive and negative poles be vertices of L. Under these conditions it is found that flipping the rotor makes no difference to the currents and potential differences in L. But K usually acquires a completely new set of currents. The authors of [1] found this fact of electrical rotor-invariance surprisingly difficult to prove. But they exploited it for the construction of planar networks with an overall resistance, between specified poles, of unity. By way of a simple geometrical correspondence some of these networks determined dissections of a square into unequal squares [1, 4].

The *tree-number* $T(G)$ of a graph G is the number of its spanning trees. The electrical theorem just mentioned can be summed up in the graph-theoretical statement that the tree-number of a rotor-stator combination is *rotor-invariant*,, that is not altered by flipping the rotor.

It is natural to ask just what functions of a general graph G are rotor-invariant. Apart from $T(G)$ and a few trivialities such as the number of edges, no rotor-invariant function seems to be known at present. But there is an interesting case of partial invariance. It occurs with the polynomial known as the dichromate of G. This polynomial is discussed in [5] and [7]. We take note here of a few of its basic properties.

The *dichromate* of a graph G is a polynomial $\chi(G; x, y)$ in two variables x and y. It takes the value 1 when G is edgeless. For graphs of one edge only it takes the value x or y according as that edge is a link or a loop. Moreover the dichromate satisfies the three following general rules.

(*i*) *The dichromate of a disconnected graph is the product of the dichromates of its components.*

(*ii*) *The dichromate of a connected but separable graph is the product of the dichromates of its blocks.*

Consider an edge A of G with two distinct ends x and y. Let $D_A G$ be the graph obtained from G by deleting A, and let $C_A G$ be the graph obtained from G by contracting A, with its two ends, to a single vertex. It is often convenient to consider the case in which A is not an isthmus, as well as not a loop. For then $D_A G$, like $C_A G$, has the same number of components as G. We can now state our third general rule.

(iii) *If A is an edge of G that is neither a loop nor an isthmus, then*

$$\chi(G; x, y) = \chi(D_A G; x, y) + \chi(C_A G; x, y).$$

Rules (ii) and (iii) enable us to determine recursively the dichromates of connected graphs. Rule (i) extends the determination to disconnected ones.

For connected graphs there is an expression for the dichromate in terms of spanning trees. Moreover in this case $\chi(G; 1, 1) = T(G)$ (see [5]).

It is shown in [6] that $\chi(G; x, y)$ is rotor-invariant for rotors of order not exceeding 5. We examine some details of the proof in the following section.

3. Partial rotor-invariance

Let $R = (K, \theta, J)$ be a rotor of order n and L a corresponding stator, with cementing mapping g. We say that L is *degenerate* if it has no edges and each of its vertices is the image under g of at least one vertex of J.

In the degenerate case L determines a partition π of J into disjoint non-null subsets, called the *parts* of π. Each part U of π corresponds to a vertex u of L. The members of U are those vertices of J that are mapped by g onto u. Let G be the rotor-stator combination determined by R, L and g. Then G is the graph $K(\pi)$ obtained from K by identifying vertices according to π. By this we mean that two vertices are to be identified if and only if they belong to the same part of π. Flipping the rotor changes G into G', that is into $K(\pi')$, where π' is that partition of J whose parts are the images under f_0 of the parts of π. Evidently each partition of J has a corresponding degenerate stator.

A particular partition π of J may be *reflectable*, that is transformable into itself by a reflection of J. In this case π can be transformed into π' by a rotation of K, and therefore $K(\pi)$ and $K(\pi')$ are isomorphic. It is easy to verify that for $n \leq 5$ all partitions of J are reflectable.

We can now discuss the inductive proof of the following theorem.

Theorem 3.1. *Let G be a connected rotor-stator combination on which the order of the rotor is at most* 5. *Then* $\chi(G; x, y)$ *is not altered by flipping the rotor.*

As usual we denote the rotor by $R = (K, \theta, J)$ and the stator by L. We proceed by induction over the number of edges of L. The induction starts with the degenerate case in which L is edgeless. The stator L then corresponds to a partition π of J. Since π is reflectable the graphs G and G', that is $K(\pi)$ and $K(\pi')$, are isomorphic. It is therefore trivially true that

$$\chi(G; x, y) = \chi(G'; x, y).$$

If L has at least one edge we can reduce to the case of fewer edges by using Rules (ii) and (iii), and so complete the induction. The details are given in [6].

We can extend Theorem 3.1 to disconnected combinations by using Rule (i).

The chromatic polynomial $P(G, \lambda)$ of a graph G is the number of colourings of G in λ colours. For a connected graph of k vertices, it is related to the dichromate by the following equation.

$$P(G, \lambda)=(-1)^{k+1}\lambda\chi(G; 1-\lambda, 0) \tag{3}$$

(see [7, Eq. (38)]).

Foldes [2] has found a rotor-stator combination in which rotor-flipping changes the chromatic polynomial and therefore the dichromate. The combination is non-planar and the order of the rotor is 6.

I know of no planar combination in which the chromatic polynomial changes. Lee [3] has shown that in the planar case the chromatic polynomial is not altered by flipping a rotor of order 6, provided that its principal orbit is the vertex-set of some circuit of the combination.

4. The rotor-invariance of $T(G)$

In this section we discuss why $T(G)$ should be rotor-invariant while $\chi(G; x, y)$ is not. We can reduce the problem of the rotor-invariance of $T(G)$ to the case of degenerate stators. For we can go from there to the general connected case by using an induction like that of Theorem 3.1 and [6].

Consider the rotor $R=(K, \theta, J)$. We have to show that $K(\pi)$ and $K(\pi')$ have the same tree-number for each partition π of J.

Let p and q be any two distinct vertices of J. Let σ be the partition of J in which $\{p, q\}$ is one part, and each other part consists of one vertex only. Such a partition we call *elementary*. An elementary partition σ is clearly reflectable. Hence $K(\sigma)$ and $K(\sigma')$ are isomorphic and so, trivially, they have the same tree-number.

In [1] the tree-number of the graph obtained from a basic graph G by identifying two distinct vertices x and y is denoted by $[xy, xy]$. This is a special case of a function $[xy, rs]$ of four vertices, defined as a second cofactor in the Kirchhoff matrix of G. It satisfies the following identity.

$$2[xy, rs]=[xs, xs]+[yr, yr]-[ys, ys]-[xr, xr] \tag{4}$$

(see [1, Eq. (2.25)]).

As a useful convention we write $[xx, xx]=0$, for each vertex x.

Suppose that we make two vertex-identifications in G, x with y and p with q. Let H be the resulting graph. There is a determinantal identity that enables us to express $T(H)$ in terms of $T(G)$ and the tree-numbers of the graphs obtainable from G by identifying two vertices only. It appears in [1] as Eq. (2.34), and we can write it as follows.

$$T(H)T(G) = [xy, xy][pq, pq] - [xy, pq]^2. \tag{5}$$

Repeated application of this rule enables us to express $T(K(\pi))$, for any partition π of J in terms of $T(G)$ and the tree-numbers of the graphs $K(\sigma)$ defined by elementary partitions σ. We find likewise that $T(K(\pi'))$ is given by the same expression, but with each σ replaced by the corresponding σ'. Since elementary partitions are reflectable it follows that $T(K(\pi)) = T(K(\pi'))$.

This completes the proof of the rotor-invariance of the tree-number. It seems that the trouble with $\chi(G; x, y)$ is that it lacks quadratic identities like (5).

The theory of rotors and spanning trees is generalized to directed graphs in [8].

References

[1] R.L. Brooks, C.A.B. Smith, A.H. Stone and W.T. Tutte, The dissection of rectangles into squares, Duke Math. J. 7 (1940) 312–340.

[2] S. Foldes, Symmetries, Thesis, Univ. of Waterloo, Waterloo, Ont. (1977).

[3] L.A. Lee, On chromatically equivalent graphs, Thesis, George Washington Univ., 1975.

[4] W.T. Tutte, Squaring the square, Can. J. Math. 2 (1950) 197–209.

[5] W.T. Tutte, A contribution to the theory of chromatic polynomials, Can. J. Math. 6 (1954) 80–91.

[6] W.T. Tutte, Codichromatic graphs, J. Combin. Theory 16 (1974) 168–174.

[7] W.T. Tutte, The dichromatic polynomial, Proc. 5th British Combinatorial Conference, 1975. Congressus Numerantium XV.

[8] W.T. Tutte, The rotor effect with generalized electrical flows, Ars Combinatoria 1 (1976) 3–31.

Annals of Discrete Mathematics 6 (1980) 349–384

EMBEDDING GRAPHS IN THE PLANE — ALGORITHMIC ASPECTS

S.G. WILLIAMSON*

Department of Mathematics, University of California, San Diego, La Jolla, CA 92093, USA

An exposition of ideas relating to graph planarity, including the construction of planar embeddings when they exist, is given. The notion of the family of "segment graphs" of a graph is discussed. It is shown how to construct a family of spanning forests for the segment graphs of a planar graph $G=(V, E)$ in linear time in $|V|$.

1. Introduction

We consider the question of, given a graph $G=(V, E)$, how does one decide whether or not a graph is planar and, if planar, construct or specify in an efficient manner a planar embedding. Our starting point will be the general ideas relating to the work of Hopcroft and Tarjan [9, 18] in which they specify an algorithm that decides whether or not a graph is planar in time linear in $|V|$. Their work is in turn based on a recursive method originally proposed by Auslander and Parter [2] and correctly formulated by Goldstein [8]. A similar method was also introduced, apparently independently, by Demoucron *et al.* [6]. The latter authors introduce (implicit also in [2]) the notion of "les graphes auxiliares", closely related to what we call below the *segment graphs* of a graph. We shall combine these various ideas below. First we give a general outline.

The recursive method used by Hopcroft and Tarjan [9] is defined by the construction of a set of cycles in G indexing the various levels of recursion. Without loss of generality G is assumed to be biconnected. The recursion has a tree structure associated with it (defined by the various states of the "stack" or recursive calls) in the usual manner. Thus this tree structure induces a tree structure on the cycles which index the recursion. The edge sets of these cycles form a partition of the edge set E of G. Given any cycle C in this tree, the union of all cycles descendant (in the tree) to C with C itself forms what we shall call the *segment of* C or SEG(C). If C is not a terminal vertex (leaf) in the tree of cycles, define the *segment graph of* C (SEGGR(C)) as follows:

The graphs SEG(D), where D is a son of C in the tree of cycles, are the vertices of SEGGR(C). Two such "vertices" SEG(D) and SEG(D') form an edge in SEGGR(C) if they must go on opposite sides of C in any planar embedding of $C \cup \text{SEG}(D) \cup \text{SEG}(D')$. If it is possible to color SEGGR(C) with 2-colors, say I and O (for "inside" and "outside"), then the vertices of SEGGR(C), as subgraphs

* Research was supported by the National Science Foundation Grant No. GJ43333.

of G, may be oriented in the plane relative to C such that if each SEG(D) is itself planar, then C together with these segments is planar. Certain segments, SEG(D), must, a priori, be colored I. Call this set of segments $\mathcal{I}$. Call SEGGR(C) *$\mathcal{I}$-bichromatic* if there exists a coloring of SEGGR(C) with I and O such that every segment of $\mathcal{I}$ is colored I. It is not difficult to see that G is planar if and only if every segment, SEGGR(C), is *$\mathcal{I}$-bichromatic*. Thus the condition of planarity for G is transferred to the condition of $\mathcal{I}$-bicolorability for every graph in a family of graphs associated with G. This family of graphs has cardinality at most $|E|-|V|$. The total number of vertices (segments) in all graphs in this collection is exactly $|E|-|V|$. Thus this family of graphs is considerably smaller than that of [6, Theorem 1].

The purpose of this paper is twofold. Firstly, we wish to give an exposition, for a mathematical audience, of the interesting ideas associated with efficient planarity algorithms. Secondly, we wish to begin a study of certain aspects of the family of segment graphs associated with a graph in order to facilitate our subsequent description of the following related problems:

(1) Linear time algorithms for the ranking, unranking, and random generation of isotopy classes of planar graphs [5].

(2) Characterization of minimal cycles of the segment graphs of a graph and linear time algorithms for finding such cycles of odd length in any nonplanar graph.

(3) A linear time algorithm for finding a Kuratowski subgraph of any nonplanar graph [12, 13, 21].

As we shall show below, the set of segment graphs of a planar graph $G=(V, E)$ cannot be constructed in time linear in $|V|$. In Section 5, however, we show that a set of spanning forests for these graphs can be constructed in linear time in $|V|$. These spanning forests provide a means for the specification of all planar embeddings of G up to isotopy and thus relate specifically to (1) above. In the first four sections we give an exposition of the basic ideas related to planarity testing and embedding.

Basic ideas in graph algorithms may be found in [1, Chapter 5; 16, Chapter 8] We refer the reader also to two basic articles by Tarjan [19, 20]. For a statement of the problem of graph planarity we refer the reader to [9, 16, Chapter 8].

2. Preliminary notions

Let V be a set and let $\mathcal{P}_2(V)$ denote the set of all subsets of V of size two. For our purposes a graph will be defined as a pair $G=(V, E)$, V a set ("vertex set") and E ("edge set") a subset of $\mathcal{P}_2(V)$. A *directed graph* is a pair $G=(V, E)$ where E is a subset of $V\times V-D$, $D=\{(x, x): x\in V\}$. We denote elements of $P_2(V)$ by $\{x, y\}$ and elements of $V\times V$ by (x, y). An undirected edge $\{x, y\}$ is *incident* on a vertex $z\in V$ if $x=z$ or $y=z$. A directed edge (x, y) is *incident* on z in V if $x=z$. In either case, let E_z denote the set of all edges of G incident on z. A graph G

will be called *ordered* if for each $z \in V$ a linear order is specified on E_z. Technically, an ordered graph G is a triple $(V, E, \{E_z, \leqslant_z): z \in V\})$ where the last component is a family of linear orders. We avoid this cumbersome notation. A connected acyclic graph $T = (V, E)$ is a *tree*. If in addition to T we specify a particular vertex $x \in V$, then the triple (V, E, x) is a *rooted tree* with *root* x. If we delete x from V and delete all incident edges E_x from E, then the remaining graph (if not empty) will decompose into one or more connected components, themselves trees, which we call the *subtrees of the root* x. If $T' = (V', E')$ is such a subtree, then there is a unique $x' \in V'$ such that $\{x, x'\} \in E_x$. We make the convention that this x' is the *root of* T'. Any rooted tree $T = (V, E, x)$ has associated with it a natural *directed* root tree: to each edge $\{x, x'\}$ in E_x we assign the directed edge (x, x') (i.e., directed "away from" the root). Recursively, we direct the edges of the subtrees of the root x to obtain a directed rooted tree which we also denote by $T = (V, E, x)$.

Using this convention we may regard any rooted tree as a directed graph, and we do so whenever it is convenient. When we speak of *a directed rooted tree* we mean directed in the above sense unless otherwise specified. The most basic structure for describing algorithms is the *ordered directed rooted tree.* One such tree is shown in Fig. 2.1 (the order on subtrees is "left to right").

Given an ordered directed rooted tree $T = (V, E, x)$ let $T_1 = (V_1, E_1, x_1), \ldots, T_k = (V_k, E_k, x_k)$ be the rooted subtrees of x in order. We define a sequence of edges DF(E) ("depth first" sequence of edges) and a sequence of vertices DF(V) ("depth first" sequence of vertices) as follows:

(1) If $V = \{x\}$, then $\text{DF}(V) = x$ and if $E = \{(x, y)\}$, then $\text{DF}(E) = (x, y), (x, y)$.

(2) *Otherwise*

$$\text{DF}(E) = (x, x_1), \text{DF}(E_1), (x, x_1), \ldots, (x, x_k), \text{DF}(E_k), (x, x_k),$$

$$\text{DF}(V) = x, \text{DF}(V_1), x \cdots x, \text{DF}(V_k), x.$$

These lists, for the directed ordered rooted tree of Fig. 2.1A, are given in Fig. 2.1. The intuitive interpretation of these lists comes from Fig. 2.1B as follows:

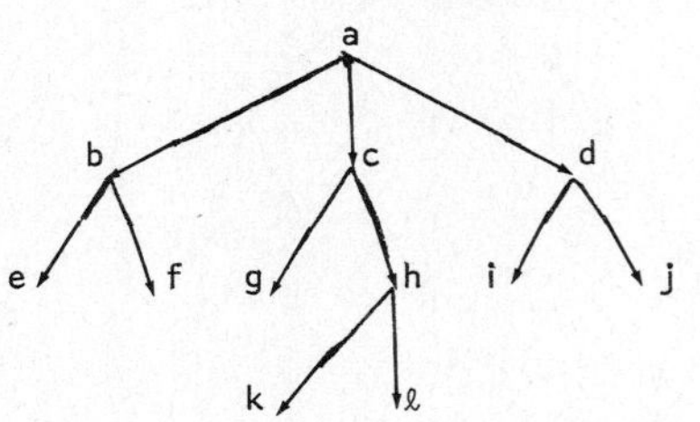

A.

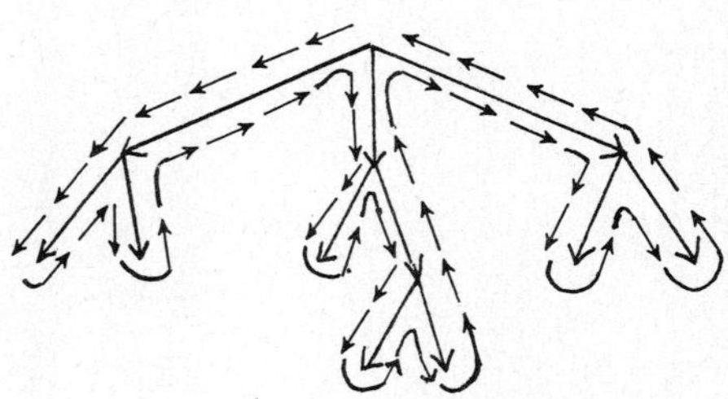

B.

DF(E) = (a, b), (b, e), (b, e), (b, f), (b, f), (a, b), (a, c), (c, g), (c, g), (c, h), (h, k), (h, k), (h, l), (h, l), (c, h), (a, c), (a, d), (d, i), (d, i), (d, j), (d, j), (a, d).

DF(V) = $a, b, e, b, f, b, a, c, g, c, h, k, h, l, h, c, a, d, i, d, j, d, a$.

Fig. 2.1. $T = (V, E, a)$.

Follow the arrows around the tree as shown. To construct DF(E) list each edge as it is traversed (forward or backward). To construct DF(V) start with the root, a, and list each vertex when it is encountered (thus "a" is listed four times in Fig. 2.1).

A subgraph $F=(V', E')$ of G is a *spanning forest* for G if F is a forest (components of F are trees) and $V'=V$. If F is a tree T, then T is called a *spanning tree* of G. A spanning forest is *rooted* if each component is rooted.

Given any rooted tree $T=(V', E', x')$, we say that two vertices $w_1, w_2 \in V'$ are lineal with respect to T if $w_1 = w_2$ or they lie on the same directed path of T. If the path leads from w_1 to w_2 (or $w_1 = w_2$) we say that w_1 is an *ancestor* of w_2 and w_2 is a *descendant* of w_1. A rooted spanning tree $T=(V, E_T, x)$ is a *lineal* spanning tree for $G=(V, E)$ if $\{s, t\} \in E - E_T$ implies that s and t are lineal in T. Given any connected graph $G=(V, E)$ and any $x \in V$, then there always exists a lineal spanning tree $T=(V, E_T, x)$ for G. Let $\{x, y\} \in E_x$. Referring to Fig. 2.2, let $\tilde{V}_y$ denote the set of all $s \in V$ for which no simple path from s to x includes y. Let $V_y = V - \tilde{V}_y$. We assume $x \in \tilde{V}_y$. Let G_y denote the restriction of G to V_y (all $\{s, t\}$ of G with $\{s, t\} \subseteq V_y$) and $\tilde{G}_y$ the restriction of G to $\tilde{V}_y$. From the definition of $\tilde{V}_y$, the only possible edges of G not in either G_y or $\tilde{G}_y$ are edges of the form $\{x, z\}$ for $z \in V_y$. Recursively we may construct lineal spanning trees T_y for G_y and $\tilde{T}_y$ for $\tilde{G}_y$. Adding the edge $\{x, y\}$ to T_y union $\tilde{T}_y$ clearly produces a lineal spanning tree T for G rooted at x. Let $T=(V, E_T, x)$ and let $E_B = E - E_T$. The subgraph of G with edge set E_B will be called B. The edges E_B are called "backedges" [1] or "fronds" [19] and the edges E_T are called "tree edges". Henceforth we shall regard T and B as directed graphs with the following conventions:

(1) Edges of T are directed away from the root in the canonical fashion.

(2) If $(s, t) \in E_B$, then s is a descendant of t in T. An example is given in Fig. 2.2(B).

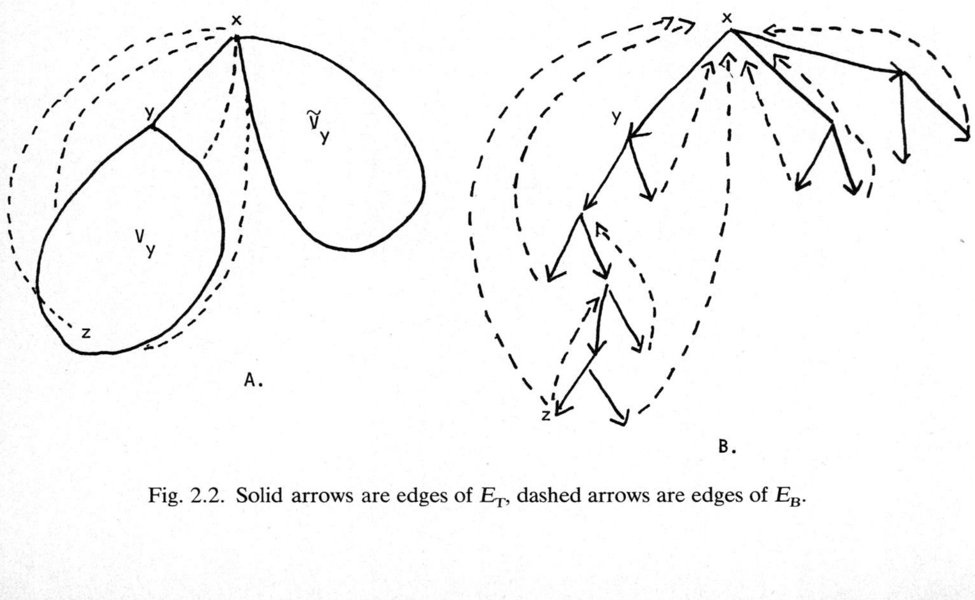

Fig. 2.2. Solid arrows are edges of E_T, dashed arrows are edges of E_B.

The above recursive description of the lineal spanning tree T and directed graph of backedges B suggests a simple recursive algorithm for constructing T and B. First some preliminary remarks. Given an ordered directed rooted tree $T=(V, E_T, x)$ we define two functions FATHER and PREORDER on V. If $t \neq x$, then FATHER(t) is the unique $s \in V$ such that $(s, t) \in E_T$. The *preorder* linear order on V is the order specified by the sequence of first occurrences of the elements of V in the list DF(V). For example, in the sequence DF(V) of Fig. 2.1 we extract just those entries z which represent a first occurrence of z to get $a\,b\,e\,f\,c\,g\,h\,k\,l\,d\,i\,j$, which is the preorder linear order of V for this example. The function which to each $z \in V$ assigns its position in preorder on V will be called PREORDER (NUMBER in [1, 9, 19]). Thus for the example

$$\text{PREORDER} = \begin{pmatrix} a & b & c & d & e & f & g & h & i & j & k & l \\ 1 & 2 & 5 & 10 & 3 & 4 & 6 & 7 & 11 & 12 & 8 & 9 \end{pmatrix}.$$

Similarly we define POSTORDER where "first occurrences" is replaced by "last occurrences".

Procedure 2.3 starts with a given ordered connected graph $G=(V, E)$ and $x \in V$. We wish to produce a lineal spanning tree $T=(V, E_T, x)$, DF(E_T), the graph $B=(V_B, E_B)$, and the functions FATHER, PREORDER, and POSTORDER. Both T and B are produced as directed graphs. $PT=(PV, PE_T, x)$ and $PB=(PV_B, PE_B)$ are used to store the parts of T and B as they are being constructed. "CARD" denotes cardinality and v is the last vertex added to PV (maximum in PREORDER on PV). For each $v \in V$, $V(v)$ initially is the list of vertices of G adjacent to v. $V(v)$ is ordered by the order on edges E_v incident to v in G. DF(E_T) is a list of edges (the depth first sequence of edges of T when $v=x$). POST is an integer.

Procedure 2.3. CONSTRUCT T AND B DESCENDANT TO v

```
begin
  while V(v) ≠ ∅ do
    begin
      w ← first element of V(v) and delete w from V(v)
    if w ∈ PV and PREORDER(w) < PREORDER(FATHER(v)) then
      add (v, w) to PE_B and v and w to PV_B
    if w ∉ PV then
      begin
        add (v, w) to PE_T and to DF(E_T);
        add w to PV and set CARD(PV) ← CARD(PV) + 1;
        FATHER(w) ← v;
        PREORDER(w) ← CARD(PV);
        CONSTRUCT T AND B DESCENDANT TO w;
      end
    end
  POST ← POST + 1 and POSTORDER(v) ← POST;
  if v ≠ x then add (FATHER(v), v) to DF(E_T);
end
```

To construct T and B we set $PT = (\{x\}, \emptyset, x)$, $PB = (\emptyset, \emptyset)$, PREORDER$(x) = 1$, POST $\leftarrow 0$, CARD $= 1$, DF$(E_T) = \emptyset$ and execute CONSTRUCT T AND B DESCENDANT TO x. At the end $T = PT$, $B = PB$.

Comments. (1) One can show that the above algorithm can be implemented in linear time in $(|V| + |E|)$. See, for example, [1, 16, 19] for the basic ideas.

(2) The statement "*if* $w \in PV$ and PREORDER$(w)<$PREORDER-(FATHER(v)) *then*..." tests that (v, w) is a backedge not yet added to PB. The condition, if true, assures us that w is an ancestor of v other than the father of v.

(3) If $G = (V, E)$ is planar, then it is a consequence of Euler's theorem on planar graphs that $|E| \leqslant 3|V| - 6$. Suppose G is nonplanar and we apply the above procedure to each component G' of G, keeping a count of all edges in $PB' \cup PT'$ at each stage of the computation for that component. If this count ever gets to 3 CARD$(PV') - 5$, then we have a connected nonplanar subgraph of that component. Suppose we stop the procedure at that point and go to the next component. In this way we can determine a nonplanar subgraph $\tilde{G} = (\tilde{V}, \tilde{E})$ of G with $|\tilde{E}| \leqslant 3|\tilde{V}| - 5$. This subgraph of G can be determined in time linear in $|\tilde{V}|$. This observation is sometimes useful in dealing efficiently with nonplanar graphs (for example in finding a Kuratowski subgraph of a given nonplanar graph [21]).

Another basic notion we shall require is that of a *biconnected* component of a graph $G = (V, E)$. We give a brief description of this idea, referring the interested reader to [19; 1, Chapter 5; 16, Chapter 8] for the algorithmic aspects of finding biconnected components. We define a relation $\sim$ in E by $e_1 \sim e_2$ if $e_1 = e_2$ or if there is an elementary (not self-intersecting) cycle of G containing both e_1 and e_2. Clearly $\sim$ is reflexive and symmetric. Less clear, but easily shown, is transitivity: $e_1 \sim e_2$ and $e_2 \sim e_3$ implies $e_1 \sim e_3$. Thus, $\sim$ is an equivalence relation. The subgraphs of G determined by the equivalence classes are called the *biconnected components* ("bicomponents") of G. If $G_1 = (V_1, E_1)$ and $G_2 = (V_2, E_2)$ are two biconnected components of G, then either $V_1 \cap V_2 = \emptyset$ or $|V_1 \cap V_2| = 1$. If $x \in V_1 \cap V_2$, then x is called an *articulation point* of G.

We indicate how one finds the biconnected components of a graph efficiently. Given an ordered connected graph $G = (V, E)$ and $x \in V$, construct the ordered directed graphs $T = (V, E_T, x)$ and $B = (V_B, E_B)$ as above. Given any $T = (V, E_T, x)$ we define the *postorder* linear order on E_T to be the order specified by the sequence of last occurrences of edges in the sequence DF(E_T). For any $e = (a, b)$ in E_T define LOW1(e) (the "first lowpoint [9, 19] of e) to be

(1) the minimum of PREORDER(b), and

(2) the minimum value of PREORDER(t) over all $(s, t) \in E_B$ with s a descendant of b (including $s = b$) in T.

If we exclude from (2) above all values t such that PREORDER$(t) =$ LOW1(e) and again minimize over (1) and (2) we obtain LOW2(e) (the "second low-point" [9, 16] of e). Note that PREORDER$(b) \geqslant$ LOW2$(e) >$ LOW1(e) unless LOW1$(e) =$ PREORDER(b).

The LOW1 function may be computed recursively by examining E_T in postorder. Given $f=(p, q)$ in E_T, the edges $f_1, \ldots, f_s$ incident to q (directed away from q) are less than f in postorder. We assume inductively that LOW1 has been computed for all $f'<f=(p, q)$ in postorder. Let $m_1=\text{MIN}\{\text{LOW1}(f_i); i=1, \ldots, s\}$ and $m_2=\text{MIN}\{\text{PREORDER}(z)\text{: } (q, z)\in E_B\}$. Then $\text{LOW1}(f)=\text{MIN}\{m_1, m_2, \text{PREORDER}(q)\}$. It is easily seen that LOW1 can be computed in linear time in $|V|+|E|$. Similarly LOW2 can be computed in linear time in $|V|+|E|$. *For* $e=(a, b)\in E_B$ *we adopt the convention that* $\text{LOW1}(e)=b$ *and* $\text{LOW2}(e)=a$.

Let $e=(a, b)$ be an edge of E_T. We say that an edge $e'=(a', b')$ is *descendant to e in T* if $e'=e$ or a' is a descendant of b (as defined above for vertices).

To compute the biconnected components ("bicomponents") of a connected graph $G=(V, E)$, scan E_T in postorder until the first edge $f=(p, q)$ is found with $\text{LOW1}(f)\geqslant\text{PREORDER}(p)$. The set of tree edges of the first bicomponent is the set of all f' in E_T descendant to f (a consecutive sequence of edges prior to f in postorder). The set of backedges of the first bicomponent consists of all edges of E_B incident on vertices p' descendant to q (including by convention $p'=q$). Delete this list of edges from $E_T\cup E_B$ and continue through E_T in postorder until the next edge $f=(p, q)$ is found such that $\text{LOW1}(f)\geqslant p$, etc. We refer the reader to [1, 19] for more detailed descriptions and references.

3. The tree of paths of a graph

Thus far we have, given a connected, ordered graph $G=(V, E)$, discussed the decomposition of G into an ordered rooted, directed spanning tree $T=(V, E_T, x)$ and a directed graph of backedges $B=(V_B, E_B)$. The order on T and B is that inherited from G. *We assume henceforth that G is biconnected.* We have discussed the functions PREORDER, LOW1, and LOW2. We now consider another ordering of T and B that greatly facilitates subsequent computations [9]. Using the notation

$$X(\textit{statement})=\begin{cases}0 & \text{if } \textit{statement}=\text{false},\\ 1 & \text{if } \textit{statement}=\text{true},\end{cases}$$

we associate with each $f=(p, q)\in E_T\cup E_B$ the triple (PREORDER(p), $LOW1(f)$, $X(\text{LOW2}(f)<\text{PREORDER}(p))$). We can, in time linear in $|V|+|E|$ (using a "radix" or "lexicographic bucket sort" [1, 16]) sort these triples in lexicographic order. First in this list will be all edges incident to the vertex $v=x$ (the root of T), then all those edges f incident to the vertex v with PREORDER(v) $=2$, etc. For a fixed vertex p, the incident edges f of $T\cup B$ will be sorted first according to LOW1(f). Next, given $\text{LOW1(f)}=\text{LOW1}(f')$ for two edges $f=(p, q)$, $f'=(p, q')$ incident to p, f will come before f' if $\text{LOW2}(f)\geqslant\text{PREORDER}(p)$ but $\text{LOW2}(f')<\text{PREORDER}(p)$. Note that, due to biconnectedness, if $p\neq x$, then $\text{LOW1}(f)<\text{PREORDER}(p)$. Thus, considering all edges of $T\cup B$ incident to a fixed vertex p and having the same LOW1 value, all edges $f=(p, q)$ with LOW2$(f) \geqslant$ PREORDER(p) come first in the sorted list of edges. Within this class of edges (incident to p, same LOW1 value and LOW2$\geqslant p$) and within the complement of this class (incident to p, same LOW1 value, LOW2$<p$) the order is arbitrary depending on exactly how the bucket sort was carried out [9].

Using the above linear order on edges we convert the ordered, directed, rooted graph $T \cup B$ into another such graph $\tilde{T} \cup \tilde{B}$ with only the order changed. We call this graph, together with T and B a *properly ordered decomposition* of G. Thus to specify such a decomposition we need $T=(V, E_T, x)$, an ordered, directed, rooted, lineal, spanning tree, $B=(V_B, E_B)$ the associated ordered directed graph of backedges, and $\tilde{T} \cup \tilde{B}$, the above described ordered, rooted, directed graph.

Fig. 3.1.

The orders on T and B are inherited from the order on G and the order on $\tilde{T} \cup \tilde{B}$ is as just described above. The functions PREORDER and POSTORDER, still refer to T, not the new ordered tree $\tilde{T}$.

By using the function PREORDER (computed in time linear in $|V|+|E|$) we can replace each vertex $v \in V$ by the integer PREORDER(v). *Henceforth we assume that* $v =$ PREORDER(v) (relative to T).

We now discuss an example. Consider the graph of Fig. 3.1. The vertices have been given the PREORDER labelling relative to the ordered spanning tree T. T and B are not shown explicitly in Fig. 3.1 but can be reconstructed immediately from the PREORDER numbers. The graph $\tilde{T} \cup \tilde{B}$ is shown in Fig. 3.2. The edges

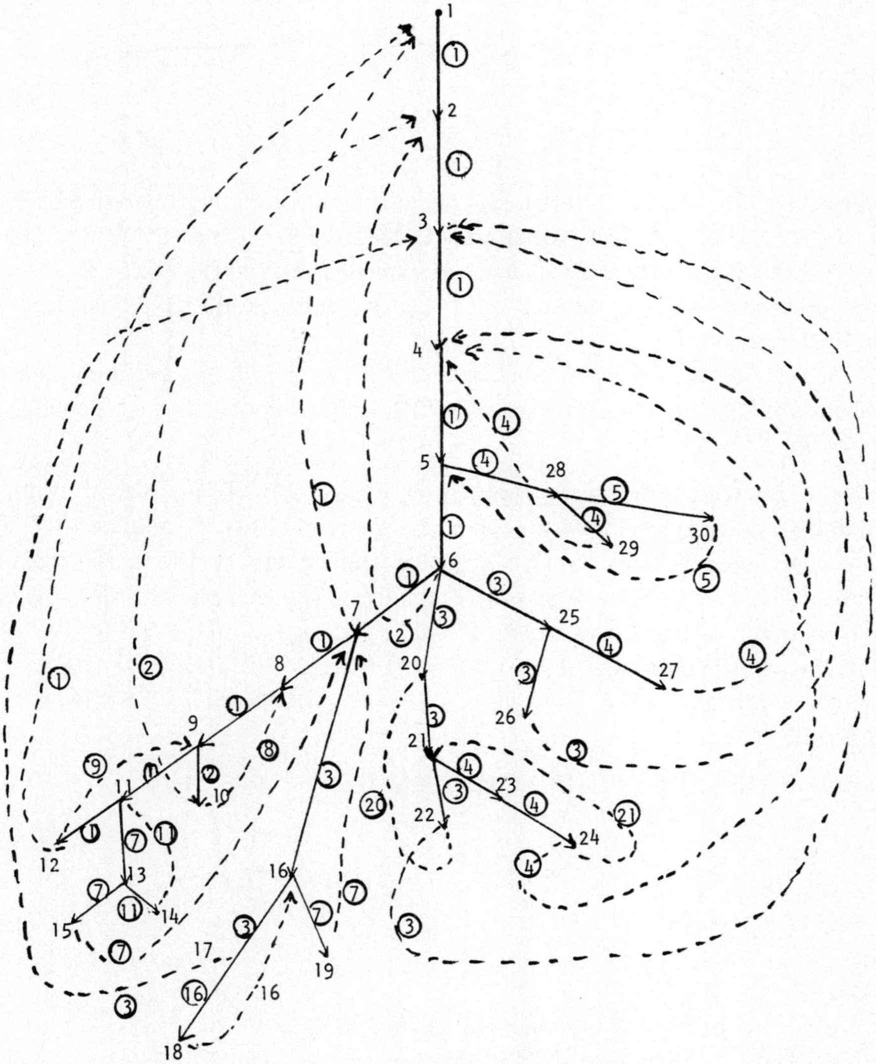

Fig. 3.2. $\tilde{T} \cup \tilde{B}$ (for Fig. 3.1).

E_B are dashed arrows, the edges E_T are solid arrows. The order edges (of $\tilde{T} \cup \tilde{B}$) incident on a vertex p is obtained by reading counterclockwise starting at 12 o'clock. The order on edges incident to p is the order in which they are encountered locally at p when read in this manner. Table 3.3 shows the adjacency table for the ordered directed graph $\tilde{T} \cup \tilde{B}$.

Table 3.3. Adjacency table for $\tilde{T} \cup \tilde{B}$

1	2	11	12, 13	21	22, 23
2	3	12	1, 9	22	3, 30
3	4	13	15, 14	23	24
4	5	14	11	24	4, 21
5	6, 28	15	7	25	3, 4
6	7, 2, 20, 25	16	3, 7	26	3
7	1, 8, 16	17	3, 18	27	4
8	9	18	16	28	29, 30
9	11, 10	19	7	29	4
10	2, 8	20	21	30	5

Remarks. (1) The order of (directed) edges incident on the vertex 6 in Fig. 3.2 is (6, 7), (6, 2), (6, 20), (6, 25). Notice that LOW2(6, 20) = 4 and LOW2(6, 25) = 4. Thus $X(\text{LOW2}(6, 20) < \text{PREORDER}(6) = 6) = 1$ and $X(\text{LOW2}(6, 25) < 6) = 1$. By our previous remarks, the order of these two edges could have been reversed in the ordering on $\tilde{T}$.

(2) At vertex 9 the order on edges is $\tilde{T} \cup \tilde{B}$ is (9, 11), (9, 10). Both edges are in $\tilde{T}$. According to the definition of PREORDER, the order of these edges is reversed in T.

Given a properly ordered decomposition T, B, $\tilde{T} \cup \tilde{B}$ of an ordered graph G, we define a new tree PATR(G, T) called the *tree of paths* of G relative to T. This tree will be needed later to give a global description of the basic recursive structure that we shall use to construct the embeddings of a planar graph. To each edge $e = (a, b)$ in G we assign a path, PATH(e). If $e \in E_B$, then PATH(e) $\leftarrow$ (e). If $e \notin E_B$, then PATH(e) will be a sequence of edges PATH(e) = $(e_1, e_2, \ldots, e_k)$ defined by the following procedure (if $e = (a, b)$, define $a = \text{TAIL}(e)$, $b = \text{HEAD}(e)$):

procedure PATH(e) [e = (TAIL(e), HEAD(e))]

```
begin
  if e ∈ E_B then
    PATH(e) ← (e)
  else
    begin
      e' ← first edge of T̃ ∪ B̃ incident to HEAD(e);
      PATH(e) ← (e, PATH(e')) [e added as first edge to PATH(e')];
    end
  end
```

Thus to obtain PATH(e), $e \in T$, one follows the sequence of first available edges (with respect to the ordering of $\tilde{T} \cup \tilde{B}$) until one encounters an edge in B. If PATH(e) $= (e_1, \ldots, e_k)$, with $e_i = (a_i, b_i)$, then notice that $b_k = \mathrm{LOW1}(e)$ (as a consequence of the ordering on $\tilde{T} \cup \tilde{B}$) and $\mathrm{LOW1}(e) < a_1$ if $a_1 \neq x$ (by biconnectedness of G). The reader should recall that we have assumed that for all $v \in V$, $v = \mathrm{PREORDER}(v)$. If we addend to each PATH(e) the unique sequence of tree edges from b_k to a_1 we obtain a cycle which we call CYCLE(e). CYCLE(e) may be computed easily from the function FATHER discussed above, once PATH(e) is known.

We now define the directed ordered rooted tree PATR(G, T). The vertices of PATR(G, T) will be a *subset* of the set {PATH(e): $e \in G$}. PATR(G, T) will be an ordered rooted tree. Since T is lineal and G is biconnected, there is only one edge d of T incident on the root x of T. The root of PATR(G, T) *is* PATH(d). In general, if PATH(e) is a vertex of PATR(G, T), then let PATH(e) $= (e_1, e_2, \ldots, e_k)$ where $e_i = (a_i, b_i)$. For each $i = 1, \ldots, k$, let $\tilde{E}_{a_i}$ denote the ordered (in $\tilde{T} \cup \tilde{B}$) list of edges incident to a_i and let $\tilde{E}[e_i] = \tilde{E}_{a_i} - \{e_i\}$ be this list with e_i (the first entry) removed. Let $\mathscr{E}(e)$ denote the list of edges obtained by concatenating $\tilde{E}[e_i]$, $i \geq 2$, in reverse order: $\tilde{E}[e_k], \tilde{E}[e_{k-1}], \ldots, \tilde{E}[e_2]$. In particular, if PATH(e) $= (e)$, then $\mathscr{E}(e)$ is empty. The ordered list of edges incident on PATH(e) in PATR(G, T) is {(PATH(e), PATH(f)): $f \in \mathscr{E}(e)$} where the order is that induced by the order on $\mathscr{E}(e)$. In this manner we recursively construct PATR(G, T). (It is easily verified that the directed graph constructed by this process is a tree.) Fig. 3.4 gives PATR(G, T) for the example of Fig. 3.1. If PATH(e) $= (e_1, \ldots, e_k)$ with $e_i = (a_i, b_i)$, then the sequence $a_1, \ldots, a_k b_k$ specifies PATH(e). In each vertex of Fig. 3.4, this sequence is written in reverse order to facilitate the planar representation of the tree. (The problem being that we have shown edges between vertices connecting the starting value a_1 of each son to the occurrence of a_1 in the string representing its father. This is intended as an aid to reading the tree and is not a part of the definition of PATR(G, T)). Note that

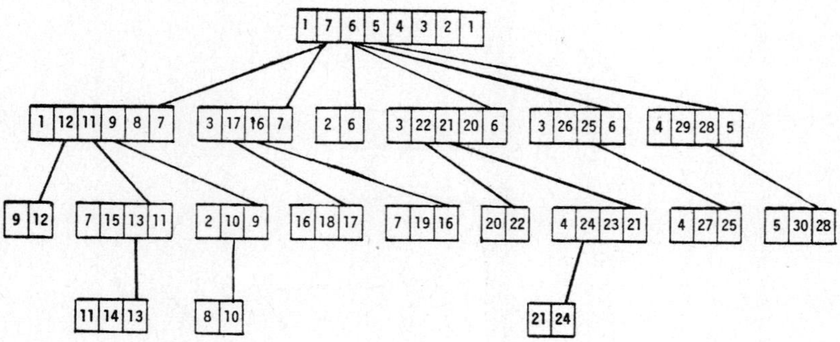

Fig. 3.4. PATR(G, T) for Fig. 3.2.

Note. PATH(e) $= (a_1, b_1)(a_2, b_2) \cdots (a_k, b_k)$ with $b_1 = a_2$, $b_2 = a_3, \ldots,$ is given by the sequence

b_k	a_k	a_{k-1}	$\cdots$	a_1

each backedge occurs in exactly one path so the number of vertices of PATR(G, T) is $|E_B| = |E| - |V| + 1$.

4. The basic recursion

Consider PATR(G, T) discussed in Section 3. $G = (V, E)$ is as before with PREORDER(v) $= v$ for all $v \in V$. Observe that the set {PATH(e): PATH(e) a vertex of PATR(G, T)} defines a (set) partition of the set of edges E of G (the "blocks" of the partition are the sets of edges of the various paths, PATH(e)). The union of all subgraphs of $\tilde{T} \cup \tilde{B}$ of the form PATH(f) for PATH(f) a descendant of PATH(e) (including PATH(e)) in PATR(G, T) is a subgraph of $\tilde{T} \cup \tilde{B}$ which we call the *segment* of e in $\tilde{T} \cup \tilde{B}$. We denote this subgraph by SEG(e). The edges of SEG(e) contained in E_B will be called the *backedges* of SEG(e) and those contained in E_T will be called the *tree edges*. SEG(e) consists of all edges of T descendant to $e = (a, b)$ and all backedges incident on vertices of T descendant to b. SEG(e) is ordered as a subgraph of $\tilde{T} \cup \tilde{B}$. When we speak of planar embeddings of SEG(e) we do not require that they reflect the ordering of $\tilde{T} \cup \tilde{B}$ (such as in Fig. 3.2). From the definition of PATR(G, T) it is apparent that if (s, t) is a backedge of SEG(e) and PATH(e') is the father of PATH(e) in PATR(G, T), then $t \geq$ LOW1(e) $\geq$ LOW1(e'). Thus t is either a vertex of SEG(e) or a vertex of CYCLE(e') intersected with CYCLE(e). See Fig. 4.1.

Let $e = (a, b)$ and consider all backedges (s, t) of SEG(e) with LOW1(e) $< t < a$. Call such backedges of SEG(e) *proper*. Observe from Fig. 4.1 that if G is planar, then the graph CYCLE(e) $\cup$ SEG(e) must be planar, and in any embedding all proper backedges must lie *inside* CYCLE(e) (as CYCLE(e') $\cup$ CYCLE(e) $\cup$ SEG(e) is planar). We call a planar embedding of CYCLE(e) $\cup$ SEG(e) *consistent* if all proper backedges of SEG(e) lie inside CYCLE(e). Fig. 4.2 shows an example where CYCLE(e) $\cup$ SEG(e) is planar but has no consistent embedding. If $e = (x, b)$ is the (unique by biconnectedness) edge of $\tilde{T}$ incident on the root x of $\tilde{T}$, then SEG(e) has no proper backedges

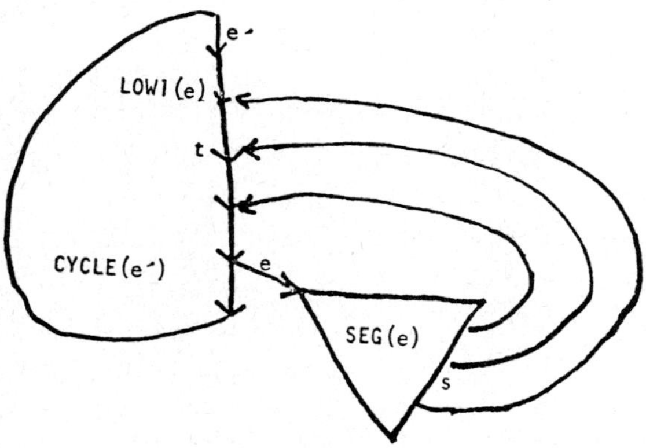

Fig. 4.1.

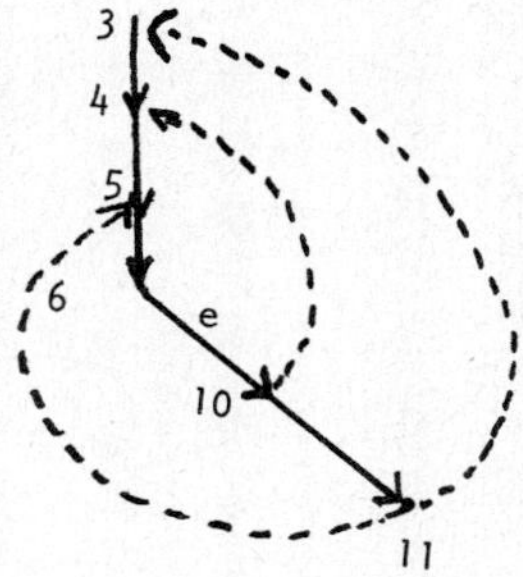

Fig. 4.2. Planar but not consistent. CYCLE(e) = (6, 10, 11, 3, 4, 5, 6); Proper backedges: (10.4), (11, 5); e = (6, 10); LOW1(e) = 3.

(LOW1(e) = x). Thus in this case $\tilde{T} \cup \tilde{B}$ = SEG(e) has a consistent embedding if and only if G is planar. Thus we have the following basic observation: *G is planar if and only if for every vertex* PATH(e) *of* PATR(G, T) *the graph* CYCLE(e) $\cup$ SEG(e) *has a consistent embedding.*

We introduce some additional basic terminology. Given PATH(e), a vertex of PATR(G, T), let SEGLST(e) (called the *list of segments of* e) denote the set {SEG(f): PATH(f) a son of PATH(e) in PATR(G, T)}. Regard SEGLST(e) as linearly ordered by the order on PATR(G, T). Referring to Fig. 3.4, let

PATH(e) = PATH((7, 8)) =

1	12	11	9	8	7

Then SEGLST((7, 8)) is (SEG((12, 9)), SEG((11, 13)), SEG((9, 10))). These graphs may be read off directly from Fig. 3.2. Let X, Y be segments in SEGLST(e). We say that X and Y are *directly linked* if, in any planar embedding of $X \cup Y \cup$ CYCLE(e), X and Y must be on opposite sides of CYCLE(e). We define an undirected graph SEGGR(e) with vertex set SEGLST(e) and edge set {{X, Y}: directly linked to Y}. SEGGR(e) is called the *segment graph* of e. In general, the number of edges of the segment graph may be quadratic in $|V|$. Fig. 4.3 shows an example where the segment graph is the complete bipartite graph $K_{3,3}$. An obvious extension gives $K_{n,n}$ as a segment graph for $|V| = 2n+3$.

It is clear that if G is planar, then every SEGGR(e) associated with PATR(T, G) must be bipartite or bichromatic. We "color" each vertex of SEGGR(e) with I if the segment defined by that vertex is inside CYCLE(e) in a given planar embedding or with O if it is outside.

Due to the above remarks about consistent embeddings, the converse is not quite true. Let us define a segment X in SEGLST(e), $e = (a, b)$, to be *internal* if it has at least one backedge (s, t) with LOW1(e) $< t < a$. Let $\mathscr{I}$ denote the set of internal segments in SEGLST(e). We say that SEGGR(e) is *$\mathscr{I}$-bichromatic* if there exists a 2-coloring of the vertices of SEGGR(e) where all vertices in $\mathscr{I}$ have the same color. Such a coloring will be called an $\mathscr{I}$-coloring. The following result is valid.

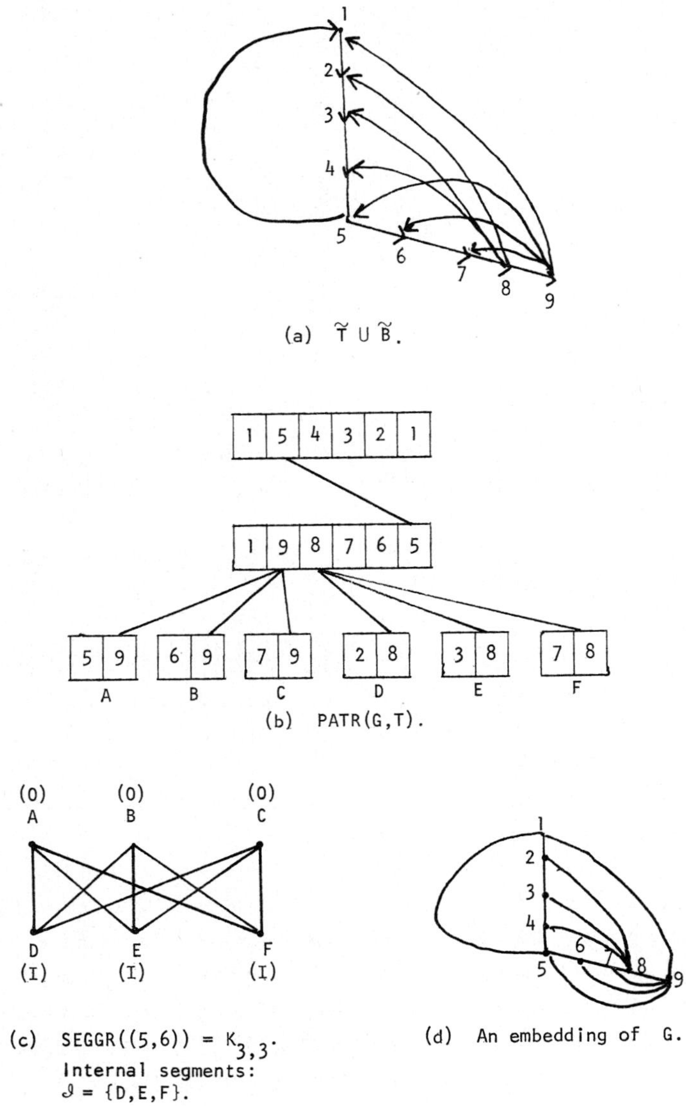

Fig. 4.3. (a) $\tilde{T} \cup \tilde{B}$. (b) PATR(G, T). (c) SEGGR((5, 6)) = $K_{3,3}$. Internal segments: $\mathcal{J} = \{D, E, F\}$. (d) An embedding of G.

G is planar if and only if every SEGGR(e) *associated with* PATR(G, T) *is $\mathcal{J}$-bichromatic where $\mathcal{J}$ is the set of internal segments of* SEGLST(e).

For example, consider Fig. 4.3. If $e = (5, 6)$ and $\tilde{T} \cup \tilde{B}$ is as shown in Fig. 4.3(a), then there are six segments in SEGLST(e). Each segment is a simple backedge as indicated in Fig. 4.3(b). We label these segments A through F. D, E, and F are the internal segments so $\mathcal{J} = \{D, E, F\}$. We make the convention that when we color a segment graph, SEGGR(e), we use the two "colors" I and O. Elements of

$\mathscr{I}$ will always be colored with I. Fig. 4.3(c) shows an $\mathscr{I}$-coloring of SEGGR(e) for this example, so SEGGR(e) is $\mathscr{I}$-bichromatic. SEGGR(1, 2) consists of one vertex, SEG(5, 6), and no edges. $\mathscr{I} = \emptyset$ for this case. Thus all segment graphs are $\mathscr{I}$-bichromatic. An embedding of G (defined by $\tilde{T} \cup \tilde{B}$) is shown in Fig. 4.3(d).

To specify a planar embedding for $\tilde{T} \cup \tilde{B}$ (order in $\tilde{T} \cup \tilde{B}$ ignored in the embedding) it suffices to construct SEGGR(e) for each vertex of PATR(G, T) and $\mathscr{I}$-color each graph (let us say with I and O, vertices in $\mathscr{I}$ colored I). Assuming recursively, that for all vertices (i.e., all SEG(e) $\in$ SEGLST(e)) of SEGGR(e) we have specified a consistent embedding, then the $\mathscr{I}$-colored SEGGR(e) specifies how to orient these embeddings relative to CYCLE(e) to construct a consistent embedding of SEG(e) relative to CYCLE(e). If at any stage in constructing the graphs SEGGR(e) we find one that is not $\mathscr{I}$-bichromatic, then G is nonplanar. As remarked above, we cannot in general compute the SEGGR(e) in linear time in $|V|$. Nevertheless, this process provides a quite effective means for constructing planar embeddings of moderately complex graphs by "pencil and paper" computation.

We conclude this section by illustrating this process in detail for the graph of Fig. 3.1. We then state two key lemmas of Hopcroft and Tarjan [9] which allow this process to be modified to construct a linear time planarity test. The reader will observe that these lemmas facilitate the "pencil and paper" computation that we shall describe. Observe that we do not really need to construct and $\mathscr{I}$-color SEGGR(e) in each case, but that we only need to construct and $\mathscr{I}$-color a *spanning forest* SEGFO(e) for each SEGGR(e). These graphs SEGFO(e) can be constructed in linear time in $|V|$. The last section (Section 5) of this paper is devoted to a careful description of this process.

Example 4.4. (Embedding G of Fig. 3.1). Consider $\tilde{T} \cup \tilde{B}$ of Fig. 3.2 and PATR(G, T) shown in Fig. 3.4. Inspect the vertices of PATR(G, T) in the depth first sequence (i.e., generate DF(vertices PATR(G, T))) Fig. 4.5 shows PATR(G, T) with the vertices replaced by their PREORDER numbers. We use these numbers to refer to the vertices of PATR(G, T). The vertices of Fig. 4.5 listed in POSTORDER are 3, 5, 4, 7, 6, 2, 9, 10, 8, 11, 13, 15, 14, 12, 17, 16, 19,

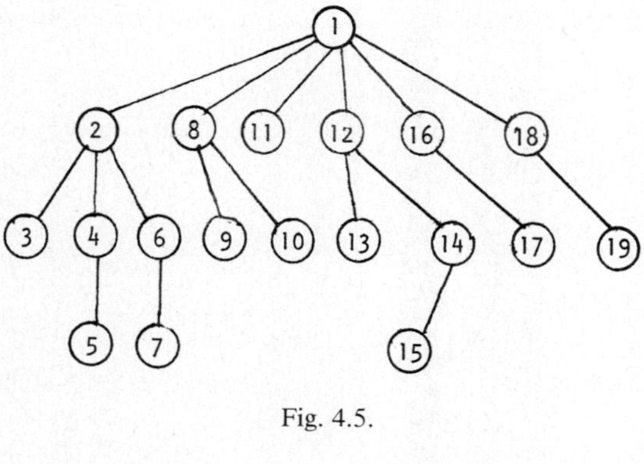

Fig. 4.5.

18, 1. For each such integer k, let SEG(k), CYCLE(k), PATH(k), SEGGR(k), etc. denote the segment, cycle, path, segment graph, etc. associated with the vertex PATH(e) of PATR(G, T) having preorder number k. We attempt to construct an $\mathscr{I}$-coloration of SEGGR(k) and a consistent embedding of SEG(k) for the values of k as they occur in POSTORDER:

k = 3

SEG(3)

SEGGR(k) = φ

k = 5

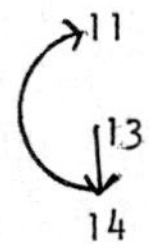

SEG(5)

SEGGR(5) = φ

k = 4

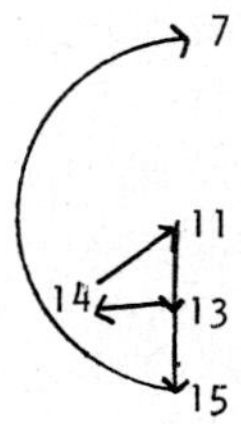

SEG(4)

SEGGR(4)

Remark. "5" refers to "SEG(5)" in SEGGR(4). The manner of embedding SEG(5) relative to CYCLE(4) is left to "trial and error" at this stage, $\mathscr{I} = \emptyset$ here.

k = 7

SEG(7)

SEGGR(7) = φ

k = 6

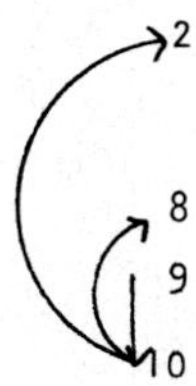

SEG(6)

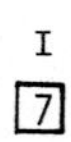

SEGGR(6)

Remark. SEG(7) is internal. This is indicated by □ in SEGGR(6).

k = 2

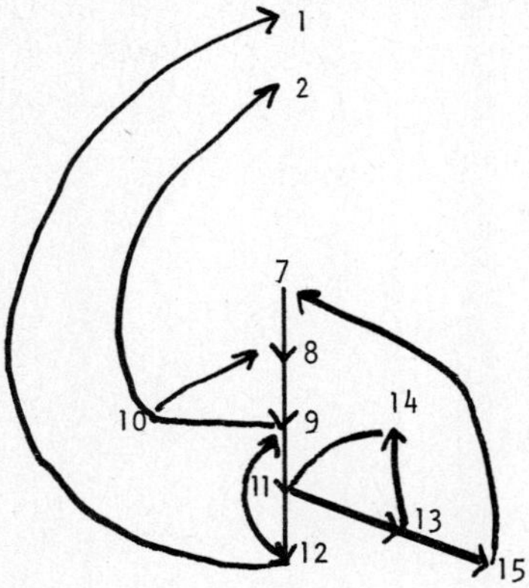

SEG(2)

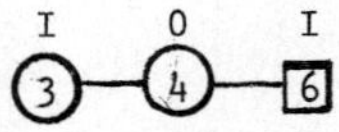

SEGGR(2)

Remarks. $\mathcal{I} = \{\text{SEG}(6)\}$.

$$\text{SEGLST}(2) = \{\text{SEG}(3), \text{SEG}(4), \text{SEG}(6)\}.$$

We have already (because of postorder computation) given consistent embeddings of each element of SEGLST(2). SEG(3) and SEG(4) are *directly linked* (by inspection), as are SEG(4) and SEG(6).

k = 9

SEG(9)

SEGGR(9) = ϕ

k = 10

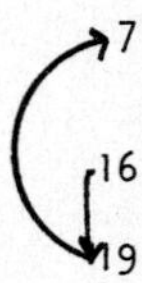

SEG(10)

SEGGR(10) = ϕ

k = 8

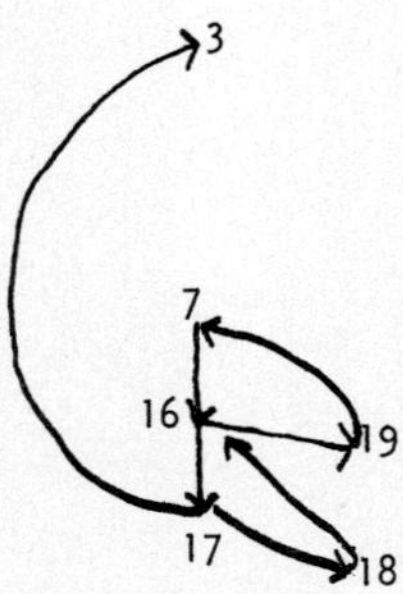

SEG(8)

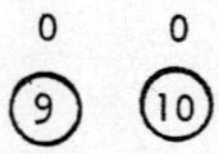

SEGGR(8)

Remark. The two vertices can be colored I or O independently, so there are three other possible $\mathscr{I}$-colorations of SEGGR(8) ($\mathscr{I} = \emptyset$).

k = 11

SEG(11)

SEGGR(11) = ϕ

k = 13

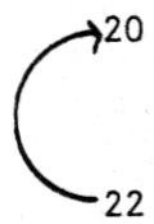

SEG(13)

SEGGR(13) = ϕ

k = 15

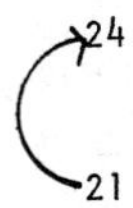

SEG(15)

SEGGR(15) = ϕ

k = 14

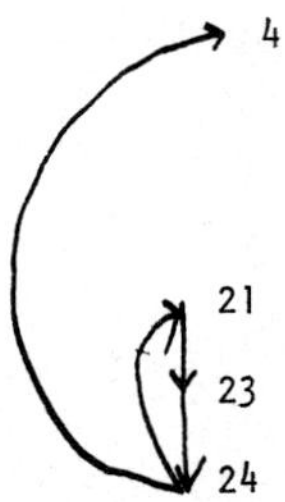

SEG(14)

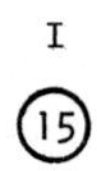

SEGGR(14)

k = 12

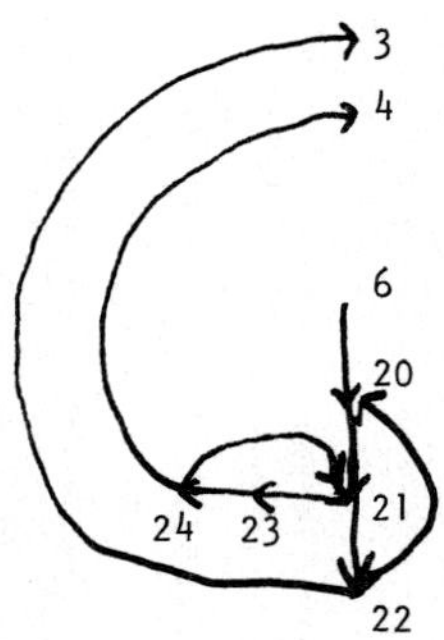

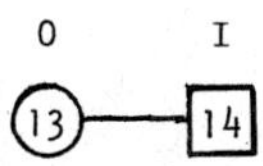

SEGGR(12)

Remark. $\mathcal{I} = \{\text{SEG}(14)\}$. The $\mathcal{I}$-coloration of SEGGR(12) is unique as SEG(13) and SEG(14) are directly linked (determined again by inspection).

k = 17

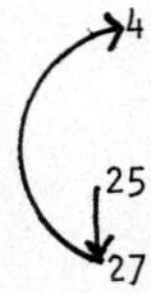

SEG (17)

SEGGR (17) = ϕ

k = 16

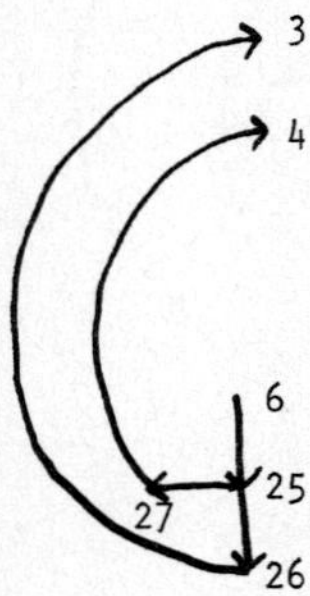

SEG (16)

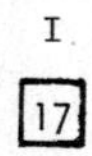

SEGGR (17)

k = 19

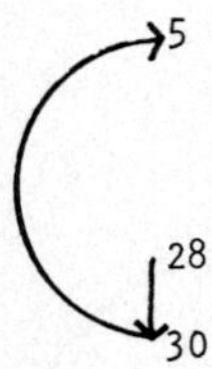

SEG (19)

SEGGR (19) = ϕ

k = 18

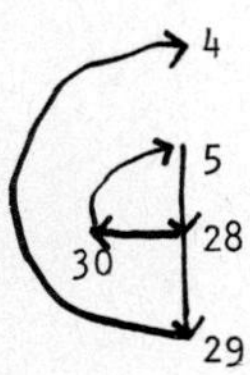

SEG (18)

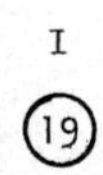

SEGGR (18)

k = 1

Remark. Sons of 1 in PATR(6, 7): 2, 8, 11, 12, 16, 18 are underlined above. Only the *proper* backedges of these segments effect the construction of SEGGR(1) and the subsequent embedding.

Schematically we have:

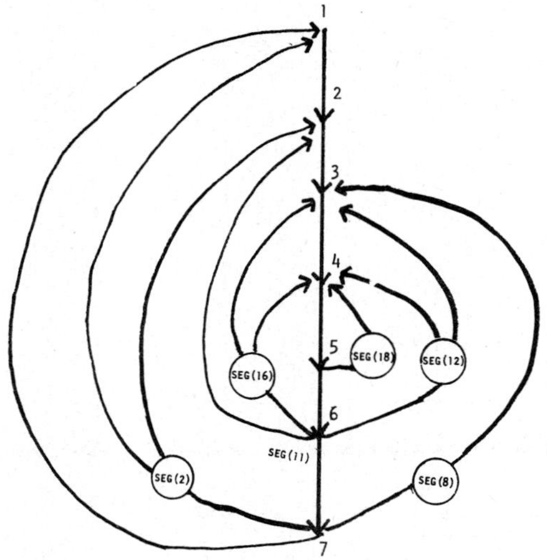

SEG(1)

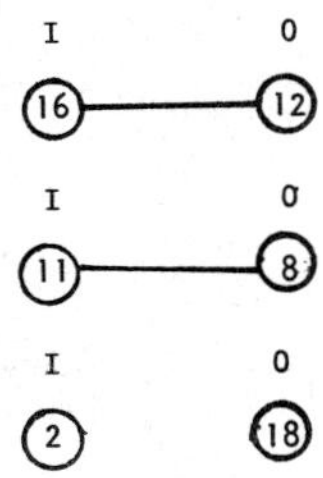

SEGGR(1)

This yields the following embedding for G (see Fig. 4.6(a)).

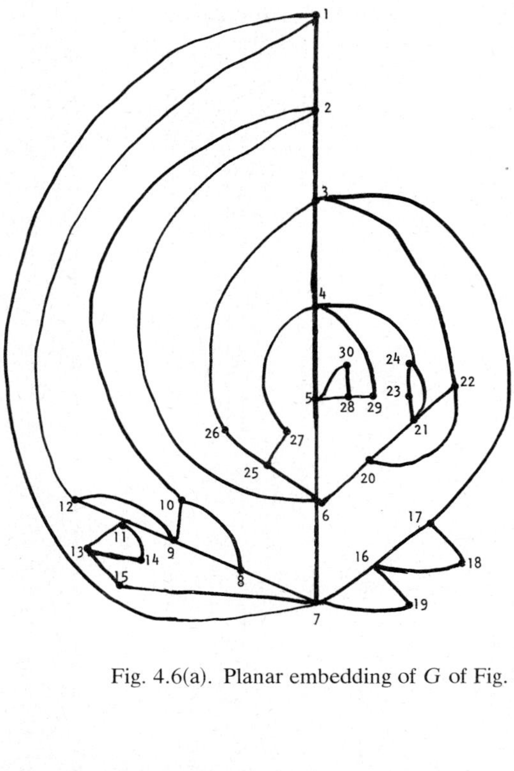

Fig. 4.6(a). Planar embedding of G of Fig. 3.1.

The reader should observe that the bicolorations of the segment graphs specify a labeling (with I, O) of PATR(G, T) as shown in Fig. 4.6(b) (referring to Fig. 4.5):

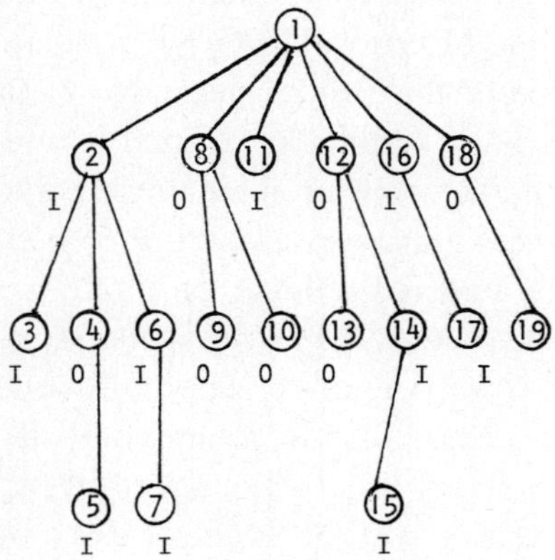

Fig. 4.6(b).

By traversing this labeled tree and, using the labels I = Inside, O = Outside, to embed the vertices of PATR(G) (as paths in G) in the plane as they are encountered in PREORDER (refer to Fig. 3.4) we again obtain Fig. 4.6(a). Thus Fig. 4.6(b) may be regarded as a specification of an embedding. We consider the construction of such labelings of PATR(G, T) in the next section (Section 5). Then we shall obtain more information than Fig. 4.6(b) by actually constructing spanning forests for all of the segment graphs associated with G.

We now state the above mentioned lemmas of Hopcroft and Tarjan [9] that result in a linear time planarity algorithm (a version of which is given in the next section). Recall that if SEG(f_1) and SEG(f_2) are in SEGLST(e), then we say SEG(f_1) and SEG(f_2) are *directly linked* if SEG(f_1) and SEG(f_2) must be embedded on opposite sides of CYCLE(e) in any planar embedding of CYCLE(e) $\cup$ SEG(f_1) $\cup$ SEG(f_2). We write SEG(f_1) dl SEG(f_2). Recall also that if $e = (a, b)$ is a directed edge, then TAIL(e) $= a$, HEAD(e) $= b$.

Lemma 4.7. *Let* SEG(f_1) $<$ SEG(f_2) *be two segments in* SEGLST(e). *Then* SEG(f_1) *is directly linked to* SEG(f_2) *if and only if there is a backedge* (s, t) *of* SEG(f_1) *with* LOW1(f_2) $< t <$ TAIL(f_2).

This lemma is proved by considering cases [9, 16]. We note only the case where TAIL(f_1) = TAIL(f_2) and LOW1(f_1) = LOW1(f_2). In this case LOW1(f_1) $< t <$ TAIL(f_1) so (s, t) is a *proper* backedge of SEG(f_1) (as defined in connection with Fig. 4.2 above). Recalling the definition of the order on SEGLST(e) as derived from the order on PATR(G, T) and the sorting of the triple

$$(\text{PREORDER}(p), \text{LOW1}(f), X(\text{LOW2}(f) < \text{PREORDER}(p)),$$

we see that SEG(f_2) must also have a proper backedge. That is,

$$X(\text{LOW2}(f_2) < \text{PREORDER}(\text{TAIL}(f_2)) = 1).$$

This is the only place where the third component of the triple plays a role in what we have done. The fact that SEG(f_1) dl SEG(f_2) in this case is an immediate consequence of the existence of this proper backedge of SEG(f_2).

We now state a second basic lemma of Hopcroft and Tarjan in a slightly generalized form. First we make some preliminary remarks.

Let S be a finite set of positive integers and let $\mathcal{P}(S)$ be the "power set" or set of all subset of S. We define a partial order $\preccurlyeq$ on $\mathcal{P}(S)$: if A and B are in $\mathcal{P}(S)$, then say $A \preccurlyeq B$ if $A = B$ or if MAX(A)$\leq$MIN(B). We assume the empty set $\emptyset \preccurlyeq A$ for all $A \in \mathcal{P}(S)$. Let $\mathcal{P}_{\preccurlyeq}(S)$ denote this partially ordered set and let $\mathcal{P}_{\trianglelefteq}(S)$ denote its dual or reverse order: $A \trianglelefteq B$ if and only if $B \preccurlyeq A$. Let $\mathcal{P}^*_{\preccurlyeq}(S) = \mathcal{P}_{\preccurlyeq}(S) \times \mathcal{P}_{\trianglelefteq}(S)$. For example $(\{1, 3, 5\}, \{4, 6\}) \leq (\{5, 6\}, \{2, 4\})$ in $\mathcal{P}^*_{\preccurlyeq}(S)$.

Let $X \in$ SEGLST(e). Define SEGLST(e, X) to be the set $\{Y: Y \in$ SEGLST(e), $Y \leq X\}$. Let SEGGR(e, X) denote the restriction of SEGGR(e) to SEGLST(e, X). For any $Z \in$ SEGLST(e), $Z =$ SEG(f), define TAIL(Z) = TAIL(f). In Example 4.4 TAIL(SEG(4)) = 11, TAIL(SEG(2)) = 7. (SEG(4) = SEG((11, 13)), SEG(2) = SEG((7, 8))). For any $Z \in$ SEGLST(e, X) define RET(Z, X) (the *return set of Z less than* TAIL(X)) to be the set of $t <$ TAIL(X) such that $t =$ HEAD(f) for some backedge f of Z. We denote RET(X, X) by simply RET(X). Let $\mathcal{C}(e, X)$ denote the set of components of SEGGR(e, X). For $C \in \mathcal{C}(e, X)$ define RET(C, X) = $\bigcup_Y$ RET(Y, X) where the union is over all Y that are vertices of C. Define the set STR(C) (for *starting set* of C) to be the set of all TAIL(Z), Z a vertex of C. Finally, let S be the vertices of CYCLE(e). We have for any graph G (planar or not):

Lemma 4.8. *The set* $\{(\text{RET}(C, X), \text{STR}(C): C \in \mathcal{C}(e, X), \text{RET}(C, X) \neq \emptyset\}$ *is a chain in* $\mathcal{P}^*_{\preccurlyeq}(S)$.

An example is given in Fig. 4.9. The components are C_1, C_2, C_3. RET(C_2, I) = ϕ. RET(C_1, I) = $\{1, 2, 3\}$, RET(C_3, I) = $\{3, 4, 5, 6, 8\}$, STR(C_1) = $\{18, 19, 20\}$, and STR(C_3) = $\{9, 10, 11, 12\}$. Thus

$$\{(\text{RET}(C_i, I), \text{STR}(C_i)): i = 1, 3\}$$
$$= \{(\{1, 2, 3\}, \{18, 19, 20\}), (\{3, 4, 5, 6, 8\}, \{9, 10, 11, 12\})\}$$

is a chain (linearly ordered subset) of $\mathcal{P}^*_{\preccurlyeq}(S)$, $s = \{1, \ldots, 20\}$.

The reader should note that Lemma 4.8 is false if RET(C, X) is replaced by RET(C) = $\bigcup$ RET(Y), Y a vertex of C. Fig. 4.10 illustrates this ($X = B$).

The proof of Lemma 4.8 is a straightforward induction on the linearly ordered set SEGLST(e). If X is the first element of SEGLST(e) the result is trivial. In general, assume the result is valid for $\mathcal{C}(e, X)$ and let X^+ be the successor of X in SEGLST(e). The structure of $\mathcal{C}(e, X^+)$ can easily be obtained using Lemma 4.7. *Note:* repeated sets of the chain of 4.8 must have size one.

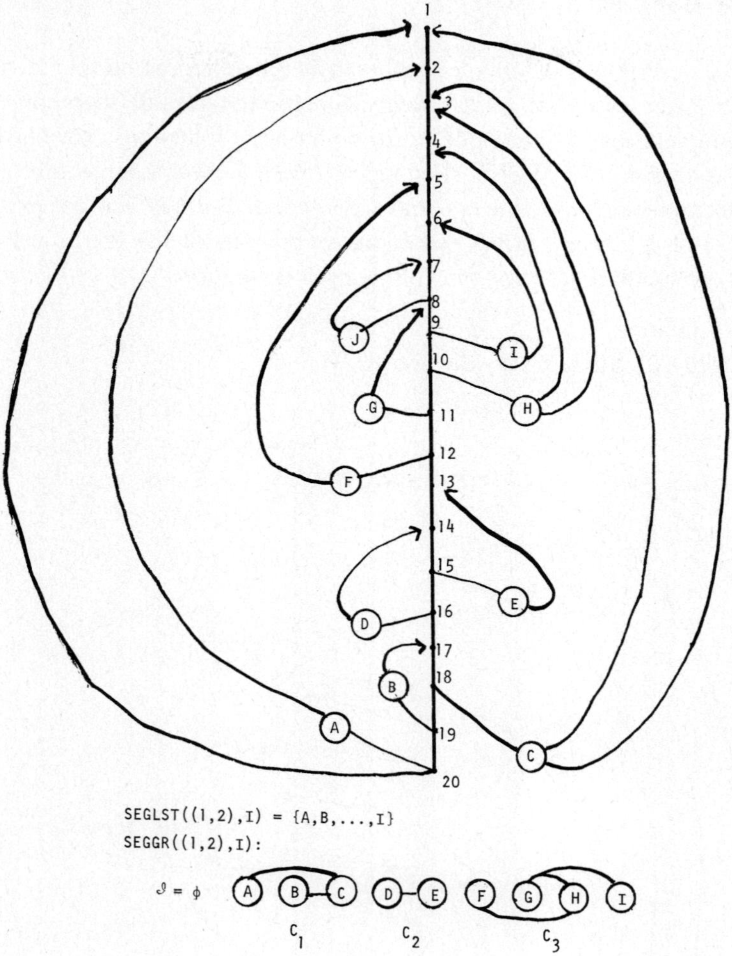

Fig. 4.9.

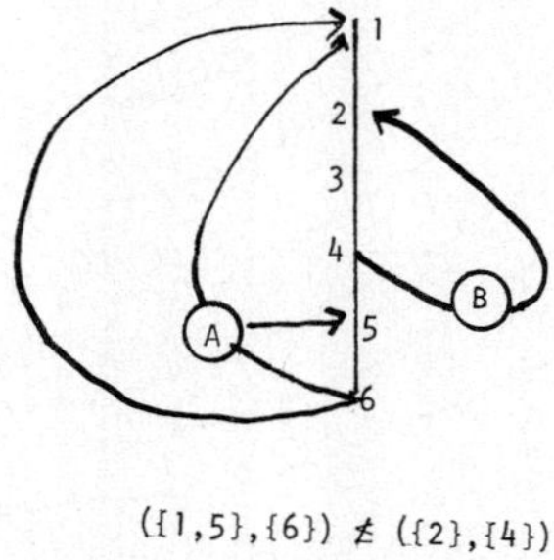

Fig. 4.10. ({1, 5}, {6}) ≢ ({2}, {4}).

5. Spanning forest for SEGGR(G)

We give a description of how a graph can be classified as nonplanar or planar and, in the latter case, spanning forests constructed for all segment graphs in linear time in vertices. As indicated in the remark (3) following Procedure 2.3, we may assume that if $G=(V, E)$, then $|E| \leqslant 3|V|-5$. We assume also that $G=(V, E)$ is biconnected and that we have constructed PATR(G, T). We traverse PATR(G, T) in depth first order. As we arrive at a vertex PATH(e) in POSTORDER we assume that we have the following information:

(1) SEGLST(e),
(2) for each $X \in$ SEGLST(e) the list RET(X).

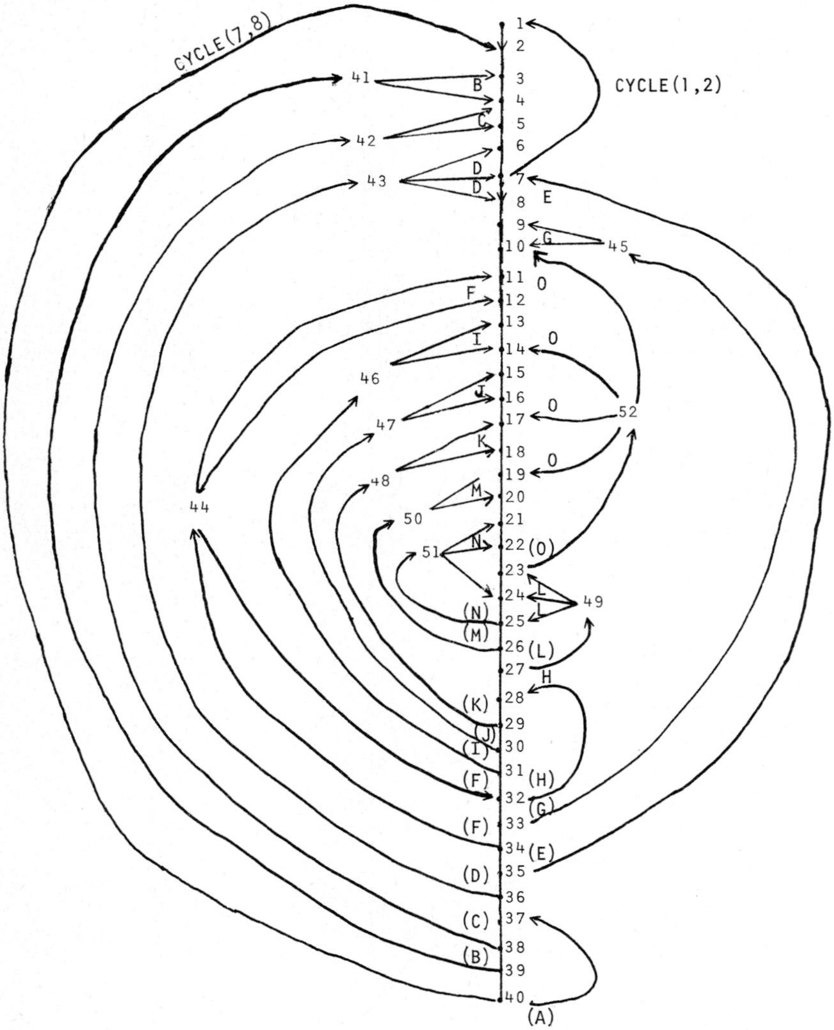

Fig. 5.1(a). Segments $A, \ldots, O$ of CYCLE(7, 8) ((X) denotes TAIL(X)).

We make the assumption that all underlying lists are "doubly linked" so that we can, in constant time, concatenate lists and find successors and predecessors of elements in a list. The elements of SEGLST(e) are represented by adjacency tables. Consider, for example, Fig. 5.1(a) and Table 5.1(b).

In Table 5.1(b) we display SEGLST(e) where $e = (7, 8)$ of Fig. 5.1(a). We are given pointers $A, B, \ldots, O$ to pairs of pointers (A_1, A_2), $(B_1, B_2) \cdots (O_1, O_2)$. At locations $A_1, B_1, \ldots, O_1$ we have the lists RET(A), RET(B), ..., RET(O) (using the pointers as symbols to represent the segments). At locations $A_2, \ldots, O_2$ we have the adjacency tables for the segments $A, \ldots, O$.

In general at each vertex PATH(e) we shall compute a spanning forest, SEGFO(e) for SEGGR(e). This will be done inductively on SEGLST(e). At any $X \in$ SEGLST(e) we will construct a spanning forest for SEGGR(e, X), denoted by SEGFO(e, X). The vertices of SEGFO(e, X) will be the elements U of SEGLST(e, X), represented by pointers U to (U_1, U_2) as in Table 5.1(b). Initially, U_1 points to RET(U) and U_2 to an adjacency table for the segment associated with U. When SEGFO(e, X) has been constructed the various lists associated with U_1 will have been modified. It will be true at this stage that U_1 points to RET(U, X) if $U \in$ I(C, X) or O(C, X) for any C in COMPONENTS(e, X) (defined below). The data associated with U_2 is never changed in computing SEGFO(e).

Table 5.1(b)

A: $(A_1, A_2)z_1$ A_1 (37)	A_2 40: 37	I: $(I_1, I_2)z_9$ I_1 13, 14	I_2 31: 46 46: 13, 14
B: $(B_1, B_2)z_2$ B_1 3, 4	B_2 39: 41 41: 3, 4	J: $(J_1, J_2)z_{10}$ J_1 15, 16	J_2 30: 47 47: 15, 16
C: $(C_1, C_2)z_3$ C_1 4, 5	C_2 38: 42 42: 4, 5	K: $(K_1, K_2)z_{11}$ K_1 17, 18	K_2 29: 48
D: $(D_1, D_2)z_4$ D_1 6, 7, 8	D_2 36: 43 43: 6, 7, 8	L: $(L_1, L_2)z_{12}$ L_1 23, 24, (25)	L_2 27: 49 49: 23, 24, 25
E: $(E_1, E_2)z_5$ E_1 7	E_2 35: 7	M: $(M_1, M_2)z_{13}$ M_1 19, 20	M_2 26: 50 50: 19, 20
F: $(F_1, F_2)z_6$ F_1 11, 12, (32)	F_2 34: 44 44: 11, 12, 32	N: $(N_1, N_2)z_{14}$ N_1 21, 22, [24]	N_2 25: 51 51: 21, 22, 24
G: $(G_1, G_2)z_z$ G_1 9, 10	G_2 33: 45 45: 9, 10	O: (O_1, O_2)	(z_{15} to be added when SEGFO(e, O) is constructed)
H: $(H_1, H_2)z_8$ H_1 (28)	H_2 32: 28	O_1 10, 14, 17, 19	O_2 23: 52 52: 10, 14, 17, 19

Associated with each such SEGFO(e, X) we shall keep certain basic data as follows:

COMPONENTS(e, x): This will be a list of pointers $C_1, \ldots, C_p$. Each C_i points to a 6-tuple ($g^-(C_i, X)$, $g^+(C_i, X)$, ROOT(C_i, X), I(C_i, X), O(C_i, X), STATUS(C_i, X)). These components are defined as follows:

$g^-(C_i, X) = \mathrm{MIN}(\mathrm{RET}(C_i, X))$,

$g^+(C_i, X) = \mathrm{MAX}(\mathrm{RET}(C_i, X))$,

ROOT(C_i, X) is a pointer to the location of the root of the tree C_i in SEGFO(e, X),

I(C_i, X) is a list of pointers to all vertices U in C_i with $\mathrm{RET}(U, X) \neq \emptyset$ that are colored I in an $\mathscr{I}$-coloration of $\tilde{C}_i$ (where $\tilde{C}_i$ is the component of SEGGR(e, X) corresponding to C_i),

O(C_i, X) is a list of pointers to all vertices U in C_i with $\mathrm{RET}(U, X) \neq \emptyset$ and colored O in the same $\mathscr{I}$-coloration,

STATUS(C_i, X) is INTERNAL if any vertex of C_i is internal (i.e., in $\mathscr{I}$), otherwise FREE.

Remarks. (1) We assume that $\mathrm{RET}(C_1, X) \preccurlyeq \mathrm{RET}(C_2, X) \preccurlyeq \cdots \preccurlyeq \mathrm{RET}(C_p, X)$ as guaranteed possible by Lemma 4.8. Thus $g^-(C_1, X) \leq g^+(C_1, X) \leq g^-(C_2, X) \leq \cdots$. (Using the remark about repeated sets in Lemma 4.8.)

(2) ROOT(C_i, X) will always be either at the top of either I(C_i, X) or O(C_i, X). This will be apparent from what follows.

(3) Note that we need know only one entry of either I(C_i, X) or O(C_i, X) to determine completely the $\mathscr{I}$-coloration of $\tilde{C}_i$ (given C_i).

(4) We assume that the sequence of entries $U_1, U_2, U_3, \ldots, U_q$ of I(C_i, X) are in order relative to SEGLST(e). Thus $\mathrm{RET}(U_1) \preccurlyeq \mathrm{RET}(U_2) \preccurlyeq \cdots \preccurlyeq \mathrm{RET}(U_q)$ as a consequence of Lemma 4.7. The same situation holds for O(C_i, X).

(5) Recall that a segment Z of SEGLST(e), $e = (a, b)$, is *internal* if it has at least one backedge (s, t) with $\mathrm{LOW1}(e) < t < a$. See Section 4, in particular Fig. 4.3.

If X is the first element of SEGLST(e), then SEGFO(e, X) = ($\{X\}, \emptyset$). In this case we have $p = 1$ and $g^-(C_1, X) = \mathrm{MIN}(\mathrm{RET}(X))$, $g^+(C_1, X) = \mathrm{MAX}(\mathrm{RET}(X))$, ROOT($C_1, X$) = ρ_1, I(C_1, X) = X, O(C_1, X) = $\emptyset$, STATUS(C_i, X) = STATUS(X). At location ρ_1 we have stored the trivial list of vertices: X. For notational convenience we put (for any X) a pointer C_0 at the bottom of the list COMPONENTS(e, X). Here $g^-(C_0, X) = 0$, $g^+(C_0, X) = 0$, I(C_0, X) = $\emptyset$, O(C_0, X) = $\emptyset$.

As an example, consider again Fig. 5.1. Let $X = N$. Then COMPONENTS(e, N), $e = (7, 8)$, is shown in Table 5.2.

Table 5.2. COMPONENTS(e, N), $e = (7, 8)$

	C_0	C_1	C_2	C_3	C_4	C_5
$g^-(C, N)$	0	3	6	9	13	19
$g^+(C, N)$	0	5	8	12	18	24
ROOT(C, N)		z_3	z_5	z_7	z_{11}	z_{14}
I(C, N)	$\emptyset$	x_1	x_2	x_3	x_4	x_5
O(C, N)	$\emptyset$	y_1	y_2	y_3	y_4	y_5
STATUS(C, N)	I	I	F	F	F	F

$X_1 B, C$ $Y_1 \emptyset$	$X_2 D$ $Y_2 E$	$X_3 F$ $y_3 G$	$X_4 I, J, K$ $y_4 \emptyset$
$x_5 M, N$ $y_5 L$			

The pointers ROOT(C, N) refer to the adjacency structure for SEGFO(e, N) = z_1. The adjacency structure and picture of this forest is shown in Fig. 5.3.

The vertices of SEGFO(e, N) are the elements of SEGLST(e, N). By the time we have produced the forest of Fig. 5.3, the initial representation of SEGLST(e) given in Table 5.1(b) is modified so that for each $U \in$ I(C, N) or O(C, N), RET(U) is replaced by RET(U, N). In L, for example, 25 has been removed from the list (indicated by parentheses) in Table 5.1(b).

We now describe the general process of going from SEGFO(e, X) to SEGFO(e, Y) where Y is the successor of X. We shall associate with each segment $Z \in$ SEGLST(e) an integer COST(Z). Also for each vertex v of PATH(e) we associate an integer COST(v). If [COST(Z)] or [COST(v)] follows a statement, then we increment these values by the time required to execute that statement. The general idea is that costs will be attributed to segments or vertices when they are removed or considered for the last time during a computation. If

```
                z1   A: B,C
                z2   B: A
ROOT(C1)    =   z3   C: A
                z4   D: E
ROOT(C2)    =   z5   E: D
                z6   F: G
ROOT(C3)    =   z7   G: F
                z8   H: I,J,K
                z9   I: H
                z10  J: H
ROOT(C4)    =   z11  K: H
                z12  L: N,M
                z13  M: L
                z14  N: L
```

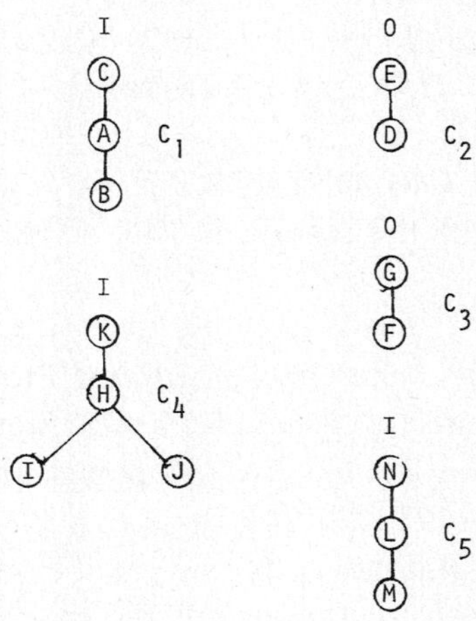

Fig. 5.3. SEGFO(e, N), $e = (7, 8)$.

no vertex or segment is removed during the computation, then the cost goes to COST(Y). We only explicitly reference those costs that might need clarification vis-a-vis linearity.

In general we will have TAIL(Y)<TAIL(X). It is necessary to remove all backedges that return to vertices of PATH(e) between TAIL(Y) and TAIL(X) as they have nothing to do with the further construction of SEGFO(e). If $\mathcal{S}$ is a list, then we use $x \Leftarrow \mathcal{S}$ or $x \Rightarrow \mathcal{S}$ to mean remove last element of $\mathcal{S}$ and call it x or make x the last element of $\mathcal{S}$. Let $\nu = \text{TAIL}(Y)$.

Procedure ERASEFROM(X to ν)

```
begin
  (1) until g⁻(C) < ν do
          C = COMPONENTS(e, X); [COST(ROOT(C))]
  (2) if COMPONENTS(e, X) ≠ C₀ then C ⇐ COMPONENTS(e, X); [COST(Y)]
  (3) if I(C, X) ≠ ∅ [COST(Y)] then until MIN(RET(S, X)) < ν
          do S ⇐ I(C, X); [COST(S)]
  (4) if I(C, X) ≠ ∅ then
          begin
            S ⇐ I(C, X); [COST(Y)]
            until t < ν do t ⇐ RET(S, X); [COST(t)]
            S ⇒ I(C, X); [COST(Y)]
          end
  (5) if O(C, X) ≠ ∅ [COST(Y)] then until MIN(RET(S, X)) < ν
          do S ⇐ O(C, X); [COST(S)]
  (6) if O(C, X) ≠ ∅ then
          begin
            S ⇐ O(C, X); [COST(Y)]
            until t < ν do t ⇐ RET(S, X); [COST(t)]
            S ⇒ O(C, X); [COST(Y)]
          end
  (7) C ⇐ COMPONENTS(e, X):
  (8) COMPONENTS(e, X, Y) ← COMPONENTS(e, X)
end
```

We consider the effect of ERASEFROM(N to 23) in our example: Statement (1) has no effect as $g^-(C_5) = 19 < 23$. Statement (2) replaces C by C_5. Referring to Table 5.2 we see that I(C, N) points to the list M, N. The pointer N_1 points to RET(N, N) = 21, 22, 24 (Table 5.1(b)). As 21<23, statement (3) does nothing. Statement (4) removes 24 (indicated by brackets in Table 5.1(b)) from RET(N, N) and returns N to the end of the list. Statement (5) removes L from O(C, N). Statement (5) removes L from O(C, N). Statement (8) simply changes the name of the list to reflect the dependence on Y.

We now are ready to add Y as a vertex to SEGFO(e), producing SEGFO(e, Y). The general situation is pictured in Fig. 5.4(a).

Consider the list COMPONENTS(e, X, Y) as it now stands (after applying ERASEFROM(X to v), v = TAIL(Y)). If C is such that LOW1(Y) < $g^-(C)$, then we call C a *critical component.* The sublist of critical components (a block of entries at the end of COMPONENTS(e, X)) will be called CRITICAL COMPONENTS(e, X, Y). A component C in COMPONENTS(e, X, Y) will be called *two sided* if both I(C, X) and O(C, X) are nonempty. Otherwise C will be called *one sided.* Note that if any C in CRITICAL COMPONENTS(e, X, Y) is

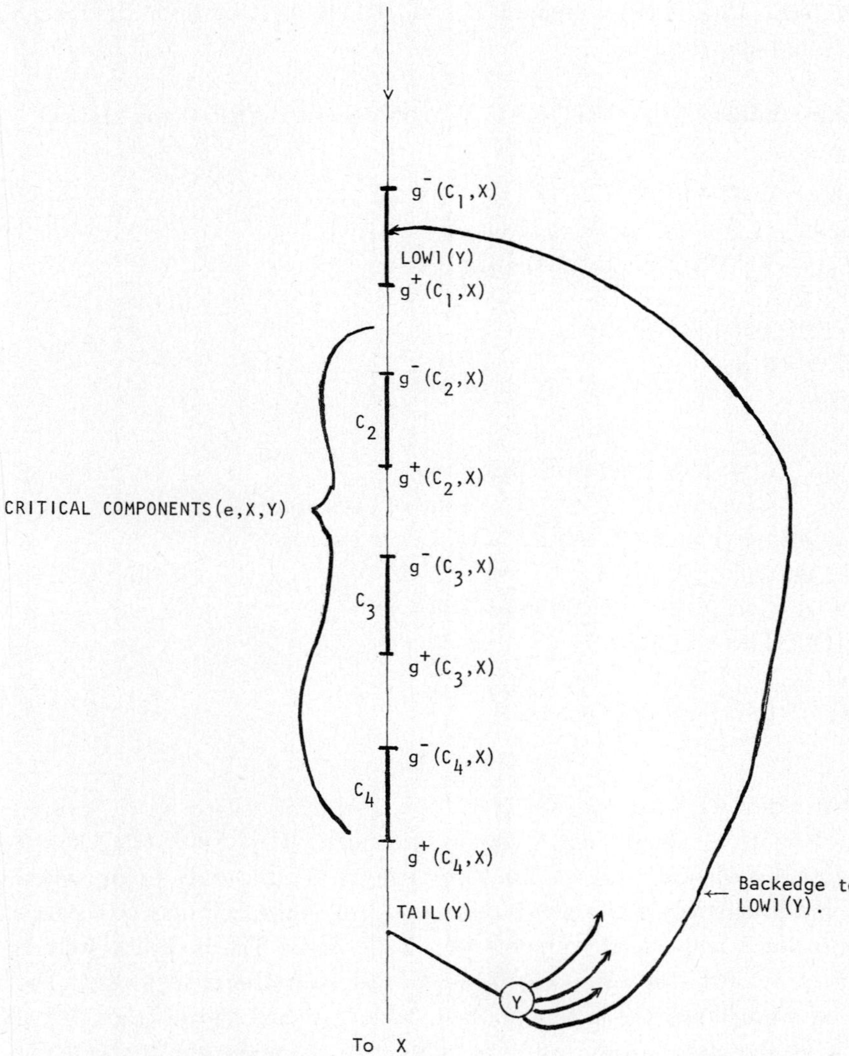

Fig. 5.4(a). Critical components (not the example of Fig. 5.1(a)).
Note. C_2 is the first element of CRITICAL COMPONENTS(e, X, Y) and C_4 is the last element.

two sided, the graph G is nonplanar (by Lemma 4.7). This follows since, by Lemma 4.7, Y would be directly linked to two segments of opposite color in the component C of SEGGR(e). The SEGGR(e), with Y added, would no longer be $\mathcal{I}$-bichromatic. If C is in COMPONENTS(e, X, Y), then either the last segment of I(C, X) or O(C, X) must have a backedge terminating at $g^+(C, X)$. In the former case we call MAXSEG(C) the last of I(C, X) otherwise MAXSEG(C) is the last of O(C, X). Clearly MAXSEG(C) can be found in constant time.

Procedure 5.4(b) returns TRUE if all critical components are sided, otherwise returns FALSE. In the former case pointers are changed, if necessary, to make $I(C, X) \neq \emptyset$ for all critical components C, and a list called CRITICAL COMPONENTS(e, X, Y) is created. $\mathcal{S}$ and CRITICAL COMPONENTS(e, X, Y) are lists, initially empty.

Procedure 5.4(b). ALL CRITICAL COMPONENTS ARE ONE SIDED

```
begin
  while g⁻(C, X) > LOW1(Y) do
    begin
      C ⇐ COMPONENTS(e, X, Y);
      C ⇒ 𝒮;
    end [COST(MAXSEG(C))]
  while 𝒮 ≠ ∅ do
    begin
      C ⇐ 𝒮
      if C is two sided return FALSE;
                                          [end of procedure].
      else interchange I(C, X) and O(C, X) if necessary to
        make I(C, X) ≠ ∅;
      C ⇒ CRITICAL COMPONENTS(e, X);
      [COST(MAXSEG(C)]
    end
  return TRUE; [end of procedure]
end
```

At this point we have a list CRITICAL COMPONENTS(e, X, Y). Every entry is a (pointer to a) component C that is one sided; $I(C, X) \neq \emptyset$, $O(C, X) = \emptyset$.

This (and a similar situation to occur below) are the only points where the COST is a little tricky with regard to linearity. MAXSEG(C) may be a part of the data associated with many segments $W > Y$ (in SEGLST(e)). It also may be the maximum segment in components associated with these segments. The key thing to note, however, is that if Y is added to SEGFO(e), then Y will be colored differently from all segments of components of CRITICAL COMPONENTS(e, X, Y). Thus (as will be apparent below) MAXSEG(C) will never again be a part of a *one sided* component (relative to a $W > Y$).

The next procedure is trivial, but we list it for the sake of completeness. The remarks above regarding COST apply here also. This procedure makes a list of all segments occurring in the lists I(C, X) associated with critical components. We assume all critical components are one sided. If $\mathcal{S}_1$ and $\mathcal{S}_2$ are lists, then $\mathcal{S}_1 \Leftarrow \mathcal{S}_2$ denotes the list obtained by putting $\mathcal{S}_1$ at the beginning of $\mathcal{S}_2$ (i.e., the "concatenation" of $\mathcal{S}_1$ with $\mathcal{S}_2$). This is done in constant time by changing pointers.

Procedure. CONSTRUCT CRITICAL SEGMENT LIST
[Initially CRITICAL SEGMENT LIST(e, X, Y) = $\emptyset$]

begin
  for C = FIRST to LAST of CRITICAL COMPONENTS(e, X, Y) *do*
    CRITICAL SEGMENT LIST(e, X, Y) $\Leftarrow$ I(C, X);
    [COST(MAXSEG(C)]
end

At this point we need to know if any of the critical components C have STATUS(C) = INTERNAL. We define a procedure called STATUS CRITICAL COMPONENTS(e, X, Y) which returns INTERNAL if one of the critical components is INTERNAL otherwise returns FREE (thus FREE if CRITICAL COMPONENTS(e, X, Y) = $\emptyset$). We omit the details of this as it is trivially just one pass through on the list of critical components [COST(MAXSEG(C)]. We also need to know STATUS(Y) = INTERNAL or FREE. To determine this, It is sufficient to check *only the two largest values of* RET(Y). Thus, this is again a trivial procedure, the details of which we omit [COST(Y)].

If all critical components are one sided and have been removed from COMPONENTS(e, X, Y), then the last element C of this latter list satisfies $g^-(C) \leq$ LOW1(Y). We must consider the relationship between this component C and Y. To do so, it will be helpful first to elaborate somewhat on the partial order introduced at the end of Section 4. Let VC denote the set of vertices of CYCLE(e). Consider $\mathcal{P}_{\leqslant}(VC)$ with partial order defined as at the end of Section 4. Let $\mathcal{A}$, $\mathcal{B}$, $\mathcal{A}'$, $\mathcal{B}'$ be chains in $\mathcal{P}_{\leqslant}(VC)$. That is $\mathcal{A} = \mathcal{A}_1 \leqslant \cdots \leqslant \mathcal{A}_p$, $\mathcal{B} = \mathcal{B}_1 \leqslant \mathcal{B}_2 \leqslant \cdots \leqslant \mathcal{B}_q$, $\mathcal{A}' = \mathcal{A}'_1 \leqslant \cdots \leqslant \mathcal{A}'_{p'}$, and $\mathcal{B}' = \mathcal{B}'_1 \leqslant \cdots \leqslant \mathcal{B}'_{q'}$. We say that $\mathcal{A} \leq \mathcal{B}$ if $\mathcal{A}_p \leqslant \mathcal{B}_1$ and the pair $(\mathcal{A}, \mathcal{B}) \leq (\mathcal{A}', \mathcal{B}')$ if $\mathcal{A} \leq \mathcal{A}'$ and $\mathcal{B} \leq \mathcal{B}'$. Alternatively $(\mathcal{A}, \mathcal{B}) \leq (\mathcal{A}, \mathcal{B})$ if $\mathcal{A}$ concatenated with $\mathcal{A}'$ ($\mathcal{A} \Leftarrow \mathcal{A}'$) and $\mathcal{B}$ concatenated with $\mathcal{B}'$ ($\mathcal{B} \Leftarrow \mathcal{B}'$) are also chains.

Return now to the last element C of COMPONENTS(e, X, Y). Let CRITICAL SEGMENT LIST(e, X, Y) = $U_1, \ldots, U_p$ and let $\mathcal{A}' = \mathcal{A}'_1 \leqslant \cdots \leqslant \mathcal{A}'_p$ where $\mathcal{A}'_i$ = RET(U_i, X), $\mathcal{B}'$ = RET(Y), $\mathcal{A}$ = I(C, X), $\mathcal{B}$ = O(C, X). First note that if STATUS CRITICAL COMPONENTS(e, X, Y) is INTERNAL and also STATUS(Y) is INTERNAL, then G is NONPLANAR by Lemma 4.7 (SEGGR(e) is not $\mathcal{I}$-bichromatic). Let us say, more generally, that STATUS($\mathcal{A}$), $\mathcal{A} = \mathcal{A}_1 \leqslant \cdots \leqslant \mathcal{A}_p$ a chain, is INTERNAL if some $t \in \mathcal{A}_1 \cup \cdots \cup \mathcal{A}_p$ satisfies LOW1(e) < t <

TAIL(e), otherwise STATUS($\mathcal{A}$) is FREE. STATUS ($\mathcal{A}$, $\mathcal{B}$) is internal if STATUS($\mathcal{A}$) or STATUS($\mathcal{B}$) is INTERNAL. We call ($\mathcal{A}'$, $\mathcal{B}'$) and ($\mathcal{A}$, $\mathcal{B}$) *incompatible* if $(\mathcal{A}', \mathcal{B}')\tilde{\geq}(\mathcal{A}, \mathcal{B})$ and $(\mathcal{A}', \mathcal{B}')\tilde{\geq}(\mathcal{B}, \mathcal{A})$. We say they are *strongly compatible* if $(\mathcal{A}', \mathcal{B}')\mathbf{e}(\mathcal{A}, \mathcal{B})$ and $(\mathcal{A}', \mathcal{B}')\geq(\mathcal{B}, \mathcal{A})$ (otherwise *weakly compatible*). We shall attempt to add Y to SEGFO(e, X) to form SEGFO(e, Y). If this is possible we shall create a component C' of SEGFO(e, Y) containing Y as a vertex. The possible lists I(C', Y) and O(C', Y) can be classified into six different types as follows: Let RET(I(C', Y)) = {RET(U, Y): $U \in$ I(C', Y)} and similarly define RET(O(C', Y)). Each of these sets is a chain in $\mathcal{P}_{\leqslant}(VC)$. The pair (RET(I($C'$, Y), RET(O(C', Y)) will be one of the following:

(1) $(\mathcal{A}', \mathcal{B}')$,
(2) $(\mathcal{B}', \mathcal{A}')$,
(3) $(\mathcal{A} \Leftarrow \mathcal{A}', \mathcal{B} \Leftarrow \mathcal{B}')$,
(4) $(\mathcal{A} \Leftarrow \mathcal{B}', \mathcal{B} \Leftarrow \mathcal{A}')$,
(5) $(\mathcal{B} \Leftarrow \mathcal{A}', \mathcal{A} \Leftarrow \mathcal{B}')$,
(6) $(\mathcal{B} \Leftarrow \mathcal{B}', \mathcal{A} \Leftarrow \mathcal{A}')$.

A decision tree showing how these cases arise is given in Fig. 5.5.

Remarks (Fig. 5.5). (1) There are four different "failure" cases resulting in NONPLANAR and eight different situations resulting in Y being successfully added to SEGFO(e, Y).

(2) In the path no, no, no, yes we use the "lemma" STATUS($\mathcal{A}$, $\mathcal{B}$) = FREE$\Rightarrow$ STATUS($\mathcal{A}'$) = FREE. In this case ($\mathcal{A}'$, $\mathcal{B}'$) and ($\mathcal{A}$, $\mathcal{B}$) are weakly compatible. This implies that $g^+(C) > g^-(C)$. Thus if STATUS($\mathcal{A}'$) = INTERNAL, then LOW1(e) $< g^+(C) <$ TAIL(e) (recall MIN($\mathcal{A}'$) $\geq g^+(C)$. Thus STATUS($\mathcal{A}$) = INTERNAL implies STATUS($\mathcal{A}$, $\mathcal{B}$) = INTERNAL, hence the "lemma".

(3) The various decisions represented by Fig. 5.5 involve direct reference to data we have already accumulated and can be implemented in constant time.

(4) Consider the basic example (Fig. 5.1). Let $Y = O$. CRITICAL COMPONENTS(e, N, O) = C_4, C_5 (obtained from Table 5.2 with LOW1(O) = 10) and $C = C_3$. CRITICAL SEGMENT LIST (e, O, N) = I, J, K, M, N. Thus $\mathcal{A}' = \{13, 14\} < \{15, 16\} < \{17, 18\} < \{19, 20\} < \{21, 22\}$ and $\mathcal{B}' =$ RET(O) = $\{11, 14, 17, 19\}$. $\mathcal{A}$ = RET(F, N) = $\{11, 12\}$, $\mathcal{B}$ = RET(G, N) = $\{9, 10\}$. Following the decision tree of Fig. 5.5 we find no, no, no ($\mathcal{A} \nleq \mathcal{B}'$), yes, yes. Thus we are in case 6, and I(C', Y) = G, O and O(C', Y) = F, I, J, K, M, N (opposite of what is shown in Fig. 5.1).

In any given case 1, . . . , 6, we must now show how to construct SEGFO(e, Y) and COMPONENTS(e, Y). In case Y is the last element of SEGLST(e) we must also construct RET(SEG(e)) and SEGGR(e) and add them to the list SEGLST(e') where PATH(e') is the father of PATH(e) in PATR(G, T). Cases 1 through 6 are all similar, so we consider only case 6 as an example.

Consider first our example of Fig. 5.1. CRITICAL COMPONENTS(e, N, O) =

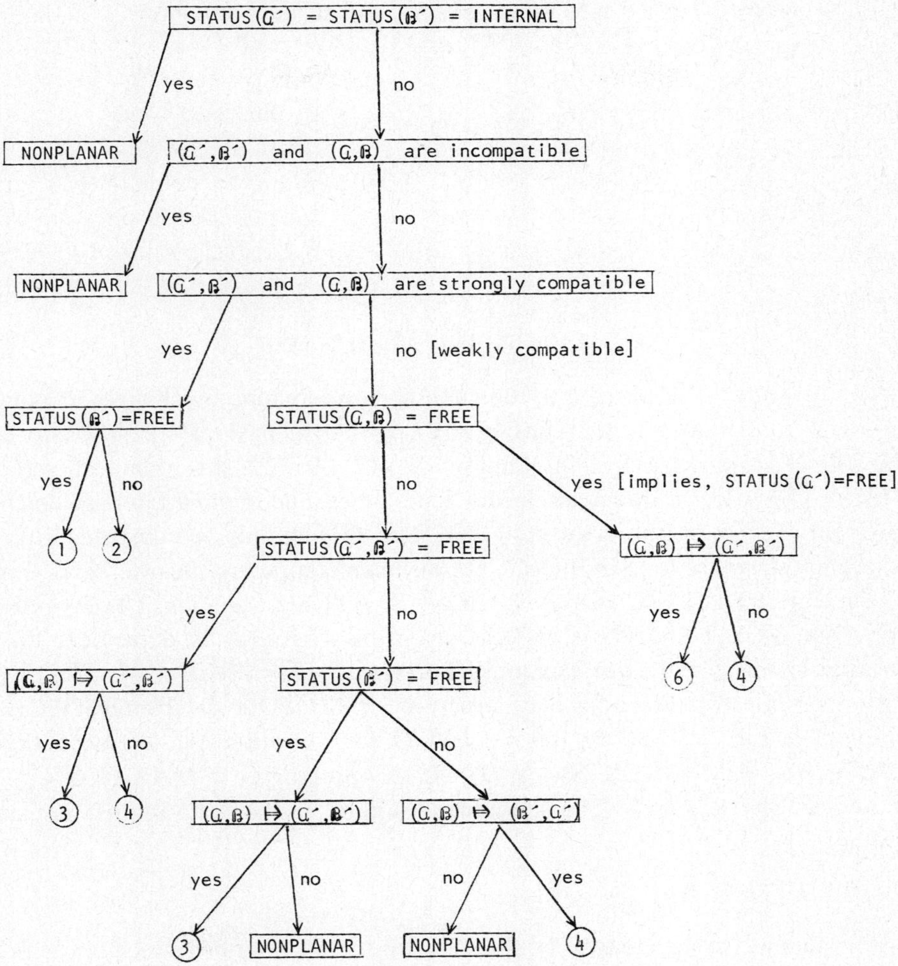

Fig. 5.5. To determine structure of SEGFO(e, Y).

C_4, C_5 and $C = C_3$ (see Table 5.2). We must join (as an edge in SEGFO(e)) O to the first entry in I(C_4, N) and I(C_5, N). Also O must be joined to the first element of I(C_3, N) (in case 6 we have $\mathscr{A} \Leftarrow \mathscr{A}'$, $\mathscr{B} \Leftarrow \mathscr{B}'$ and $\mathscr{B} = \text{RET}(O)$). In case 6 $\mathscr{B}'$ appears on the left in the pair ($\mathscr{B} \Leftarrow \mathscr{B}'$, $\mathscr{A} \Leftarrow \mathscr{A}'$) so the new root O will have color I (i.e., be in I(C', O) where C' is the new component containing O). We see the result for our example in Fig. 5.6 (compare with Fig. 5.3).

We must obtain the list COMPONENTS(e, O) and all related data. In addition to COMPONENTS(e, O) we must modify SEGFO(e, N) (see Fig. 5.3) to obtain SEGFO(e, O)).

We first discuss COMPONENTS(e, N). As described above, if C' is the new component of SEGFO(e, O), then I(C', O) = G, O and O(C', O) = F, I, J, K, M, N. These are obtained by concatenation I(C', O) = $\mathscr{B} \Leftarrow \mathscr{B}'$ and

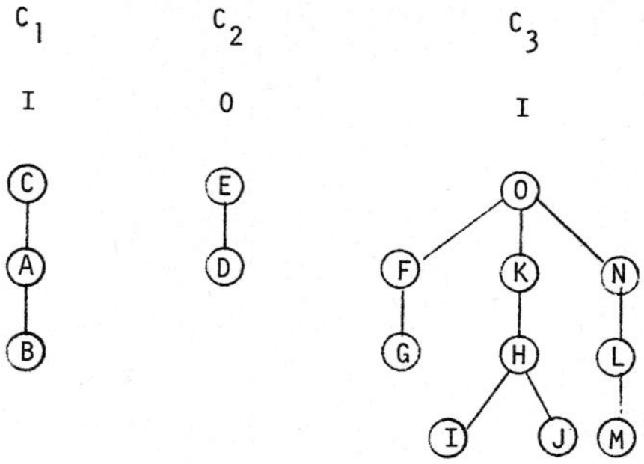

Fig. 5.6. SEGFO(e, O) = SEGFO(e).

O(C', O) = $\mathscr{A} \Leftarrow \mathscr{A}'$. This can, in the general case, be done by changes of pointers and thus in constant time. Thus $g^{-}(C', O) = 9$ (smallest of smallest entries in RET(G, O) and RET(F, O)). Similarly $g^{+}(C', O) = 22$ (from largest value of RET(N, O)). Clearly these can, in the general case, be obtained in constant time. The root of C' is O but its location ROOT(C, O) can only be computed after the adjacency structure for SEGFO(C', O) has been computed, thus we defer this for a moment. STATUS(C', O) is FREE if STATUS($\mathscr{B}'$) = STATUS(O) = FREE, otherwise STATUS(C, O) is INTERNAL (this condition is dependent on our position in Fig. 5.5. In our example STATUS(C', O) = FREE).

We now construct the adjacency structure for SEGFO(e, O) from SEGFO(e, N) as given in Fig. 5.3. To do this we add O to a location z_{15}. We scan the list CRITICAL COMPONENTS(e, N, O) = C_4, C_5, linking O to MAX(SEG(C_4)) = K and MAX(SEG(C_5)) = N. This amounts to adding K and N to the list associated with z_{15}:

$$z_{15}O\colon K, N.$$

In the general case, linearity is obtained by the same trick associated with the Procedure 5.4(b). Finally we must add a pointer from O to the last element of I(C, N) which is F, now we have

$$z_{15}O\colon K, N, F.$$

We now follow the pointers K, N, F to Table 5.1(b) to locate the values z_{11}, z_{14}, z_6, pointing to Fig. 5.3. O is then added to these lists: ($z_{11}K$: H, O, $z_{14}N$: L, O, z_6F: G, O). The pointer z_{15} is added to Table 5.1(b) after O. Finally we set ROOT(C') = z_{15}.

Thus we have now constructed SEGFO(e, O) and COMPONENTS(e, O). The earlier procedure ERASEFROM(N to 23) has changed Table 5.1(b) so that if $W \in$ I($\tilde{C}$, O) or O($\tilde{C}$, O), then W points to RET(W, O). ($\tilde{C}$ in COMPONENTS(e, O). This is valid, in general, only for W in I($\tilde{C}$, O) or $O(\tilde{C}, O)$ (due to statement (1) of ERASEFROM(X to ν)). Thus we are now ready to consider the successor to O in SEGLST(e). In this case, O is the last element of SEGLST(e), $e = (7, 8)$. Thus we have constructed SEGFO(e). If PATH(e) is not the root of PATR(G, T), then call PATH(e') the father of PATH(e) in PATR(G). We must construct SEG(e) and RET(SEG(e)) and add them to SEGLST(e') (to construct at the next level up in the recursion the data corresponding to Table 5.1(b)). To obtain SEG(e) we follow PATH(e) stopping at each vertex to set up pointers to entries in SEGLST(e) (Table 5.1(b)). This process is easily seen to be linear and to result in an adjacency structure (with linked lists) representing SEG(e). To obtain RET(SEG(e)) we perform ERASEFROM(O, TAIL(e)). The entries left in COMPONENTS(e, O) contain pointers to the values of RET(SEG(e)). In this example RET(SEG(e)) contains 2, 3, 4, 5, 6.

The general case of this process, involving all cases of Fig. 5.5 represents an algorithm of considerable complexity. The general algorithm parallels closely our example. We omit a detailed algol description and instead briefly summarize the general process:

(1) At PATH(e) in PATR(G) we have SEGLST(e) and RET(U) for all $U \in$ SEGLST(e) (as specified, for example, in Table 5.1(b)).

(2) SEGFO(e) is constructed inductively on SEGLST(e). At X on SEGLST(e) we assume that we are given SEGFO(e, X) and COMPONENTS(e, X).

(3) Let Y be the successor of X. First compute ERASEFROM(X to TAIL(Y)). If ALL CRITICAL COMPONENTS ARE ONE SIDED is FALSE the G is nonplanar, otherwise go to (4).

(4) Following the decision tree of Fig. 5.5 determine if G is NONPLANAR or, if planar, which case 1, . . . , 6 holds. If planar go to (5).

(5) Depending on which case 1, . . . , 6 holds, add Y to SEGFO(e, X) by

(a) Joining Y to one vertex of each component of CRITICAL COMPONENTS(e, X, Y) (if this list is not empty).

(b) Joining Y to the component C (if any) for which $g^-(C) <$ LOW1(Y) $< g^+(C)$.

(6) Update SEGFO(e, X) to get SEGFO(e, Y) and COMPONENTS(e, X, Y) to get COMPONENTS(e, Y).

(7) If Y is the last entry in SEGLST(e), then construct SEG(e) from PATH(e) and SEGLST(e). Also construct RET(SEG(e)) by applying ERASEFROM(Y, TAIL(e)). Add SEG(e) and RET(SEG(e)) to SEGLST(e').

We have indicated carefully how this process can be implemented in linear time in $|V|$, V the vertex set of G, up through (4) and by example for steps (5), (6), and (7). These latter steps involve repetition of the same ideas previously stated (in terms of complexity).

References

[1] A.V. Aho, J.E. Hopcroft and J.D. Ullman, The Design and Analysis of Computer Algorithms (Addison-Wesley, Reading, MA, 1974).

[2] L. Auslander and S.V. Parter, On imbedding graphs in the plane, J. Math. Mech., 10 (1961) 517–523.

[3] J.A. Bondy and U.S.R. Murty, Graph Theory with Applications (Am. Elsevier, New York, 1977).
[4] S.H. Chung and P.H. Roe, Algorithms for testing the planarity of a graph, Proc. Thirteenth Midwest Symposium on Circuit Theory, Univ. of Minnesota, Minneapolis, MN, May 1970, VII.4.1-VII.4.12.
[5] R. Cori, Un code pour les graphes planaires, Astérisque 1975 (Soc. Math. de France) 169 pp.
[6] G. Demoucron, Y. Malgrange and R. Pertuiset, Graphes planaires: Reconnaissance et construction des représentations planaires topologiques, Rev. Française Recherche Opérationelle 8 (1965) 34–47.
[7] N. Deo, Note on Hopcroft and Tarjan's planarity algorithm, J. Assoc. Comput. Mach. 23 (1976) 74–75.
[8] A.J. Goldstein, An efficient and constructive algorithm for testing whether a graph can be embedded in a plane, Graph and Combinatorics Conf., Contract no. NONR 1858-(21), Office of Naval Research Logistics Proj., Dept. of Math., Princeton Univ. May 16–18 (1963) 2 pp.
[9] J. Hopcroft and R. Tarjan, Efficient planarity testing, J. Assoc. Comput. Mach. 21 (1974) 549–568.
[10] J. Hopcroft and R. Tarjan, Isomorphism of planar graphs, in: R.E. Miller and J.W. Thatcher, Eds., Complexity of Computer Computations (Plenum Press, New York, 1972) 154–150.
[11] J. Hopcroft and R. Tarjan, Planarity testing in $V \log V$ steps: Extended abstract, in: Proc. IFIP Congr. 1971: Foundations of Information Processing, Ljubljana, Yugoslavia, Aug. 1971 (North-Holland, Amsterdam, 1972) 18–22.
[12] C. Kuratowski, Sur le problème des corbes gauches en topologie, Fund. Math. 15 (1930) 271–283.
[13] A.S. LaPaugh and R.L. Rivest, The subgraph homeomorphism problem, in: Tenth Annual ACM Symposium on Theory of Computing, San Diego, May 1978 (ACM, 1978) pp. 40–50.
[14] A. Lempel, S. Even and I. Cederbaum, An algorithm for planarity testing of graphs, in: P. Rosensteiehl, Ed., Theory of Graphs: International Symposium: Rome, July 1966 (Gordon and Breach, New York, 1967) 215–232.
[15] P. Mei and N. Gibbs, A planarity algorithm based on the Kuratowski theorem, in: Proc. AFIPS 1970 SJCC, Vol. 36 (AFIPS Press, Montvale, NJ) 91–95.
[16] E.M. Riengold, J. Nievergelt and N. Deo, Combinatorial Algorithms (Prentice-Hall, Englewood Cliffs, NJ, (1977) (Sect. 8.6, p. 364 and p. 399, Problem 51).
[17] F. Rubin, An improved algorithm for testing the planarity of a graph, IEEE Trans. Comput. C-24 (1975) 113–121.
[18] R. Tarjan, An efficient planarity algorithm, STAN-CS-244-71, Comput. Sci. Dept., Stanford Univ. (Nov. 1971).
[19] R. Tarjan, Depth-first search and linear graph algorithms, SIAM J. Comput. 1 (1972) 146–159.
[20] R. Tarjan, Complexity of combinatorial algorithms, SIAM Rev. 20 (1978) 457–492.
[21] S.G. Williamson, Finding a Kuratowski sub-graph in linear time, IRMA Strasbourg (preprint).

Annals of Discrete Mathematics 6 (1980) 385–391

HYPERCLAW DECOMPOSITION OF COMPLETE HYPERGRAPHS

Sumiyasu YAMAMOTO and Shinsei TAZAWA
Hiroshima University and Hiroshima College of Economics, Hiroshima, Japan

1. Introduction

Let $V=\{v_i \mid i=1,2,\ldots,m\}$ be a finite set and let $\mathscr{E}=\{E_i \mid i\in I\}$ be a family of nonempty subsets of V. The couple $H=(V,\mathscr{E})$ is called a *hypergraph* on V (see [2]) if

$$\bigcup_{i\in I} E_i = V.$$

The cardinality $|V|=m$ of the set V is called the *order*, the elements of V are called the *vertices* and the elements of $\mathscr{E}$ are called the *edges*. An edge E is sometimes called k-*edge* by indicating its cardinality $|E|=k$. A hypergraph H is said to be *simple* if the edges E_i are all distinct.

Two vertices are said to be *adjacent* if there is an edge that contains both vertices. Two edges are said to be *adjacent* if their intersection is not empty. An edge E and a vertex v are said to be *incident* with each other if $v\in E$. Two edges E_i and E_j are said to be *incident* with each other if either $E_i \supsetneqq E_j$ or $E_i \subsetneqq E_j$.

A simple hypergraph on V is called a k-*graph* if the cardinality of every edge is the same k. A 2-graph is a simple undirected graph without isolated vertices. A k-graph on V is said to be *complete* and denoted by $H_{m,k}$ if $\mathscr{E}$ is composed of all $\binom{m}{k}$ edges. A complete 2-graph on V is a complete graph with m points (vertices) and $\binom{m}{2}$ lines (edges).

A collection of k-edges $E_1, E_2, \ldots, E_c$ $(k\geq 2, c\geq 2)$ is called a k-*hyperclaw* (abbreviated k-HC) or simply a *hyperclaw* of degree c if

$$\left|\bigcap_{i=1}^{c} E_i\right| = k-1,$$

where $|A|$ denotes the cardinality of a set A. The set $\bigcap_{i=1}^{c} E_i$ is called the root and the vertices

$$E_i - \bigcap_{j=1}^{c} E_j, \quad i=1,2,\ldots,c,$$

are called the leaves of the hyperclaw. For a 2-graph, a hyperclaw of degree c is a claw or a star composed of c lines being incident to the same point.

A k-graph $(V, \mathscr{E})$ is said to be *k-hyperclaw decomposable* or k-HC decomposable if $\mathscr{E}$ can be decomposed into pairwise edge-disjoint k-hyperclaws of the same degree.

The problem of the claw-decomposabability of a complete graph has been raised in connection with the design of a file organization scheme by Yamamoto, *et al.* [4]. A necessary and sufficient condition for the claw-decomposabability of a complete graph $K_m = H_{m,2}$ given by Yamamoto, *et al.* [5] is (i) $\binom{m}{2}$ is an integral multiple of c, the degree of the claw, and (ii) $m \geqslant 2c$. An algorithm for the decomposition has also been given there. The problem of k-HC decomposability of a complete k-graph has, therefore, been solved completely for the case of $k = 2$.

In this paper, a necessary condition and several sufficient conditions for the k-HC decomposability of a complete k-graph $H_{m,k}$ for $k \geqslant 3$ are given. The 3-HC decomposability of $H_{m,3}$ is given in some detail.

2. A necessary condition

Consider a complete k-graph $H_{m,k}$ with m vertices and $\binom{m}{k}$ k-edges, where $k \geqslant 3$.

Theorem 2.1. *A necessary condition for a complete k-graph $H_{m,k}$ to be k-HC decomposable of degree c is*

(i) *$\binom{m}{k}$ is an integral multiple of c, and*

(ii) $$m \geqslant \frac{-kc+4k-6+(k(k+24)c^2-12kc+4)^{1/2}}{4},$$

or

$$c \leqslant \frac{m-k+3}{6} + \frac{(k(k+24)m^2-2k(k^2+21k-36)m+k(k^3+18k^2-63k+48))^{1/2}}{6k}.$$

Proof. Suppose $H_{m,k}$ has been decomposed into pairwise edge-disjoint k-HC of degree c. The condition (i) is obviously necessary. For an arbitrary selected set P of $k-2$ vertices $x_1, x_2, \ldots, x_{k-2}$, the remaining $m-k+2$ vertices can be classified into two sets $Q = \{y_1, y_2, \ldots, y_\alpha\}$ and $R = \{z_1, z_2, \ldots, z_{m-k-\alpha+2}\}$ in such a way that every set of $k-1$ vertices $P \cup \{y_i\}$ is the root of some hyperclaw of degree c whereas every $P \cup \{z_j\}$ is not.

If $|R| = m-k-\alpha+2 = 1$ or 0, then we have $\alpha = m-k+1$ or $m-k+2$. Suppose $|R| = m-k-\alpha+2 \geqslant 2$ and consider a k-edge $P \cup \{z_i, z_j\}$ for every pair of vertices in R. Since both $P \cup \{z_i\}$ and $P \cup \{z_j\}$ are not the root of any k-HC of degree c, one of the subsets, each consisting of z_i, z_j and $k-3$ vertices of P, must be the root of a k-HC of degree c. Thus, for any pair of $\{z_i, z_j\}$ there corresponds at least one k-HC of degree c. Enumerating the number of k-edges of such hyperclaws and that of k-edges which may possibly be the members of these hyperclaws, we have

$$\binom{m-k-\alpha+2}{2}c \leq \binom{m-k-\alpha+2}{2}(\alpha+1)+\binom{m-k-\alpha+2}{3}. \tag{2.1}$$

Thus α must satisfy the inequality

$$\alpha \geq \tfrac{1}{2}(3c-m+k-3). \tag{2.2}$$

The inequality (2.2) holds whatever $|R| \geq 2$ or not, since we have $m-k+1 \geq c$ from the existence of a k-HC of degree c in $H_{m,k}$. Moreover, since (2.2) holds for every selection of the set P of $k-2$ vertices, we have,

$$\binom{m}{k-2}\cdot \tfrac{1}{2}(3c-m+k-3) \leq \sum \alpha = (k-1)\binom{m}{k}\Big/ c. \tag{2.3}$$

Thus, we have the inequality

$$2m^2+(kc-4k+6)m-(c-2)k^2-3(c^2-c+2)k+4 \geq 0. \tag{2.4}$$

The condition (ii) follows immediately from (2.4).

It can be verified easily that the lower bound of m given in (ii) cannot be attained except possibly for $k=3$, since $k(k+24)c^2-12kc+4$ cannot be a perfect square except for $k=3$. Only in the case of $k=3$, the condition (ii) can be simplified as

(ii)* $\quad m \geq \tfrac{1}{2}(3c+2) \quad$ or $\quad c \leq \tfrac{2}{3}(m-1)$.

The parameter m which satisfies (i) and attains the bound in (ii)* can be expressed as $m=6n+4$ for $c=4n+2$ where n is a non-negative integer. For such a pair of m and c, $H_{m,3}$ can really be decomposed into the pairwise disjoint 3-HC of degree c as will be seen presently. In this sense, the bound of m or equivalently that of c given in (ii) is tight at least for the case $k=3$.

3. Sufficient conditions for the k-hyperclaw decomposability

With respect to the k-HC decomposability of a complete k-graph $H_{m,k}$, we have the following theorems.

Theorem 3.1. *If $H_{m,k}$ is k-HC decomposable of degree λc for positive integers λ and $c \geq 2$, then $H_{m,k}$ is k-HC decomposable of degree c.*

The proof is immediate.

Theorem 3.2. *If $m=kc+k-1$ for $c \geq 2$, then $H_{m,k}$ is k-HC decomposable of degree c.*

Proof. Since

$$\binom{kc+k-1}{k} \Big/ c = \binom{kc+k-1}{k-1},$$

the condition (i) of Theorem 2.1 is satisfied. It is easy to verify that the condition (ii) is also satisfied.

Let $\mathscr{E}^{(k-1)}$ be the family of all $\binom{kc+k-1}{k-1}$ subsets, each composed of $k-1$ vertices and $\mathscr{E}^{(k)}$ be the family of all $\binom{kc+k-1}{k}$ k-edges. Every element of $\mathscr{E}^{(k-1)}$ is incident with kc elements of $\mathscr{E}^{(k)}$ and every element of $\mathscr{E}^{(k)}$ is incident with k elements of $\mathscr{E}^{(k-1)}$. There exists, therefore, a c-ple SDR or a c-fold matching (see, e.g., [1, 3]) from $\mathscr{E}^{(k-1)}$ to $\mathscr{E}^{(k)}$. This matching implies that every set of c k-edges which corresponds to an element of $\mathscr{E}^{(k-1)}$ is a k-HC of degrees c. These k-HC's are, of course, pairwise disjoint.

Corollary 3.3. *If $m = k\lambda c + k - 1$ for positive integers λ and $c \geqslant 2$, then $H_{m,k}$ is k-HC decomposable of degree c.*

The proof is immediate.

Theorem 3.4. *If $m = kc$ for $c \geqslant 2$, then $H_{m,k}$ is k-HC decomposable of degree c.*

Proof. Let $V = \{v_1, \ldots, v_m\}$ and consider $V_0 = \{v_1, \ldots, v_{m-1}\}$. Since $m-1 = k(c-1)+k-1$, $H_{m-1,k}$ is k-HC decomposable of degree $c-1$. By adjoining v_m to every set of leaves of the k-HC of degree $c-1$, a family of pairwise disjoint k-HC's of degree c is constructed. Since $\binom{m-1}{k}/(c-1) = \binom{m}{k}/c$, $H_{m,k}$ on V is k-HC decomposable of degree c.

Corollary 3.5. *If $m = k\lambda c$ for positive integers λ and $c \geqslant 2$, then $H_{m,k}$ is k-HC decomposable of degree c.*

The proof is immediate.

Theorem 3.6. *If $H_{m-1,k}$ is k-HC decomposable of degree c and if $H_{m-1,k-1}$ is $(k-1)$-HC decomposable of degree c, then $H_{m,k}$ is k-HC decomposable of degree c.*

Proof. Let $V = \{v_1, v_2, \ldots, v_m\}$ and let $\mathscr{E}^{(k)}$ be the family of all k-edges on V. Let $V_0 = V - \{v_m\}$ and let $\mathscr{E}_0^{(k)}$ and $\mathscr{E}_0^{(k-1)}$ be the family of all k-edges and that of all $(k-1)$-edges on V_0, respectively. The family $\mathscr{E}^{(k)}$ can be partitioned into two subfamilies $\mathscr{E}_1 = \mathscr{E}_0^{(k)}$ and $\mathscr{E}_2 = \{E \mid E = E_0 \cup \{v_m\},\ E_0 \in \mathscr{E}_0^{(k-1)}\}$. The subfamily $\mathscr{E}_1$ of k-edges is, by assumption, k-HC decomposable of degree c. Since every k-edge in $\mathscr{E}_2$ is composed by adjoining v_m to $(k-1)$-edge on V_0 and since $\mathscr{E}_0^{(k-1)}$ is $(k-1)$-HC decomposable of degree c, the subfamily $\mathscr{E}_2$ is also k-HC decomposable of degree c. This completes the proof.

Corollary 3.7. *If $m=k(k-1)c+1$ for $c\geqslant 2$, then $H_{m,k}$ is k-HC decomposable of degree c.*

This is a straightforward consequence of Theorem 3.6, since it can be shown by Corollary 3.5 that $H_{k(k-1)c,k}$ is k-HC decomposable of degree c and $H_{k(k-1)c,k-1}$ is $(k-1)$-HC decomposable of degree c.

4. Miscellaneous results on 3-HC decomposability

It has been pointed out in Section 2, that a necessary condition for an $H_{m,3}$ to be 3-HC decomposable is (i) $m(m-1)(m-2)\equiv 0 \pmod{6c}$ and (ii)* $m\geqslant\frac{1}{2}(3c+2)$ or $c\leqslant\frac{2}{3}(m-1)$. The parameters m and c which satisfy (i) and attain the bound given in (ii)* can be expressed as $m=6n+4$, $c=4n+2$ for a nonnegative integer n. With respect to such series of parameters $m=6n+4$ and $c=4n+2$, we have the following theorem.

Theorem 4.1. *A complete 3-graph $H_{m,3}$ is 3-HC decomposable of degree c if $m=6n+4$ and $c=4n+2$ for a nonnegative integer n.*

Proof. Suppose $n\geqslant 2$. Let V_1 and V_2 be the two subsets of V and suppose $V_1\cup V_2=V$, $V_1\cap V_2=\emptyset$ and $|V_1|=|V_2|=3n+2$. Theorem 3.2 shows that a 3-HC decomposition of degree n exists on each of V_1 and V_2. If $3n+2$ 3-edges constructed by adjoining respectively a vertex in V_2 (V_1) to the root of a 3-HC on V_1 (V_2) are added to the 3-HC of degree n, then a 3-HC of degree $4n+2$ on V can be constructed. The number of such 3-HC is $2\binom{3n+2}{3}/n=(3n+2)(3n+1)$, which is equal to $\binom{m}{3}/c$. These hyperclaws are, of course, edge-disjoint.

Suppose $n=1$. Let $V=\{1,2,\ldots,10\}$ and let $V_1=\{1,2,3,4,5\}$ and $V_2=\{6,7,8,9,10\}$. Using the relation of incidence between the set of $\binom{5}{3}$ 3-edges and $\binom{5}{2}$ 2-edges on V_1 (V_2), a 2-edge on V_1 (V_2) is selected for each of the 3-edges on V_1 (V_2). If five 3-edges on V, constructed by adjoining respectively a vertex in V_2 (V_1) to a 2-edge on V_1 (V_2), are added to the corresponding 3-edge, then a 3-HC of degree 6 on V can be constructed. Table 1 illustrates the situation and shows that $H_{10,3}$ is 3-HC decomposable of degree 6, where (1, 2) 3 6 7 8 9 10 shows a 3-HC composed of six 3-edges $\{1,2,3\}$, $\{1,2,6\}$, $\{1,2,7\}$, $\{1,2,8\}$, $\{1,2,9\}$ and $\{1,2,10\}$.

Suppose $n=0$. The existence of two 3-HC of degree 2 on $V=\{1,2,3,4\}$, i.e., $\{\{1,2,3\},\{1,2,4\}\}$ and $\{\{1,3,4\},\{2,3,4\}\}$, shows that $H_{4,3}$ is 3-HC decomposable of degree 2.

Theorem 4.2. *Every $H_{m,3}$ can be decomposed into pairwise disjoint 3-HC's of degree c for a positive integer $c\geqslant 2$ provided $m\geqslant 8$.*

Table 1. Decomposition of $H_{10,3}$ on $V=\{1, 2, \ldots, 10\}$ into 20 3-HC's of degree 6

Roots in	Leaves in		Roots in	Leaves in	
V_1	V_1	V_2	V_2	V_2	V_1
(1, 2)	3	6 7 8 9 10	(6, 7)	8	1 2 3 4 5
(2, 3)	4	6 7 8 9 10	(7, 8)	9	1 2 3 4 5
(3, 4)	5	6 7 8 9 10	(8, 9)	10	1 2 3 4 5
(4, 5)	1	6 7 8 9 10	(9, 10)	6	1 2 3 4 5
(5, 1)	2	6 7 8 9 10	(10, 6)	7	1 2 3 4 5
(1, 3)	4	6 7 8 9 10	(6, 8)	9	1 2 3 4 5
(2, 4)	5	6 7 8 9 10	(7, 9)	10	1 2 3 4 5
(3, 5)	1	6 7 8 9 10	(8, 10)	6	1 2 3 4 5
(4, 1)	2	6 7 8 9 10	(9, 6)	7	1 2 3 4 5
(5, 2)	3	6 7 8 9 10	(10, 7)	8	1 2 3 4 5

Proof. (a) Case $m \equiv 0 \pmod 3$. Put $m=3c$, then Theorem 3.4 shows that $H_{m,3}$ is 3-HC decomposable of degree c, since $c=\frac{1}{3}m \geq 2$.

(b) Case $m \equiv 1 \pmod 3$. Two cases, namely, $m=6c+1$ for $c \geq 2$ and $m=6n+4$ for $n \geq 1$ have to be considered. In the former, it is easily seen by Corollary 3.7 that $H_{m,3}$ is 3-HC decomposable of degree c. In the latter case, $H_{m,3}$ *is* 3-HC decomposable of degree $c=4n+2$ by Theorem 4.1.

(c) Case $m \equiv 2 \pmod 3$. Put $m=3c+2$, then Theorem 3.2 shows that $H_{m,3}$ is 3-HC decomposable of degree c, since $c=\frac{1}{3}(m-2) \geq 2$.

Corollary 4.3. *For any factor c^* $(2 \leq c^* < c)$ of c given in Theorem 4.2, if it exists, $H_{m,3}$ is 3-HC decomposable of degree c^*.*

In a complete 3-graph $H_{m,3}$, a 3-HC of degree c exists if and only if $m \geq 4$. We, therefore, have to investigate the 3-HC decomposability of $H_{4,3}$, $H_{5,3}$, $H_{6,3}$ and $H_{7,3}$.

It has been shown in Theorem 4.1 that $H_{4,3}$ is 3-HC decomposable of degree 2. Theorem 3.4 shows that $H_{6,3}$ is 3-HC decomposable of degree 2. $H_{5,3}$ is 3-HC decomposable of degree 2 as is illustrated in Table 2.

Only the complete 3-graph $H_{7,3}$ cannot be decomposed into pairwise edge

Table 2. Decomposition of $H_{5,3}$ on $V=\{1, 2, 3, 4, 5\}$ into five 3-HC's of degree 2

Roots	Leaves	Roots	Leaves
(1, 2)	3 4	(4, 5)	1 2
(2, 3)	4 5	(5, 1)	2 3
(3, 4)	5 1		

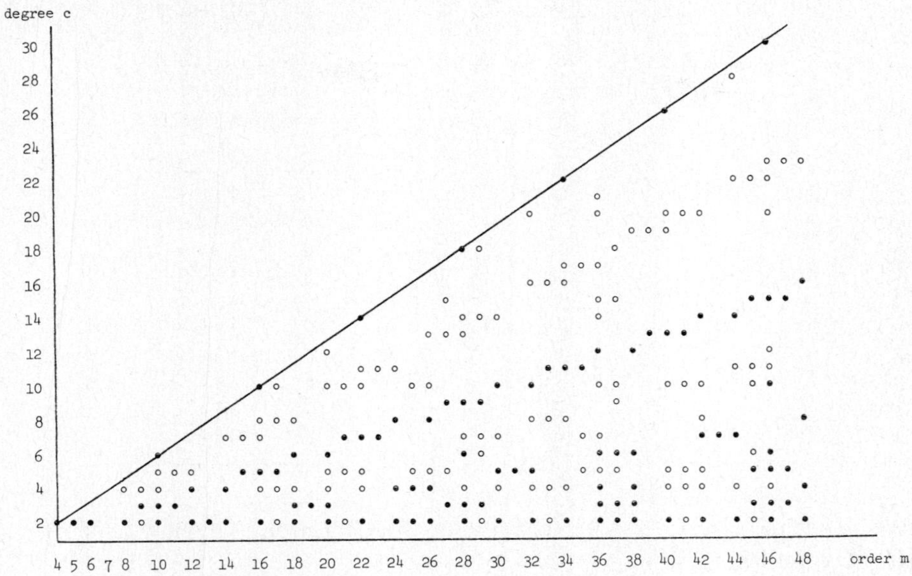

Fig. 1. The 3-HC decomposability of degree c for $H_{m,3}$.

disjoint 3-HC's in that there is no integer which satisfy the necessary condition given in Theorem 2.1.

These results obtained for $H_{m,3}$ within $4 \leq m \leq 48$ are illustrated in Fig. 1. Straight line drawn in the figure indicates the upper bound of c given in (ii)* of Theorem 2.1. The closed dots (•) indicate those pairs of parameters (m, c) which have been shown to be 3-HC decomposable by Theorem 3.1 and subsequent considerations. The open dots (◦) indicate those pairs of parameters which are not yet decided whether they are 3-HC decomposable or not, although they satisfy the necessary condition given in Theorem 2.1.

Though not yet decided completely, it might be conjectured that the necessary condition for k-HC decomposability given in Theorem 2.1 is also sufficient, since the condition (ii) is tight enough if $k = 3$.

References

[1] H. Agrawal, Some generalizations of distinct representatives with applications to statistical designs, Ann. Math. Statist. 37 (1966) 525–528.

[2] C. Berge, Graphes et Hypergraphes (Dunod, Paris, 1970).

[3] D. Raghavarao, Constructions and Combinatorial Problems in design of Experiments (Wiley, New York, 1971).

[4] S. Yamamoto, H. Ikeda, S. Shige-eda, K. Ushio and N. Hamada, Design of a new balanced file organization scheme with the least redundancy, Information and control 28 (1975) 156–175.

[5] S. Yamamoto, H. Ikeda, S. Shige-eda, K. Ushio and N. Hamada. On claw-decomposition of complete graphs and complete bigraphs, Hiroshima Math. J. 5 (1975) 33–42.